Graduate Texts in Mathematics 271

Graduate Texts in Mathematics

Series Editors:

Sheldon Axler
San Francisco State University, San Francisco, CA, USA

Kenneth Ribet
University of California, Berkeley, CA, USA

Advisory Board:

Colin Adams, *Williams College, Williamstown, MA, USA*
Alejandro Adem, *University of British Columbia, Vancouver, BC, Canada*
Ruth Charney, *Brandeis University, Waltham, MA, USA*
Irene M. Gamba, *The University of Texas at Austin, Austin, TX, USA*
Roger E. Howe, *Yale University, New Haven, CT, USA*
David Jerison, *Massachusetts Institute of Technology, Cambridge, MA, USA*
Jeffrey C. Lagarias, *University of Michigan, Ann Arbor, MI, USA*
Jill Pipher, *Brown University, Providence, RI, USA*
Fadil Santosa, *University of Minnesota, Minneapolis, MN, USA*
Amie Wilkinson, *University of Chicago, Chicago, IL, USA*

Graduate Texts in Mathematics bridge the gap between passive study and creative understanding, offering graduate-level introductions to advanced topics in mathematics. The volumes are carefully written as teaching aids and highlight characteristic features of the theory. Although these books are frequently used as textbooks in graduate courses, they are also suitable for individual study.

More information about this series at http://www.springer.com/series/136

Michele Conforti • Gérard Cornuéjols
Giacomo Zambelli

Integer Programming

 Springer

Michele Conforti
Department of Mathematics
University of Padova
Padova, Italy

Gérard Cornuéjols
Tepper School of Business
Carnegie Mellon University
Pittsburgh, PA, USA

Giacomo Zambelli
Department of Management
London School of Economics
  and Political Science
London, UK

ISSN 0072-5285   ISSN 2197-5612 (electronic)
ISBN 978-3-319-11007-3   ISBN 978-3-319-11008-0 (eBook)
DOI 10.1007/978-3-319-11008-0
Springer Cham Heidelberg New York Dordrecht London

Library of Congress Control Number: 2014952029

© Springer International Publishing Switzerland 2014
This work is subject to copyright. All rights are reserved by the Publisher, whether the whole or part of the material is concerned, specifically the rights of translation, reprinting, reuse of illustrations, recitation, broadcasting, reproduction on microfilms or in any other physical way, and transmission or information storage and retrieval, electronic adaptation, computer software, or by similar or dissimilar methodology now known or hereafter developed. Exempted from this legal reservation are brief excerpts in connection with reviews or scholarly analysis or material supplied specifically for the purpose of being entered and executed on a computer system, for exclusive use by the purchaser of the work. Duplication of this publication or parts thereof is permitted only under the provisions of the Copyright Law of the Publisher's location, in its current version, and permission for use must always be obtained from Springer. Permissions for use may be obtained through RightsLink at the Copyright Clearance Center. Violations are liable to prosecution under the respective Copyright Law.
The use of general descriptive names, registered names, trademarks, service marks, etc. in this publication does not imply, even in the absence of a specific statement, that such names are exempt from the relevant protective laws and regulations and therefore free for general use.
While the advice and information in this book are believed to be true and accurate at the date of publication, neither the authors nor the editors nor the publisher can accept any legal responsibility for any errors or omissions that may be made. The publisher makes no warranty, express or implied, with respect to the material contained herein.

Printed on acid-free paper

Springer is part of Springer Science+Business Media (www.springer.com)

Preface

Integer programming is a thriving area of optimization, which is applied nowadays to a multitude of human endeavors, thanks to high quality software. It was developed over several decades and is still evolving rapidly.

The goal of this book is to present the mathematical foundations of integer programming, with emphasis on the techniques that are most successful in current software implementations: convexification and enumeration.

This textbook is intended for a graduate course in integer programming in M.S. or Ph.D. programs in applied mathematics, operations research, industrial engineering, or computer science.

To better understand the excitement that is generated today by this area of mathematics, it is helpful to provide a historical perspective.

Babylonian tablets show that mathematicians were already solving systems of linear equations over 3,000 years ago. The eighth book of the Chinese *Nine Books of Arithmetic*, written over 2,000 years ago, describes what is now known as the Gaussian elimination method. In 1809, Gauss [160] used this method in his work, stating that it was a "standard technique." The method was subsequently named after him.

A major breakthrough occurred when mathematicians started analyzing systems of linear *inequalities*. This is a fertile ground for beautiful theories. In 1826 Fourier [145] gave an algorithm for solving such systems by eliminating variables one at a time. Other important contributions are due to Farkas [135] and Minkowski [279]. Systems of linear inequalities define *polyhedra* and it is natural to optimize a linear function over them. This is the topic of *linear programming*, arguably one of the greatest successes of computational mathematics in the twentieth century. The *simplex method*, developed by Dantzig [102] in 1951, is currently used to solve large-scale problems in all sorts of application areas. It is often desirable to find *integer solutions* to linear programs. This is the topic of this book. The first algorithm for solving pure integer linear programs was discovered in 1958 by Gomory [175].

When considering algorithmic questions, a fundamental issue is the increase in computing time when the size of the problem instance increases. In the 1960s Edmonds [123] was one of the pioneers in stressing the importance of *polynomial-time* algorithms. These are algorithms whose computing time is bounded by a polynomial function of the instance size. In particular Edmonds [125] pointed out that the Gaussian elimination method can be turned into a polynomial-time algorithm by being a bit careful with the intermediate numbers that are generated. The existence of a polynomial-time algorithm for linear programming remained a challenge for many years. This question was resolved positively by Khachiyan [235] in 1979, and later by Karmarkar [229] using a totally different algorithm. Both algorithms were (and still are) very influential, each in its own way. In integer programming, Lenstra [256] found a polynomial-time algorithm when the number of variables is fixed.

Although integer programming is NP-hard in general, the *polyhedral approach* has proven successful in practice. It can be traced back to the work of Dantzig, Fulkerson, and Johnson [103] in 1954. Research is currently very active in this area. Beautiful mathematical results related to the polyhedral approach pervade the area of integer programming. This book presents several of these results. Also very promising are nonpolyhedral approximations that can be computed in polynomial-time, such as semidefinite relaxations, see Lovász and Schrijver [264], and Goemans and Williamson [173].

We are grateful to the colleagues and students who read earlier drafts of this book and have helped us improve it. In particular many thanks to Lawrence Wolsey for carefully checking the whole manuscript. Many thanks also to Marco Di Summa, Kanstantsin Pashkovich, Teresa Provesan, Sercan Yildiz, Monique Laurent, Sebastian Pokutta, Dan Bienstock, François Margot, Giacomo Nannicini, Juan Pablo Viema, Babis Tsourakakis, Thiago Serra, Yang Jiao, and Tarek Elgindy for their excellent suggestions.

Thank you to Julie Zavon for the artwork at the end of most chapters.

This work was supported in part by NSF grant CMMI1263239 and ONR grant N000141210032.

Padova, Italy	Michele Conforti
Pittsburgh, PA, USA	Gérard Cornuéjols
London, UK	Giacomo Zambelli

Contents

1 Getting Started   **1**
    1.1 Integer Programming 1
    1.2 Methods for Solving Integer Programs 5
        1.2.1 The Branch-and-Bound Method 6
        1.2.2 The Cutting Plane Method 11
        1.2.3 The Branch-and-Cut Method 15
    1.3 Complexity 16
        1.3.1 Problems, Instances, Encoding Size 17
        1.3.2 Polynomial Algorithm 18
        1.3.3 Complexity Class NP 19
    1.4 Convex Hulls and Perfect Formulations 20
        1.4.1 Example: A Two-Dimensional Mixed Integer Set . . . 22
        1.4.2 Example: A Two-Dimensional Pure Integer Set 24
    1.5 Connections to Number Theory 25
        1.5.1 The Greatest Common Divisor 26
        1.5.2 Integral Solutions to Systems of Linear Equations . . 29
    1.6 Further Readings 36
    1.7 Exercises 38

2 Integer Programming Models   **45**
    2.1 The Knapsack Problem 45
    2.2 Comparing Formulations 46
    2.3 Cutting Stock: Formulations with Many Variables 48
    2.4 Packing, Covering, Partitioning 51
        2.4.1 Set Packing and Stable Sets 51
        2.4.2 Strengthening Set Packing Formulations 52
        2.4.3 Set Covering and Transversals 53
        2.4.4 Set Covering on Graphs: Many Constraints 55

		2.4.5 Set Covering with Many Variables: Crew Scheduling	57
		2.4.6 Covering Steiner Triples	58
	2.5	Generalized Set Covering: The Satisfiability Problem	58
	2.6	The Sudoku Game	60
	2.7	The Traveling Salesman Problem	61
	2.8	The Generalized Assignment Problem	65
	2.9	The Mixing Set	66
	2.10	Modeling Fixed Charges	66
		2.10.1 Facility Location	67
		2.10.2 Network Design	69
	2.11	Modeling Disjunctions	70
	2.12	The Quadratic Assignment Problem and Fortet's Linearization	72
	2.13	Further Readings	73
	2.14	Exercises	74

3 Linear Inequalities and Polyhedra — 85

3.1	Fourier Elimination	85
3.2	Farkas' Lemma	88
3.3	Linear Programming	89
3.4	Affine, Convex, and Conic Combinations	91
	3.4.1 Linear Combinations, Linear Spaces	91
	3.4.2 Affine Combinations, Affine Spaces	92
	3.4.3 Convex Combinations, Convex Sets	92
	3.4.4 Conic Combinations, Convex Cones	93
3.5	Polyhedra and the Theorem of Minkowski–Weyl	94
	3.5.1 Minkowski–Weyl Theorem for Polyhedral Cones	94
	3.5.2 Minkowski–Weyl Theorem for Polyhedra	95
3.6	Lineality Space and Recession Cone	97
3.7	Implicit Equalities, Affine Hull, and Dimension	98
3.8	Faces	101
3.9	Minimal Representation and Facets	104
3.10	Minimal Faces	108
3.11	Edges and Extreme Rays	109
3.12	Decomposition Theorem for Polyhedra	111
3.13	Encoding Size of Vertices, Extreme Rays, and Facets	112
3.14	Carathéodory's Theorem	113
3.15	Projections	116
3.16	Polarity	119

	3.17	Further Readings	120
	3.18	Exercises	124

4 Perfect Formulations — 129

- 4.1 Properties of Integral Polyhedra … 130
- 4.2 Total Unimodularity … 131
- 4.3 Networks … 134
  - 4.3.1 Circulations … 135
  - 4.3.2 Shortest Paths … 137
  - 4.3.3 Maximum Flow and Minimum Cut … 139
- 4.4 Matchings in Graphs … 145
  - 4.4.1 Augmenting Paths … 146
  - 4.4.2 Cardinality Bipartite Matchings … 147
  - 4.4.3 Minimum Weight Perfect Matchings in Bipartite Graphs … 149
  - 4.4.4 The Matching Polytope … 150
- 4.5 Spanning Trees … 153
- 4.6 Total Dual Integrality … 155
- 4.7 Submodular Polyhedra … 157
- 4.8 The Fundamental Theorem of Integer Programming … 159
  - 4.8.1 An Example: The Mixing Set … 161
  - 4.8.2 Mixed Integer Linear Programming is in NP … 163
  - 4.8.3 Polynomial Encoding of the Facets of the Integer Hull … 165
- 4.9 Union of Polyhedra … 166
  - 4.9.1 Example: Modeling Disjunctions … 169
  - 4.9.2 Example: All the Even Subsets of a Set … 170
  - 4.9.3 Mixed Integer Linear Representability … 171
- 4.10 The Size of a Smallest Perfect Formulation … 174
  - 4.10.1 Rectangle Covering Bound … 177
  - 4.10.2 An Exponential Lower-Bound for the Cut Polytope … 179
  - 4.10.3 An Exponential Lower-Bound for the Matching Polytope … 181
- 4.11 Further Readings … 182
- 4.12 Exercises … 187

5 Split and Gomory Inequalities — 195

- 5.1 Split Inequalities … 195
  - 5.1.1 Inequality Description of the Split Closure … 199
  - 5.1.2 Polyhedrality of the Split Closure … 202
  - 5.1.3 Split Rank … 203

		5.1.4	Gomory's Mixed Integer Inequalities	205

        5.1.4 Gomory's Mixed Integer Inequalities 205
        5.1.5 Mixed Integer Rounding Inequalities 206
  5.2 Chvátal Inequalities 207
        5.2.1 The Chvátal Closure of a Pure Integer Linear Set . . . 208
        5.2.2 Chvátal Rank 209
        5.2.3 Chvátal Inequalities for Other Forms
                of the Linear System 211
        5.2.4 Gomory's Fractional Cuts 212
        5.2.5 Gomory's Lexicographic Method for Pure
                Integer Programs 213
  5.3 Gomory's Mixed Integer Cuts 216
  5.4 Lift-and-Project 222
        5.4.1 Lift-and-Project Rank for Mixed 0,1
                Linear Programs 223
        5.4.2 A Finite Cutting Plane Algorithm for Mixed 0, 1
                Linear Programming 225
  5.5 Further Readings 227
  5.6 Exercises 230

6 Intersection Cuts and Corner Polyhedra   **235**
  6.1 Corner Polyhedron 235
  6.2 Intersection Cuts 240
        6.2.1 The Gauge Function 248
        6.2.2 Maximal Lattice-Free Convex Sets 249
  6.3 Infinite Relaxations 253
        6.3.1 Pure Integer Infinite Relaxation 256
        6.3.2 Continuous Infinite Relaxation 264
        6.3.3 The Mixed Integer Infinite Relaxation 268
        6.3.4 Trivial and Unique Liftings 270
  6.4 Further Readings 273
  6.5 Exercises 275

7 Valid Inequalities for Structured Integer Programs   **281**
  7.1 Cover Inequalities for the 0,1 Knapsack Problem 282
  7.2 Lifting 283
        7.2.1 Lifting Minimal Cover Inequalities 285
        7.2.2 Lifting Functions, Superadditivity, and Sequence
                Independent Lifting 287
        7.2.3 Sequence Independent Lifting for Minimal
                Cover Inequalities 289

	7.3	Flow Cover Inequalities	291

  7.3 Flow Cover Inequalities 291
  7.4 Faces of the Symmetric Traveling Salesman Polytope 299
       7.4.1 Separation of Subtour Elimination Constraints 302
       7.4.2 Comb Inequalities 303
       7.4.3 Local Cuts 305
  7.5 Equivalence Between Optimization
       and Separation 307
  7.6 Further Readings 311
  7.7 Exercises 315

8 Reformulations and Relaxations — 321
  8.1 Lagrangian Relaxation 321
       8.1.1 Examples 324
       8.1.2 Subgradient Algorithm 326
  8.2 Dantzig–Wolfe Reformulation 330
       8.2.1 Problems with Block Diagonal Structure 332
       8.2.2 Column Generation 334
       8.2.3 Branch-and-Price 337
  8.3 Benders Decomposition 338
  8.4 Further Readings 341
  8.5 Exercises 344

9 Enumeration — 351
  9.1 Integer Programming in Fixed Dimension 351
       9.1.1 Basis Reduction 352
       9.1.2 The Flatness Theorem and Rounding Polytopes . 358
       9.1.3 Lenstra's Algorithm 362
  9.2 Implementing Branch-and-Cut 364
  9.3 Dealing with Symmetries 373
  9.4 Further Readings 380
  9.5 Exercises 383

10 Semidefinite Bounds — 389
  10.1 Semidefinite Relaxations 389
  10.2 Two Applications in Combinatorial Optimization 391
       10.2.1 The Max-Cut Problem 391
       10.2.2 The Stable Set Problem 393
  10.3 The Lovász–Schrijver Relaxation 394
       10.3.1 Semidefinite Versus Linear Relaxations 396

| | | 10.3.2 Connection with Lift-and-Project . . . . 397 |
| | | 10.3.3 Iterating the Lovász–Schrijver Procedure . . . . 399 |

10.4 The Sherali–Adams and Lasserre Hierarchies 400
    10.4.1 The Sherali–Adams Hierarchy 400
    10.4.2 The Lasserre Hierarchy 402
10.5 Further Readings 408
10.6 Exercises 411

Bibliography   415

Index   447

The authors discussing the outline of the book

Chapter 1

Getting Started

What are integer programs? We introduce this class of problems and present two algorithmic ideas for solving them, the branch-and-bound and cutting plane methods. Both capitalize heavily on the fact that linear programming is a well-solved class of problems. Many practical situations can be modeled as integer programs, as will be shown in Chap. 2. But in Chap. 1 we reflect on the consequences of the algorithmic ideas just mentioned. Algorithms raise the question of computational complexity. Which problems can be solved in "polynomial time"? The cutting plane approach also leads naturally to the notion of "convex hull" and the so-called polyhedral approach. We illustrate these notions in the case of 2-variable integer programs. Complementary to the connection with linear programming, there is also an interesting connection between integer programming and number theory. In particular, we show how to find an integer solution to a system of linear equations. Contrary to general integer linear programming problems which involve inequalities, this problem can be solved in polynomial time.

1.1 Integer Programming

A *pure integer linear program* is a problem of the form

$$\begin{array}{rl} \max & cx \\ \text{subject to} & Ax \leq b \\ & x \geq 0 \text{ integral} \end{array} \quad (1.1)$$

where the data, usually rational, are the row vector $c = (c_1, \ldots, c_n)$, the $m \times n$ matrix $A = (a_{ij})$, and the column vector $b = \begin{pmatrix} b_1 \\ \vdots \\ b_m \end{pmatrix}$. The column vector $x = \begin{pmatrix} x_1 \\ \vdots \\ x_n \end{pmatrix}$ contains the variables to be optimized. An n-vector x is said to be *integral* when $x \in \mathbb{Z}^n$. The set $S := \{x \in \mathbb{Z}_+^n : Ax \leq b\}$ of feasible solutions to (1.1) is called a *pure integer linear set*.

In this book we also consider *mixed integer linear programs*. These are problems of the form

$$\begin{aligned} \max \quad & cx + hy \\ \text{subject to} \quad & Ax + Gy \leq b \\ & x \geq 0 \text{ integral} \\ & y \geq 0, \end{aligned} \qquad (1.2)$$

where the data are row vectors $c = (c_1, \ldots, c_n)$, $h = (h_1, \ldots, h_p)$, an $m \times n$ matrix $A = (a_{ij})$, an $m \times p$ matrix $G = (g_{ij})$ and a column vector $b = \begin{pmatrix} b_1 \\ \vdots \\ b_m \end{pmatrix}$. We will usually assume that all entries of c, h, A, G, b are rational. The column vectors $x = \begin{pmatrix} x_1 \\ \vdots \\ x_n \end{pmatrix}$ and $y = \begin{pmatrix} y_1 \\ \vdots \\ y_p \end{pmatrix}$ contain the variables to be optimized. The variables x_j are constrained to be nonnegative integers while the variables y_j are allowed to take any nonnegative real value. We will always assume that there is at least one integer variable, i.e., $n \geq 1$. The pure integer linear program (1.1) is the special case of the mixed integer linear program (1.2) where $p = 0$. For convenience, we refer to mixed integer linear programs simply as *integer programs* in this book.

The set of feasible solutions to (1.2)

$$S := \{(x, y) \in \mathbb{Z}_+^n \times \mathbb{R}_+^p : Ax + Gy \leq b\} \qquad (1.3)$$

is called a *mixed integer linear set* (Fig. 1.1).

A *mixed 0, 1 linear set* is a set of the form (1.3) in which the integer variables are restricted to take the value 0 or 1:

$$S := \{(x, y) \in \{0, 1\}^n \times \mathbb{R}_+^p : Ax + Gy \leq b\}.$$

1.1. INTEGER PROGRAMMING

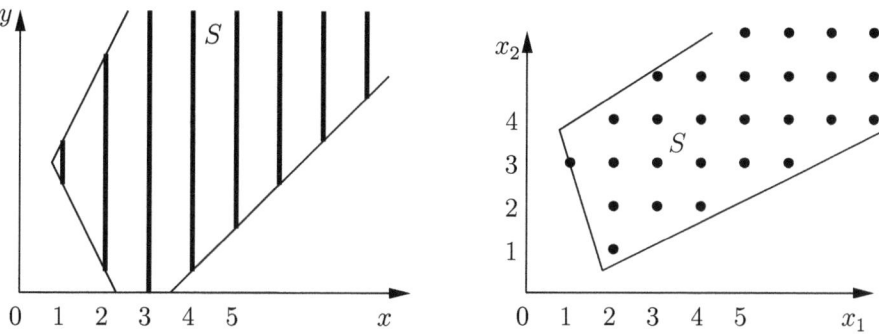

Figure 1.1: A mixed integer linear set and a pure integer linear set

A *mixed 0,1 linear program* is an integer program whose set of feasible solutions is a mixed 0, 1 linear set.

Solving integer programs is a difficult task in general. One approach that is commonly used in computational mathematics is to find a relaxation that is easier to solve numerically and gives a good approximation. In this book we focus mostly on *linear programming relaxations*.

We will use the symbols "$\subseteq$" to denote inclusion and "$\subset$" to denote strict inclusion. Given a mixed integer set $S \subseteq \mathbb{Z}^n \times \mathbb{R}^p$, a *linear relaxation* of S is a set of the form $P' := \{(x, y) \in \mathbb{R}^n \times \mathbb{R}^p : A'x + G'y \leq b'\}$ that contains S. A *linear programming relaxation* of (1.2) is a linear program $\max\{cx + hy : (x, y) \in P'\}$. Why linear programming relaxations? Mainly for two reasons. First, solving linear programs is one of the greatest successes in computational mathematics. There are algorithms that are efficient in theory and practice and therefore one can solve these relaxations in a reasonable amount of time. Second, one can generate a sequence of linear relaxations of S that provide increasingly tighter approximations of the set S.

For the mixed integer linear set S defined in (1.3), there is a *natural linear relaxation*, namely the relaxation

$$P_0 := \{(x, y) \in \mathbb{R}_+^n \times \mathbb{R}_+^p : Ax + Gy \leq b\}$$

obtained from S by discarding the integrality requirement on the vector x. The *natural linear programming relaxation* of (1.2) is the linear program $\max\{cx + hy : (x, y) \in P_0\}$.

For example, the 2-variable pure integer program

$$\begin{aligned} \max \quad & 5.5x_1 + 2.1x_2 \\ & -x_1 + x_2 \leq 2 \\ & 8x_1 + 2x_2 \leq 17 \\ & x_1, x_2 \geq 0 \\ & x_1, x_2 \text{ integer} \end{aligned}$$

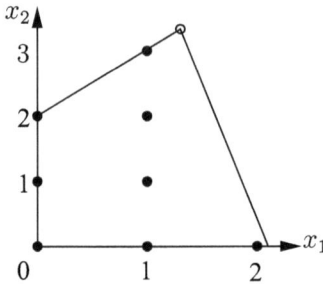

Figure 1.2: A 2-variable integer program

has eight feasible solutions represented by the dots in Fig. 1.2. One can verify that the optimal solution of this integer program is $x_1 = 1$, $x_2 = 3$ with objective value 11.8. The solution of the natural linear programming relaxation is $x_1 = 1.3$, $x_2 = 3.3$, with objective value 14.08.

A pure mathematician might dismiss integer programming as trivial. For example, how difficult can it be to solve a 0, 1 linear program $\max\{cx : x \in S\}$ when $S := \{x \in \{0,1\}^n : Ax \leq b\}$? After all, one can check whether S is empty by enumerating all vectors $x \in \{0,1\}^n$ and, if $S \neq \emptyset$, pick a vector $x \in S$ that results in the largest value of cx, again by complete enumeration. Yes, this is possible in theory, but it is not practical when n is large. The key algorithmic question is: Can an optimal solution be found more efficiently than by total enumeration? As an example, consider the following *assignment problem*. There are n jobs to be performed by n workers. We know the cost c_{ij} of assigning job i to worker j for each pair i, j. What is the cheapest way of assigning one job to each of the n workers (the same job cannot be assigned to two different workers)? The number of possible assignments is $n!$, the number of permutations of n elements, since the first job can be assigned to any of the n workers, the second job to any of the remaining $n-1$ workers and so on. However, total enumeration is only practical for small values of n as should be evident from the following table.

n	$n!$
10	3.6×10^6
100	9.3×10^{157}
1000	4×10^{2567}

Already for $n = 100$, the value of $n!$ exceeds the number of atoms in the universe, which is approximately 10^{80} according to Wolfram (http://www.wolframalpha.com).

A 0, 1 programming formulation of the assignment problem is as follows. Let $x_{ij} = 1$ if job i is assigned to worker j, and 0 otherwise. The integer program is

$$\min \sum_{i=1}^{n}\sum_{j=1}^{n} c_{ij}x_{ij}$$
$$\text{subject to} \quad \sum_{i=1}^{n} x_{ij} = 1 \quad \text{for } j = 1,\ldots,n \quad (1.4)$$
$$\sum_{j=1}^{n} x_{ij} = 1 \quad \text{for } i = 1,\ldots,n$$
$$x \in \{0,1\}^{n \times n}$$

where the objective is to minimize the overall cost of the assignments, the first constraint guarantees that each worker is assigned exactly one job, and the second constraint guarantees that each job is assigned only once. Assignment problems with $n = 1000$ can be solved in seconds on a computer, using the Hungarian method [245, 281] (This algorithm will be presented in Sect. 4.4.3). We will also show in Chap. 4 that solving the natural linear programming relaxation of (1.4) is enough to find an optimal solution of the assignment problem. This is much more efficient than total enumeration!

1.2 Methods for Solving Integer Programs

In this section we introduce two algorithmic principles that have proven successful for solving integer programs. These two approaches, the branch-and-bound and the cutting plane methods, are based on simple ideas but they are at the heart of the state-of-the-art software for integer programming. Some mathematical sophistication is necessary to make them successful.

The choice of topics in the remainder of this book is motivated by the desire to establish sound mathematical foundations for these deceptively simple algorithmic ideas.

The integer programming formulation (1.2) will be denoted by MILP here for easy reference.

$$\text{MILP}: \quad \max\{cx + hy : (x,y) \in S\}$$

where $S := \{(x,y) \in \mathbb{Z}_+^n \times \mathbb{R}_+^p : Ax + Gy \leq b\}$. For ease of exposition, we assume in this section that MILP admits a finite optimum. Let (x^*, y^*) denote an optimal solution and z^* the optimal value of MILP. These are the unknowns that we are looking for.

Let (x^0, y^0) and z_0 be, respectively, an optimal solution and the optimal value of the natural linear programming relaxation

$$\max\{cx + hy : (x,y) \in P_0\} \tag{1.5}$$

where P_0 is the natural linear relaxation of S (we will show later in the book that the existence of an optimal rational solution (x^0, y^0) to (1.5) follows from our assumption on the existence of (x^*, y^*) and the rationality of the data; here we just assume it). We will also assume that we have a linear programming solver at our disposal, thus (x^0, y^0) and z_0 are available to us. Since $S \subseteq P_0$, it follows that $z^* \leq z_0$. Furthermore, if x^0 is an integral vector, then $(x^0, y^0) \in S$ and therefore $z^* = z_0$; in this case MILP is solved. We describe two strategies that deal with the case in which at least one component of the vector x^0 is fractional.

1.2.1 The Branch-and-Bound Method

We give a formal description of the branch-and-bound algorithm at the end of this section. We first present the method informally.

Choose an index j, where $1 \leq j \leq n$, such that x_j^0 is fractional. For simplicity, let $f := x_j^0$ denote this fractional value and define the sets

$$S_1 := S \cap \{(x,y) : x_j \leq \lfloor f \rfloor\}, \quad S_2 := S \cap \{(x,y) : x_j \geq \lceil f \rceil\}$$

where $\lfloor f \rfloor$ denotes the largest integer $k \leq f$ and $\lceil f \rceil$ denotes the smallest integer $l \geq f$. Since x_j is an integer for every $(x,y) \in S$, it follows that (S_1, S_2) is a partition of S. Consider now the two following integer programs based on this partition

$$\text{MILP}_1 : \max\{cx + hy : (x,y) \in S_1\}, \quad \text{MILP}_2 : \max\{cx + hy : (x,y) \in S_2\}.$$

1.2. METHODS FOR SOLVING INTEGER PROGRAMS

The optimal solution of MILP is the best among the optimal solutions of MILP$_1$ and MILP$_2$, therefore the solution of the original problem is reduced to solving the two new subproblems.

Denote by P_1, P_2 the natural linear relaxations of S_1, S_2, that is

$$P_1 := P_0 \cap \{(x,y) : x_j \leq \lfloor f \rfloor\}, \quad P_2 := P_0 \cap \{(x,y) : x_j \geq \lceil f \rceil\},$$

and consider the two corresponding natural linear programming relaxations

$$\text{LP}_1 : \max\{cx + hy : (x,y) \in P_1\}, \quad \text{LP}_2 : \max\{cx + hy : (x,y) \in P_2\}.$$

(i) If one of the linear programs LP$_i$ is infeasible, i.e., $P_i = \emptyset$, then we also have $S_i = \emptyset$ since $S_i \subseteq P_i$. Thus MILP$_i$ is infeasible and does not need to be considered any further. We say that this problem is *pruned by infeasibility*.

(ii) Let (x^i, y^i) be an optimal solution of LP$_i$ and z_i its value, $i = 1, 2$.

  (iia) If x^i is an integral vector, then (x^i, y^i) is an optimal solution of MILP$_i$ and a feasible solution of MILP. Problem MILP$_i$ is solved, and we say that it is *pruned by integrality*. Since $S_i \subseteq S$, it follows that $z_i \leq z^*$, that is, z_i is a lower bound on the value of MILP.

  (iib) If x^i is not an integral vector and z_i is smaller than or equal to the best known lower bound on the value of MILP, then S_i cannot contain a better solution and the problem is *pruned by bound*.

  (iic) If x^i is not an integral vector and z_i is greater than the best known lower bound, then S_i may still contain an optimal solution to MILP. Let $x^i_{j'}$ be a fractional component of vector x^i. Let $f' := x^i_{j'}$, define the sets $S_{i_1} := S_i \cap \{(x,y) : x_{j'} \leq \lfloor f' \rfloor\}$, $S_{i_2} := S_i \cap \{(x,y) : x_{j'} \geq \lceil f' \rceil\}$ and repeat the above process.

To illustrate the branch-and-bound method, consider the integer program depicted in Fig. 1.2:

$$\begin{aligned}
\max \quad & 5.5x_1 + 2.1x_2 \\
& -x_1 + x_2 \leq 2 \\
& 8x_1 + 2x_2 \leq 17 \\
& x_1, x_2 \geq 0 \\
& x_1, x_2 \text{ integer.}
\end{aligned}$$

The solution of the natural linear programming relaxation is $x_1 = 1.3$, $x_2 = 3.3$ with objective value 14.08. Thus 14.08 is an upper bound on the optimal solution of the problem. Branching on variable x_1, we create two integer programs. The linear programming relaxation of the one with the additional constraint $x_1 \leq 1$ has solution $x_1 = 1$, $x_2 = 3$ with value 11.8, and it is therefore pruned by integrality in Case (iia). Thus 11.8 is a lower bound on the value of an optimal solution of the integer program. The linear programming relaxation of the subproblem with the additional constraint $x_1 \geq 2$ has solution $x_1 = 2$, $x_2 = 0.5$ and objective value 12.05. The above steps can be represented graphically in the *enumeration tree* shown in Fig. 1.3.

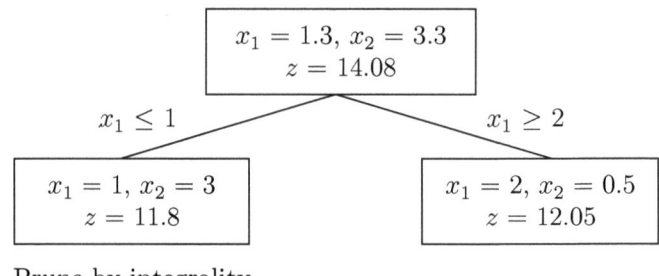

Figure 1.3: Branching on variable x_1

Note that the value of x_2 is fractional, so this solution is not feasible to the integer program. Since its objective value is higher than 11.8 (the value of the best integer solution found so far), we need to continue the search as described in Case (iic). Therefore we branch on variable x_2. We create two integer programs, one with the additional constraint $x_2 \geq 1$, the other with $x_2 \leq 0$. The linear programming relaxation of the first of these linear programs is infeasible, therefore this problem is pruned by infeasibility in Case (i). The second integer program is

$$
\begin{array}{rl}
\max & 5.5x_1 + 2.1x_2 \\
& -x_1 + x_2 \leq 2 \\
& 8x_1 + 2x_2 \leq 17 \\
& x_1 \geq 2 \\
& x_2 \leq 0 \\
& x_1, x_2 \geq 0 \\
& x_1, x_2 \text{ integer.}
\end{array}
$$

1.2. METHODS FOR SOLVING INTEGER PROGRAMS

The optimal solution of its linear relaxation is $x_1 = 2.125$, $x_2 = 0$, with objective value 11.6875. Because this value is smaller than the best lower bound 11.8, the corresponding node of the enumeration tree is pruned by bounds in Case (iib) and the enumeration is complete. The optimal solution is $x_1 = 1$, $x_2 = 3$ with value 11.8. The complete enumeration tree is shown in Fig. 1.4.

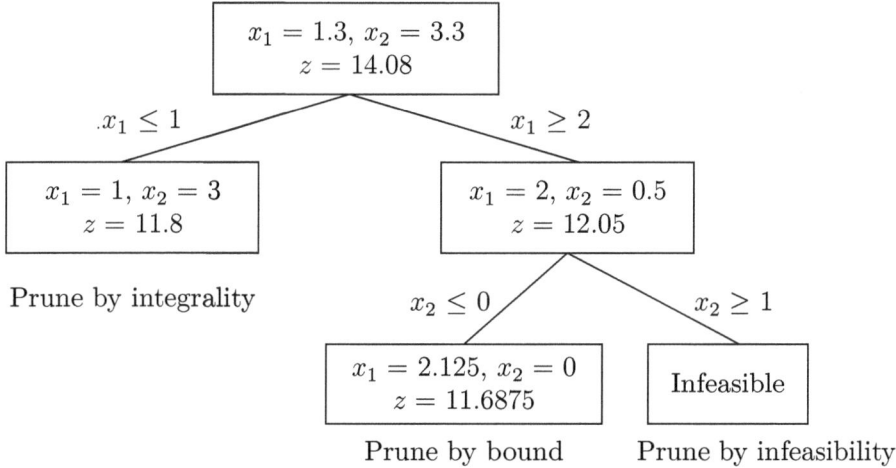

Figure 1.4: Example of a branch-and-bound tree

The procedure that we just described searches for an optimal solution by *branching*, i.e., partitioning the set S into subsets, and it attempts to prune the enumeration by *bounding* the objective value of the subproblems generated by this partition. A *branch-and-bound* algorithm is based on these two principles. In the procedure described above, the branching step creates two subproblems obtained by restricting the range of a variable. This branching strategy is called *variable branching*. Bounding the objective value of a subproblem is done by solving its natural linear programming relaxation. This bounding strategy is called *linear programming bounding*. Variable branching and linear programming bounding are widely used in state-of-the-art integer programming solvers but they are not the only way to implement a branch-and-bound algorithm. We discuss other strategies in Chaps. 8 and 9. Next we formalize the steps of a branch-and-bound algorithm based on linear programming bounding.

The branch-and-bound algorithm keeps a list of linear programming problems obtained by relaxing the integrality requirements on the variables

x_j, $j = 1, \ldots, n$, and imposing linear constraints, such as bounds on the variables $x_j \leq u_j$ or $x_j \geq l_j$. Each such linear program corresponds to a *node* of the enumeration tree. For a node N_i, let z_i denote the value of the corresponding linear program LP_i. Node N_0 is associated with the linear programming relaxation (1.5). Let $\mathcal{L}$ denote the list of nodes that must still be solved (i.e., that have not been pruned nor branched on). Let $\underline{z}$ denote a lower bound on the optimum value z^* (initially, the bound $\underline{z}$ can be derived from a heuristic solution of MILP, or it can be set to $-\infty$).

Branch-and-Bound Algorithm

0. Initialize

$\mathcal{L} := \{N_0\}$, $\underline{z} := -\infty$, $(x^*, y^*) := \emptyset$.

1. Terminate?

If $\mathcal{L} = \emptyset$, the solution (x^*, y^*) is optimal.

2. Select node

Choose a node N_i in $\mathcal{L}$ and delete it from $\mathcal{L}$.

3. Bound

Solve LP_i. If it is infeasible, go to Step 1. Else, let (x^i, y^i) be an optimal solution of LP_i and z_i its objective value.

4. Prune

If $z_i \leq \underline{z}$, go to Step 1.

If (x^i, y^i) is feasible to MILP, set $\underline{z} := z_i$, $(x^*, y^*) := (x^i, y^i)$ and go to Step 1.

Otherwise:

5. Branch

From LP_i, construct $k \geq 2$ linear programs $\text{LP}_{i_1}, \ldots, \text{LP}_{i_k}$ with smaller feasible regions whose union does not contain (x^i, y^i), but contains all the solutions of LP_i with $x \in \mathbb{Z}^n$. Add the corresponding new nodes $N_{i_1}, \ldots, N_{i_k}$ to $\mathcal{L}$ and go to Step 1.

Various choices are left open in this algorithm, such as the node selection criterion in Step 2 and the branching strategy in Step 5. We will discuss options for these choices in Chap. 9. Even more important to the success

1.2. METHODS FOR SOLVING INTEGER PROGRAMS

of branch-and-bound algorithms is the ability to prune the tree (Step 4). This will occur when $\underline{z}$ is a tight lower bound on z^* and when z_i is a tight upper bound. Good lower bounds $\underline{z}$ are obtained using heuristics. Heuristics can be designed either "ad hoc" for a specific application, or as part of the branch-and-bound algorithm (by choosing branching and node selection strategies that are more likely to produce good feasible solutions (x^i, y^i) to MILP in Step 4). We discuss such heuristics in Chap. 9 (Sect. 9.2). In order to get good upper bounds z_i, it is crucial to formulate MILP in such a way that the value of its natural linear programming relaxation z_0 is as close as possible to z^*. Formulations will be discussed extensively in Chap. 2. To summarize, four issues need attention when solving integer programs by branch-and-bound algorithms.

- Formulation (so that the gap $z_0 - z^*$ is small),
- Heuristics (to find a good lower bound $\underline{z}$),
- Branching,
- Node selection.

The formulation issue is a central one in integer programming and it will be discussed extensively in this book (Chap. 2 but also Chaps. 3–7). One way to tighten the formulation is by adding cutting planes. We introduce this idea in the next section.

1.2.2 The Cutting Plane Method

Consider the integer program

$$\text{MILP}: \quad \max\{cx + hy : (x, y) \in S\}$$

where, as earlier, $S := \{(x, y) \in \mathbb{Z}^n_+ \times \mathbb{R}^p_+ : Ax + Gy \leq b\}$. Let P_0 be the natural linear relaxation of S. Solve the linear program

$$\max\{cx + hy : (x, y) \in P_0\}. \tag{1.6}$$

Let z_0 be its optimal value and (x^0, y^0) an optimal solution. We may assume that (x^0, y^0) is a *basic* optimal solution of (1.6) (this notion will be defined precisely in Chap. 3). Optimal basic solutions can be computed by standard linear programming algorithms.

We present now a second strategy for dealing with the case when the solution (x^0, y^0) is not in S. The idea is to find an inequality $\alpha x + \gamma y \leq \beta$ that is satisfied by every point in S and such that $\alpha x^0 + \gamma y^0 > \beta$. The existence of such an inequality is guaranteed when (x^0, y^0) is a basic solution of (1.6).

An inequality $\alpha u \leq \beta$ is *valid* for a set $K \subseteq \mathbb{R}^d$ if it is satisfied by every point $\bar{u} \in K$. A valid inequality $\alpha x + \gamma y \leq \beta$ for S that is violated by (x^0, y^0) is a *cutting plane* separating (x^0, y^0) from S. Let $\alpha x + \gamma y \leq \beta$ be a cutting plane and define

$$P_1 := P_0 \cap \{(x, y) : \alpha x + \gamma y \leq \beta\}.$$

Since $S \subseteq P_1 \subset P_0$, the linear programming relaxation of MILP based on P_1 is stronger than the natural linear programming relaxation (1.5), in the sense that the optimal value of the linear program

$$\max\{cx + hy : (x, y) \in P_1\}$$

is at least as good an upper-bound on the value z^* as z_0, while the optimal solution (x^0, y^0) of the natural linear programming relaxation does not belong to P_1. The recursive application of this idea leads to the *cutting plane approach*.

Cutting Plane Algorithm

Starting with $i = 0$, repeat:

Recursive Step. Solve the linear program $\max\{cx + hy : (x, y) \in P_i\}$.

- If the associated optimal basic solution (x^i, y^i) belongs to S, stop.
- Otherwise solve the *separation problem*

  Find a cutting plane $\alpha x + \gamma y \leq \beta$ separating (x^i, y^i) from S.

  Set $P_{i+1} := P_i \cap \{(x, y) : \alpha x + \gamma y \leq \beta\}$ and repeat the recursive step.

The separation problem that needs to be solved in the cutting plane algorithm is a central issue in integer programming. If the basic solution (x^i, y^i) is not in S, there are infinitely many cutting planes separating (x^i, y^i) from S. How does one produce effective cuts? Usually, there is a tradeoff between the running time of a separation procedure and the quality of the cutting planes it produces. In practice, it may also be preferable to generate

1.2. METHODS FOR SOLVING INTEGER PROGRAMS

several cutting planes separating (x^i, y^i) from S, instead of a single cut as suggested in the above algorithm, and to add them all to P_i to create problem P_{i+1}. We will study several separation procedures in this book (Chaps. 5–7).

For now, we illustrate the cutting plane approach on our two-variable example:
$$\begin{aligned} \max \quad & 5.5x_1 + 2.1x_2 \\ & -x_1 + x_2 \leq 2 \\ & 8x_1 + 2x_2 \leq 17 \\ & x_1, x_2 \geq 0 \\ & x_1, x_2 \text{ integer.} \end{aligned} \quad (1.7)$$

We first introduce a variable z representing the objective function and slack variables x_3 and x_4 to turn the inequality constraints into equalities. The problem becomes to maximize z subject to
$$\begin{aligned} z - 5.5x_1 - 2.1x_2 & = 0 \\ -x_1 + x_2 + x_3 & = 2 \\ 8x_1 + 2x_2 + x_4 & = 17 \\ x_1, x_2, x_3, x_4 \geq 0 \text{ integer.} & \end{aligned}$$

Note that x_3 and x_4 can be constrained to be integer because the data in the constraints of (1.7) are all integers.

Solving the linear programming relaxation using standard techniques, we get the optimal tableau:
$$\begin{aligned} z \quad & +0.58x_3 + 0.76x_4 = 14.08 \\ x_2 \quad & +0.8x_3 + 0.1x_4 = 3.3 \\ x_1 \quad & -0.2x_3 + 0.1x_4 = 1.3 \\ & x_1, x_2, x_3, x_4 \geq 0. \end{aligned}$$

The corresponding basic solution is $x_3 = x_4 = 0$, $x_1 = 1.3$, $x_2 = 3.3$ with objective value $z = 14.08$. Since the values of x_1 and x_2 are not integer, this is not a solution of (1.7). We can generate a cut from the constraint $x_2 + 0.8x_3 + 0.1x_4 = 3.3$ in the above tableau by using the following reasoning. Since x_2 is an integer variable, we have
$$0.8x_3 + 0.1x_4 = 0.3 + k \quad \text{where } k \in \mathbb{Z}.$$

Since the left-hand side is nonnegative for every feasible solution of (1.7), we must have $k \geq 0$, which implies
$$0.8x_3 + 0.1x_4 \geq 0.3 \quad (1.8)$$

This is the famous Gomory fractional cut [175]. Note that it cuts off the above fractional solution $x_3 = x_4 = 0$, $x_1 = 1.3$, $x_2 = 3.3$. More generally, if nonnegative integer variables $x_1, \ldots, x_n$ satisfy the equation

$$\sum_{j=1}^{n} a_j x_j = a_0$$

where $a_0 \notin \mathbb{Z}$, the Gomory fractional cut is

$$\sum_{j=1}^{n} (a_j - \lfloor a_j \rfloor) x_j \geq a_0 - \lfloor a_0 \rfloor. \tag{1.9}$$

This inequality is satisfied by any $x \in \mathbb{Z}_+^n$ satisfying the equation $\sum_{j=1}^n a_j x_j = a_0$ because $\sum_{j=1}^n a_j x_j = a_0$ implies $\sum_{j=1}^n (a_j - \lfloor a_j \rfloor) x_j = a_0 - \lfloor a_0 \rfloor + k$ for some integer k; furthermore $k \geq 0$ since the left-hand side is nonnegative.

Let us return to our example. Since $x_3 = 2 + x_1 - x_2$ and $x_4 = 17 - 8x_1 - 2x_2$, we can express Gomory's fractional cut (1.8) in the space (x_1, x_2). This yields $x_2 \leq 3$ (see Cut 1 in Fig. 1.5).

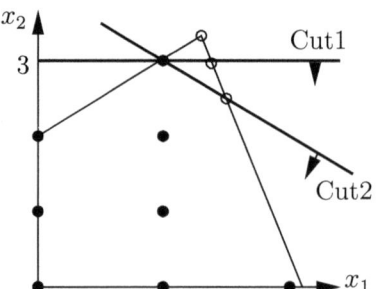

Figure 1.5: The first two cuts in the cutting plane algorithm

Adding this cut to the linear programming relaxation, we get:

$$\begin{aligned}
\max \quad & 5.5x_1 + 2.1x_2 \\
& -x_1 + x_2 \leq 2 \\
& 8x_1 + 2x_2 \leq 17 \\
& x_2 \leq 3 \\
& x_1, x_2 \geq 0.
\end{aligned}$$

1.2. METHODS FOR SOLVING INTEGER PROGRAMS

We introduce a slack variable x_5 for the inequality $x_2 \leq 3$, and observe as before that x_5 can be assumed to be integer. Solving this linear program, we find the optimal tableau

$$
\begin{array}{rlrll}
z & & +0.6875x_4 & +0.725x_5 & = 13.8625 \\
& x_3 & +0.125x_4 & -1.25x_5 & = 0.375 \\
x_1 & & +0.125x_4 & -0.25x_5 & = 1.375 \\
& x_2 & & +x_5 & = 3 \\
\end{array}
$$
$$x_1, x_2, x_3, x_4, x_5 \geq 0.$$

The corresponding basic solution in the (x_1, x_2)-space is $x_1 = 1.375$, $x_2 = 3$ with value $z = 13.8625$. Since x_1 is fractional, we need to generate another cut. From the row of the tableau $x_1 + 0.125x_4 - 0.25x_5 = 1.375$, we generate the new fractional cut according to (1.9), namely $(0.125 - \lfloor 0.125 \rfloor)x_4 + (-0.25 - \lfloor -0.25 \rfloor)x_5 \geq 1.375 - \lfloor 1.375 \rfloor$, which is $0.125x_4 + 0.75x_5 \geq 0.375$. Replacing $x_4 = 17 - 8x_1 - 2x_2$ and $x_5 = 3 - x_2$ we get (see Cut 2 in Fig. 1.5):

$$x_1 + x_2 \leq 4.$$

Adding this cut and solving the updated linear program, we find a new optimal solution $x_1 = 1.5$, $x_2 = 2.5$ with value $z = 13.5$. This solution is again fractional. Two more iterations are needed to obtain the optimal integer solution $x_1 = 1$, $x_2 = 3$ with value $z = 11.8$. We leave the last two iterations as an exercise for the reader (Exercise 1.6).

1.2.3 The Branch-and-Cut Method

In the branch-and-bound approach, the tightness of the upper bound is crucial for pruning the enumeration tree (Step 4). Tighter upper bounds can be calculated by applying the cutting plane approach to the subproblems. This leads to the *branch-and-cut* approach, which is currently the most successful method for solving integer programs. It is obtained by adding a cutting-plane step before the branching step in the branch-and-bound algorithm of Sect. 1.2.1:

Branch-and-Cut Algorithm

0. Initialize

$\mathcal{L} := \{N_0\}$, $\underline{z} := -\infty$, $(x^*, y^*) := \emptyset$.

1. **Terminate?**

   If $\mathcal{L} = \emptyset$, the solution (x^*, y^*) is optimal.

2. **Select node**

   Choose a node N_i in $\mathcal{L}$ and delete it from $\mathcal{L}$.

3. **Bound**

   Solve LP_i. If it is infeasible, go to Step 1. Else, let (x^i, y^i) be an optimal solution of LP_i and z_i its objective value.

4. **Prune**

   If $z_i \leq \underline{z}$, go to Step 1.

   If (x^i, y^i) is feasible to MILP, set $\underline{z} := z_i$, $(x^*, y^*) := (x^i, y^i)$ and go to Step 1.

   Otherwise:

5. **Add Cuts?**

   Decide whether to strengthen the formulation LP_i or to branch.

   In the first case, strengthen LP_i by adding cutting planes and go back to Step 3.

   In the second case, go to Step 6.

6. **Branch**

   From LP_i, construct $k \geq 2$ linear programs $\text{LP}_{i_1}, \ldots, \text{LP}_{i_k}$ with smaller feasible regions whose union does not contain (x^i, y^i), but contains all the solutions of LP_i with $x \in \mathbb{Z}^n$. Add the corresponding new nodes $N_{i_1}, \ldots, N_{i_k}$ to $\mathcal{L}$ and go to Step 1.

The decision of whether to add new cuts in Step 5 is made empirically based on the success of previously added cuts and characteristics of the new cuts such as their density (the fraction of nonzero coefficients in the cut). Typically several rounds of cuts are added at the root node N_0, while fewer or no cuts might be generated deeper in the enumeration tree.

1.3 Complexity

When analyzing an algorithm to solve a problem, a key question is the computing time needed to find a solution.

1.3.1 Problems, Instances, Encoding Size

By *problem*, we mean a question to be answered for any given set of data. An example is the *linear programming* problem: Given positive integers m, n, an $m \times n$ matrix A, m-vector b, and n-vector c, find an n-vector x that solves
$$\begin{aligned} \max \quad & cx \\ & Ax \leq b \\ & x \geq 0. \end{aligned}$$

An *instance* of a problem is a specific data set. For example,
$$\begin{aligned} \max \quad & 5.5x_1 + 2.1x_2 \\ & -x_1 + x_2 \leq 2 \\ & 8x_1 + 2x_2 \leq 17 \\ & x_1, x_2 \geq 0 \end{aligned}$$
is an instance of a linear program. One measures the *size* of an instance by the space required to write down the data. We assume that all instances of a problem are provided in a "standard" way. For example, we may assume that all integers are written in binary encoding. With this assumption, to encode an integer n we need one bit to represent the sign and $\lceil \log_2(|n|+1) \rceil$ to write $|n|$ in base 2. Thus the *encoding size* of an integer n is $1 + \lceil \log_2(|n| + 1) \rceil$. For a rational number $\frac{p}{q}$ where p is an integer and q is a positive integer, the encoding size is $1 + \lceil \log_2(|p| + 1) \rceil + \lceil \log_2(q + 1) \rceil$. The encoding size of a rational vector or matrix is the sum of the encoding sizes of its entries.

Remark 1.1. *Given a rational n-dimensional vector $v = (\frac{p_1}{q_1}, \ldots, \frac{p_n}{q_n})$, let L be its encoding size and D the least common multiple of $q_1, \ldots, q_n$. Then Dv is an integral vector whose encoding size is at most nL.*

Proof. The absolute value of the ith component of the vector Dv is bounded above by $|p_i|q_1 \cdots q_n$. The encoding size of this integer number is bounded above by L. □

In this book we will consider systems of linear inequalities and equations with rational coefficients. Since multiplying any such system by a positive number does not change the set of solutions, the above remark shows that we can consider an equivalent system with integer coefficients whose encoding size is polynomially bounded by the size of the original system. By "polynomially bounded" we mean the following. Function $f : S \to \mathbb{R}_+$ is *polynomially bounded* by $g : S \to \mathbb{R}_+$ if there exists a polynomial $\phi : \mathbb{R} \to \mathbb{R}$ such that $f(s) \leq \phi(g(s))$ for every $s \in S$. For example, s^2 is polynomially bounded by s since $\phi(x) = x^2$ is a polynomial that satisfies the definition.

In Remark 1.1, the encoding size of Dv is at most L^2 since $L \geq n$. This shows that the encoding size of Dv is polynomially bounded by the encoding size of v.

We will also need the following more general definition. Given real valued functions $f, g_1, \ldots, g_t$ defined on some set S, we say that f is *polynomially bounded* by $g_1, \ldots, g_t$ if there exists a polynomial $\phi : \mathbb{R}^t \to \mathbb{R}$ such that $f(s) \leq \phi(g_1(s), \ldots, g_t(s))$ for every $s \in S$.

Let $f : S \to \mathbb{R}_+$ and $g : S \to \mathbb{R}_+$ be two functions, where S is an unbounded subset of $\mathbb{R}_+$. One writes $f(x) = O(g(x))$ if there exists a positive real number M and an $x_0 \in S$ such that $f(x) \leq Mg(x)$ for every $x > x_0$. This notion can be extended to several real variables: $f(x_1, \ldots, x_k) = O(g(x_1, \ldots, x_k))$ if there exist positive real numbers M and x_0 such that $f(x_1, \ldots, x_k) \leq Mg(x_1, \ldots, x_k)$ for every $x_1 > x_0, \ldots, x_k > x_0$. For example, considering the set S of all rational vectors v in Remark 1.1, we can write that the encoding size of Dv is $O(nL)$.

1.3.2 Polynomial Algorithm

An *algorithm* for solving a problem is a procedure that, given any possible instance, produces the correct answer in a finite amount of time. For example, Dantzig's simplex method with an anti-cycling rule is an algorithm for solving the linear programming problem [102]. Karmarkar's interior point algorithm is another. An algorithm is said to solve a problem in *polynomial time* if its running time, measured as the number of arithmetic operations carried out by the algorithm (a function f defined on the set S of instances), is polynomially bounded by the encoding size of the input (a function g defined on S). Instead of polynomial-time algorithm, we often just say *polynomial algorithm*. Karmarkar's interior point algorithm is a polynomial algorithm for linear programming but the simplex method is not (Klee and Minty [240] gave a family of instances where the simplex method with the largest cost pivoting rule (Dantzig's rule) takes an exponential number of steps as a function of the instance size). A problem is said to be in the *complexity class* P if there exists a polynomial algorithm to solve it. For example linear programming is in P since Karmarkar's algorithm solves it in polynomial time. Specifically, if L denotes the encoding size of a linear program with n variables, Karmarkar's algorithm performs $O(n^{3.5}L)$ arithmetic operations, where the sizes of numbers during the entire execution is $O(L)$. The running time was later improved to $O(n^3 L)$ by Renegar [314].

1.3. COMPLEXITY

Another issue in computational complexity is the encoding size of the output as a function of the encoding size of the input. For example, given some input A, b, what is the encoding size of a solution to $Ax = b$?

Proposition 1.2. *Let A be a nonsingular $n \times n$ rational matrix and b a rational n-vector. The encoding size of the unique solution of $Ax = b$ is polynomially bounded by the encoding size of (A, b).*

Proof. By Remark 1.1, we may assume that (A, b) has integer entries. Let θ be the largest absolute value of an entry in (A, b). The absolute value of the determinant of any square $n \times n$ submatrix of (A, b) is at most $n!\theta^n$ (by using the standard formula for computing determinants as the sum of $n!$ products of n entries). The encoding size of $n!\theta^n$ is $O(n(\log n + \log(1 + \theta)))$, thus it is polynomially bounded by n and the encoding size of θ. By Cramer's rule, each entry of $A^{-1}b$ is the ratio of two such determinants. Now the proposition follows by observing that the encoding size of (A, b) is at least $n + \lceil \log_2(1 + \theta) \rceil$. □

1.3.3 Complexity Class NP

A *decision problem* is a problem whose answer is either "yes" or "no." An important complexity class is NP, which stands for "nondeterministic polynomial-time." Intuitively, NP is the class of all decision problems for which the "yes"-answer has a *certificate* that can be checked in polynomial time. The complexity class Co-NP is the class of all decision problems for which the "no"-answer has a certificate that can be checked in polynomial time, see Exercises 1.10, 1.11 and 1.12.

Recall that a set of the form $S := \{(x, y) \in \mathbb{Z}_+^n \times \mathbb{R}_+^p : Ax + Gy \leq b\}$ is called a mixed integer linear set. We assume the data to be rational. We will show in Chap. 4 that the question "given a mixed integer linear set, is it nonempty?" is in NP. Indeed, we will show that, if S contains a solution, then it contains one whose encoding size is polynomial in the encoding size of the input (A, G, b). Given such a solution, one can therefore verify in polynomial time that it satisfies the constraints. On the other hand, the question "given a mixed integer linear set, is it empty?" does not have any obvious certificate that can be checked in polynomial time and, in fact, it is conjectured that this problem is not in NP.

A decision problem Q in NP is said to be *NP-complete* if all other problems D in NP are *reducible* to Q in polynomial time. That is there exists a polynomial algorithm that, for every instance I of D, produces an instance of Q whose answer is "yes" if and only if the answer to I is yes. This implies that, up to a polynomial factor, solving Q requires at least as much computing time as solving any problem in NP. A fundamental result of Cook [84] is that the question "given a mixed integer linear set, is it nonempty?" is NP-complete. Analogously, a decision problem Q in co-NP is said to be *co-NP-complete* if all other problems D in co-NP are reducible to Q in polynomial time.

A problem Q (not necessarily in NP or a decision problem) is said to be *NP-hard* if all problems D in NP are reducible to Q in polynomial time. That is, there exist two polynomial algorithms such that the first produces an instance $J(I)$ of Q for any given instance I of D, and the second produces the correct answer for I given a solution for $J(I)$. In particular, integer programming is NP-hard.

In this book, we will focus on integer programs with linear objective and constraints. Nonlinear constraints can make the problem much harder. In fact, Jeroslow [210] showed that integer programming with quadratic constraints is *undecidable*, i.e., there cannot exist an algorithm to solve this problem. In particular, if no solution is found after a very large amount of time, it may still not be possible to conclude that none exists.

1.4 Convex Hulls and Perfect Formulations

A set $S \subseteq \mathbb{R}^n$ is *convex* if, for any two distinct points in S, the line segment joining them is also in S, i.e., if $x, y \in S$ then $\lambda x + (1 - \lambda)y \in S$ for all $0 \leq \lambda \leq 1$.

It follows from the definition that the intersection of an arbitrary family of convex sets is a convex set. In particular, given a set $S \subseteq \mathbb{R}^n$, the intersection of all convex sets containing S is itself a convex set containing S, and it is therefore the inclusionwise minimal convex set containing S. The inclusionwise minimal convex set containing S is the *convex hull* of $S \subseteq \mathbb{R}^n$, and it is denoted by $\mathrm{conv}(S)$. Figure 1.6 illustrates the notion of convex hull.

1.4. CONVEX HULLS AND PERFECT FORMULATIONS

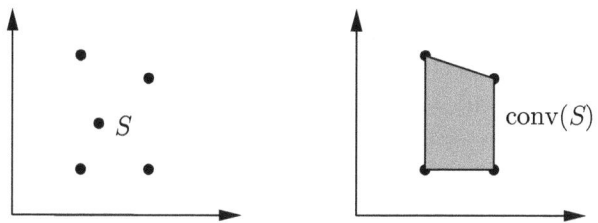

Figure 1.6: Convex hull of a set S of five points in $\mathbb{R}^2$

Next we give a characterization of conv(S) that will be useful in the remainder. A point $x \in \mathbb{R}^n$ is a *convex combination* of points in S if there exists a finite set of points $x^1, \ldots, x^p \in S$ and scalars $\lambda_1, \ldots, \lambda_p$ such that

$$x = \sum_{j=1}^{p} \lambda_j x^j, \quad \sum_{j=1}^{p} \lambda_j = 1, \quad \lambda_1, \ldots, \lambda_p \geq 0.$$

Note that the word "finite" is important in the above definition.

By convexity, every convex set containing S must contain all convex combinations of points in S. Conversely, it is easy to verify that the set of all convex combinations of points in S is itself a convex set containing S (Exercise 1.17). It follows that

$$\text{conv}(S) = \{x \in \mathbb{R}^n : x \text{ is a convex combination of points in } S\}. \quad (1.10)$$

Next, we show that optimizing a linear function cx over $S \subseteq \mathbb{R}^n$ is equivalent to optimizing cx over conv(S).

Lemma 1.3. *Let $S \subset \mathbb{R}^n$ and $c \in \mathbb{R}^n$. Then $\sup\{cx : x \in S\} = \sup\{cx : x \in \text{conv}(S)\}$.*
Furthermore, the supremum of cx is attained over S if and only if it is attained over conv(S).

Proof. Since $S \subseteq \text{conv}(S)$, we have $\sup\{cx : x \in S\} \leq \sup\{cx : x \in \text{conv}(S)\}$. We prove $\sup\{cx : x \in S\} \geq \sup\{cx : x \in \text{conv}(S)\}$. Let $z^* := \sup\{cx : x \in S\}$. We assume that $z^* < +\infty$, otherwise the statement is trivially satisfied. Let $H := \{x \in \mathbb{R}^n : cx \leq z^*\}$. Note that H is convex and by definition of z^* it contains S. By definition conv(S) is contained in every convex set containing S, therefore conv(S) $\subseteq H$, i.e., $\sup\{cx : x \in \text{conv}(S)\} \leq z^*$.

We now show the second part of the statement. For the "only if" direction, assume $\sup\{cx : x \in S\} = c\bar{x}$ for some $\bar{x} \in S$. Then clearly $\bar{x} \in \text{conv}(S)$ and it follows from the first part that $\sup\{cx : x \in \text{conv}(S)\} = c\bar{x}$.

For the "if" direction, assume there exists $\bar{x} \in \text{conv}(S)$ such that $\sup\{cx : x \in \text{conv}(S)\} = c\bar{x}$. By (1.10), $\bar{x} = \sum_{i=1}^{k} \lambda_i x^i$ for some finite number of points $x^1, \ldots, x^k \in S$ and $\lambda_1, \ldots, \lambda_k > 0$ such that $\sum_{i=1}^{k} \lambda_i = 1$. By definition of $\bar{x}$, $c\bar{x} \geq cx^i$ for $i = 1, \ldots, k$, thus $c\bar{x} = \sum_{i=1}^{k} \lambda_i(cx^i) \leq c\bar{x} \sum_{i=1}^{k} \lambda_i = c\bar{x}$. Since $\lambda_i > 0$, it follows that $c\bar{x} = cx^i$ for $i = 1, \ldots, n$. Thus the supremum over S is achieved by $x^1, \ldots, x^k$. □

In this book we will be concerned with the case where S is a mixed integer linear set, that is, a set of the form $S := \{(x,y) \in \mathbb{Z}_+^n \times \mathbb{R}_+^p : Ax + Gy \leq b\}$ where A, G, b have rational entries. A fundamental result of Meyer, which will be proved in Chap. 4, states that in this case $\text{conv}(S)$ is a set of the form $\{(x,y) : A'x + G'y \leq b'\}$ where A', G', b' have rational entries. The first part of Lemma 1.3 shows that, for any objective function cx, the integer program $\max\{cx + hy : (x,y) \in S\}$ and the linear program $\max\{cx + hy : A'x + G'y \leq b'\}$ have the same value. A well-known fact in the theory of linear programming implies that this linear program is infeasible, or unbounded or admits a finite optimal solution, i.e., $\sup\{cx + hy : A'x + G'y \leq b'\} = \max\{cx + hy : A'x + G'y \leq b'\}$. Therefore, the second part of Lemma 1.3 implies that the integer program $\max\{cx + hy : (x,y) \in S\}$ admits an optimal solution whenever the linear program is feasible and bounded.

The above discussion illustrates that, in principle, in order to solve the integer program $\max\{cx + hy : Ax + Gy \leq b, x \geq 0 \text{ integral}, y \geq 0\}$ it is sufficient to solve the linear program $\max\{cx + hy : A'x + G'y \leq b'\}$. A central question in integer programming is the constructive aspect of this linear program: Given A, G, b, how does one compute A', G', b'?

Clearly the system of inequalities $A'x + G'y \leq b'$ also provides a formulation for the mixed integer set $S := \{(x,y) \in \mathbb{Z}_+^n \times \mathbb{R}_+^p : Ax + Gy \leq b\}$. In other words, we can also write $S := \{(x,y) \in \mathbb{Z}_+^n \times \mathbb{R}_+^p : A'x + G'y \leq b'\}$. This new formulation has the property that for every objective function $cx + hy$, the integer program can be solved as a linear program, disregarding the integrality requirement on the vector x. We call such a formulation a *perfect formulation*. When there are no continuous variables y, the set $\{x \in \mathbb{R}^n : A'x \leq b'\}$ defined by a perfect formulation $A'x \leq b'$ is called an *integral polyhedron*.

1.4.1 Example: A Two-Dimensional Mixed Integer Set

In this section we describe the convex hull of the following mixed integer linear set (see Fig. 1.7):

$$S := \{(x,y) \in \mathbb{Z} \times \mathbb{R}_+ : x - y \leq \beta\}.$$

1.4. CONVEX HULLS AND PERFECT FORMULATIONS

Lemma 1.4. *Consider the 2-variable mixed integer linear set $S := \{(x, y) \in \mathbb{Z} \times \mathbb{R}_+ : x - y \leq \beta\}$. Let $f := \beta - \lfloor \beta \rfloor$. Then*

$$x - \frac{1}{1-f} y \leq \lfloor \beta \rfloor \qquad (1.11)$$

is a valid inequality for S.

Proof. We prove that the inequality (1.11) is satisfied by every $(x, y) \in S$. Note that, since $x \in \mathbb{Z}$, either $x \leq \lfloor \beta \rfloor$ or $x \geq \lfloor \beta \rfloor + 1$. If $x \leq \lfloor \beta \rfloor$, then adding this inequality to the inequality $-y \leq 0$ multiplied by $\frac{1}{1-f}$ yields (1.11). If $x \geq \lfloor \beta \rfloor + 1$, then summing the inequality $-x \leq -\lfloor \beta \rfloor - 1$ multiplied by $\frac{f}{1-f}$ and the inequality $x - y \leq \beta$ multiplied by $\frac{1}{1-f}$ yields (1.11). □

Note that the assumption $y \geq 0$ is critical in the above derivation, whereas we can have indifferently $x \in \mathbb{Z}$ or $x \in \mathbb{Z}_+$.

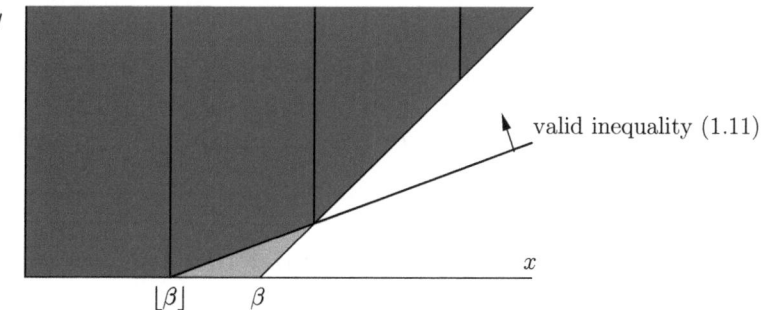

Figure 1.7: Illustration of Lemma 1.4

The three inequalities

$$\begin{aligned} x - y &\leq \beta \\ x - \frac{1}{1-f} y &\leq \lfloor \beta \rfloor \\ y &\geq 0 \end{aligned}$$

are valid for S and we show in the next proposition that they provide a perfect formulation for conv(S), that is, no other inequality is needed to describe conv(S).

Proposition 1.5. *Let $S := \{(x, y) \in \mathbb{Z} \times \mathbb{R}_+ : x - y \leq \beta\}$. Then*

$$\mathrm{conv}(S) = \left\{ (x, y) \in \mathbb{R}^2 : x - y \leq \beta,\ x - \frac{1}{1-f} y \leq \lfloor \beta \rfloor,\ y \geq 0 \right\}.$$

Proof. Let $Q := \{(x,y) \in \mathbb{R}^2 : x - y \leq \beta, \ x - \frac{1}{1-f}y \leq \lfloor\beta\rfloor, \ y \geq 0\}$. We need to show that $\text{conv}(S) = Q$. By Lemma 1.4, $\text{conv}(S) \subseteq Q$. To show the reverse inclusion, we show that every point $(\bar{x}, \bar{y}) \in Q$ is a convex combination of two points (x^1, y^1) and (x^2, y^2) in S.

We may assume $\bar{x} \notin \mathbb{Z}$, since otherwise the result trivially holds with $(x^1, y^1) = (x^2, y^2) = (\bar{x}, \bar{y})$. Let $\lambda = \bar{x} - \lfloor\bar{x}\rfloor$. Note that $0 < \lambda < 1$. We consider three possible cases.

If $\bar{x} < \lfloor\beta\rfloor$, let $(x^1, y^1) = (\lfloor\bar{x}\rfloor, \bar{y})$ and $(x^2, y^2) = (\lceil\bar{x}\rceil, \bar{y})$. Clearly $(\bar{x}, \bar{y}) = (1-\lambda)(x^1, y^1) + \lambda(x^2, y^2)$, and it is immediate to verify that $(x^1, y^1), (x^2, y^2)$ are both in S.

If $\lfloor\beta\rfloor < \bar{x} < \lceil\beta\rceil$, let $(x^1, y^1) = (\lfloor\bar{x}\rfloor, \bar{y} - \lambda(1-f))$ and $(x^2, y^2) = (\lceil\bar{x}\rceil, \bar{y} + (1-\lambda)(1-f))$. Note that $(\bar{x}, \bar{y}) = (1-\lambda)(x^1, y^1) + \lambda(x^2, y^2)$. We next show that $(x^1, y^1), (x^2, y^2)$ are both in S. Indeed, $y^2 \geq 0$ and $y^1 = -(1-f)(\bar{x} - \frac{\bar{y}}{1-f} - \lfloor\bar{x}\rfloor) \geq -(1-f)(\lfloor\beta\rfloor - \lfloor\bar{x}\rfloor) = 0$, while $x^1 - y^1 = \lfloor\bar{x}\rfloor - \bar{y} + \lambda(1-f) = \bar{x} - \bar{y} - \lambda f \leq \bar{x} - \bar{y} \leq \beta$ and $x^2 - y^2 = (1-f)(x^2 - \frac{y^2}{1-f}) + fx^2 = (1-f)(\bar{x} - \frac{\bar{y}}{1-f}) + f\lceil\beta\rceil \leq (1-f)\lfloor\beta\rfloor + f\lceil\beta\rceil = \beta$.

If $\bar{x} > \lceil\beta\rceil$, let $(x^1, y^1) = (\lfloor\bar{x}\rfloor, \bar{y} - \lambda)$ and $(x^2, y^2) = (\lceil\bar{x}\rceil, \bar{y} + 1 - \lambda)$. Note that $(\bar{x}, \bar{y}) = (1-\lambda)(x^1, y^1) + \lambda(x^2, y^2)$. We next show that $(x^1, y^1), (x^2, y^2)$ are both in S. Indeed, $y^2 \geq 0$ and $y^1 = \bar{y} - \lambda = \lfloor\bar{x}\rfloor + \bar{y} - \bar{x} \geq \lceil\beta\rceil - \beta \geq 0$, while $x^1 - y^1 = x^2 - y^2 = \bar{x} - \bar{y} \leq \beta$. $\square$

1.4.2 Example: A Two-Dimensional Pure Integer Set

Consider the following region in the plane:

$$P := \{(x_1, x_2) \in \mathbb{R}^2 : -a_1 x_1 + a_2 x_2 \leq 0, \ x_1 \leq u, \ x_2 \geq 0\}, \tag{1.12}$$

where a_1, a_2 and u are positive integers. Let $S := P \cap \mathbb{Z}^2$. Our goal is to characterize the inequalities that define $\text{conv}(S)$.

Figure 1.8 illustrates this problem when $a_1 = 19$, $a_2 = 12$ and $u = 5$.

Two of the inequalities that describe $\text{conv}(S)$ go through the origin. One of them is $x_2 \geq 0$. The other has a positive slope, which is determined by an integral point $(x_1^*, x_2^*) \neq (0,0)$ such that

$$\frac{x_2^*}{x_1^*} = \max\left\{\frac{x_2}{x_1} : (x_1, x_2) \in S\right\}. \tag{1.13}$$

This problem can be solved in polynomial time using continued fractions, see [237]. We do not present this algorithm here (we refer the interested reader to [196, 202]).

1.5. CONNECTIONS TO NUMBER THEORY

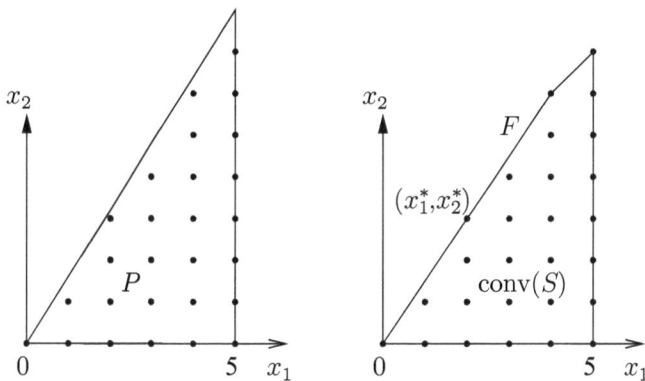

Figure 1.8: Illustration of problem (1.12)

The solution (x_1^*, x_2^*) of problem (1.13) is a point on the boundary segment F of conv(S) that starts at 0 and has positive slope. Once we have (x_1^*, x_2^*), it is easy to compute the vertex of conv(S) at the other end of the boundary segment F: Assuming that the fraction x_2^*/x_1^* is irreducible, this vertex is the point (kx_1^*, kx_2^*), where $k = \lfloor u/x_1^* \rfloor$.

If $kx_1^* < u$ the next step is to determine the other boundary segment of conv(S) that contains the point (kx_1^*, kx_2^*). If we move the origin to (kx_1^*, kx_2^*) and set $u := u - kx_1^*$, the problem is again to find a fraction as in (1.13). This process is iterated until the bound u is reached.

We now discuss the number of sides that conv(S) can have. Note that at each iteration $kx_1^* > u/2$ (by definition of k), thus the new value of u is at most half the previous value. This implies that the number of sides of conv(S) (including the horizontal and vertical sides) is at most $3 + \log_2 u$, so it is at most linear in the binary encoding of the data. Thus conv(S) can be computed in polynomial time when a_1, a_2, u are given.

1.5 Connections to Number Theory

Integer programs tend to come in two different flavors. Some, like the assignment problem introduced in Sect. 1.1, fall into the category of combinatorial optimization. Other integer programs are closely related to number theory. We illustrate the ties to number theory with two examples. The first one is the computation of the greatest common divisor of two integers, the second is the solution of Diophantine linear equations, that is, finding integer solutions to linear systems of equations.

1.5.1 The Greatest Common Divisor

Let $a, b \in \mathbb{Z}$, $(a, b) \neq (0, 0)$. The *greatest common divisor* of a and b, $\gcd(a, b)$, is the largest positive integer d that divides both a and b, i.e., $a = dx$ and $b = dy$ for some integers x, y. That is,

$$\gcd(a, b) := \max\{d \in \mathbb{Z}_+ \setminus \{0\} : d \mid a, \, d \mid b\},$$

where the notation $d \mid c$ means d divides c. The following proposition shows that the problem of computing $\gcd(a, b)$ can be formulated as a pure integer program.

Proposition 1.6. *Let $a, b \in \mathbb{Z}$, $(a, b) \neq (0, 0)$. Then*

$$\gcd(a, b) = \min\{ax + by : x, y \in \mathbb{Z}, \, ax + by \geq 1\}. \tag{1.14}$$

Proof. Let $m := \min\{ax + by : x, y \in \mathbb{Z}, \, ax + by \geq 1\}$ (it is clear that this set is not empty, as $(a, b) \neq (0, 0)$, thus this minimum exists). Clearly, if $d \mid a$ and $d \mid b$ then $d \mid ax + by$ for every $x, y \in \mathbb{Z}$ and so $d \mid m$. Thus $\gcd(a, b) \mid m$. We show that $m \mid \gcd(a, b)$. We need to show that $m \mid a$ and $m \mid b$. It suffices to show that $m \mid a$. Suppose not, then there exist $q, r \in \mathbb{Z}$ with $1 \leq r < m$ such that $a = qm + r$. Let $\bar{x}, \bar{y} \in \mathbb{Z}$ such that $m = a\bar{x} + b\bar{y}$. We have $r = a - q(a\bar{x} + b\bar{y}) = a(1 - q\bar{x}) - bq\bar{y}$ and so $r \in \{ax + by : x, y \in \mathbb{Z}, \, ax + by \geq 1\}$, a contradiction since $r < m$. □

The problem of finding the greatest common divisor of two given integers $(a, b) \neq (0, 0)$ can be solved by the *Euclidean algorithm*. Since $\gcd(a, b) = \gcd(|a|, |b|)$, we can assume that $a \geq b \geq 0$ and $a > 0$.

The Euclidean Algorithm

Input. $a, b \in \mathbb{Z}_+$ such that $a \geq b$ and $a > 0$.
Output. $\gcd(a, b)$.

Iterative Step.
If $b = 0$, return $\gcd(a, b) = a$.
Otherwise, let $r := a - \left\lfloor \dfrac{a}{b} \right\rfloor b$.
Let $a := b$, $b := r$ and repeat the iterative step.

Proposition 1.7. *Given $a, b \in \mathbb{Z}_+$ such that $a \geq b$ and $a > 0$, the Euclidean algorithm runs in polynomial time and correctly returns $\gcd(a, b)$.*

1.5. CONNECTIONS TO NUMBER THEORY

Proof. At each iteration the pair (a, b) is replaced by (b, r), where $r = a - \lfloor a/b \rfloor b$. Therefore $r < a/2$. Furthermore, in the pair (a, b), r replaces a after two iterations. Therefore the Euclidean algorithm terminates after at most $1 + 2 \log_2 a$ iterations. Noting that the size of the input a, b is $2 + \lceil \log_2(a+1) \rceil + \lceil \log_2(b+1) \rceil$ (using binary encoding of integers as discussed in Sect. 1.3.1) and that the work in each iteration is polynomial in this size, we conclude that the Euclidean algorithm runs in polynomial time.

Note that $d \mid a$ and $d \mid b$ if and only if $d \mid b$ and $d \mid r$. It follows that the set of common divisors of (a, b) is precisely the set of common divisors of (b, r). Thus $\gcd(a, b) = \gcd(b, r)$. Since $\gcd(a, 0) = a$, the algorithm correctly computes $\gcd(a, b)$. $\square$

Integral Solutions of a Linear Equation

We now show that the Euclidean algorithm can be used to find all integral solutions of the equation $ax + by = c$ where $a, b, c \in \mathbb{Z}$ and $(a, b) \neq (0, 0)$. We may assume without loss of generality that $a \geq b \geq 0$ and $a > 0$. At each iteration, whenever $b \neq 0$, the Euclidean algorithm replaces the vector (a, b) with the vector (b, r), where $r = a - \lfloor a/b \rfloor b$. This exchange can be performed in terms of matrices as follows:

$$(a, b) \begin{pmatrix} 0 & 1 \\ 1 & -\lfloor a/b \rfloor \end{pmatrix} = (b, r).$$

Let $T_{(a,b)}$ be the above 2×2 matrix. Note that its inverse is also an integral matrix, namely

$$T_{(a,b)}^{-1} = \begin{pmatrix} \lfloor a/b \rfloor & 1 \\ 1 & 0 \end{pmatrix}.$$

Let $(a_1, b_1), \ldots, (a_k, b_k)$ be the sequence produced by the Euclidean algorithm, where $(a_1, b_1) := (a, b)$, $(a_k, b_k) := (\gcd(a, b), 0)$, and $(a_i, b_i) := (b_{i-1}, a_{i-1} - \lfloor a_{i-1}/b_{i-1} \rfloor b_{i-1})$ for $i = 2, \ldots, k$. Define

$$T := T_{(a_1, b_1)} T_{(a_2, b_2)} \cdots T_{(a_k, b_k)}.$$

It follows that

$$(a, b) T = (\gcd(a, b), 0),$$

and $T^{-1} = T_{(a_k, b_k)}^{-1} T_{(a_{k-1}, b_{k-1})}^{-1} \cdots T_{(a_1, b_1)}^{-1}$ is an integral matrix.

Theorem 1.8. *Let a, b be integers not both 0, and let $g := \gcd(a, b)$. Equation $ax + by = c$ admits an integral solution if and only if c is an integer and g divides c.*

Furthermore, there exists a 2×2 integral matrix T whose inverse is also integral such that $(a, b)T = (g, 0)$. All integral solutions of $ax + by = c$ are of the form

$$T \begin{pmatrix} \frac{c}{g} \\ z \end{pmatrix}, \qquad z \in \mathbb{Z}.$$

Proof. We may assume without loss of generality that $a \geq b \geq 0$ and $a > 0$. We already established the existence of a matrix T as claimed in the theorem. Equation $(a\ b)\begin{pmatrix} x \\ y \end{pmatrix} = c$ can be rewritten as $(a\ b)TT^{-1}\begin{pmatrix} x \\ y \end{pmatrix} = c$. This latter equation is $gt + 0z = c$ where $\begin{pmatrix} t \\ z \end{pmatrix} := T^{-1}\begin{pmatrix} x \\ y \end{pmatrix}$. Since both T and T^{-1} are integral matrices, vector $\begin{pmatrix} x \\ y \end{pmatrix} = T\begin{pmatrix} t \\ z \end{pmatrix}$ is integral if and only if $\begin{pmatrix} t \\ z \end{pmatrix}$ is integral. Therefore the integral solutions of $ax + by = c$ are the vectors $\begin{pmatrix} x \\ y \end{pmatrix} = T\begin{pmatrix} t \\ z \end{pmatrix}$ such that $\begin{pmatrix} t \\ z \end{pmatrix}$ is an integral solution to $gt + 0z = c$. The latter equation admits an integral solution if and only if c is an integer and g divides c. If this is the case, the integral solutions of $gt + 0z = c$ are $\begin{pmatrix} \frac{c}{g} \\ z \end{pmatrix}$, for all $z \in \mathbb{Z}$. $\square$

The above theorem has a straightforward extension to more than two integers. For $(a_1, \ldots, a_n) \in \mathbb{Z}^n \setminus \{0\}$, $\gcd(a_1, \ldots, a_n)$ denotes the largest positive integer d that divides each of the integers a_i for $i = 1, \ldots, n$. When $\gcd(a_1, \ldots, a_n) = 1$, the integers $a_1, \ldots, a_n$ are said to be *relatively prime*.

Corollary 1.9. *Given $a \in \mathbb{Z}^n \setminus \{0\}$, let $g := \gcd(a_1, \ldots, a_n)$. Equation $ax = c$ admits an integral solution if and only if c is an integer and g divides c.*

Furthermore, there exists an $n \times n$ integral matrix T whose inverse is also integral such that $aT = (g, 0, \ldots, 0)$. All integral solutions of $ax = c$ are of the form

$$T \begin{pmatrix} \frac{c}{g} \\ z \end{pmatrix}, \qquad z \in \mathbb{Z}^{n-1}.$$

Proof. Let $g_i := \gcd(a_i, \ldots, a_n)$, $i = 1, \ldots, n$. For $i = 1, \ldots, n-1$, let T_i' be the 2×2 integral matrix whose inverse is also integral such that $(a_i, g_{i+1})T_i' = (\gcd(a_i, g_{i+1}), 0)$, and let T_i be the matrix obtained from the $n \times n$ identity matrix I by substituting T_i' for the 2×2 submatrix of I with row and column indices i and $i+1$. Clearly T_i and T_i^{-1} are both

1.5. CONNECTIONS TO NUMBER THEORY

integral. Let $T := T_{n-1}T_{n-2}\ldots T_1$. Since $g = g_1$ and $g_i = \gcd(a_i, g_{i+1})$, $i = 1,\ldots,n-1$, we have that $aT = (g, 0\ldots, 0)$. Equation $(g, 0\ldots, 0)y = c$ admits an integral solution if and only if c is an integer and g divides c. If this is the case, all integral solutions are of the form $y = \begin{pmatrix} \frac{c}{g} \\ z \end{pmatrix}$, $z \in \mathbb{Z}^{n-1}$.

Since T and T^{-1} are both integral, it follows that $x = Ty$ is integral if and only if y is integral, therefore all integral solutions of $ax = c$ are of the form $T\begin{pmatrix} \frac{c}{g} \\ z \end{pmatrix}$, $z \in \mathbb{Z}^{n-1}$. □

The Convex Hull of the Integer Points in a Halfspace

Theorem 1.10. *Given $a \in \mathbb{Z}^n \setminus \{0\}$ and $c \in \mathbb{R}$, let $S := \{x \in \mathbb{Z}^n : ax \leq c\}$, and let $g := \gcd(a_1,\ldots,a_n)$. Then*

$$\operatorname{conv}(S) = \left\{x \in \mathbb{R}^n : \frac{a}{g}x \leq \left\lfloor \frac{c}{g} \right\rfloor\right\}.$$

Proof. By Corollary 1.9 there exists an integral $n \times n$ matrix T whose inverse is integral such that $aT = (g, 0, \ldots, 0)$. Let $S' := \{y \in \mathbb{R}^n : Ty \in S\}$. Note that $\operatorname{conv}(S) = \{x \in \mathbb{R}^n : T^{-1}x \in \operatorname{conv}(S')\}$, since a set $C \subseteq \mathbb{R}^n$ is a convex set containing S if and only if $C = \{x \in \mathbb{R}^n : T^{-1}x \in C'\}$ for some convex set C' containing S'.

We first compute $\operatorname{conv}(S')$. Note that $Ty \in S$ if and only if $aTy \leq c$ and $Ty \in \mathbb{Z}^n$. Since T and T^{-1} are both integral, $Ty \in \mathbb{Z}^n$ if and only if $y \in \mathbb{Z}^n$. Thus $S' = \{y \in \mathbb{Z}^n : y_1 \leq \frac{c}{g}\}$, because $aT = (g, 0, \ldots, 0)$. It follows that $\operatorname{conv}(S') = \{y \in \mathbb{R}^n : y_1 \leq \lfloor \frac{c}{g} \rfloor\}$.

From $\operatorname{conv}(S) = \{x \in \mathbb{R}^n : T^{-1}x \in \operatorname{conv}(S')\}$ and the above form of $\operatorname{conv}(S')$, we have that $x \in \operatorname{conv}(S)$ if and only if $(1, 0, \ldots, 0)(T^{-1}x) \leq \lfloor \frac{c}{g} \rfloor$. Since $(1, 0, \ldots, 0)T^{-1} = \frac{a}{g}$, we get $\operatorname{conv}(S) = \{x \in \mathbb{R}^n : \frac{a}{g}x \leq \lfloor \frac{c}{g} \rfloor\}$. □

1.5.2 Integral Solutions to Systems of Linear Equations

In this section we present an algorithm to solve the following problem:

Given a rational matrix $A \in \mathbb{Q}^{m \times n}$ and a rational vector $b \in \mathbb{Q}^m$, find a vector x satisfying

$$Ax = b, \quad x \in \mathbb{Z}^n. \tag{1.15}$$

With a little care, the algorithm can be made to run in polynomial time. We note that this is in contrast to the problem of finding a solution to

$$Ax = b, \quad x \geq 0, \quad x \in \mathbb{Z}^n,$$

which is an NP-hard problem. The idea of the algorithm is to reduce problem (1.15) to a form for which the solution is immediate.

A matrix $A \in \mathbb{Q}^{m \times n}$ is in *Hermite normal form* if $A = \begin{pmatrix} D & 0 \end{pmatrix}$ where 0 is the $m \times (n-m)$ matrix of zeroes and

- D is an $m \times m$ lower triangular nonnegative matrix,
- $d_{ii} > 0$ for $i = 1, \ldots, m$, and $d_{ij} < d_{ii}$ for every $1 \leq j < i \leq m$.

Since $d_{ii} > 0$, $i = 1, \ldots, m$, a matrix in Hermite normal form has full row rank.

Remark 1.11. *If A is an $m \times n$ matrix in Hermite normal form $\begin{pmatrix} D & 0 \end{pmatrix}$, the set $\{x \in \mathbb{Z}^n : Ax = b\}$ can be easily described. Since D is a nonsingular matrix, the system $Dy = b$ has a unique solution $\bar{y}$. Furthermore $\begin{pmatrix} \bar{y} \\ 0 \end{pmatrix}$ is a solution to $Ax = b$, and the solutions of $Ax = 0$ are of the form $\begin{pmatrix} 0 \\ k \end{pmatrix}$ for all $k \in \mathbb{R}^{n-m}$. Thus the solutions of $Ax = b$ are of the form $\begin{pmatrix} \bar{y} \\ k \end{pmatrix}$, where $k \in \mathbb{R}^{n-m}$. In particular, $Ax = b$ has integral solutions if and only if $\bar{y}$ is integral, in which case*

$$\{x \in \mathbb{Z}^n : Ax = b\} = \left\{ \begin{pmatrix} \bar{y} \\ k \end{pmatrix} : k \in \mathbb{Z}^{n-m} \right\}.$$

Consider the three following matrix operations, called *unimodular operations*:

- Interchange two columns.
- Add an integer multiple of a column to another column.
- Multiply a column by -1.

Theorem 1.12. *Every rational matrix with full row rank can be brought into Hermite normal form by a finite sequence of unimodular operations.*

Proof. Let $A \in \mathbb{Q}^{m \times n}$ be a rational matrix with full row rank. Let M be a positive integer such that MA is an integral matrix. We prove the result by induction on the number of rows of A. Assume that A has been transformed with unimodular operations into the form $\begin{pmatrix} D & 0 \\ B & C \end{pmatrix}$, where $\begin{pmatrix} D & 0 \end{pmatrix}$ is in Hermite normal form.

1.5. CONNECTIONS TO NUMBER THEORY

Permuting columns and multiplying columns by -1, we transform C so that $c_{11} \geq c_{12} \geq \cdots \geq c_{1k} \geq 0$. If $c_{1j} > 0$ for some $j = 2, \ldots, k$, we subtract the jth column of C from the first and we repeat the previous step. Note that, at each iteration, $c_{11} + \cdots + c_{1k}$ decreases by at least $1/M$, thus after a finite number of iterations $c_{12} = \ldots = c_{1k} = 0$.

When $c_{12} = \ldots = c_{1k} = 0$, we add or subtract integer multiples of the first column of C to the columns of B, so that $0 \leq b_{1j} < c_{11}$ for $j = 1, \ldots, n-k$. Then the matrix $(D \ \ 0)$ can be extended by one row by adding the first row of (B, C). □

The rationality assumption is critical in Theorem 1.12. For example the matrix $A = (\sqrt{5} \ 1)$ cannot be brought into Hermite normal form using unimodular operations (Exercise 1.21).

We leave it as an exercise to show that, for every rational matrix A with full row rank, there is a *unique* matrix H in Hermite normal form that can be obtained from A by a sequence of unimodular operations (Exercise 1.22).

Remark 1.13. *The statement in Theorem 1.12 can be strengthened: There is a polynomial algorithm to transform a rational matrix with full row rank into Hermite normal form. In particular every rational matrix with full row rank can be brought into Hermite normal form using a polynomial number of unimodular operations [228]. We do not provide the details of this polynomial algorithm here. The interested reader is referred to [228] or [325] (see also [217], p. 513).*

An $m \times n$ matrix A is *unimodular* if it has rank m, it is integral and $\det(B) = 0, \pm 1$ for every $m \times m$ submatrix B of A. In particular, a square matrix U is unimodular if it is integral and $\det(U) = \pm 1$.

Remark 1.14. *If U is a matrix obtained from the identity matrix by performing a unimodular operation, then U is unimodular.*

Indeed, if U is obtained by interchanging two columns of the identity or by multiplying a column of I by -1 then $\det(U) = -1$, whereas if U is obtained by adding to a column an integer multiple of another column then U differs from the identity only in one component, and it is therefore a triangular matrix with all ones in the diagonal, thus $\det(U) = 1$.

Remark 1.15. *If matrix H is obtained from an $m \times n$ matrix A by a single unimodular operation, then $H = AU$, where U is the $n \times n$ matrix obtained from the identity matrix by performing the same unimodular operation. In particular, by Theorem 1.12, if H is the Hermite normal form of A, then $H = AU$ for some square unimodular matrix U.*

Lemma 1.16. *Let U be an $n \times n$ nonsingular matrix. The following are equivalent.*

(i) *U is unimodular,*

(ii) *U and U^{-1} are both integral,*

(iii) *U^{-1} is unimodular,*

(iv) *For all $x \in \mathbb{R}^n$, Ux is integral if and only if x is integral,*

(v) *U is obtained from the identity matrix by a sequence of unimodular operations.*

Proof. (i)$\Rightarrow$ (ii) Assume U is unimodular. By standard linear algebra, U^{-1} equals the adjugate matrix of U divided by $\det(U)$. Since U is integral, its adjugate is integral as well, thus U^{-1} is integral because $\det(U) = \pm 1$.

(ii)$\Rightarrow$ (i) If U and U^{-1} are both integral, then $\det(U)$ and $\det(U^{-1})$ are both integer numbers. Since $\det(U) = 1/\det(U^{-1})$, it follows that $\det(U) = \pm 1$.

(i)$\Leftrightarrow$ (iii) follows from (i)$\Leftrightarrow$ (ii).

(ii)$\Rightarrow$ (iv) Assume U and U^{-1} are both integral matrices and let $x \in \mathbb{R}^n$. If x is an integral vector, then Ux is integral because U is integral. Conversely, if $y = Ux$ is integral then $x = U^{-1}y$ is integral because U^{-1} is integral.

(iv)$\Rightarrow$ (ii) This is immediate.

(v)$\Rightarrow$ (i) Suppose that U is obtained from the identity matrix by a sequence of unimodular operations. It follows from Remark 1.15 that $U = U_1 \cdots U_k$ where $U_1, \ldots, U_k$ are matrices obtained from the identity matrix by performing a single unimodular operation. By Remark 1.14 $U_1, \ldots, U_k$ are unimodular matrices, therefore U is unimodular, since it is integral and $\det(U) = \det(U_1) \cdots \det(U_k) = \pm 1$.

(i)$\Rightarrow$ (v) Suppose U is unimodular. Let H be the Hermite normal form of U^{-1}. By Theorem 1.12 and Remark 1.15, $H = U^{-1}U'$, where U' is obtained from the identity matrix by a sequence of unimodular operations. It follows from the implication (v)$\Rightarrow$ (i) that U' is unimodular, and it follows from (i)$\Rightarrow$ (iii) that U^{-1} is unimodular. Therefore H is unimodular as well. Since H is diagonal, H is the identity matrix. Therefore $U = U'$, which shows that U can be obtained from the identity matrix by a sequence of unimodular operations. □

1.5. CONNECTIONS TO NUMBER THEORY

The above result easily implies the following characterization of the solutions of problem (1.15).

Theorem 1.17. *Let A be a rational $m \times n$ matrix with full row-rank, and let $b \in \mathbb{R}^m$. Let $H = \begin{pmatrix} D & 0 \end{pmatrix}$ be the Hermite normal form of A, and U be a unimodular matrix such that $H = AU$. Then $Ax = b$, $x \in \mathbb{Z}^n$, has a solution if and only if $\bar{y} := D^{-1}b \in \mathbb{Z}^m$. In this case, all solutions are of the form*

$$\left\{ U \begin{pmatrix} \bar{y} \\ k \end{pmatrix} : k \in \mathbb{Z}^{n-m} \right\}.$$

Proof. By Lemma 1.16, for all $y \in \mathbb{R}^n$ we have that $y \in \mathbb{Z}^n$ if and only if $Uy \in \mathbb{Z}^n$, therefore y is an integral solution of $Hy = AUy = b$ if and only if $x := Uy$ is an integral solution of $Ax = b$. By Remark 1.11, all solutions of $Hx = b$ are of the form $\begin{pmatrix} \bar{y} \\ k \end{pmatrix}$, $k \in \mathbb{Z}^{n-m}$, and the result follows. □

This theorem suggests a natural algorithm to solve systems of linear Diophantine equations.

Algorithm for Solving the Linear Diophantine Equation Problem (1.15)

Step 1. Check whether A has full row rank. If not, either the system $Ax = b$ is infeasible and (1.15) has no solution or it contains redundant equations that can be removed.

Step 2. Transform A into a matrix H in Hermite normal form by a sequence of unimodular operations (this is possible by Theorem 1.12). By Remark 1.15 and Lemma 1.16, $H = AU$ for some square unimodular matrix U.

Step 3. Solve $Hy = b$, $y \in \mathbb{Z}^n$ as in Remark 1.11. If this problem is infeasible, (1.15) has no solution. Otherwise let $\bar{y}$ be a solution. Then $\bar{x} = U\bar{y}$ is a solution of (1.15).

Step 1 of the algorithm can be performed by Gaussian elimination. If an equation $0x = \alpha$ is produced and $\alpha \neq 0$, then $Ax = b$ is infeasible and (1.15) has no solution. If $\alpha = 0$, the equation is redundant.

Example 1.18. Solve the system

$$\begin{pmatrix} 10 & 4 & 3 & 0 \\ 58 & 24 & 19 & 2 \\ 3 & 2 & 0 & 0 \end{pmatrix} x = \begin{pmatrix} 3 \\ 5 \\ 5 \end{pmatrix} \quad \text{with } x \in \mathbb{Z}^4. \quad (1.16)$$

Solution: We first transform the matrix $A = \begin{pmatrix} 10 & 4 & 3 & 0 \\ 58 & 24 & 19 & 2 \\ 3 & 2 & 0 & 0 \end{pmatrix}$ into Hermite normal form by performing unimodular operations. We start by subtracting 3 times column 3 from column 1. We get

$$A_1 = \begin{pmatrix} 1 & 4 & 3 & 0 \\ 1 & 24 & 19 & 2 \\ 3 & 2 & 0 & 0 \end{pmatrix} = \begin{pmatrix} 10 & 4 & 3 & 0 \\ 58 & 24 & 19 & 2 \\ 3 & 2 & 0 & 0 \end{pmatrix} \begin{pmatrix} 1 & 0 & 0 & 0 \\ 0 & 1 & 0 & 0 \\ -3 & 0 & 1 & 0 \\ 0 & 0 & 0 & 1 \end{pmatrix}.$$

We now use column 1 of A_1 to cancel out the nonzero entries in the first row of columns 2 and 3.

$$A_2 = \begin{pmatrix} 1 & 0 & 0 & 0 \\ 1 & 20 & 16 & 2 \\ 3 & -10 & -9 & 0 \end{pmatrix} = \begin{pmatrix} 1 & 4 & 3 & 0 \\ 1 & 24 & 19 & 2 \\ 3 & 2 & 0 & 0 \end{pmatrix} \begin{pmatrix} 1 & -4 & -3 & 0 \\ 0 & 1 & 0 & 0 \\ 0 & 0 & 1 & 0 \\ 0 & 0 & 0 & 1 \end{pmatrix}.$$

Next we interchange columns 2 and 4 of A_2. We get

$$A_3 = \begin{pmatrix} 1 & 0 & 0 & 0 \\ 1 & 2 & 16 & 20 \\ 3 & 0 & -9 & -10 \end{pmatrix} = \begin{pmatrix} 1 & 0 & 0 & 0 \\ 1 & 20 & 16 & 2 \\ 3 & -10 & -9 & 0 \end{pmatrix} \begin{pmatrix} 1 & 0 & 0 & 0 \\ 0 & 0 & 0 & 1 \\ 0 & 0 & 1 & 0 \\ 0 & 1 & 0 & 0 \end{pmatrix}.$$

Then we cancel out the nonzero entries in the second row of columns 3 and 4 using column 2 of A_3.

$$A_4 = \begin{pmatrix} 1 & 0 & 0 & 0 \\ 1 & 2 & 0 & 0 \\ 3 & 0 & -9 & -10 \end{pmatrix} = \begin{pmatrix} 1 & 0 & 0 & 0 \\ 1 & 2 & 16 & 20 \\ 3 & 0 & -9 & -10 \end{pmatrix} \begin{pmatrix} 1 & 0 & 0 & 0 \\ 0 & 1 & -8 & -10 \\ 0 & 0 & 1 & 0 \\ 0 & 0 & 0 & 1 \end{pmatrix}.$$

To get the matrix in lower triangular form, we subtract column 4 of A_4 from column 3, and then add 10 times column 3 to column 4. This gives

$$A_5 = \begin{pmatrix} 1 & 0 & 0 & 0 \\ 1 & 2 & 0 & 0 \\ 3 & 0 & 1 & 0 \end{pmatrix} = \begin{pmatrix} 1 & 0 & 0 & 0 \\ 1 & 2 & 0 & 0 \\ 3 & 0 & -9 & -10 \end{pmatrix} \begin{pmatrix} 1 & 0 & 0 & 0 \\ 0 & 1 & 0 & 0 \\ 0 & 0 & 1 & 0 \\ 0 & 0 & -1 & 1 \end{pmatrix} \begin{pmatrix} 1 & 0 & 0 & 0 \\ 0 & 1 & 0 & 0 \\ 0 & 0 & 1 & 10 \\ 0 & 0 & 0 & 1 \end{pmatrix}.$$

Finally, in order to satisfy the condition $d_{ij} < d_{ii}$ for $j < i$, we subtract 3 times column 3 of A_5 from column 1.

$$H = \begin{pmatrix} 1 & 0 & 0 & 0 \\ 1 & 2 & 0 & 0 \\ 0 & 0 & 1 & 0 \end{pmatrix} = \begin{pmatrix} 1 & 0 & 0 & 0 \\ 1 & 2 & 0 & 0 \\ 3 & 0 & 1 & 0 \end{pmatrix} \begin{pmatrix} 1 & 0 & 0 & 0 \\ 0 & 1 & 0 & 0 \\ -3 & 0 & 1 & 0 \\ 0 & 0 & 0 & 1 \end{pmatrix}.$$

1.5. CONNECTIONS TO NUMBER THEORY

Matrix H is the Hermite normal form of A. Note that $H = AU$ where

$$U = \begin{pmatrix} -2 & 0 & 1 & 6 \\ 3 & 0 & -1 & -9 \\ 3 & 0 & -2 & -8 \\ -6 & 1 & 2 & 10 \end{pmatrix}$$

is obtained by multiplying the 4×4 matrices that were used above to perform the column operations.

We now return to question (1.16), namely to solve $Ax = b$ with $x \in \mathbb{Z}^4$. This is equivalent to solving $HU^{-1}x = b$ with $x \in \mathbb{Z}^4$. Set $y := U^{-1}x$. Since U is unimodular, this is equivalent to solving $Hy = b$ with $y \in \mathbb{Z}^4$. If D denotes the 3×3 submatrix of H induced by the first 3 columns of H, the system is equivalent to $(D\ 0)y = b$ with $y \in \mathbb{Z}^4$. The answer is

$$\begin{pmatrix} y_1 \\ y_2 \\ y_3 \end{pmatrix} = D^{-1}b, \qquad y_4 \in \mathbb{Z}.$$

Going back to x through $x = Uy$ we get $x = U_1 D^{-1} b + U_2 \mathbb{Z}$ where U_1 and U_2 are the column submatrices of U associated with the first 3 columns, and the last one respectively. Here

$$D = \begin{pmatrix} 1 & 0 & 0 \\ 1 & 2 & 0 \\ 0 & 0 & 1 \end{pmatrix}, \quad D^{-1} = \begin{pmatrix} 1 & 0 & 0 \\ -\frac{1}{2} & \frac{1}{2} & 0 \\ 0 & 0 & 1 \end{pmatrix}, \quad U_1 = \begin{pmatrix} -2 & 0 & 1 \\ 3 & 0 & -1 \\ 3 & 0 & -2 \\ -6 & 1 & 2 \end{pmatrix}, \quad U_2 = \begin{pmatrix} 6 \\ -9 \\ -8 \\ 10 \end{pmatrix}.$$

Therefore the solutions of (1.16) are

$$x = \begin{pmatrix} -1 \\ 4 \\ -1 \\ -7 \end{pmatrix} + \begin{pmatrix} 6 \\ -9 \\ -8 \\ 10 \end{pmatrix} k \qquad \text{with } k \in \mathbb{Z}.$$

∎

There is a simple characterization of when a system of linear equations does not have a solution in integers. The characterization is similar to the Fredholm alternative.

Theorem 1.19 (Fredholm Alternative). *A system of linear equations $Ax = b$ is infeasible if and only if there exists a vector $u \in \mathbb{R}^m$ such that $uA = 0$, $ub \neq 0$.*

A constructive proof of Theorem 1.19 is straightforward using Gaussian elimination on the system $Ax = b$.

Theorem 1.20 (Integer Farkas Lemma or Kronecker Approximation Theorem). *Let A be a rational matrix and b a rational vector. The system $Ax = b$ admits no integral solution if and only if there exists a vector $u \in \mathbb{R}^m$ such that $uA \in \mathbb{Z}^n$, $ub \notin \mathbb{Z}$.*

Proof. Assume that $Ax = b$ admits an integral solution x. Then, for any vector u such that uA is integral, $ub = uAx$ is an integer.

Suppose now that $Ax = b$ does not have an integral solution. If system $Ax = b$ is infeasible, Theorem 1.19 shows that there exists u such that $uA = 0$ and $ub \notin \mathbb{Z}$ (note that u may have to be scaled appropriately). So the theorem holds in this case, and we may assume that $Ax = b$ is feasible and that A has full row rank.

By Theorem 1.12 and Remark 1.14, there exists a matrix in Hermite normal form $(D \ \ 0)$ and a square unimodular matrix U such that $(D \ \ 0) = AU$. By Theorem 1.17, $D^{-1}b \notin \mathbb{Z}^m$, say, the ith component of $D^{-1}b$ is fractional. Let u be the ith row of D^{-1}. Then the ith component of the vector $D^{-1}b$ is ub, so $ub \notin \mathbb{Z}$. We conclude the proof by showing that uA is integral. Indeed
$$uA = u\begin{pmatrix}D & 0\end{pmatrix}U^{-1} = e^i U^{-1},$$
where e^i denotes the ith unit row-vector in $\mathbb{R}^n$. Since U^{-1} is integral, it follows that uA is integral. □

1.6 Further Readings

There are several textbooks on integer programming. Garfinkel and Nemhauser [159] is an important early reference, Schrijver [325], Nemhauser and Wolsey [285], Wolsey [353], Bertsimas and Weismantel [51] are more recent. We also recommend the book "50 Years of Integer Programming 1958–2008" edited by Jünger et al. [217]. There are several monographs that deal with central topics in integer programming, see Grötschel, Lovász, and Schrijver [188] and De Loera, Hemmecke, and Köppe [110]. A good reference on computational integer programming is the book edited by Jünger and Naddef [218].

Integer programming is closely related to combinatorial optimization. Many excellent references can be found under this heading; we just mention Korte and Vygen [243] and the encyclopedic volumes of Schrijver [327].

1.6. FURTHER READINGS

A point of view that is popular in theoretical computer science is to ask, for an optimization problem, how well can its optimum value be approximated in polynomial time. A polynomial algorithm that, for all instances, finds a solution with value within a ratio α of the optimum value is called an *α-approximation algorithm*. We refer to the books of Williamson and Shmoys [349] and Vazirani [345]. There is also a vast literature on metaheuristics (local search, tabu search, genetic algorithms, etc.) that we do not cover. Our focus in this book is on exact methods. We will only discuss heuristics briefly in the context of the branch-and-bound algorithm in Chap. 9.

The first cutting plane algorithm was devised in 1954 by Dantzig, Fulkerson, and Johnson [103] to solve the traveling salesman problem, using special purpose cuts. The first general cutting plane algorithm for pure integer programming was devised by Gomory [175] in 1958. Around the same time Bellman [46] wrote his book on dynamic programming, a technique that can be useful for certain types of integer programs.

The first branch-and-bound algorithm was proposed in 1960 by Land and Doig [246]. Another early reference is Dakin [100]. The term branch-and-cut was coined in the 80s by Padberg and Rinaldi [301, 302]. Extensive computational experience on the progress in software for integer programming up to 2004 is reported in Bixby, Fenelon, Gu, Rothberg, and Wunderling [57].

Complexity issues were stressed by Edmonds [123, 125] and put in sharp focus by the work of Cook [84] on NP-completeness. Papadimitriou [305] showed that integer programming is in NP. Rather innocent-looking problems can be NP-complete, such as the following: Given $n+1$ points in $\mathbb{Q}^n$, does their convex hull contain an integral point? Arora and Barak [14] present a modern treatment of complexity theory.

Edmonds [123] was also a pioneer in developing the polyhedral approach in combinatorial optimization. We will give examples of this approach, which is known as polyhedral combinatorics, in Chap. 4.

Kannan and Bachem [228] presented polynomial algorithms for computing the Smith and Hermite normal forms of an integer matrix.

Lenstra [256] gave a polynomial algorithm to solve integer programs with a fixed number of integer variables. We will present Lenstra's algorithm in Chap. 9.

Bachem and von Randow [21] proved a mixed integer Farkas lemma in the spirit of Theorem 1.20 (see also [242]).

Primal Methods

This book focuses on the methods that have been most successful in solving a wide range of integer programs in practice, such as branch-and-bound and cutting plane algorithms. Other approaches, such as primal methods, might be more appropriate for certain classes of integer programs. Like the simplex method, primal methods start with a feasible solution and seek a new feasible solution with better objective function value until optimality is achieved. Consider an integer program of the form

$$\min\{cx \ : \ Ax = b, \ l \leq x \leq u, \ x \in \mathbb{Z}^n\} \qquad (1.17)$$

where all data are assumed to be integral. Primal methods address the following question:

Let x_0 be a feasible solution to (1.17). Either prove that x_0 is optimal, or find a vector z such that $x_0 + z$ is feasible for (1.17) and $c(x_0 + z) < cx_0$. Such a vector z is called an improving direction.

The main result is the following:

Given an integral matrix A, there is a finite set of vectors $\mathcal{Z}(A)$ such that, for every integral vectors c, b, l, u and feasible solution x_0 to (1.17) that is not optimal, there is an improving direction $z \in \mathcal{Z}(A)$.

The set $\mathcal{Z}(A)$ is called a *Graver test set*. As an example where primal methods might be appropriate, we mention stochastic integer programs, which often have a fixed constraint matrix but need to be solved for a large number of different right-hand sides (see Tayur, Thomas, and Natraj [336] for an application). We refer the reader to the books of Bertsimas and Weismantel [51] and De Loera, Hemmecke, and Köppe [110] for a treatment of primal methods.

1.7 Exercises

Exercise 1.1. Consider the objective function: $f(y, x) = 8y + 3x_1 + 4x_2 + 2x_3$ and the following inequalities:

$$y + x_1 \geq 0.2 \qquad (1.18)$$
$$y + x_2 \geq 0.8 \qquad (1.19)$$
$$y + x_3 \geq 1.5 \qquad (1.20)$$
$$y, x_1, x_2, x_3 \geq 0 \qquad (1.21)$$

Define the following programs:

1.7. EXERCISES

- LP: $\min\{f(y,x) : (y,x) \text{ satisfies } (1.18)-(1.21)\}$.
- ILP: $\min\{f(y,x) : (y,x) \text{ satisfies } (1.18)-(1.21), (y,x) \text{ integral}\}$.
- MILP1: $\min\{f(y,x) : (y,x) \text{ satisfies } (1.18)-(1.21), y \text{ integer}\}$.
- MILP2: $\min\{f(y,x) : (y,x) \text{ satisfies } (1.18)-(1.21), x \text{ integral}\}$.

Show that:

- The unique optimum of LP is attained at $y = 0.2$.
- The unique optimum of ILP is attained at $y = 1$.
- The unique optimum of MILP1 is attained at $y = 0$.
- The unique optimum of MILP2 is attained at $y = 0.8$.

Exercise 1.2. Consider the program with $n+1$ variables:

$$\begin{aligned}
\min \quad & hy + \sum_{i=1}^{n} c_i x_i \\
& y + x_i \geq b_i, \ 1 \leq i \leq n \\
& y \geq 0 \\
& x_i \geq 0, \ 1 \leq i \leq n
\end{aligned}$$

where $0 \leq b_1 \leq b_2 \leq \ldots \leq b_n$ and $0 \leq c_i \leq h$ for $i = 1, \ldots, n$. Design the most efficient algorithms that you can to solve the above program under the additional conditions:

- LP: y, x_i continuous.
- ILP: y, x_i integer.
- MILP: x_i continuous, y integer.

In particular, can you give a closed-form formula for the optimal values of y, x_i in LP and ILP?

Exercise 1.3. Solve the following integer program by branch-and-bound, using variable branching and linear programming bounding.

$$\begin{aligned}
\min \quad & 8y + 3x_1 + 4x_2 + 2x_3 \\
& y + x_1 \geq 0.2 \\
& y + x_2 \geq 0.8 \\
& y + x_3 \geq 1.5 \\
& y \geq 0 \\
& x_1, x_2, x_3 \geq 0 \text{ integer}.
\end{aligned}$$

Exercise 1.4. Solve the following pure integer program by branch-and-bound.

$$\begin{array}{rrrl} \max & 2x_1 & +x_2 & \\ & -x_1 & +x_2 & \leq 0 \\ & 6x_1 & +2x_2 & \leq 21 \\ & x_1, & x_2 & \geq 0 \text{ integer.} \end{array}$$

Exercise 1.5. When the branch-and-bound algorithm takes too much time to find an optimal solution, one may be willing to settle for an approximate solution. Assume that the optimum value z^* of (1.2) is positive. Modify Step 4 of the branch-and-bound algorithm presented in Sect. 1.2.1 when the goal is to find a feasible solution $(\bar{x}, \bar{y})$ whose objective value $\bar{z}$ is within p % of the optimum, i.e., $\frac{\bar{z}}{z^*} \geq 1 - p/100$.

Exercise 1.6. Finish solving the 2-variable integer program of Sect. 1.2.2 using fractional cuts.

Exercise 1.7. Solve the following pure integer program using fractional cuts.

$$\begin{array}{rrrl} \max & 2x_1 & +x_2 & \\ & -x_1 & +x_2 & \leq 0 \\ & 6x_1 & +2x_2 & \leq 21 \\ & x_1, & x_2 & \geq 0 \text{ integer.} \end{array}$$

Exercise 1.8. Let (x^0, y^0) be an optimal solution of (1.5) that is not in S. Suppose that $(x^0, y^0) = \frac{1}{2}(x^1, y^1) + \frac{1}{2}(x^2, y^2)$ where $(x^i, y^i) \in S$ for $i = 1, 2$. Show that there is no cutting plane separating (x^0, y^0) from S.

Exercise 1.9.

(i) Let A be a square nonsingular rational matrix. Show that the encoding size of A^{-1} is polynomially bounded by the encoding size of A.
(ii) Let $A \in \mathbb{Q}^{m \times n}$ be a rational matrix and $b \in \mathbb{Q}^m$ a rational vector. Show that if the system $Ax = b$ has a solution $\bar{x} \in \mathbb{R}^n$, it has one whose encoding size is polynomially bounded by the encoding size of (A, b).

Exercise 1.10. Show that the following decision problems belong to the class NP.

- Given two graphs $G_1 = (V_1, E_1)$, $G_2 = (V_2, E_2)$, determine if G_1 and G_2 are isomorphic.

- Given positive integers m and n, determine if m has a factor less than n and greater than 1 (Integer factorization).

Exercise 1.11. Show that the following decision problems belong to the class Co-NP.

- Determine whether $\{x \in \{0,1\}^n : Ax \leq b\}$ and $\{x \in \{0,1\}^n : A'x \leq b'\}$ have the same set of solutions.

- Given a graph $G = (V, E)$ and a positive integer k, determine whether the chromatic number of G is greater than k.

Exercise 1.12. Show that the following decision problems belong to the class NP and to the class Co-NP. (Some answers are a consequence of basic results in linear algebra, linear programming or graph theory. Please cite any such results that you use.)

- Does a rational system of equations $Ax = b$ admit a solution?

- Does a rational system of inequalities $Ax \leq b$ admit a solution?

- Is $\bar{x}$ an optimal solution to the linear program $\max\{cx : Ax \leq b\}$?

- Does a rational system of linear equations $Ax = b$ admit an integral solution?

- Is a graph $G = (V, E)$ Eulerian? (i.e., does G contain a closed walk traversing each edge exactly once?)

- Is a graph $G = (V, E)$ connected?

- Is a directed graph $D = (V, A)$ strongly connected?

- Does a directed graph $D = (V, A)$ contain a directed cycle?

Exercise 1.13. Draw the convex hull of the following 2-variable integer sets.

- $S := \{x \in \mathbb{Z}_+^2 : x_1 + x_2 \leq 2,\ x_1 - x_2 \leq 1,\ x_2 - x_1 \leq 1\}$
- $S := \{(x, y) \in \mathbb{Z}_+ \times \mathbb{R}_+ : x + y \geq 1.6,\ x \leq 2,\ y \leq 2\}$.

Exercise 1.14. Let $c \in \mathbb{R}_+$. Generate the valid inequalities needed to describe the convex hull of $S := \{(x, y) \in \mathbb{R}^2 : x - cy \leq b,\ x \text{ integer},\ y \geq 0\}$.

Exercise 1.15. Generate the valid inequalities needed to describe the convex hull of the following sets S.

- $S := \{(x, y) : x + y \geq b,\ x \geq 0 \text{ integer},\ y \geq 0\}$
- $S := \{(x, y) : x + y \geq b,\ x \geq 0 \text{ integer}\}$
- $S := \{(x, y) : x + y \geq b,\ x \geq d \text{ integer}\}$
- $S := \{(x, y) : x + y \geq b,\ x \geq d \text{ integer},\ y \geq 0\}$

Exercise 1.16. Generate the valid inequalities needed to describe the convex hull of the set $S := \{(x, y_1, y_2) : x + y_1 \geq b_1,\ x + y_2 \geq b_2,\ x \geq 0 \text{ integer},\ y_1, y_2 \geq 0\}$.

Exercise 1.17. Consider a (possibly infinite) set $S \subseteq \mathbb{R}^n$. Prove that the set of all convex combinations of points in S is a convex set containing S.

Exercise 1.18. Modify the Euclidean algorithm so that it has the following input and output.
Input. $a, b \in \mathbb{Z}_+$ such that $a \geq b$ and $a > 0$.
Output. $\gcd(a, b)$ and $x, y \in \mathbb{Z}$ such that $ax + by = \gcd(a, b)$.
Prove the correctness of your algorithm, and show that it runs in polynomial time.

Exercise 1.19. A rectangle is subdivided into an $m \times n$ checkerboard. How many of the mn cells are traversed by a diagonal of the rectangle (A cell is *traversed* if the diagonal intersects its interior).

Exercise 1.20. Let $a_1, \ldots, a_n \in \mathbb{Z}$ be relatively prime integers. Show that the hyperplane $a_1 x_1 + \ldots a_n x_n = b$ contains integer points $x \in \mathbb{Z}^n$ for every $b \in \mathbb{Z}$.

Exercise 1.21. Let $A = (\sqrt{5}\ 1)$. Show that A cannot be put in Hermite normal form using unimodular operations.

Exercise 1.22. Let A be a rational matrix with full row rank. Show that there is a *unique* matrix H in Hermite normal form that can be obtained from A by a sequence of unimodular operations.

Exercise 1.23. Put the following matrix in Hermite normal form.
$$A = \begin{pmatrix} 4 & 4 & 3 & 3 & 3 \\ 2 & 4 & 4 & 8 & 0 \\ 5 & 7 & 0 & 0 & 5 \end{pmatrix}.$$

1.7. EXERCISES

Exercise 1.24. Solve the system of Diophantine equations

$$\begin{aligned} 4x_1 + 4x_2 + 3x_3 + 3x_4 + 3x_5 &= 3 \\ 2x_1 + 4x_2 + 4x_3 + 8x_4 &= 2 \\ 5x_1 + 7x_2 + 5x_5 &= 2 \\ x &\in \mathbb{Z}^5. \end{aligned}$$

(Hint: Use Exercise 1.23.)

Exercise 1.25. Let $a_1, \ldots, a_n \in \mathbb{Z}$ be relatively prime integers. Show that there exists an $n \times n$ unimodular matrix whose first row is the vector $(a_1, \ldots, a_n)$.

Exercise 1.26. Prove the following statement: Let A, G be rational matrices and b a rational vector. The problem $Ax + Gy = b$, $x \in \mathbb{Z}^n$ is infeasible if and only if there exists a vector $u \in \mathbb{R}^m$ such that $uA \in \mathbb{Z}^n$, $uG = 0$, $ub \notin \mathbb{Z}$. (Hint: prove that $\{x : Ax + Gy = b \text{ for some } y\} = \{x : (u^i A)x = u^i b, i = 1, \ldots, k\}$ where $u^1, \ldots, u^k$ is a basis of the linear subspace $\{u : uG = 0\}$. Then apply Theorem 1.20.)

Exercise 1.27. Let A be an $n \times n$ nonsingular integral matrix.

(i) Show that there exist $n \times n$ unimodular matrices R and C such that RAC is a diagonal matrix D with diagonal entry $d_i \in \mathbb{Z}_+ \setminus \{0\}$ in row i and column i, where d_i divides d_{i+1} for $i = 1, \ldots, n-1$.
(ii) Show that the diagonal matrix D found in (i) is unique.

Exercise 1.28. Susie meets Roberto at a candy store. Susie tells Roberto: "I want to buy chocolates for my friends. Unfortunately all the boxes contain either 7 or 9 chocolates, so it is impossible for me to buy exactly one chocolate for each of my friends." How many friends does Susie have, knowing that she has more than 40?

Exercise 1.29. Let $a_1, \ldots, a_n$ be relatively prime positive integers. Show that there exists a positive integer k such that, for every integer $m \geq k+1$ the equation $a_1 x_1 + \cdots + a_n x_n = m$ has a nonnegative integral solution.

This is a fantastic proof!

Chapter 2
Integer Programming Models

The importance of integer programming stems from the fact that it can be used to model a vast array of problems arising from the most disparate areas, ranging from practical ones (scheduling, allocation of resources, etc.) to questions in set theory, graph theory, or number theory. We present here a selection of integer programming models, several of which will be further investigated later in this book.

2.1 The Knapsack Problem

We are given a knapsack that can carry a maximum weight b and there are n types of items that we could take, where an item of type i has weight $a_i > 0$. We want to load the knapsack with items (possibly several items of the same type) without exceeding the knapsack capacity b. To model this, let a variable x_i represent the number of items of type i to be loaded. Then the *knapsack set*

$$S := \left\{ x \in \mathbb{Z}^n : \sum_{i=1}^{n} a_i x_i \leq b,\ x \geq 0 \right\}$$

contains precisely all the feasible loads.

If an item of type i has value c_i, the problem of loading the knapsack so as to maximize the total value of the load is called the *knapsack problem*. It can be modeled as follows:

$$\max\left\{\sum_{i=1}^n c_i x_i \; : \; x \in S\right\}.$$

If only one unit of each item type can be selected, we use binary variables instead of general integers. The $0,1$ *knapsack set*

$$K := \left\{ x \in \{0,1\}^n \; : \; \sum_{i=1}^n a_i x_i \leq b \right\}$$

can be used to model the $0,1$ *knapsack problem* $\max\{cx \; : \; x \in K\}$.

2.2 Comparing Formulations

Given scalars $b > 0$ and $a_j > 0$ for $j = 1, \ldots, n$, consider the $0,1$ knapsack set $K := \{x \in \{0,1\}^n \; : \; \sum_{i=1}^n a_i x_i \leq b\}$. A subset C of indices is a *cover* for K if $\sum_{i \in C} a_i > b$ and it is a *minimal cover* if $\sum_{i \in C \setminus \{j\}} a_i \leq b$ for every $j \in C$. That is, C is a cover if the knapsack cannot contain all items in C, and it is minimal if every proper subset of C can be loaded. Consider the set

$$K^C := \left\{ x \in \{0,1\}^n \; : \; \sum_{i \in C} x_i \leq |C| - 1 \text{ for every minimal cover } C \text{ for } K \right\}.$$

Proposition 2.1. *The sets K and K^C coincide.*

Proof. It suffices to show that (i) if C is a minimal cover of K, the inequality $\sum_{i \in C} x_i \leq |C| - 1$ is valid for K and (ii) the inequality $\sum_{i=1}^n a_i x_i \leq b$ is valid for K^C. The first statement follows from the fact that the knapsack cannot contain all the items in a minimal cover.

Let $\bar{x}$ be a vector in K^C and let $J := \{j : \bar{x}_j = 1\}$. Suppose $\sum_{i=1}^n a_i \bar{x}_i > b$ or equivalently $\sum_{i \in J} a_i > b$. Let C be a minimal subset of J such that $\sum_{i \in C} a_i > b$. Then obviously C is a minimal cover and $\sum_{i \in C} \bar{x}_i = |C|$. This contradicts the assumption $\bar{x} \in K^C$ and the second statement is proved. □

So the $0,1$ knapsack problem $\max\{cx \; : \; x \in K\}$ can also be formulated as $\max\{cx \; : \; x \in K^C\}$. The constraints that define K and K^C look quite different. The set K is defined by a single inequality with nonnegative integer coefficients whereas K^C is defined by many inequalities (their number may be exponential in n) whose coefficients are $0,1$. Which of the two formulations is "better"? This question has great computational relevance and

2.2. COMPARING FORMULATIONS

the answer depends on the method used to solve the problem. In this book we focus on algorithms based on linear programming relaxations (remember Sect. 1.2) and for these algorithms, the answer can be stated as follows:

Assume that $\{(x,y) : A_1x + G_1y \leq b_1, x \text{ integral}\}$ and $\{(x,y) : A_2x + G_2y \leq b_2, x \text{ integral}\}$ represent the same mixed integer set S and consider their linear relaxations $P_1 = \{(x,y) : A_1x + G_1y \leq b_1\}$, $P_2 = \{(x,y) : A_2x + G_2y \leq b_2\}$. If $P_1 \subset P_2$ the first representation is better. If $P_1 = P_2$ the two representations are equivalent and if $P_1 \setminus P_2$ and $P_2 \setminus P_1$ are both nonempty, the two representations are incomparable.

Next we discuss how to compare the two linear relaxations P_1 and P_2. If, for every inequality $a_2 x + g_2 y \leq \beta_2$ in $A_2 x + G_2 y \leq b_2$, the system

$$uA_1 = a_2, \ uG_1 = g_2, \ ub_1 \leq \beta_2, \ u \geq 0$$

admits a solution $u \in \mathbb{R}^m$, where m is the number of components of b_1, then every inequality defining P_2 is implied by the set of inequalities that define P_1 and therefore $P_1 \subseteq P_2$. Indeed, every point in P_1 satisfies the inequality $(uA_1)x + (uG_1)y \leq ub_1$ for every nonnegative $u \in \mathbb{R}^m$; so in particular every point in P_1 satisfies $a_2 x + g_2 y \leq \beta_2$ whenever u satisfies the above system.

Farkas's lemma, an important result that will be proved in Chap. 3, implies that the converse is also true if P_1 is nonempty. That is

Assume $P_1 \neq \emptyset$. $P_1 \subseteq P_2$ if and only if for every inequality $a_2 x + g_2 y \leq \beta_2$ in $A_2 x + G_2 y \leq b_2$ the system $uA_1 = a_2$, $uG_1 = g_2$, $ub_1 \leq \beta_2$, $u \geq 0$ is feasible.

This fact is of fundamental importance in comparing the tightness of different linear relaxations of a mixed integer set. These conditions can be checked by solving linear programs, one for each inequality in $A_2 x + G_2 y \leq b_2$.

We conclude this section with two examples of $0, 1$ knapsack sets, one where the minimal cover formulation is better than the knapsack formulation and another where the reverse holds. Consider the following $0, 1$ knapsack set

$$K := \{x \in \{0,1\}^3 : 3x_1 + 3x_2 + 3x_3 \leq 5\}.$$

Its minimal cover formulation is

$$K^C := \left\{ x \in \{0,1\}^3 : \begin{array}{rrrr} x_1 & +x_2 & & \leq 1 \\ x_1 & & +x_3 & \leq 1 \\ & x_2 & +x_3 & \leq 1 \end{array} \right\}.$$

The corresponding linear relaxations are the sets

$$P := \{x \in [0,1]^3 : 3x_1 + 3x_2 + 3x_3 \leq 5\}, \text{ and}$$

$$P^C := \left\{ x \in [0,1]^3 : \begin{array}{lll} x_1 + x_2 & & \leq 1 \\ x_1 & +x_3 & \leq 1 \\ x_2 & +x_3 & \leq 1 \end{array} \right\}.$$

respectively. By summing up the three inequalities in P^C we get

$$2x_1 + 2x_2 + 2x_3 \leq 3$$

which implies $3x_1 + 3x_2 + 3x_3 \leq 5$. Thus $P^C \subseteq P$. The inclusion is strict since, for instance $(1, \frac{2}{3}, 0) \in P \setminus P^C$. In other words, the minimal cover formulation is strictly better than the knapsack formulation in this case.

Now consider a slightly modified example. The knapsack set $K := \{x \in \{0,1\}^3 : x_1 + x_2 + x_3 \leq 1\}$ has the same minimal cover formulation K^C as above, but this time the inclusion is reversed: We have $P := \{x \in [0,1]^3 : x_1+x_2+x_3 \leq 1\} \subseteq P^C$. Furthermore $(\frac{1}{2}, \frac{1}{2}, \frac{1}{2}) \in P^C \setminus P$. In other words, the knapsack formulation is strictly better than the minimal cover formulation in this case.

One can also construct examples where neither formulation is better (Exercise 2.2). In Sect. 7.2.1 we will show how to improve minimal cover inequalities through a procedure called *lifting*.

2.3 Cutting Stock: Formulations with Many Variables

A paper mill produces large rolls of paper of width W, which are then cut into rolls of various smaller widths in order to meet demand. Let m be the number of different widths that the mill produces. The mill receives an order for b_i rolls of width w_i for $i = 1, \ldots, m$, where $w_i \leq W$. How many of the large rolls are needed to meet the order?

To formulate this problem, we may assume that an upper bound p is known on the number of large rolls to be used. We introduce variables $j = 1, \ldots, n$, which take value 1 if large roll j is used and 0 otherwise. Variables z_{ij}, $i = 1, \ldots, m$, $j = 1, \ldots, p$, indicate the number of small rolls of width w_i to be cut out of roll j. Using these variables, one can formulate the cutting stock problem as follows:

2.3. CUTTING STOCK: FORMULATIONS WITH MANY...

$$\begin{aligned}
\min \quad & \sum_{j=1}^{p} y_j \\
& \sum_{i=1}^{m} w_i z_{ij} \leq W y_j && j = 1, \ldots, p \\
& \sum_{j=1}^{p} z_{ij} \geq b_i && i = 1, \ldots, m \\
& y_j \in \{0, 1\} && j = 1, \ldots, p \\
& z_{ij} \in \mathbb{Z}_+ && i = 1, \ldots, m, \ j = 1, \ldots, p.
\end{aligned} \quad (2.1)$$

The first set of constraints ensures that the sum of the widths of the small rolls cut out of a large roll does not exceed W, and that a large roll is used whenever a small roll is cut out of it. The second set ensures that the numbers of small rolls that are cut meets the demands. Computational experience shows that this is not a strong formulation: The bound provided by the linear programming relaxation is rather distant from the optimal integer value.

A better formulation is needed. Let us consider all the possible different *cutting patterns*. Each pattern is represented by a vector $s \in \mathbb{Z}^m$ where component i represents the number of rolls of width w_i cut out of the big roll. The set of cutting patterns is therefore $\mathcal{S} := \{s \in \mathbb{Z}^n : \sum_{i=1}^{m} w_i s_i \leq W, s \geq 0\}$. Note that $\mathcal{S}$ is a knapsack set. For example, if $W = 5$, and the order has rolls of 3 different widths $w_1 = 2.1$, $w_2 = 1.8$ and $w_3 = 1.5$, a possible cutting pattern consists of 3 rolls of width 1.5, i.e., $\begin{pmatrix} 0 \\ 0 \\ 3 \end{pmatrix}$, another consists of one roll of width 2.1 and one of width 1.8, i.e., $\begin{pmatrix} 1 \\ 1 \\ 0 \end{pmatrix}$, etc.

If we introduce integer variables x_s representing the number of rolls cut according to pattern $s \in \mathcal{S}$, the cutting stock problem can be formulated as

$$\begin{aligned}
\min \quad & \sum_{s \in \mathcal{S}} x_s \\
& \sum_{s \in \mathcal{S}} s_i x_s \geq b_i \quad i = 1, \ldots, m \\
& x \geq 0 \quad \text{integral}.
\end{aligned} \quad (2.2)$$

This is an integer programming formulation in which the columns of the constraint matrix are all the feasible solutions of a knapsack set. The number of these columns (i.e., the number of possible patterns) is typically enormous, but this is a strong formulation in the sense that the bound provided by the linear programming relaxation is usually close to the optimal value of the integer program. A good solution to the integer program can typically be obtained by simply rounding the linear programming solution. However, solving the linear programming relaxation of (2.2) is challenging. This is best done using column generation, as first proposed by Gilmore and Gomory [168]. We briefly outline this technique here. We will return to it in Sect. 8.2.2. We suggest that readers not familiar with linear programming duality (which will be discussed later in Sect. 3.3) skip directly to Sect. 2.4.

The dual of the linear programming relaxation of (2.2) is:

$$\max \sum_{i=1}^{m} b_i u_i$$
$$\sum_{i=1}^{m} s_i u_i \leq 1 \quad s \in \mathcal{S} \quad (2.3)$$
$$u \geq 0.$$

Let $\mathcal{S}'$ be a subset of $\mathcal{S}$, and consider the cutting stock problem (2.2) restricted to the variables indexed by $\mathcal{S}'$. The dual is the problem defined by the inequalities from (2.3) indexed by $\mathcal{S}'$. Let $\bar{x}$, $\bar{u}$ be optimal solutions to the linear programming relaxations of (2.2) and (2.3) restricted to $\mathcal{S}'$. By setting $\bar{x}_s = 0$, $s \in \mathcal{S} \setminus \mathcal{S}'$, $\bar{x}$ can be extended to a feasible solution of the linear relaxation of (2.2). By strong duality $\bar{x}$ is an optimal solution of the linear relaxation of (2.2) if $\bar{u}$ provides a feasible solution to (2.3) (defined over $\mathcal{S}$). The solution $\bar{u}$ is feasible for (2.3) if and only if $\sum_{i=1}^{m} s_i \bar{u}_i \leq 1$ for every $s \in \mathcal{S}$ or equivalently if and only if the value of the following knapsack problem is at most equal to 1.

$$\max\{\sum_{i=1}^{m} \bar{u}_i s_i : s \in \mathcal{S}\}$$

If the value of this knapsack problem exceeds 1, let s^* be an optimal solution. Then s^* corresponds to a constraint of (2.3) that is most violated by $\bar{u}$, and s^* is added to $\mathcal{S}'$, thus enlarging the set of candidate patterns.

This is the *column generation* scheme, where variables of a linear program with exponentially many variables are generated on the fly when strong duality is violated, by solving an optimization problem (knapsack, in our case).

2.4 Packing, Covering, Partitioning

Let $E := \{1, \ldots, n\}$ be a finite set and $\mathcal{F} := \{F_1, \ldots, F_m\}$ a family of subsets of E. A set $S \subseteq E$ is said to be a *packing, partitioning* or *covering* of the family $\mathcal{F}$ if S intersects each member of $\mathcal{F}$ at most once, exactly once, or at least once, respectively. Representing a set $S \subseteq E$ by its *characteristic vector* $x^S \in \{0,1\}^n$, i.e., $x_j^S = 1$ if $j \in S$, and $x_j^S = 0$ otherwise, the families of packing, partitioning and covering sets have the following formulations.

$$S^P := \{x \in \{0,1\}^n : \sum_{j \in F_i} x_j \leq 1, \forall F_i \in \mathcal{F}\},$$
$$S^T := \{x \in \{0,1\}^n : \sum_{j \in F_i} x_j = 1, \forall F_i \in \mathcal{F}\},$$
$$S^C := \{x \in \{0,1\}^n : \sum_{j \in F_i} x_j \geq 1, \forall F_i \in \mathcal{F}\}.$$

Given weights w_j on the elements $j = 1, \ldots, n$, the *set packing* problem is $\max\{\sum_{j=1}^n w_j x_j : x \in S^P\}$, the *set partitioning* problem is $\min\{\sum_{j=1}^n w_j x_j : x \in S^T\}$, and the *set covering* problem is $\min\{\sum_{j=1}^n w_j x_j : x \in S^C\}$.

Given $E := \{1, \ldots, n\}$ and a family $\mathcal{F} := \{F_1, \ldots, F_m\}$ of subsets of E, the *incidence matrix* of $\mathcal{F}$ is the $m \times n$ 0,1 matrix in which $a_{ij} = 1$ if and only if $j \in F_i$. Then $S^P = \{x \in \{0,1\}^n : Ax \leq \mathbf{1}\}$, where $\mathbf{1}$ denotes the column vector in $\mathbb{R}^m$ all of whose components are equal to 1. Similarly the sets S^T, S^C can be expressed in terms of A. Conversely, given any 0,1 matrix A, one can define a *set packing family* $S^P(A) := \{x \in \{0,1\}^n : Ax \leq \mathbf{1}\}$. The families $S^T(A)$, $S^C(A)$ are defined similarly.

Numerous practical problems and several problems in combinatorics and graph theory can be formulated as set packing or covering. We illustrate some of them.

2.4.1 Set Packing and Stable Sets

Let $G = (V, E)$ be an undirected graph and let $n := |V|$. A *stable set* in G is a set of nodes no two of which are adjacent. Therefore $S \subseteq V$ is a stable set if and only if its characteristic vector $x \in \{0,1\}^n$ satisfies $x_i + x_j \leq 1$ for every edge $ij \in E$. If we consider E as a family of subsets of V, the characteristic vectors of the stable sets in G form a set packing family, namely

$$\text{stab}(G) := \{x \in \{0,1\}^n : x_i + x_j \leq 1 \text{ for all } ij \in E\}.$$

We now show the converse: Every set packing family is the family of characteristic vectors of the stable sets of some graph. Given an $m \times n$ 0,1 matrix A, the *intersection graph* of A is an undirected simple graph

$G_A = (V, E)$ on n nodes, corresponding to the columns of A. Two nodes u, v are adjacent in G_A if and only if $a_{iu} = a_{iv} = 1$ for some row index i, $1 \leq i \leq m$. In Fig. 2.1 we show a matrix A and its intersection graph.

$$A := \begin{pmatrix} 1 & 1 & 0 & 0 & 1 \\ 0 & 1 & 1 & 0 & 0 \\ 0 & 0 & 1 & 1 & 0 \\ 0 & 0 & 0 & 1 & 1 \\ 0 & 1 & 0 & 1 & 0 \end{pmatrix}$$

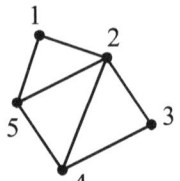

Figure 2.1: A $0, 1$ matrix A and its intersection graph G_A

We have $S^P(A) = \text{stab}(G_A)$ since a vector $x \in \{0,1\}^n$ is in $S^P(A)$ if and only if $x_j + x_k \leq 1$ whenever $a_{ij} = a_{ik} = 1$ for some row i.

All modern integer programming solvers use intersection graphs to model logical conditions among the binary variables of integer programming formulations. Nodes are often introduced for the complement of binary variables as well: This is useful to model conditions such as $x_i \leq x_j$, which can be reformulated in set packing form as $x_i + (1 - x_j) \leq 1$. In this context, the intersection graph is called a *conflict graph*. We refer to the paper of Atamtürk, Nemhauser, and Savelsbergh [16] for the use of conflict graphs in integer programming. This paper stresses the practical importance of strengthening set packing formulations.

2.4.2 Strengthening Set Packing Formulations

Given a $0, 1$ matrix A, the correspondence between $S^P(A)$ and $\text{stab}(G_A)$ can be used to strengthen the formulation $\{x \in \{0,1\}^n : Ax \leq 1\}$.

A *clique* in a graph is a set of pairwise adjacent nodes. Since a clique K in G_A intersects any stable set in at most one node, the inequality

$$\sum_{j \in K} x_j \leq 1$$

is valid for $S^P(A) = \text{stab}(G_A)$. This inequality is called a *clique inequality*. Conversely, given $Q \subseteq V$, if $\sum_{j \in Q} x_j \leq 1$ is a valid inequality for $\text{stab}(G_A)$, then every pair of nodes in Q must be adjacent, that is, Q is a clique of G_A.

A clique is *maximal* if it is not properly contained in any other clique. Note that, given two cliques K, K' in G_A such that $K \subset K'$, inequality $\sum_{j \in K} x_j \leq 1$ is implied by the inequalities $\sum_{j \in K'} x_j \leq 1$ and $x_j \geq 0$, $j \in K' \setminus K$.

2.4. PACKING, COVERING, PARTITIONING

On the other hand, the following argument shows that no maximal clique inequality is implied by the other clique inequalities and the constraints $0 \leq x_j \leq 1$. Let K be a maximal clique of G_A. We will exhibit a point $\bar{x} \in [0,1]^V$ that satisfies all clique inequalities except for the one relative to K. Let $\bar{x}_j := \frac{1}{|K|-1}$ for all $j \in K$ and $\bar{x}_j := 0$ otherwise. Since K is a maximal clique, every other clique K' intersects it in at most $|K|-1$ nodes, therefore $\sum_{j \in K'} \bar{x}_j \leq 1$. On the other hand, $\sum_{j \in K} \bar{x}_j = 1 + \frac{1}{|K|-1} > 1$. We have shown the following.

Theorem 2.2. *Given an $m \times n$ 0,1 matrix A, let $\mathcal{K}$ be the collection of all maximal cliques of its intersection graph G_A. The strongest formulation for $S^P(A) = \text{stab}(G_A)$ that only involves set packing constraints is*

$$\{x \in \{0,1\}^n : \sum_{j \in K} x_j \leq 1, \forall K \in \mathcal{K}\}.$$

Example 2.3. In the example of Fig. 2.1, the inequalities $x_2 + x_3 + x_4 \leq 1$ and $x_2 + x_4 + x_5 \leq 1$ are clique inequalities relative to the cliques $\{2,3,4\}$ and $\{2,4,5\}$ in G_A. Note that the point $(0, 1/2, 1/2, 1/2, 0)$ satisfies $Ax \leq 1$, $0 \leq x \leq 1$ but violates $x_2 + x_3 + x_4 \leq 1$. A better formulation of $Ax \leq 1$, $x \in \{0,1\}^n$ is obtained by replacing the constraint matrix A by the maximal clique versus node incidence matrix A_c of the intersection graph of A. For the example of Fig. 2.1, $A_c := \begin{pmatrix} 1 & 1 & 0 & 0 & 1 \\ 0 & 1 & 1 & 1 & 0 \\ 0 & 1 & 0 & 1 & 1 \end{pmatrix}$. The reader can verify that this formulation is perfect, as defined in Sect. 1.4. ∎

Note that the strongest set packing formulation described in Theorem 2.2 may contain exponentially many inequalities. If $\mathcal{K}$ denotes the collection of all maximal cliques of a graph G, the $|\mathcal{K}| \times n$ incidence matrix of $\mathcal{K}$ is called the *clique matrix* of G. Exercise 2.10 gives a characterization of clique matrices due to Gilmore [167]. Theorem 2.2 prompts the following question:

> *For which graphs G is the formulation defined in Theorem 2.2 a perfect formulation of $\text{stab}(G)$?*

This leads to the theory of perfect graphs, see Sect. 4.11 for references on this topic. In Chap. 10 we will discuss a semidefinite relaxation of $\text{stab}(G)$.

2.4.3 Set Covering and Transversals

We have seen the equivalence between general packing sets and stable sets in graphs. Covering sets do not seem to have an equivalent graphical representation. However some important questions in graph theory regarding connectivity, coloring, parity, and others can be formulated using covering sets. We first introduce a general setting for these formulations.

Given a finite set $E := \{1, \ldots, n\}$, a family $\mathcal{S} := \{S_1, \ldots, S_m\}$ of subsets E is a *clutter* if it has the following property:

For every pair S_i, $S_j \in \mathcal{S}$, both $S_i \setminus S_j$ and $S_j \setminus S_i$ are nonempty.

A subset T of E is a *transversal* of $\mathcal{S}$ if $T \cap S_i \neq \emptyset$ for every $S_i \in \mathcal{S}$. Let $\mathcal{T} := \{T_1, \ldots, T_q\}$ be the family of all inclusionwise minimal transversals of $\mathcal{S}$. The family $\mathcal{T}$ is a clutter as well, called the *blocker* of $\mathcal{S}$. The following set-theoretic property, due to Lawler [251] and Edmonds and Fulkerson [127], is fundamental for set covering formulations.

Proposition 2.4. *Let $\mathcal{S}$ be a clutter and $\mathcal{T}$ its blocker. Then $\mathcal{S}$ is the blocker of $\mathcal{T}$.*

Proof. Let $\mathcal{Q}$ be the blocker of $\mathcal{T}$. We need to show that $\mathcal{Q} = \mathcal{S}$. By definition of clutter, it suffices to show that every member of $\mathcal{S}$ contains some member of $\mathcal{Q}$ and every member of $\mathcal{Q}$ contains some member of $\mathcal{S}$.

Let $S_i \in \mathcal{S}$. By definition of $\mathcal{T}$, $S_i \cap T \neq \emptyset$ for every $T \in \mathcal{T}$. Therefore S_i is a transversal of $\mathcal{T}$. Because $\mathcal{Q}$ is the blocker of $\mathcal{T}$, this implies that S_i contains a member of $\mathcal{Q}$.

We now show the converse, namely every member of $\mathcal{Q}$ contains a member of $\mathcal{S}$. Suppose not. Then there exists a member Q of $\mathcal{Q}$ such that $(E \setminus Q) \cap S \neq \emptyset$ for every $S \in \mathcal{S}$. Therefore $E \setminus Q$ is a transversal of $\mathcal{S}$. This implies that $E \setminus Q$ contains some member $T \in \mathcal{T}$. But then $Q \cap T = \emptyset$, a contradiction to the assumption that Q is a transversal of $\mathcal{T}$. □

In light of the previous proposition, we call the pair of clutters $\mathcal{S}$ and its blocker $\mathcal{T}$ a *blocking pair*.

Given a vector $x \in \mathbb{R}^n$, the *support* of x is the set $\{i \in \{1, \ldots, n\} : x_i \neq 0\}$. Proposition 2.4 yields the following:

Observation 2.5. *Let $\mathcal{S}$, $\mathcal{T}$ be a blocking pair of clutters and let A be the incidence matrix of $\mathcal{T}$. The vectors with minimal support in the set covering family $\mathcal{S}^C(A)$ are the characteristic vectors of the family $\mathcal{S}$.*

Consider the following problem, which arises often in combinatorial optimization (we give three examples in Sect. 2.4.4).

Let $E := \{1, \ldots, n\}$ be a set of elements where each element $j = 1, \ldots, n$ is assigned a nonnegative weight w_j, and let $\mathcal{R}$ be a family of subsets of E. Find a member $S \in \mathcal{R}$ having minimum weight $\sum_{j \in S} w_j$.

2.4. PACKING, COVERING, PARTITIONING

Let $\mathcal{S}$ be the clutter consisting of the minimal members of $\mathcal{R}$. Note that, since the weights are nonnegative, the above problem always admit an optimal solution that is a member of $\mathcal{S}$.

Let $\mathcal{T}$ be the blocker of $\mathcal{S}$ and A the incidence matrix of $\mathcal{T}$. In light of Observation 2.5 an integer programming formulation for the above problem is given by

$$\min\{wx : x \in S^C(A)\}.$$

2.4.4 Set Covering on Graphs: Many Constraints

We now apply the technique introduced above to formulate some optimization problems on an undirected graph $G = (V, E)$ with nonnegative edge weights w_e, $e \in E$.

Given $S \subseteq V$, let $\delta(S) := \{uv \in E : u \in S, v \notin S\}$. A *cut* in G is a set F of edges such that $F = \delta(S)$ for some $S \subseteq V$. A cut F is *proper* if $F = \delta(S)$ for some $\emptyset \neq S \subset V$. For every node v, we will write $\delta(v) := \delta(\{v\})$ to denote the set of edges containing v. The *degree* of node v is $|\delta(v)|$.

Minimum Weight s,t-Cuts

Let s, t be two distinct nodes of a connected graph G. An *s,t-cut* is a cut of the form $\delta(S)$ such that $s \in S$ and $t \notin S$. Given nonnegative weights on the edges, w_e for $e \in E$, the *minimum weight s,t-cut problem* is to find an s,t-cut F that minimizes $\sum_{e \in F} w_e$.

An *s,t-path* in G is a path between s and t in G. Equivalently, an s,t-path is a minimal set of edges that induce a connected graph containing both s and t. Let $\mathcal{S}$ be the family of inclusionwise minimal s,t-cuts. Note that its blocker $\mathcal{T}$ is the family of s,t-paths. Therefore the minimum weight s,t-cut problem can be formulated as follows:

$$\begin{array}{ll} \min & \sum_{e \in E} w_e x_e \\ & \sum_{e \in P} x_e \geq 1 \quad \text{for all } s,t\text{-paths } P \\ & x_e \in \{0, 1\} \quad e \in E. \end{array}$$

Fulkerson [156] showed that the above formulation is a perfect formulation. Ford and Fulkerson [146] gave a polynomial-time algorithm for the minimum weight of an s,t-cut problem, and proved that the minimum weight of an s,t-cut is equal to the maximum value of an s,t-flow. This will be discussed in Chap. 4.

Let A be the incidence matrix of a clutter and B the incidence matrix of its blocker. Lehman [254] proved that $S^C(A)$ is a perfect formulation if and only if $S^C(B)$ is a perfect formulation. Lehman's theorem together with Fulkerson's theorem above imply that the following linear program solves the shortest s,t-path problem when $w \geq 0$:

$$\begin{array}{ll} \min & \sum_{e \in E} w_e x_e \\ & \sum_{e \in C} x_e \geq 1 \quad \text{for all } s,t\text{-cuts } C \\ & 0 \leq x_e \leq 1 \quad e \in E. \end{array}$$

We give a more traditional formulation of the shortest s,t-path problem in Sect. 4.3.2.

Minimum Cut

In the *min-cut problem* one wants to find a proper cut of minimum total weight in a connected graph G with nonnegative edge weights w_e, $e \in E$. An edge set $T \subseteq E$ is a *spanning tree* of G if it is an inclusionwise minimal set of edges such that the graph (V,T) is connected.

Let $\mathcal{S}$ be the family of inclusionwise minimal proper cuts. Note that the blocker of $\mathcal{S}$ is the family of spanning trees of G, hence one can formulate the min-cut problem as

$$\begin{array}{ll} \min & \sum_{e \in E} w_e x_e \\ & \sum_{e \in T} x_e \geq 1 \quad \text{for all spanning trees } T \\ & x_e \in \{0,1\} \quad e \in E. \end{array} \qquad (2.4)$$

This is not a perfect formulation (see Exercise 2.13). Nonetheless, the min-cut problem is polynomial-time solvable. A solution can be found by fixing a node $s \in V$, computing a minimum weight s,t-cut for every choice of t in $V \setminus \{s\}$, and selecting the cut of minimum weight among the $|V|-1$ cuts computed.

Max-Cut

Given a graph $G=(V,E)$ with edge weights w_e, $e \in E$, the *max-cut problem* asks to find a set $F \subseteq E$ of maximum total weight in G such that F is a cut of G. That is, $F = \delta(S)$, for some $S \subseteq V$.

Given a graph $G=(V,E)$ and $C \subseteq E$, let $V(C)$ denote the set of nodes that belong to at least one edge in C. A set of edges $C \subseteq E$ is a *cycle* in G if the graph $(V(C),C)$ is connected and all its nodes have degree two. A cycle C in G is an *odd cycle* if C has an odd number of edges. A basic fact in graph theory states that a set $F \subseteq E$ is contained in a cut of G if and only if $(E \setminus F) \cap C \neq \emptyset$ for every odd cycle C of G (see Exercise 2.14).

2.4. PACKING, COVERING, PARTITIONING

Therefore when $w_e \geq 0$, $e \in E$, the max-cut problem in the graph $G = (V, E)$ can be formulated as the problem of finding a set $E' \subseteq E$ of minimum total weight such that $E' \cap C \neq \emptyset$ for every odd cycle C of G.

$$\begin{aligned} \min \ & \sum_{e \in E} w_e x_e \\ & \sum_{e \in C} x_e \geq 1 \quad \text{for all odd cycles } C \\ & x_e \in \{0, 1\} \quad e \in E. \end{aligned} \tag{2.5}$$

Given an optimal solution $\bar{x}$ to (2.5), the optimal solution of the max-cut problem is the cut $\{e \in E : \bar{x}_e = 0\}$.

Unlike the two previous examples, the max-cut problem is NP-hard. However, Goemans and Williamson [173] show that a near-optimal solution can be found in polynomial time, using a semidefinite relaxation that will be discussed in Sect. 10.2.1.

2.4.5 Set Covering with Many Variables: Crew Scheduling

An airline wants to operate its daily flight schedule using the smallest number of crews. A crew is on duty for a certain number of consecutive hours and therefore may operate several flights. A feasible crew schedule is a sequence of flights that may be operated by the same crew within its duty time. For instance it may consist of the 8:30–10:00 am flight from Pittsburgh to Chicago, then the 11:30 am–1:30 pm Chicago–Atlanta flight and finally the 2:45–4:30 pm Atlanta–Pittsburgh flight.

Define $A = \{a_{ij}\}$ to be the 0, 1 matrix whose rows correspond to the daily flights operated by the company and whose columns correspond to all the possible crew schedules. The entry a_{ij} equals 1 if flight i is covered by crew schedule j, and 0 otherwise. The problem of minimizing the number of crews can be formulated as

$$\min\{\sum_j x_j : x \in S^C(A)\}.$$

In an optimal solution a flight may be covered by more than one crew: One crew operates the flight and the other occupies passenger seats. This is why the above formulation involves covering constraints. The number of columns (that is, the number of possible crew schedules) is typically enormous. Therefore, as in the cutting stock example, column generation is relevant in crew scheduling applications.

2.4.6 Covering Steiner Triples

Fulkerson, Nemhauser, and Trotter [157] constructed set covering problems of small size that are notoriously difficult to solve. A *Steiner triple system* of order n (denoted by $STS(n)$) consists of a set E of n elements and a collection $\mathcal{S}$ of triples of E with the property that every pair of elements in E appears together in a unique triple in $\mathcal{S}$. It is known that a Steiner triple system of order n exists if and only if $n \equiv 1$ or $3 \mod 6$. A subset C of E is a covering of the Steiner triple system if $C \cap T \neq \emptyset$ for every triple T in $\mathcal{S}$. Given a Steiner triple system, the problem of computing the smallest cardinality of a cover is

$$\min\{\sum_j x_j : x \in S^C(A)\}$$

where A is the $|\mathcal{S}| \times n$ incidence matrix of the collection $\mathcal{S}$. Fulkerson, Nemhauser, and Trotter constructed an infinite family of Steiner triple systems in 1974 and asked for the smallest cardinality of a cover. The question was solved 5 years later for $STS(45)$, it took another 15 years for $STS(81)$, and the current record is the solution of $STS(135)$ and $STS(243)$ [292].

2.5 Generalized Set Covering: The Satisfiability Problem

We generalize the set covering model by allowing constraint matrices whose entries are $0, \pm 1$ and we use it to formulate problems in propositional logic.

Atomic propositions $x_1, \ldots, x_n$ can be either *true* or *false*. A *truth assignment* is an assignment of "true" or "false" to every atomic proposition. A *literal* L is an atomic proposition x_j or its negation $\neg x_j$. A *conjunction* of two literals $L_1 \wedge L_2$ is true if both literals are true and a *disjunction* of two literals $L_1 \vee L_2$ is true if at least one of L_1, L_2 is true. A *clause* is a disjunction of literals and it is *satisfied* by a given truth assignment if at least one of its literals is true.

For example, the clause with three literals $x_1 \vee x_2 \vee \neg x_3$ is satisfied if "x_1 is true or x_2 is true or x_3 is false." In particular, it is satisfied by the truth assignment $x_1 = x_2 = x_3 =$ "false."

2.5. GENERALIZED SET COVERING: THE SATISFIABILITY... 59

It is usual to identify truth assignments with 0,1 vectors: $x_i = 1$ if $x_i =$ "true" and $x_i = 0$ if $x_i =$ "false." A truth assignment satisfies the clause

$$\bigvee_{j \in P} x_j \vee (\bigvee_{j \in N} \neg x_j)$$

if and only if the corresponding 0, 1 vector satisfies the inequality

$$\sum_{j \in P} x_j - \sum_{j \in N} x_j \geq 1 - |N|.$$

For example the clause $x_1 \vee x_2 \vee \neg x_3$ is satisfied if and only if the corresponding 0, 1 vector satisfies the inequality $x_1 + x_2 - x_3 \geq 0$.

A logic statement consisting of a conjunction of clauses is said to be in *conjunctive normal form*. For example the logical proposition $(x_1 \vee x_2 \vee \neg x_3) \wedge (x_2 \vee x_3)$ is in conjunctive normal form. Such logic statements can be represented by a system of m linear inequalities, where m is the number of clauses in the conjunctive normal form. This can be written in the form:

$$Ax \geq 1 - n(A) \tag{2.6}$$

where A is an $m \times n$ $0, \pm 1$ matrix and the ith component of $n(A)$ is the number of -1's in row i of A. For example the logical proposition $(x_1 \vee x_2 \vee \neg x_3) \wedge (x_2 \vee x_3)$ corresponds to the system of constraints

$$\begin{aligned} x_1 + x_2 - x_3 &\geq 0 \\ x_2 + x_3 &\geq 1 \\ x_i &\in \{0, 1\}^3. \end{aligned}$$

In this example $A = \begin{pmatrix} 1 & 1 & -1 \\ 0 & 1 & 1 \end{pmatrix}$ and $n(A) = \begin{pmatrix} 1 \\ 0 \end{pmatrix}$.

Every logic statement can be written in conjunctive normal form by using rules of logic such as $L_1 \vee (L_2 \wedge L_3) = (L_1 \vee L_2) \wedge (L_1 \vee L_3)$, $\neg(L_1 \wedge L_2) = \neg L_1 \vee \neg L_2$, etc. This will be illustrated in Exercises 2.24, 2.25.

We present two classical problems in logic.

The *satisfiability problem* (SAT) for a set S of clauses, asks for a truth assignment satisfying all the clauses in S or a proof that none exists. Equivalently, SAT consists of finding a 0, 1 solution x to (2.6) or showing that none exists.

Logical inference in propositional logic consists of a set S of clauses (the premises) and a clause C (the conclusion), and asks whether every truth assignment satisfying all the clauses in S also satisfies the conclusion C. To the clause C, we associate the inequality

$$\sum_{j \in P(C)} x_j - \sum_{j \in N(C)} x_j \geq 1 - |N(C)|. \tag{2.7}$$

Therefore the conclusion C cannot be deduced from the premises S if and only if (2.6) has a 0, 1 solution that violates (2.7).

Equivalently C cannot be deduced from S if and only if the integer program

$$\min \left\{ \sum_{j \in P(C)} x_j - \sum_{j \in N(C)} x_j \; : \; Ax \geq \mathbf{1} - n(A), \; x \in \{0,1\}^n \right\}$$

has a solution with value $-|n(C)|$.

2.6 The Sudoku Game

The game is played on a 9×9 grid which is subdivided into 9 blocks of 3×3 contiguous cells. The grid must be filled with numbers $1, \ldots, 9$ so that all the numbers between 1 and 9 appear in each row, in each column and in each of the nine blocks. A game consists of an initial assignment of numbers in some cells (Fig. 2.2).

8				2	6			
						7		4
			7					5
			1				3	6
	1			8			4	
9	8				3			
3					1			
7		5						
			2	5				8

Figure 2.2: An instance of the Sudoku game

This is a decision problem that can be modeled with binary variables x_{ijk}, $1 \leq i, j, k \leq 9$ where $x_{ijk} = 1$ if number k is entered in position with coordinates i, j of the grid, and 0 otherwise.

The constraints are:

$$\sum_{j=1}^{9} x_{ijk} = 1, \quad 1 \leq j, k \leq 9 \quad \text{(each number } k \text{ appears once in column } j\text{)}$$
$$\sum_{j=1}^{9} x_{ijk} = 1, \quad 1 \leq i, k \leq 9 \quad \text{(each } k \text{ appears once in row } i\text{)}$$
$$\sum_{q,r=0}^{2} x_{i+q,j+r,k} = 1, \quad i,j = 1, 4, 7, \ 1 \leq k \leq 9 \quad \text{(each } k \text{ appears once in a block)}$$
$$\sum_{k=1}^{9} x_{ijk} = 1, \quad 1 \leq i, j \leq 9 \quad \text{(each cell contains exactly one number)}$$
$$x_{ijk} \in \{0, 1\}, \quad 1 \leq i, j, k \leq 9$$
$$x_{ijk} = 1, \quad \text{when the initial assignment has number } k \text{ in cell } i, j.$$

In *constraint programming*, variables take values in a specified domain, which may include data that are non-quantitative, and constraints restrict the space of possibilities in a way that is more general than the one given by linear constraints. We refer to the book "Constraint Processing" by R. Dechter [108] for an introduction to constraint programming. One of these constraints is \alldifferent$\{z_1, \ldots, z_n\}$ which forces variables $z_1, \ldots, z_n$ to take distinct values in the domain. Using the \alldifferent{} constraint, we can formulate the Sudoku game using 2-index variables, instead of the 3-index variables used in the above integer programming formulation. Variable x_{ij} represents the value in the cell of the grid with coordinates (i, j). Thus x_{ij} take its values in the domain $\{1, \ldots, 9\}$ and there is an \alldifferent{} constraint that involves the set of variables in each row, each column and each of the nine blocks.

2.7 The Traveling Salesman Problem

This section illustrates the fact that several formulations may exist for a given problem, and it is not immediately obvious which is the best for branch-and-cut algorithms.

A traveling salesman must visit n cities and return to the city he started from. We will call this a *tour*. Given the cost c_{ij} of traveling from city i to city j, for each $1 \leq i, j \leq n$ with $i \neq j$, in which order should the salesman visit the cities to minimize the total cost of his tour? This problem is the famous *traveling salesman problem*. If we allow costs c_{ij} and c_{ji} to be different for any given pair of cities i, j, then the problem is referred to as the *asymmetric traveling salesman problem*, while if $c_{ij} = c_{ji}$ for every pair of cities i and j, the problem is known as the *symmetric traveling salesman problem*. In Fig. 2.3, the left diagram represents eight cities in the plane. The cost of traveling between any two cities is assumed to be proportional to the Euclidean distance between them. The right diagram depicts the optimal tour.

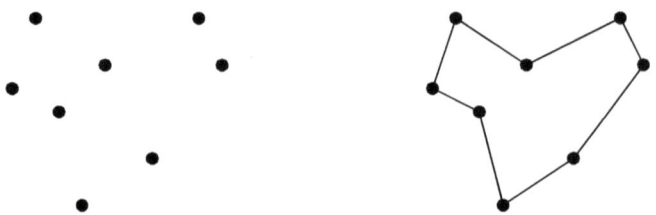

Figure 2.3: An instance of the symmetric traveling salesman problem in the Euclidean plane, and the optimal tour

It will be convenient to define the traveling salesman problem on a graph (directed or undirected). Given a *digraph* (a directed graph) $D = (V, A)$, a *(directed) Hamiltonian tour* is a circuit that traverses each node exactly once. Given costs c_a, $a \in A$, the asymmetric traveling salesman problem on D consists in finding a Hamiltonian tour in D of minimum total cost. Note that, in general, D might not contain any Hamiltonian tour. We give three different formulations for the asymmetric traveling salesman problem.

The first formulation is due to Dantzig, Fulkerson, and Johnson [103]. They introduce a binary variable x_{ij} for all $ij \in A$, where $x_{ij} = 1$ if the tour visits city j immediately after city i, and 0 otherwise. Given a set of cities $S \subseteq V$, let $\delta^+(S) := \{ij \in A : i \in S, j \notin S\}$, and let $\delta^-(S) := \{ij \in A : i \notin S, j \in S\}$. For ease of notation, for $v \in V$ we use $\delta^+(v)$ and $\delta^-(v)$ instead of $\delta^+(\{v\})$ and $\delta^-(\{v\})$. The Dantzig–Fulkerson–Johnson formulation of the traveling salesman problem is as follows.

$$\min \sum_{a \in A} c_a x_a \qquad (2.8)$$

$$\sum_{a \in \delta^+(i)} x_a = 1 \quad \text{for } i \in V \qquad (2.9)$$

$$\sum_{a \in \delta^-(i)} x_a = 1 \quad \text{for } i \in V \qquad (2.10)$$

$$\sum_{a \in \delta^+(S)} x_a \geq 1 \quad \text{for } \emptyset \subset S \subset V \qquad (2.11)$$

$$x_a \in \{0, 1\} \quad \text{for } a \in A. \qquad (2.12)$$

Constraints (2.9)–(2.10), known as *degree constraints*, guarantee that the tour visits each node exactly once and constraints (2.11) guarantee that the solution does not decompose into several subtours. Constraints (2.11) are known under the name of *subtour elimination constraints*. Despite the

2.7. THE TRAVELING SALESMAN PROBLEM

exponential number of constraints, this is the formulation that is most widely used in practice. Initially, one solves the linear programming relaxation that only contains (2.9)–(2.10) and $0 \leq x_{ij} \leq 1$. The subtour elimination constraints are added later, on the fly, only when needed. This is possible because the so-called separation problem can be solved efficiently for such constraints (see Chap. 4).

Miller, Tucker and Zemlin [278] found a way to avoid the subtour elimination constraints (2.11). Assume $V = \{1, \ldots, n\}$. The formulation has extra variables u_i that represent the position of node $i \geq 2$ in the tour, assuming that the tour starts at node 1, i.e., node 1 has position 1. Their formulation is identical to (2.8)–(2.12) except that (2.11) is replaced by

$$u_i - u_j + 1 \leq n(1 - x_{ij}) \quad \text{for all } ij \in A,\ i,j \neq 1. \qquad (2.13)$$

It is not difficult to verify that the Miller–Tucker–Zemlin formulation is correct. Indeed, if x is the incident vector of a tour, define u_i to be the position of node i in the tour, for $i \geq 2$. Then constraint (2.13) is satisfied. Conversely, if $x \in \{0,1\}^E$ satisfies (2.9)–(2.10) but is not the incidence vector of a tour, then (2.9)–(2.10) and (2.12) imply that there is at least one subtour $C \subseteq A$ that does not contain node 1. Summing the inequalities (2.13) relative to every $ij \in C$ gives the inequality $|C| \leq 0$, a contradiction. Therefore, if (2.9)–(2.10), (2.12), (2.13) are satisfied, x must represent a tour. Although the Miller–Tucker–Zemlin formulation is correct, we will show in Chap. 4 that it produces weaker bounds for branch-and-cut algorithms than the Dantzig–Fulkerson–Johnson formulation. It is for this reason that the latter is preferred in practice.

It is also possible to formulate the traveling salesman problem using variables x_{ak} for every $a \in A$, $k \in V$, where $x_{ak} = 1$ if arc a is the kth leg of the Hamiltonian tour, and $x_{ak} = 0$ otherwise. The traveling salesman problem can be formulated as follows.

$$\begin{aligned}
\min \quad & \sum_{a \in A} \sum_k c_a x_{ak} \\
& \sum_{a \in \delta^+(i)} \sum_k x_{ak} = 1 \quad \text{for } i = 1, \ldots, n \\
& \sum_{a \in \delta^-(i)} \sum_k x_{ak} = 1 \quad \text{for } i = 1, \ldots, n
\end{aligned} \qquad (2.14)$$

$$\sum_{a \in A} x_{ak} = 1 \quad \text{for } k = 1, \ldots, n$$

$$\sum_{a \in \delta^-(i)} x_{ak} = \sum_{a \in \delta^+(i)} x_{a,k+1} \quad \text{for } i = 1, \ldots, n \text{ and } k = 1, \ldots, n-1$$

$$\sum_{a \in \delta^-(1)} x_{an} = \sum_{a \in \delta^+(1)} x_{a1} = 1$$

$$x_{ak} = 0 \text{ or } 1 \quad \text{for } a \in A, \ k = 1, \ldots, n.$$

The first three constraints impose that each city is entered once, left once, and each leg of the tour contains a unique arc. The next constraint imposes that if leg k brings the salesman to city i, then he leaves city i on leg $k+1$. The last constraint imposes that the first leg starts from city 1 and the last returns to city 1. The main drawback of this formulation is its large number of variables.

The Dantzig–Fulkerson–Johnson formulation has a simple form in the case of the symmetric traveling salesman problem. Given an undirected graph $G = (V, E)$, a *Hamiltonian tour* is a cycle that goes exactly once through each node of G. Given costs c_e, $e \in E$, the symmetric traveling salesman problem is to find a Hamiltonian tour in G of minimum total cost. The Dantzig–Fulkerson–Johnson formulation for the symmetric traveling salesman problem is the following.

$$\begin{aligned}
\min \ & \sum_{e \in E} c_e x_e \\
& \sum_{e \in \delta(i)} x_e = 2 \quad \text{for } i \in V \\
& \sum_{e \in \delta(S)} x_e \geq 2 \quad \text{for } \emptyset \subset S \subset V \\
& x_e \in \{0,1\} \quad \text{for } e \in E.
\end{aligned} \qquad (2.15)$$

In this context $\sum_{e \in \delta(i)} x_e = 2$ for $i \in V$ are the *degree constraints* and $\sum_{e \in \delta(S)} x_e \geq 2$ for $\emptyset \subset S \subset V$ are the *subtour elimination constraints*. Despite its exponential number of constraints, the formulation (2.15) is very effective in practice. We will return to this formulation in Chap. 7.

Kaibel and Weltge [224] show that the traveling salesman problem cannot be formulated with polynomially many inequalities in the space of variables x_e, $e \in E$.

2.8 The Generalized Assignment Problem

The *generalized assignment problem* is the following 0,1 program, defined by coefficients c_{ij} and t_{ij}, and capacities T_j, $i = 1, \ldots, m$, $j = 1, \ldots, n$,

$$\max \quad \sum_{i=1}^{m} \sum_{j=1}^{n} c_{ij} x_{ij}$$
$$\sum_{j=1}^{n} x_{ij} = 1 \quad i = 1, \ldots, m \qquad (2.16)$$
$$\sum_{i=1}^{m} t_{ij} x_{ij} \leq T_j \quad j = 1, \ldots, n$$
$$x \in \{0,1\}^{m \times n}.$$

The following example is a variation of this model. In hospitals, operating rooms are a scarce resource that needs to be utilized optimally. The basic problem can be formulated as follows, acknowledging that each hospital will have its own specific additional constraints. Suppose that a hospital has n operating rooms. During a given time period T, there may be m surgeries that could potentially be scheduled. Let t_{ij} be the estimated time of operating on patient i in room j, for $i = 1, \ldots, m$, $j = 1, \ldots, n$. The goal is to schedule surgeries during the given time period so as to waste as little of the operating rooms' capacity as possible.

Let x_{ij} be a binary variable that takes the value 1 if patient i is operated on in operating room j, and 0 otherwise. The basic operating rooms scheduling problem is as follows:

$$\max \quad \sum_{i=1}^{m} \sum_{j=1}^{n} t_{ij} x_{ij}$$
$$\sum_{j=1}^{n} x_{ij} \leq 1 \quad i = 1, \ldots, m \qquad (2.17)$$
$$\sum_{i=1}^{m} t_{ij} x_{ij} \leq T \quad j = 1, \ldots, n$$
$$x \in \{0,1\}^{m \times n}.$$

The objective is to maximize the utilization time of the operating rooms during the given time period (this is equivalent to minimizing wasted capacity). The first constraints guarantee that each patient i is operated on at most once. If patient i *must* be operated on during this period, the inequality constraint is changed into an equality. The second constraints are the capacity constraints on each of the operating rooms.

A special case of interest is when all operating rooms are identical, that is, $t_{ij} := t_i$, $i = 1, \ldots, m$, $j = 1, \ldots, n$, where the estimated time t_i of operation i is independent of the operating room. In this case, the above formulation admits numerous symmetric solutions, since permuting operating rooms does not modify the objective value. Intuitively, symmetry in the problem seems helpful but, in fact, it may cause difficulties in the context of a standard branch-and-cut algorithm. This is due to the creation of a potentially very large number of isomorphic subproblems in the enumeration tree, resulting in a duplication of the computing effort unless the isomorphisms are discovered. Special techniques are available to deal with symmetries, such as isomorphism pruning, which can be incorporated in branch-and-cut algorithms. We will discuss this in Chap. 9.

The operating room scheduling problem is often complicated by the fact that there is also a limited number of surgeons, each surgeon can only perform certain operations, and a support team (anesthesiologist, nurses) needs to be present during the operation. To deal with these aspects of the operating room scheduling problem, one needs new variables and constraints.

2.9 The Mixing Set

We now describe a mixed integer linear set associated with a simple make-or-buy problem. The demand for a given product takes values $b_1, \ldots, b_n \in \mathbb{R}$ with probabilities $p_1, \ldots, p_n$. Note that the demand values in this problem need not be integer. Today we produce an amount $y \in \mathbb{R}$ of the product at a unit cost h, before knowing the actual demand. Tomorrow the actual demand b_i is experienced; if $b_i > y$ then we purchase the extra amount needed to meet the demand at a unit cost c. However, the product can only be purchased in unit batches, that is, in integer amounts. The problem is to describe the production strategy that minimizes the expected total cost. Let x_i be the amount purchased tomorrow if the demand takes value b_i. Define the mixing set
$$MIX := \left\{(y,x) \in \mathbb{R}_+ \times \mathbb{Z}_+^n : y + x_i \geq b_i, 1 \leq i \leq n\right\}.$$
Then the above problem can be formulated as
$$\begin{array}{l} \min \quad hy + c\sum_{i=1}^n p_i x_i \\ (y,x) \in MIX. \end{array}$$

2.10 Modeling Fixed Charges

Integer variables naturally represent entities that come in discrete amounts. They can also be used to model:

- logical conditions such as implications or dichotomies;
- nonlinearities, such as piecewise linear functions;
- nonconvex sets that can be expressed as a union of polyhedra.

2.10. MODELING FIXED CHARGES

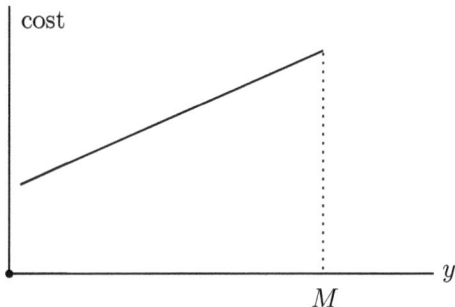

Figure 2.4: Fixed and variable costs

We introduce some of these applications. Economic activities frequently involve both fixed and variable costs. In this case, the cost associated with a certain variable y is 0 when the variable y takes value 0, and it is $c + hy$ whenever y takes positive value (see Fig. 2.4). For example, variable y may represent a production quantity that incurs both a fixed cost if anything is produced at all (e.g., for setting up the machines), and a variable cost (e.g., for operating the machines). This situation can be modeled using a binary variable x indicating whether variable y takes a positive value. Let M be some upper bound, known a priori, on the value of variable y. The (nonlinear) cost of variable y can be written as the linear expression

$$cx + hy$$

where we impose

$$\begin{aligned} y &\leq Mx \\ x &\in \{0,1\} \\ y &\geq 0. \end{aligned}$$

Such "big M" formulations should be used with caution in integer programming because their linear programming relaxations tend to produce weak bounds in branch-and-bound algorithms. Whenever possible, one should use the tightest known bound, instead of an arbitrarily large M. We give two examples.

2.10.1 Facility Location

A company would like to set up facilities in order to serve geographically dispersed customers at minimum cost. The m customers have known annual demands d_i, for $i = 1, \ldots, m$. The company can open a facility of capacity u_j and fixed annual operating cost f_j in location j, for $j = 1, \ldots, n$. Knowing the variable cost c_{ij} of transporting one unit of goods from location j to customer i, where should the company locate its facilities in order to minimize its annual cost

To formulate this problem, we introduce variables x_j that take the value 1 if a facility is opened in location j, and 0 if not. Let y_{ij} be the fraction of the demand d_i transported annually from j to i.

$$\min \sum_{i=1}^{m}\sum_{j=1}^{n} c_{ij} d_i y_{ij} + \sum_{j=1}^{n} f_j x_j$$

$$\sum_{j=1}^{n} y_{ij} = 1 \qquad i=1,\ldots,m$$

$$\sum_{i=1}^{m} d_i y_{ij} \leq u_j x_j \qquad j=1,\ldots,n$$

$$y \geq 0$$

$$x \in \{0,1\}^n.$$

The objective function is the total yearly cost (transportation plus operating costs). The first set of constraints guarantees that the demand is met, the second type of constraints are capacity constraints at the facilities. Note that the capacity constraints are fixed charge constraints, since they force $x_j = 1$ whenever $y_{ij} > 0$ for some i.

A classical special case is the uncapacitated facility location problem, in which $u_j = +\infty$, $j = 1,\ldots,n$. In this case, it is always optimal to satisfy all the demand of client i from the closest open facility, therefore y_{ij} can be assumed to be binary. Hence the problem can be formulated as

$$\begin{array}{lll} \min & \sum\sum c_{ij} d_i y_{ij} + \sum f_j x_j & \\ & \sum_j y_{ij} = 1 & i=1,\ldots,m \\ & \sum_i y_{ij} \leq m x_j & j=1,\ldots,n \\ & y \in \{0,1\}^{m\times n}, \quad x \in \{0,1\}^n. & \end{array} \qquad (2.18)$$

Note that the constraint $\sum_i y_{ij} \leq m x_j$ forces $x_j = 1$ whenever $y_{ij} > 0$ for some i. The same condition could be enforced by the disaggregated set of constraints $y_{ij} \leq x_j$, for all i, j.

$$\begin{array}{lll} \min & \sum\sum c_{ij} d_i y_{ij} + \sum f_j x_j & \\ & \sum_j y_{ij} = 1 & i=1,\ldots,m \\ & y_{ij} \leq x_j & i=1,\ldots,m,\ j=1,\ldots,n \\ & y \in \{0,1\}^{m\times n}, \quad x \in \{0,1\}^n. & \end{array} \qquad (2.19)$$

2.10. MODELING FIXED CHARGES

The disaggregated formulation (2.19) is stronger than the aggregated one (2.18), since the constraint $\sum_i y_{ij} \leq mx_i$ is just the sum of the constraints $y_{ij} \leq x_i$, $i = 1, \ldots, m$. According to the paradigm presented in Sect. 2.2 in this chapter, the disaggregated formulation is better, because it yields tighter bounds in a branch-and-cut algorithm. In practice it has been observed that the difference between these two bounds is typically enormous. It is natural to conclude that formulation (2.19) is the one that should be used in practice. However, the situation is more complicated. When the aggregated formulation (2.18) is given to state-of-the-art solvers, they are able to detect and generate disaggregated constraints on the fly, whenever these constraints are violated by the current feasible solution. So, in fact, it is preferable to use the aggregated formulation because the size of the linear relaxation is much smaller and faster to solve.

Let us elaborate on this interesting point. Nowadays, state-of-the-art solvers automatically detect violated minimal cover inequalities (this notion was introduced in Sect. 2.2), and the disaggregated constraints in (2.19) happen to be minimal cover inequalities for the aggregated constraints. More formally, let us write the aggregated constraint relative to facility j as

$$mz_j + \sum_{j=1}^{m} y_{ij} \leq m$$

where $z_j = 1 - x_j$ is also a $0, 1$ variable. This is a knapsack constraint. Note that any minimal cover inequality is of the form $z_j + y_{ij} \leq 1$. Substituting $1 - x_j$ for z_j, we get the disaggregated constraint $y_{ij} \leq x_j$. We will discuss the separation of minimal cover inequalities in Sect. 7.1.

2.10.2 Network Design

Network design problems arise in the telecommunication industry. Let N be a given set of nodes. Consider a directed network $G = (N, A)$ consisting of arcs that could be constructed. We need to select a subset of arcs from A in order to route commodities. Commodity k has a source $s_k \in N$, a destination $t_k \in N$, and volume v_k for $k = 1, \ldots, K$. Each commodity can be viewed as a flow that must be routed through the network. Each arc $a \in A$ has a construction cost f_a and a capacity c_a. If we select arc a, the sum of the commodity flows going through arc a should not exceed its capacity c_a. Of course, if we do not select arc a, no flow can be routed through a. How should we design the network in order to route all the demand at minimum cost?

Let us introduce binary variables x_a, for $a \in A$, where $x_a = 1$ if arc a is constructed, 0 otherwise. Let y_a^k denote the amount of commodity k flowing through arc a. The formulation is

$$\min \sum_{a \in A} f_a x_a$$

$$\sum_{a \in \delta^+(i)} y_{ij}^k - \sum_{a \in \delta^-(i)} y_{ji}^k = \begin{cases} v_k & \text{for } i = s_k \\ -v_k & \text{for } i = t_k \\ 0 & \text{for } i \in N \setminus \{s_k, t_k\} \end{cases} \quad \text{for } k = 1, \ldots K$$

$$\sum_{k=1}^{K} y_a^k \leq c_a x_a \quad \text{for } a \in A$$

$$y \geq 0$$

$$x_a \in \{0,1\} \quad \text{for } a \in A.$$

The first set of constraints are conservation of flow constraints: For each commodity k, the amount of flow out of node i equals to the amount of flow going in, except at the source and destination. The second constraints are the capacity constraints that need to be satisfied for each arc $a \in A$. Note that they are fixed-charge constraints.

2.11 Modeling Disjunctions

Many applications have disjunctive constraints. For example, when scheduling jobs on a machine, we might need to model that either job i is scheduled before job j or vice versa; if p_i and p_j denote the processing times of these two jobs on the machine, we then need a constraint stating that the starting times t_i and t_j of jobs i and j satisfy $t_j \geq t_i + p_i$ or $t_i \geq t_j + p_j$. In such applications, the feasible solutions lie in the union of two or more polyhedra.

In this section, the goal is to model that a point belongs to the union of k polytopes in $\mathbb{R}^n$, namely bounded sets of the form

$$\begin{aligned} A_i y &\leq b_i \\ 0 \leq y &\leq u_i, \end{aligned} \quad (2.20)$$

for $i = 1, \ldots, k$. The same modeling question is more complicated for unbounded polyhedra and will be discussed in Sect. 4.9.

A way to model the union of k polytopes in $\mathbb{R}^n$ is to introduce k variables $x_i \in \{0,1\}$, indicating whether y is in the ith polytope, and k vectors of variables $y_i \in \mathbb{R}^n$. The vector $y \in \mathbb{R}^n$ belongs to the union of the k polytopes (2.20) if and only if

2.11. MODELING DISJUNCTIONS

$$\begin{aligned}
\sum_{i=1}^{k} y_i &= y \\
A_i y_i &\leq b_i x_i & i &= 1, \ldots, k \\
0 \leq y_i &\leq u_i x_i & i &= 1, \ldots, k \\
\sum_{i=1}^{k} x_i &= 1 \\
x &\in \{0,1\}^k.
\end{aligned} \qquad (2.21)$$

The next proposition shows that formulation (2.21) is perfect in the sense that the convex hull of its solutions is simply obtained by dropping the integrality restriction.

Proposition 2.6. *The convex hull of solutions to (2.21) is*

$$\begin{aligned}
\sum_{i=1}^{k} y_i &= y \\
A_i y_i &\leq b_i x_i & i &= 1, \ldots, k \\
0 \leq y_i &\leq u_i x_i & i &= 1, \ldots, k \\
\sum_{i=1}^{k} x_i &= 1 \\
x &\in [0,1]^k.
\end{aligned}$$

Proof. Let $P \subset \mathbb{R}^n \times \mathbb{R}^{kn} \times \mathbb{R}^k$ be the polytope given in the statement of the proposition. It suffices to show that any point $\bar{z} := (\bar{y}, \bar{y}_1, \ldots, \bar{y}_k, \bar{x}_1, \ldots, \bar{x}_k)$ in P is a convex combination of solutions to (2.21). For t such that $\bar{x}_t \neq 0$, define the point $z^t = (y^t, y_1^t, \ldots, y_k^t, x_1^t, \ldots, x_k^t)$ where

$$y^t := \frac{\bar{y}_t}{\bar{x}_t}, \qquad y_i^t := \begin{cases} \frac{\bar{y}_t}{\bar{x}_t} & \text{for } i = t, \\ 0 & \text{otherwise,} \end{cases} \qquad x_i^t := \begin{cases} 1 & \text{for } i = t, \\ 0 & \text{otherwise.} \end{cases}$$

The z^ts are solutions of (2.21). We claim that $\bar{z}$ is a convex combination of these points, namely $\bar{z} = \sum_{t : \bar{x}_t \neq 0} \bar{x}_t z^t$. To see this, observe first that $\bar{y} = \sum \bar{y}_i = \sum_{t : \bar{x}_t \neq 0} \bar{y}_t = \sum_{t : \bar{x}_t \neq 0} \bar{x}_t y^t$. Second, note that when $\bar{x}_i \neq 0$ we have $\bar{y}_i = \sum_{t : \bar{x}_t \neq 0} \bar{x}_t y_i^t$. This equality also holds when $\bar{x}_i = 0$ because then $\bar{y}_i = 0$ and $y_i^t = 0$ for all t such that $\bar{x}_t \neq 0$. Finally $\bar{x}_i = \sum_{t : \bar{x}_t \neq 0} \bar{x}_t x_i^t$ for $i = 1, \ldots, k$. □

2.12 The Quadratic Assignment Problem and Fortet's Linearization

In this book we mostly deal with *linear* integer programs. However, *nonlinear* integer programs (in which the objective function or some of the constraints defining the feasible region are nonlinear) are important in some applications. The *quadratic assignment problem (QAP)* is an example of a nonlinear 0, 1 program that is simple to state but notoriously difficult to solve. Interestingly, we will show that it can be linearized.

We have to place n facilities in n locations. The data are the amount $f_{k\ell}$ of goods that has to be shipped from facility k to facility ℓ, for $k = 1, \ldots, n$ and $\ell = 1, \ldots, n$, and the distance d_{ij} between locations i, j, for $i = 1, \ldots, n$ and $j = 1, \ldots, n$.

The problem is to assign facilities to locations so as to minimize the total cumulative distance traveled by the goods. For example, in the electronics industry, the quadratic assignment problem is used to model the problem of placing interconnected electronic components onto a microchip or a printed circuit board.

Let x_{ki} be a binary variable that takes the value 1 if facility k is assigned to location i, and 0 otherwise. The quadratic assignment problem can be formulated as follows:

$$\max \quad \sum_{i,j} \sum_{k,\ell} d_{ij} f_{k\ell} x_{ki} x_{\ell j}$$
$$\sum_k x_{ki} = 1 \qquad i = 1, \ldots, n$$
$$\sum_i x_{ki} = 1 \qquad k = 1, \ldots, n$$
$$x \in \{0, 1\}^{n \times n}.$$

The quadratic assignment problem is an example of a *0,1 polynomial program*

$$\begin{aligned} \min \; z = \;& f(x) \\ & g_i(x) = 0 \quad i = 1, \ldots, m \\ & x_j \in \{0, 1\} \quad j = 1, \ldots, n \end{aligned} \qquad (2.22)$$

where the functions f and g_i ($i = 1, \ldots, m$) are polynomials. Fortet [144] observed that such nonlinear functions can be linearized when the variables only take value 0 or 1.

Proposition 2.7. *Any 0,1 polynomial program (2.22) can be formulated as a pure 0,1 linear program by introducing additional variables.*

Proof. Note that, for any integer exponent $k \geq 1$, the 0,1 variable x_j satisfies $x_j^k = x_j$. Therefore we can replace each expression of the from x_j^k with x_j, so that no variable appears in f or g_i with exponent greater than 1.

The product $x_i x_j$ of two 0,1 variables can be replaced by a new 0,1 variable y_{ij} related to x_i, x_j by linear constraints. Indeed, to guarantee that $y_{ij} = x_i x_j$ when x_i and x_j are binary variables, it suffices to impose the linear constraints $y_{ij} \leq x_i$, $y_{ij} \leq x_j$ and $y_{ij} \geq x_i + x_j - 1$ in addition to the 0,1 conditions on x_i, x_j, y_{ij}. □

As an example, consider f defined by $f(x) = x_1^5 x_2 + 4 x_1 x_2 x_3^2$. Applying Fortet's linearization sequentially, function f is initially replaced by $z = x_1 x_2 + 4 x_1 x_2 x_3$ for 0,1 variables x_j, $j = 1, 2, 3$. Subsequently, we introduce 0,1 variables y_{12} in place of $x_1 x_2$, and y_{123} in place of $y_{12} x_3$, so that the objective function is replaced by the linear function $z = y_{12} + 4 y_{123}$, where we impose

$$y_{12} \leq x_1, \quad y_{12} \leq x_2, \quad y_{12} \geq x_1 + x_2 - 1,$$
$$y_{123} \leq y_{12}, \quad y_{123} \leq x_3, \quad y_{123} \geq y_{12} + x_3 - 1,$$
$$y_{12}, y_{123}, x_1, x_2, x_3 \in \{0, 1\}.$$

2.13 Further Readings

The book "Applications of Optimization with Xpress" by Guéret, Prins, and Servaux [193], which can also be downloaded online, provides an excellent guide for constructing integer programming formulations in various areas such as planning, transportation, telecommunications, economics, and finance. The book "Production Planning by Mixed-Integer Programming" by Pochet and Wolsey [309] contains several optimization models in production planning and an accessible exposition of the theory of mixed integer linear programming. The book "Optimization Methods in Finance" by Cornuéjols and Tütüncü [96] gives an application of integer programming to modeling index funds. Several formulations in this chapter are defined on graphs. We refer to Bondy and Murty [62] for a textbook on graph theory.

The knapsack problem is one of the most widely studied models in integer programming. A classic book for the knapsack problem is the one of Martello and Toth [268], which is downloadable online. A more recent textbook is [234]. In Sect. 2.2 we introduced alternative formulations (in the context of 0, 1 knapsack set) and discussed the strength of different formulations. This topic is central in integer programming theory and applications. In fact, a strong formulation is a key ingredient to solving integer programs even of moderate size: A weak formulation may prove to be unsolvable by state-of-the-art solvers even for small-size instances. Formulations can be strengthened a priori or dynamically, by adding cuts and this will be discussed at length in this book. Strong formulations can also be obtained with the use of additional variables, that model properties of a mixed integer set to be optimized and we will develop this topic. The book "Integer Programming" by Wolsey [353] contains an accessible exposition of this topic.

There is a vast literature on the traveling salesman problem: This problem is easy to state and it has been popular for testing the methods exposed in this book. The book edited by Lawler, Lenstra, Rinnooy Kan, and Shmoys [253] contains a series of important surveys; for instance the chapters on polyhedral theory and computations by Grötschel and Padberg. The book by Applegate, Bixby, Chvátal, and Cook [13] gives a detailed account of the theory and the computational advances that led to the solution of traveling salesman instances of enormous size. The recent book "In the pursuit of the traveling salesman" by Cook [86] provides an entertaining account of the traveling salesman problem, with many historical insights.

Vehicle routing is related to the traveling salesman problem and refers to a class of problems where goods located at a central depot need to be delivered to customers who have placed orders for such goods. The goal is to minimize the cost of delivering the goods. There are many references in this area. We just cite the monograph of Toth and Vigo [337].

Constraint programming has been mentioned while introducing formulations for the Sudoku game. The interaction between integer programming and constraint programming is a growing area of research, see, e.g., Hooker [206] and Achterberg [5].

For machine scheduling we mention the survey of Queyranne and Schulz [312].

2.14 Exercises

Exercise 2.1. Let

$$S := \{x \in \{0,1\}^4 : \quad 90x_1 \quad +35x_2 \quad +26x_3 \quad +25x_4 \quad \leq 138\}.$$

(i) Show that

$$S = \{x \in \{0,1\}^4 : \quad 2x_1 \quad +x_2 \quad +x_3 \quad +x_4 \quad \leq 3\},$$

and
$$S = \{x \in \{0,1\}^4 : 2x_1 + x_2 + x_3 + x_4 \leq 3\\ x_1 + x_2 + x_3 \leq 2\\ x_1 + x_2 + x_4 \leq 2\\ x_1 + x_3 + x_4 \leq 2\}.$$

(ii) Can you rank these three formulations in terms of the tightness of their linear relaxations, when $x \in \{0,1\}^4$ is replaced by $x \in [0,1]^4$? Show any strict inclusion.

Exercise 2.2. Give an example of a 0, 1 knapsack set where both $P \setminus P^C \neq \emptyset$ and $P^C \setminus P \neq \emptyset$, where P and P^C are the linear relaxations of the knapsack and minimal cover formulations respectively.

Exercise 2.3. Produce a family of 0, 1 knapsack sets (having an increasing number n of variables) whose associated family of minimal covers grows exponentially with n.

Exercise 2.4. (Constraint aggregation) Given a finite set E and a clutter $\mathcal{C}$ of subsets of E, does there always exist a 0, 1 knapsack set K such that $\mathcal{C}$ is the family of all minimal covers of K? Prove or disprove.

Exercise 2.5. Show that any integer linear program of the form

$$\min cx\\ Ax = b\\ 0 \leq x \leq u\\ x \text{ integral}$$

can be converted into a 0,1 knapsack problem.

Exercise 2.6. The pigeonhole principle states that the problem

(P) Place $n+1$ pigeons into n holes so that no two pigeons share a hole

has no solution.

Formulate (P) as an integer linear program with two kinds of constraints:

(a) those expressing the condition that every pigeon must get into a hole;

(b) those expressing the condition that, for each pair of pigeons, at most one of the two birds can get into a given hole.

Show that there is no integer solution satisfying (a) and (b), but that the linear program with constraints (a) and (b) is feasible.

Exercise 2.7. Let A be a $0,1$ matrix and let A^{max} be the row submatrix of A containing one copy of all the rows of A whose support is not included in the support of another row of A. Show that the packing sets $S^P(A)$ and $S^P(A^{max})$ coincide and that their linear relaxations are equivalent.

Similarly let A^{min} be the row submatrix of A containing one copy of all the rows of A whose support does not include the support of another row of A. Show that $S^C(A)$ and $S^C(A^{min})$ coincide and that their linear relaxations are equivalent.

Exercise 2.8. We use the notation introduced in Sect. 2.4.2. Given the matrix

$$A = \begin{bmatrix} 1 & 1 & 0 & 0 & 0 & 0 \\ 0 & 1 & 1 & 0 & 0 & 0 \\ 0 & 0 & 1 & 1 & 0 & 0 \\ 0 & 0 & 0 & 1 & 1 & 0 \\ 1 & 0 & 0 & 0 & 1 & 0 \\ 1 & 0 & 0 & 0 & 0 & 1 \\ 0 & 1 & 0 & 0 & 0 & 1 \\ 0 & 0 & 1 & 0 & 0 & 1 \\ 0 & 0 & 0 & 1 & 0 & 1 \\ 0 & 0 & 0 & 0 & 1 & 1 \end{bmatrix}$$

- What is G_A?
- What is A_c?
- Give a formulation for $S^P(A)$ that is better than $A_c x \leq 1$, $0 \leq x \leq 1$.

Exercise 2.9. Let A be a matrix with two 1's per row. Show that the sets $S^P(A)$ and $S^C(A)$ have the same cardinality.

Exercise 2.10. Given a clutter $\mathcal{F}$, let A be the incidence matrix of the family $\mathcal{F}$ and G_A the intersection graph of A. Prove that A is the clique matrix of G_A if and only if the following holds:

For every F_1, F_2, F_3 in $\mathcal{F}$, there is an $F \in \mathcal{F}$ that contains $(F_1 \cap F_2) \cup (F_1 \cap F_3) \cup (F_2 \cap F_3)$.

Exercise 2.11. Let T be a minimal transversal of $\mathcal{S}$ and $e_j \in T$. Then $T \cap S_i = \{e_j\}$ for some $S_i \in \mathcal{S}$.

Exercise 2.12. Prove that, for an undirected connected graph $G = (V, E)$, the following pairs of families of subsets of edges of G are blocking pairs:

2.14. EXERCISES

- Spanning trees and minimal cuts
- st-paths and minimal st-cuts.
- Minimal postman sets and minimal odd cuts. (A set $E' \subseteq E$ is a *postman set* if $G = (V, E \setminus E')$ is an Eulerian graph and a cut is *odd* if it contains an odd number of edges.) Assume that G is not an Eulerian graph.

Exercise 2.13. Construct an example showing that the formulation (2.4) is not perfect.

Exercise 2.14. Show that a graph is bipartite if and only if it contains no odd cycle.

Exercise 2.15. (Chromatic number) The following is (a simplified version of) a *frequency assignment problem* in telecommunications. Transmitters $1, \ldots, n$ broadcast different signals using preassigned frequencies. Transmitters that are geographically close might interfere and they must therefore use distinct frequencies. The problem is to determine the minimum number of frequencies that need to be assigned to the transmitters so that interference is avoided.

This problem has a natural graph-theoretic counterpart: The *chromatic number* $\chi(G)$ of an undirected graph $G = (V, E)$ is the minimum number of colors to be assigned to the nodes of G so that adjacent nodes receive distinct colors. Equivalently, the chromatic number is the minimum number of (maximal) stable sets whose union is V.

Define the *interference graph* of a frequency assignment problem to be the undirected graph $G = (V, E)$ where V represents the set of transmitters and E represents the set of pairs of transmitters that would interfere with each other if they were assigned the same frequency. Then the minimum number of frequencies to be assigned so that interference is avoided is the chromatic number of the interference graph.

Consider the following integer programs. Let $\mathcal{S}$ be the family of all maximal stable sets of G. The first one has one variable x_S for each maximal stable set S of G, where $x_S = 1$ if S is used as a color, $x_S = 0$ otherwise.

$$\chi_1(G) = \min \sum_{S \in \mathcal{S}} x_S$$

$$\sum_{S \supseteq \{v\}} x_S \geq 1 \quad v \in V$$

$$x_S \in \{0, 1\} \quad S \in \mathcal{S}.$$

The second one has one variable $x_{v,c}$ for each node v in V and color c in a set C of available colors (with $|C| \geq \chi(G)$), where $x_{v,c} = 1$ if color c is assigned to node v, 0 otherwise. It also has color variables, $y_c = 1$ if color c used, 0 otherwise.

$$\chi_2(G) = \min \sum_{c \in C} y_c$$
$$x_{u,c} + x_{v,c} \leq 1 \quad \forall uv \in E \text{ and } c \in C$$
$$x_{v,c} \leq y_c \quad v \in V, c \in C$$
$$\sum_{c \in C} x_{v,c} = 1 \quad v \in V$$
$$x_{v,c} \in \{0,1\}, y_c \geq 0 \quad v \in V, c \in C.$$

- Show that $\chi_1(G) = \chi_2(G) = \chi(G)$.

- Let $\chi_1^*(G)$, $\chi_2^*(G)$ be the optimal values of the linear programming relaxations of the above integer programs. Prove that $\chi_1^*(G) \geq \chi_2^*(G)$ for all graphs G. Prove that $\chi_1^*(G) > \chi_2^*(G)$ for some graph G.

Exercise 2.16 (Combinatorial auctions). A company sets an auction for N objects. Bidders place their bids for some subsets of the N objects that they like. The auction house has received n bids, namely bids b_j for subset S_j, for $j = 1, \ldots, n$. The auction house is faced with the problem of choosing the winning bids so that profit is maximized and each of the N objects is given to at most one bidder. Formulate the optimization problem faced by the auction house as a set packing problem.

Exercise 2.17. (Single machine scheduling) Jobs $\{1, \ldots, n\}$ must be processed on a single machine. Each job is available for processing after a certain time, called release time. For each job we are given its release time r_i, its processing time p_i and its weight w_i. Formulate as an integer linear program the problem of sequencing the jobs without overlap or interruption so that the sum of the weighted completion times is minimized.

Exercise 2.18. (Lot sizing) The demand for a product is known to be d_t units in periods $t = 1, \ldots, n$. If we produce the product in period t, we incur a machine setup cost f_t which does not depend on the number of units produced plus a production cost p_t per unit produced. We may produce any number of units in any period. Any inventory carried over from period t to period $t + 1$ incurs an inventory cost i_t per unit carried over. Initial inventory is s_0. Formulate a mixed integer linear program in order to meet the demand over the n periods while minimizing overall costs.

2.14. EXERCISES

Exercise 2.19. A firm is considering project $A, B, \ldots, H$. Using binary variables $x_a, \ldots, x_h$ and linear constraints, model the following conditions on the projects to be undertaken.

1. At most one of $A, B, \ldots, H$.
2. Exactly two of $A, B, \ldots, H$.
3. If A then B.
4. If A then not B.
5. If not A then B.
6. If A then B, and if B then A.
7. If A then B and C.
8. If A then B or C.
9. If B or C then A.
10. If B and C then A.
11. If two or more of B, C, D, E then A.
12. If m or more than n projects $B, \ldots, H$ then A.

Exercise 2.20. Prove or disprove that the formulation $\mathcal{F} = \{x \in \{0,1\}^{n^2}, \sum_{i=1}^{n} x_{ij} = 1 \text{ for } 1 \leq j \leq n, \sum_{j=1}^{n} x_{ij} = 1 \text{ for } 1 \leq i \leq n\}$ describes the set of $n \times n$ permutation matrices.

Exercise 2.21. For the following subsets of edges of an undirected graph $G = (V, E)$, find an integer linear formulation and prove its correctness:

- The family of Hamiltonian paths of G with endnodes u, v. (A *Hamiltonian path* is a path that goes exactly once through each node of the graph.)

- The family of all Hamiltonian paths of G.

- The family of edge sets that induce a triangle of G.

- Assuming that G has $3n$ nodes, the family of n node-disjoint triangles.

- The family of odd cycles of G.

Exercise 2.22. Consider a connected undirected graph $G = (V, E)$. For $S \subseteq V$, denote by $E(S)$ the set of edges with both ends in S. For $i \in V$, denote by $\delta(i)$ the set of edges incident with i. Prove or disprove that the following formulation produces a spanning tree with maximum number of leaves.

$$\begin{aligned}
\max \quad & \sum_{i \in V} z_i \\
& \sum_{e \in E} x_e = |V| - 1 \\
& \sum_{e \in E(S)} x_e \leq |S| - 1 && S \subset V, |S| \geq 2 \\
& \sum_{e \in \delta(i)} x_e + (|\delta(i)| - 1) z_i \leq |\delta(i)| && i \in V \\
& x_e \in \{0, 1\} && e \in E \\
& z_i \in \{0, 1\} && i \in V.
\end{aligned}$$

Exercise 2.23. One sometimes would like to maximize the sum of nonlinear functions $\sum_{i=1}^{n} f_i(x_i)$ subject to $x \in P$, where $f_i : \mathbb{R} \to \mathbb{R}$ for $i = 1, \ldots, n$ and P is a polytope. Assume $P \subset [l, u]$ for $l, u \in \mathbb{R}^n$. Show that, if the functions f_i are piecewise linear, this problem can be formulated as a mixed integer linear program. For example a utility function might be approximated by f_i as shown in Fig. 2.5 (risk-averse individuals dislike more a monetary loss of y than they like a monetary gain of y dollars).

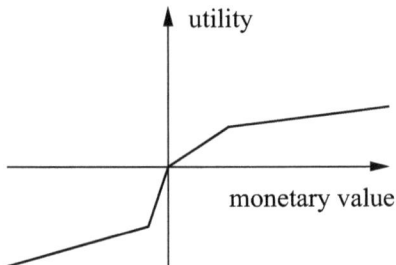

Figure 2.5: Example of a piecewise linear utility function

Exercise 2.24.

(i) Write the logic statement $(x_1 \wedge x_2 \wedge \neg x_3) \vee (\neg(x_1 \wedge x_2) \wedge x_3)$ in conjunctive normal form.

(ii) Formulate the following logical inference problem as an integer linear program. "Does the proposition $(x_1 \wedge x_2 \wedge \neg x_3) \vee (\neg(x_1 \wedge x_2) \wedge x_3)$ imply $x_1 \vee x_2 \vee x_3$?"

Exercise 2.25. Let $x_1, \ldots, x_n$ be atomic propositions and let A and B be two logic statements in CNF. The logic statement $A \Longrightarrow B$ is satisfied if any truth assignment that satisfies A also satisfies B. Prove that $A \Longrightarrow B$ is satisfied if and only if the logic statement $\neg A \vee B$ is satisfied.

2.14. EXERCISES

Exercise 2.26. Consider a 0,1 set $S := \{x \in \{0,1\}^n : Ax \leq b\}$ where $A \in \mathbb{R}^{m \times n}$ and $b \in \mathbb{R}^m$. Prove that S can be written in the form $S = \{x \in \{0,1\}^n : Dx \leq d\}$ where D is a matrix all of whose entries are $0, +1$ or -1 (Matrices D and A may have a different number of rows).

Exercise 2.27 (Excluding $(0,1)$-vectors). Find integer linear formulations for the following integer sets (Hint: Use the generalized set covering inequalities).

- The set of all $(0,1)$-vectors in $\mathbb{R}^4$ except $\begin{pmatrix} 0 \\ 1 \\ 1 \\ 0 \end{pmatrix}$.

- The set of all $(0,1)$-vectors in $\mathbb{R}^6$ except $\begin{pmatrix} 0 \\ 1 \\ 1 \\ 0 \\ 1 \\ 1 \end{pmatrix} \begin{pmatrix} 0 \\ 1 \\ 0 \\ 1 \\ 1 \\ 0 \end{pmatrix} \begin{pmatrix} 1 \\ 1 \\ 1 \\ 1 \\ 1 \\ 1 \end{pmatrix}$.

- The set of all $(0,1)$-vectors in $\mathbb{R}^6$ except all the vectors having exactly two 1s in the first 3 components and one 1 in the last 3 components.

- The set of all $(0,1)$-vectors in $\mathbb{R}^n$ with an even number of 1s.

- The set of all $(0,1)$-vectors in $\mathbb{R}^n$ with an odd number of 1's.

Exercise 2.28. Show that if $P = \{x \in \mathbb{R}^n : Ax \leq b\}$ is such that $P \cap \mathbb{Z}^n$ is the set of 0–1 vectors with an even number of 1's, then $Ax \leq b$ contains at least 2^{n-1} inequalities.

Exercise 2.29. Given a Sudoku game and a solution $\bar{x}$, formulate as an integer linear program the problem of certifying that $\bar{x}$ is the unique solution.

Exercise 2.30 (Crucipixel Game). Given a $m \times n$ grid, the purpose of the game is to darken some of the cells so that in every row (resp. column) the darkened cells form distinct strings of the lengths and in the order prescribed by the numbers on the left of the row (resp. on top of the column).

Two strings are distinct if they are separated by at least one white cell. For instance, in the figure below the tenth column must contain a string of length 6 followed by some white cells and then a sting of length 2. The game consists in darkening the cells to satisfy the requirements.

										1	
			1	1	1	1		3		2	
			1	1	1	2	3	2	3	1	1	6
			3	1	1	1	3	3	4	1	2	2
4	1	2									
	3	1									
1	3	2									
1	1	1	2								
1	1	1	1								
1	1	1	1								
1	1	1	4								
	1	3									
1	4	2									
	2	2									

- Formulate the game as an integer linear program.

- Formulate the problem of certifying that a given solution is unique as an integer linear program.

- Play the game in the figure.

Exercise 2.31. Let $P = \{A_1 x \leq b_1\}$ be a polytope and $S = \{A_2 x < b_2\}$. Formulate the problem of maximizing a linear function over $P \setminus S$ as a mixed 0,1 program.

Exercise 2.32. Consider continuous variables y_j that can take any value between 0 and u_j, for $j = 1, \ldots, k$. Write a set of mixed integer linear constraints to impose that at most ℓ of the k variables y_j can take a nonzero value. [Hint: use k binary variables $x_j \in \{0, 1\}$.] Either prove that your formulation is perfect, in the spirit of Proposition 2.6, or give an example showing that it is not.

Exercise 2.33. Assume $c \in \mathbb{Z}^n$, $A \in \mathbb{Z}^{m \times n}$, $b \in \mathbb{Z}^m$. Give a polynomial transformation of the 0,1 linear program

$$\begin{array}{ll} \max & cx \\ & Ax \leq b \\ & x \in \{0,1\}^n \end{array}$$

into a quadratic program

$$\begin{array}{ll} \max & cx - Mx^T(1-x) \\ & Ax \leq b \\ & 0 \leq x \leq 1, \end{array}$$

i.e., show how to choose the scalar M as a function of A, b and c so that an optimal solution of the quadratic program is always an optimal solution of the 0,1 linear program (if any).

The authors working on Chap. 2

Giacomo Zambelli at the US border. Immigration Officer: What is the purpose of your trip? Giacomo: Visiting a colleague; I am a mathematician. Immigration Officer: What do mathematicians do? Giacomo: Sit in a chair and think.

Chapter 3

Linear Inequalities and Polyhedra

The focus of this chapter is on the study of systems of linear inequalities $Ax \leq b$. We look at this subject from two different angles. The first, more algebraic, addresses the issue of solvability of $Ax \leq b$. The second studies the geometric properties of the set of solutions $\{x \in \mathbb{R}^n : Ax \leq b\}$ of such systems. In particular, this chapter covers Fourier's elimination procedure, Farkas' lemma, linear programming, the theorem of Minkowski–Weyl, polarity, Carathéorory's theorem, projections and minimal representations of the set $\{x \in \mathbb{R}^n : Ax \leq b\}$.

3.1 Fourier Elimination

The most basic question concerning a system of linear inequalities is whether or not it has a solution. Fourier [145] devised a simple method to address this problem. Fourier's method is similar to Gaussian elimination, in that it performs row operations to eliminate one variable at a time.

Let $A \in \mathbb{R}^{m \times n}$ and $b \in \mathbb{R}^m$, and suppose we want to determine if the system $Ax \leq b$ has a solution. We first reduce this question to one about a system with $n-1$ variables. Namely, we determine necessary and sufficient conditions for which, given a vector $(\bar{x}_1, \ldots, \bar{x}_{n-1}) \in \mathbb{R}^{n-1}$, there exists $\bar{x}_n \in \mathbb{R}$ such that $(\bar{x}_1, \ldots, \bar{x}_n)$ satisfies $Ax \leq b$. Let $I := \{1, \ldots, m\}$ and define

$$I^+ := \{i \in I : a_{in} > 0\}, \quad I^- := \{i \in I : a_{in} < 0\}, \quad I^0 := \{i \in I : a_{in} = 0\}.$$

Dividing the ith row by $|a_{in}|$ for each $i \in I^+ \cup I^-$, we obtain the following system, which is equivalent to $Ax \leq b$:

$$\begin{array}{lll} \sum_{j=1}^{n-1} a'_{ij} x_j + x_n & \leq b'_i, & i \in I^+ \\ \sum_{j=1}^{n-1} a'_{ij} x_j - x_n & \leq b'_i, & i \in I^- \\ \sum_{j=1}^{n-1} a_{ij} x_j & \leq b_i, & i \in I^0 \end{array} \quad (3.1)$$

where $a'_{ij} = a_{ij}/|a_{in}|$ and $b'_i = b_i/|a_{in}|$ for $i \in I^+ \cup I^-$.

For each pair $i \in I^+$ and $k \in I^-$, we sum the two inequalities indexed by i and k, and we add the resulting inequality to the system (3.1). Furthermore, we remove the inequalities indexed by I^+ and I^-. This way, we obtain the following system:

$$\begin{array}{ll} \sum_{j=1}^{n-1} (a'_{ij} + a'_{kj}) x_j \leq b'_i + b'_k, & i \in I^+,\ k \in I^-, \\ \sum_{j=1}^{n-1} a_{ij} x_j \leq b_i, & i \in I^0. \end{array} \quad (3.2)$$

If $(\bar{x}_1, \ldots, \bar{x}_{n-1}, \bar{x}_n)$ satisfies $Ax \leq b$, then $(\bar{x}_1, \ldots, \bar{x}_{n-1})$ satisfies (3.2). The next theorem states that the converse also holds.

Theorem 3.1. *A vector $(\bar{x}_1, \ldots, \bar{x}_{n-1})$ satisfies the system (3.2) if and only if there exists $\bar{x}_n$ such that $(\bar{x}_1, \ldots, \bar{x}_{n-1}, \bar{x}_n)$ satisfies $Ax \leq b$.*

Proof. We already remarked the "if" statement. For the converse, assume there is a vector $(\bar{x}_1, \ldots, \bar{x}_{n-1})$ satisfying (3.2). Note that the first set of inequalities in (3.2) can be rewritten as

$$\sum_{j=1}^{n-1} a'_{kj} x_j - b'_k \leq b'_i - \sum_{j=1}^{n-1} a'_{ij} x_j, \quad i \in I^+,\ k \in I^-. \quad (3.3)$$

Let $l := \max_{k \in I^-} \{\sum_{j=1}^{n-1} a'_{kj} \bar{x}_j - b'_k\}$ and $u := \min_{i \in I^+} \{b'_i - \sum_{j=1}^{n-1} a'_{ij} \bar{x}_j\}$, where we define $l := -\infty$ if $I^- = \emptyset$ and $u := +\infty$ if $I^+ = \emptyset$. Since $(\bar{x}_1, \ldots, \bar{x}_{n-1})$ satisfies (3.3), we have that $l \leq u$. Therefore, for any $\bar{x}_n$ such that $l \leq \bar{x}_n \leq u$, the vector $(\bar{x}_1, \ldots, \bar{x}_n)$ satisfies the system (3.1), which is equivalent to $Ax \leq b$. □

Therefore, the problem of finding a solution to $Ax \leq b$ is reduced to finding a solution to (3.2), which is a system of linear inequalities in $n-1$ variables. *Fourier's elimination method* is:

Given a system of linear inequalities $Ax \leq b$, let $A^n := A$, $b^n := b$;
For $i = n, \ldots, 1$, eliminate variable x_i from $A^i x \leq b^i$ with the above procedure to obtain system $A^{i-1} x \leq b^{i-1}$.

3.1. FOURIER ELIMINATION

System $A^1 x \leq b^1$, which involves variable x_1 only, is of the type, $x_1 \leq b_p^1$, $p \in P$, $-x_1 \leq b_q^1$, $q \in N$, and $0 \leq b_i^1$, $i \in Z$.

System $A^0 x \leq b^0$ has the following inequalities: $0 \leq b_{pq}^0 := b_p^1 + b_q^1$, $p \in P$, $q \in N$, $0 \leq b_i^0 := b_i^1$, $i \in Z$.

Applying Theorem 3.1, we obtain that $Ax \leq b$ is feasible if and only if $A^0 x \leq b^0$ is feasible, and this happens when the b_{pq}^0 and b_i^0 are all nonnegative.

Remark 3.2.

(i) At each iteration, Fourier's method removes $|I^+| + |I^-|$ inequalities and adds $|I^+| \times |I^-|$ inequalities, hence the number of inequalities may roughly be squared at each iteration. Thus, after eliminating p variables, the number of inequalities may be exponential in p.

(ii) If matrix A and vector b have only rational entries, then all coefficients in (3.2) are rational.

(iii) Every inequality of $A^i x \leq b^i$ is a nonnegative combination of inequalities of $Ax \leq b$.

Example 3.3. Consider the system $A^3 x \leq b^3$ of linear inequalities in three variables

$$\begin{aligned}
-x_1 & & & \leq -1 \\
& -x_2 & & \leq -1 \\
& & -x_3 & \leq -1 \\
-x_1 & -x_2 & & \leq -3 \\
-x_1 & & -x_3 & \leq -3 \\
& -x_2 & -x_3 & \leq -3 \\
x_1 & +x_2 & +x_3 & \leq 6
\end{aligned}$$

Applying Fourier's procedure to eliminate variable x_3, we obtain the system $A^2 x \leq b^2$:

$$\begin{aligned}
-x_1 & & \leq -1 \\
& -x_2 & \leq -1 \\
-x_1 & -x_2 & \leq -3 \\
x_1 & +x_2 & \leq 5 \\
& x_2 & \leq 3 \\
x_1 & & \leq 3
\end{aligned}$$

where the last three inequalities are obtained from $A^3 x \leq b^3$ by summing the third, fifth, and sixth inequality, respectively, with the last inequality. Eliminating variable x_2, we obtain $A^1 x \leq b^1$

$$
\begin{array}{rcr}
-x_1 & \leq & -1 \\
x_1 & \leq & 3 \\
x_1 & \leq & 4 \\
0 & \leq & 2 \\
0 & \leq & 2 \\
-x_1 & \leq & 0
\end{array}
$$

Finally $A^0 x \leq b^0$ is

$$
\begin{array}{rcr}
0 & \leq & 3-1 \\
0 & \leq & 4-1 \\
0 & \leq & 3 \\
0 & \leq & 4 \\
0 & \leq & 2 \\
0 & \leq & 2
\end{array}
$$

Therefore $A^0 x \leq b^0$ is feasible. A solution can now be found by *backward substitution*. System $A^1 x \leq b^1$ is equivalent to $1 \leq x_1 \leq 3$. Since x_1 can take any value in this interval, choose $\bar{x}_1 = 3$. Substituting $x_1 = 3$ in $A^2 x \leq b^2$, we obtain $1 \leq x_2 \leq 2$. If we choose $\bar{x}_2 = 1$ and substitute $x_2 = 1$ and $x_1 = 3$ in $A^3 x \leq b^3$, we finally obtain $x_3 = 2$. This gives the solution $\bar{x} = (3, 1, 2)$. ∎

3.2 Farkas' Lemma

Next we present Farkas' lemma, which gives a simple necessary and sufficient condition for the existence of a solution to a system of linear inequalities. Farkas' lemma is the analogue of the Fredholm alternative for a system of linear equalities (Theorem 1.19).

Theorem 3.4 (Farkas' Lemma). *A system of linear inequalities $Ax \leq b$ is infeasible if and only if the system $uA = 0$, $ub < 0$, $u \geq 0$ is feasible.*

Proof. Assume $uA = 0$, $ub < 0$, $u \geq 0$ is feasible. Then $0 = uAx \leq ub < 0$ for any x satisfying $Ax \leq b$. It follows that $Ax \leq b$ is infeasible and this proves the "if" part.

We now prove the "only if" part. Assume that $Ax \leq b$ has no solution. Apply the Fourier elimination method to $Ax \leq b$ to eliminate all variables $x_n, \ldots, x_1$. System $A^0 x \leq b^0$ is of the form $0 \leq b^0$, and the system $Ax \leq b$ has a solution if and only if all the entries of b^0 are nonnegative. Since $Ax \leq b$ has no solution, it follows that b^0 has a negative entry, say $b_i^0 < 0$.

By Remark 3.2(iii), every inequality of the system $0 \leq b^0$ is a nonnegative combination of inequalities of $Ax \leq b$. In particular, there exists some vector $u \geq 0$ such that the inequality $0 \leq b_i^0$ is identical to $uAx \leq ub$. That is, $u \geq 0$, $uA = 0$, $ub = b_i^0 < 0$ is feasible. □

Farkas' lemma is sometimes referred to as a *theorem of the alternative* because it can be restated as follows.

Exactly one among the system $Ax \leq b$ and the system $uA = 0$, $ub < 0$, $u \geq 0$ is feasible.

The following is Farkas' lemma for systems of equations in nonnegative variables.

Theorem 3.5. *The system $Ax = b$, $x \geq 0$ is feasible if and only if $ub \leq 0$ for every u satisfying $uA \leq 0$.*

Proof. If $Ax = b$, $x \geq 0$ is feasible, then $ub \leq 0$ for every u satisfying $uA \leq 0$. For the converse, suppose that $Ax = b$, $x \geq 0$ is infeasible. Then the system $Ax \leq b$, $-Ax \leq -b$, $-x \leq 0$ is infeasible. By Theorem 3.4 there exists $(v, v', w) \geq 0$ such that $vA - v'A - w = 0$ and $vb - v'b < 0$. The vector $u := v' - v$ satisfies $ub > 0$ and since $w \geq 0$, u satisfies $uA \leq 0$. □

We finally present a more general, yet still equivalent, form of Farkas' lemma.

Theorem 3.6. *The system $Ax + By \leq f$, $Cx + Dy = g$, $x \geq 0$ is feasible if and only if $uf + vg \geq 0$ for every (u, v) satisfying $uA + vC \geq 0$, $uB + vD = 0$, $u \geq 0$.*

Theorem 3.6 can be derived from Theorem 3.4. We leave this proof as an exercise.

3.3 Linear Programming

Linear programming is the problem of maximizing a linear function subject to a finite number of linear constraints. Given a matrix $A \in \mathbb{R}^{m \times n}$ and vectors $c \in \mathbb{R}^n$, $b \in \mathbb{R}^m$, the *dual* of the linear programming problem $\max\{cx : Ax \leq b\}$ is the problem $\min\{ub : uA = c, u \geq 0\}$. Next we derive the fundamental theorem of linear programming, stating that the optimum values of the primal and dual problems coincide whenever both problems have a feasible solution. This property is called *strong duality*.

Theorem 3.7 (Linear Programming Duality). *Given a matrix $A \in \mathbb{R}^{m \times n}$ and vectors $c \in \mathbb{R}^n$, $b \in \mathbb{R}^m$, let $P := \{x : Ax \leq b\}$ and $D := \{u : uA = c, u \geq 0\}$. If P and D are both nonempty, then*

$$\max\{cx : Ax \leq b\} = \min\{ub : uA = c, u \geq 0\}, \qquad (3.4)$$

and there exist $x^* \in P$ and $y^* \in D$ such that $cx^* = u^*b$.

Proof. For every $x \in P$ and $u \in D$, we have $cx = uAx \leq ub$, where the equality follows from $uA = c$ and the inequality follows from $u \geq 0$, $Ax \leq b$. Hence $\max\{cx : x \in P\} \leq \min\{ub : u \in D\}$. Since $D \neq \emptyset$, this also implies that $\max\{cx : x \in P\}$ is bounded.

Note that $\max\{cx : x \in P\} = \max\{z : z - cx \leq 0, Ax \leq b\}$. Apply Fourier's method to the system $z - cx \leq 0$, $Ax \leq b$, to eliminate the variables $x_1, \ldots, x_n$. The result is a system $\bar{a}z \leq \bar{b}$ in the variable z, where $\bar{a}$ and $\bar{b}$ are vectors. We may assume that the entries of $\bar{a}$ are $0, \pm 1$. By Theorem 3.1, $\max\{z : z - cx \leq 0, Ax \leq b\} = \max\{z : \bar{a}z \leq \bar{b}\}$, and there exists $x^* \in P$ such that cx^* achieves the maximum. Since $\max\{z : \bar{a}z \leq \bar{b}\}$ is bounded, at least one entry of $\bar{a}$ equals 1, and $\max\{z : \bar{a}z \leq \bar{b}\} = \min_{i : \bar{a}_i = 1} \bar{b}_i$. Let h be an index achieving the minimum in the previous equation. By Remark 3.2(iii), inequality $z \leq \bar{b}_h$ is a nonnegative combination of the $m+1$ inequalities $z - cx \leq 0$, $Ax \leq b$. Thus there exists a nonnegative vector $(u_0, u^*) \in \mathbb{R}_+ \times \mathbb{R}_+^m$ such that $u_0 = 1$, $(u_0, u^*)\binom{-c}{A} = 0$ and $u^*b = \bar{b}_h$. It follows that $u^* \in D$ and $u^*b = \max\{cx : x \in P\}$. $\square$

Theorem 3.8 (Complementary Slackness). *Given a matrix $A \in \mathbb{R}^{m \times n}$ and vectors $c \in \mathbb{R}^n$, $b \in \mathbb{R}^m$, let $P := \{x : Ax \leq b\}$ and $D := \{u : uA = c, u \geq 0\}$. Given $x^* \in P$ and $u^* \in D$, x^* and u^* are optimal solutions for the primal and dual problem $\max\{cx : x \in P\}$ and $\min\{ub : u \in D\}$, respectively, if and only if the following complementary slackness conditions hold*

$$u_i^*(a^i x^* - b_i) = 0 \text{ for } i = 1, \ldots, m.$$

Proof. We have that $cx^* = u^* A x^* \leq u^* b$, and by Theorem 3.7 equality holds if and only if x^* and u^* are optimal solutions for $\max\{cx : x \in P\}$ and $\min\{ub : u \in D\}$. Since $a^i x^* \leq b_i$ and $u_i^* \geq 0$, equality holds if and only if, for $i = 1, \ldots, m$, $u_i^*(a^i x^* - b_i) = 0$. $\square$

Here is another consequence of Farkas' lemma:

Proposition 3.9. *Let $P := \{x : Ax \leq b\}$ and $D := \{u : uA = c, u \geq 0\}$, and suppose $P \neq \emptyset$. Then $\max\{cx : x \in P\}$ is unbounded if and only if $D = \emptyset$. Equivalently, $\max\{cx : x \in P\}$ is unbounded if and only if there exists a vector $\bar{y}$ such that $A\bar{y} \leq 0$ and $c\bar{y} > 0$.*

3.4. AFFINE, CONVEX, AND CONIC COMBINATIONS

Proof. By Farkas' lemma (Theorem 3.5), $D = \emptyset$ if and only if there exists a vector $\bar{y}$ such that $A\bar{y} \leq 0$ and $c\bar{y} > 0$. If $D \neq \emptyset$, then by Theorem 3.7 $\max\{cx : x \in P\} = \min\{ub : u \in D\}$, therefore $\max\{cx : x \in P\}$ is bounded. Conversely, assume $D = \emptyset$. Given $\bar{x} \in P$ and $\bar{y}$ such that $A\bar{y} \leq 0$ and $c\bar{y} > 0$, it follows that $\bar{x} + \lambda \bar{y} \in P$ for every $\lambda \geq 0$ and $\lim_{\lambda \to +\infty} c(\bar{x} + \lambda \bar{y}) = +\infty$. Thus $\max\{cx : x \in P\}$ is unbounded. □

Remark 3.10. *Let $P := \{x : Ax \leq b\}$. Define $\max\{cx : x \in P\}$ to be $-\infty$ when $P = \emptyset$ and $+\infty$ when the problem is unbounded; similarly $\min\{cx : x \in P\} = +\infty$ when $P = \emptyset$ and $-\infty$ when this minimization problem is unbounded. Then, for $P := \{x : Ax \leq b\}$ and $D := \{u : uA = c, u \geq 0\}$, the duality equation $\max\{cx : x \in P\} = \min\{ub : u \in D\}$ holds in all cases except when P and D are both empty.*

3.4 Affine, Convex, and Conic Combinations

3.4.1 Linear Combinations, Linear Spaces

Vector $x \in \mathbb{R}^n$ is a *linear combination* of the vectors $x^1, \ldots, x^q \in \mathbb{R}^n$ if there exist scalars $\lambda_1, \ldots, \lambda_q$ such that

$$x = \sum_{j=1}^{q} \lambda_j x^j.$$

Vectors $x^1, \ldots, x^q \in \mathbb{R}^n$ are *linearly independent* if $\lambda_1 = \ldots = \lambda_q = 0$ is the unique solution to the system $\sum_{j=1}^{q} \lambda_j x^j = 0$.

A nonempty subset $\mathcal{L}$ of $\mathbb{R}^n$ is a *linear space* if $\mathcal{L}$ is closed under taking linear combinations, i.e., every linear combination of vectors in $\mathcal{L}$ belongs to $\mathcal{L}$. A subset $\mathcal{L}$ of $\mathbb{R}^n$ is a linear space if and only if $\mathcal{L} = \{x \in \mathbb{R}^n : Ax = 0\}$ for some matrix A (Exercise 3.6).

A *basis* of a linear space $\mathcal{L}$ is a maximal set of linearly independent vectors in $\mathcal{L}$. All bases have the same cardinality (Exercise 3.5). This cardinality is called the *dimension* of $\mathcal{L}$. If $\mathcal{L} = \{x \in \mathbb{R}^n : Ax = 0\}$, then the dimension of $\mathcal{L}$ is $n - \text{rank}(A)$ (Exercise 3.6).

The inclusionwise minimal linear space containing a set $S \subseteq \mathbb{R}^n$ is the *linear space generated* by S, and is denoted by $\langle S \rangle$. Given any maximal set S' of linearly independent vectors in S, we have that $\langle S \rangle = \langle S' \rangle$.

3.4.2 Affine Combinations, Affine Spaces

A point $x \in \mathbb{R}^n$ is an *affine combination* of $x^1, \ldots, x^q \in \mathbb{R}^n$ if there exist scalars $\lambda_1, \ldots, \lambda_q$ such that

$$x = \sum_{j=1}^{q} \lambda_j x^j, \quad \sum_{j=1}^{q} \lambda_j = 1.$$

Points $x^0, x^1, \ldots, x^q \in \mathbb{R}^n$ are *affinely independent* if $\lambda_0 = \lambda_1 = \ldots = \lambda_q = 0$ is the unique solution to the system

$$\sum_{j=0}^{q} \lambda_j x^j = 0, \quad \sum_{j=0}^{q} \lambda_j = 0.$$

Equivalently, $x^0, x^1, \ldots, x^q \in \mathbb{R}^n$ are affinely independent if and only if no point in $x^0, \ldots, x^q$ can be written as an affine combination of the others.

A subset $\mathcal{A}$ of $\mathbb{R}^n$ is an *affine space* if $\mathcal{A}$ is closed under taking affine combinations. A *basis* of an affine subspace $\mathcal{A} \subseteq \mathbb{R}^n$ is a maximal set of affinely independent points in $\mathcal{A}$. All bases of $\mathcal{A}$ have the same cardinality.

Equivalently, $\mathcal{A} \subseteq \mathbb{R}^n$ is an affine subspace if and only if, for every distinct $x, y \in \mathcal{A}$, the line $\{\lambda x + (1-\lambda)y\}$ passing through x and y belongs to $\mathcal{A}$. Furthermore a subset $\mathcal{A}$ of $\mathbb{R}^n$ is an affine space if and only if $\mathcal{A} = \{x \in \mathbb{R}^n : Ax = b\}$ for some matrix A and vector b (Exercise 3.8). Note that the linear subspaces are precisely the affine subspaces containing the origin.

The *dimension* of a set $S \subseteq \mathbb{R}^n$, denoted by $\dim(S)$, is the maximum number of affinely independent points in S minus one. So the dimension of the empty set is -1, the dimension of a point is 0 and the dimension of a segment is 1. If $\mathcal{A} = \{x \in \mathbb{R}^n : Ax = b\}$ is nonempty, then $\dim(\mathcal{A}) = n - \text{rank}(A)$ (Exercise 3.8).

The inclusionwise minimal affine space containing a set $S \subseteq \mathbb{R}^n$ is called the *affine hull* of S and is denoted by $\text{aff}(S)$. Since the intersection of affine spaces is an affine space, the affine hull is well defined. Note that $\dim(S) = \dim(\text{aff}(S))$.

3.4.3 Convex Combinations, Convex Sets

A point x in $\mathbb{R}^n$ is a *convex combination* of the points $x^1, \ldots, x^q \in \mathbb{R}^n$ if there exist nonnegative scalars $\lambda_1, \ldots, \lambda_q$ such that

$$x = \sum_{j=1}^{q} \lambda_j x^j, \quad \sum_{j=1}^{q} \lambda_j = 1.$$

3.4. AFFINE, CONVEX, AND CONIC COMBINATIONS

A set $C \subseteq \mathbb{R}^n$ is *convex* if C contains all convex combinations of points in C. Equivalently, $C \subseteq \mathbb{R}^n$ is convex if for any two points $x, y \in C$, the line segment $\{\lambda x + (1-\lambda)y, \ 0 \leq \lambda \leq 1\}$ with endpoints x, y is contained in C. (This is the definition given in Sect. 1.4).

Given a set $S \subseteq \mathbb{R}^n$, the *convex hull* of S, denoted by $\text{conv}(S)$, is the inclusionwise minimal convex set containing S. As the intersection of convex sets is a convex set, $\text{conv}(S)$ exists. As observed in Sect. 1.4, it is the set of all points that are convex combinations of points in S. That is

$$\text{conv}(S) = \{\sum_{j=1}^{q} \lambda_j x^j \ : \ x^1, \ldots, x^q \in S, \ \lambda_1, \ldots, \lambda_q \geq 0, \ \sum_{j=1}^{q} \lambda_j = 1\}.$$

3.4.4 Conic Combinations, Convex Cones

A vector $x \in \mathbb{R}^n$ is a *conic combination* of vectors $x^1, \ldots, x^q \in \mathbb{R}^n$ if there exist scalars $\lambda_j \geq 0$, $j = 1, \ldots, q$, such that

$$x = \sum_{j=1}^{q} \lambda_j r^j.$$

A set $C \subseteq \mathbb{R}^n$ is a *cone* if $0 \in C$ and for every $x \in C$ and $\lambda \geq 0$, λx belongs to C. In other words, C is a cone if and only if $0 \in C$ and, for every $x \in C \setminus \{0\}$, C contains the half line starting from the origin in the direction x.

A cone C is a *convex cone* if C contains every conic combination of vectors in C. A convex cone is a convex set, since by definition every convex combination of points is also a conic combination.

Given a nonempty set $S \subseteq \mathbb{R}^n$, the *cone of S*, denoted by $\text{cone}(S)$, is the inclusionwise minimal convex cone containing S. As the intersection of convex cones is a convex cone, $\text{cone}(S)$ exists. It is the set of all conic combinations of vectors in S. We say that $\text{cone}(S)$ is the cone generated by S. For convenience, we define $\text{cone}(\emptyset) := \{0\}$.

Given a cone C and a vector $r \in C \setminus \{0\}$, the half line $\text{cone}(r) = \{\lambda r, \lambda \geq 0\}$ is called a *ray* of C. We will often simply refer to a vector $r \in C \setminus \{0\}$ as a ray of C to denote the corresponding ray $\text{cone}(r)$. Since $\text{cone}(\lambda r) = \text{cone}(r)$ for every $\lambda > 0$, we say that two rays r and r' of a cone are distinct when there is no $\mu > 0$ such that $r = \mu r'$.

3.5 Polyhedra and the Theorem of Minkowski–Weyl

A subset P of $\mathbb{R}^n$ is a *polyhedron* if there exists a positive integer m, an $m \times n$ matrix A, and a vector $b \in \mathbb{R}^m$, such that

$$P = \{x \in \mathbb{R}^n : Ax \leq b\}.$$

For example $\mathbb{R}^n$ is a polyhedron because it is obtained as $0x \leq 0$. If $a^i \neq 0$ is a row of A, the corresponding inequality $a^i x \leq b_i$ defines a half-space of $\mathbb{R}^n$. Therefore a polyhedron is the intersection of a finite number of half-spaces. It follows immediately from the definition that the intersection of a finite number of polyhedra is again a polyhedron.

A polyhedron P is said to be a *rational polyhedron* if there exists a rational matrix $A \in \mathbb{Q}^{m \times n}$ and a rational vector $b \in \mathbb{Q}^m$ such that $P = \{x \in \mathbb{R}^n : Ax \leq b\}$.

A set $C \subseteq \mathbb{R}^n$ is a *polyhedral cone* if C is the intersection of a finite number of half-spaces containing the origin on their boundaries. That is,

$$C := \{x \in \mathbb{R}^n : Ax \leq 0\}$$

for some $m \times n$ matrix A.

3.5.1 Minkowski–Weyl Theorem for Polyhedral Cones

A set $C \subseteq \mathbb{R}^n$ is a *finitely generated cone* if C is the convex cone generated by a finite set of vectors $r^1, \ldots, r^k \in \mathbb{R}^n$, for $k \geq 1$. We write $C = \text{cone}(r^1, \ldots, r^k)$, and we say that $r^1, \ldots, r^k$ are the *generators* of C. If R is the $n \times k$ matrix with columns $r^1, \ldots, r^k$, we sometimes write $\text{cone}(R)$ instead of $\text{cone}(r^1, \ldots, r^k)$. Note that

$$\text{cone}(R) = \{x \in \mathbb{R}^n : \exists \mu \geq 0 \text{ s.t. } x = R\mu\}.$$

Theorem 3.11 (Minkowski–Weyl Theorem for Cones). *A subset of $\mathbb{R}^n$ is a finitely generated cone if and only if it is a polyhedral cone.*

Proof. We first show that if $C \subseteq \mathbb{R}^n$ is a finitely generated cone, then C is polyhedral. Let R be an $n \times k$ matrix such that $C = \{x \in \mathbb{R}^n : \exists \mu \geq 0 \text{ s.t. } x = R\mu\}$. We need to show that there exists a matrix A such that $C = \{x \in \mathbb{R}^n : Ax \leq 0\}$. By applying Fourier's elimination method k times to the system $x - R\mu = 0, \mu \geq 0$ to eliminate all variables $\mu_1, \ldots, \mu_k$, we obtain a system of linear inequalities involving only the variables $x_1, \ldots, x_n$.

Note that, if we apply an iteration of Fourier's method to a homogeneous system, we again obtain a homogeneous system, therefore the output of the k iterations of Fourier's method is a system of the form $Ax \leq 0$. By Theorem 3.1, we have $C = \{x \in \mathbb{R}^n : Ax \leq 0\}$.

We now prove that, if $C \subseteq \mathbb{R}^n$ is a polyhedral cone, then C is finitely generated. Let A be an $m \times n$ matrix such that $C = \{x \in \mathbb{R}^n : Ax \leq 0\}$. We need to show that there exists a matrix R such that $C = \text{cone}(R)$. Consider the finitely generated cone $C^* := \{y \in \mathbb{R}^n : \exists \nu \geq 0 \text{ s.t. } y = \nu A\}$. Since we have shown that every finitely generated cone is polyhedral, there exists a matrix R such that $C^* = \{y \in \mathbb{R}^n : yR \leq 0\}$. We will show that $C = \text{cone}(R)$.

To show "$C \supseteq \text{cone}(R)$," we need to prove that, for every $\mu \geq 0$, the point $x := R\mu$ satisfies $Ax \leq 0$. Note that every row of A is in C^*, since it is the product a unit row vector in $\mathbb{R}^m$ and A. Therefore, by definition of R, $AR \leq 0$. We have that $Ax = AR\mu \leq 0$ because $AR \leq 0$ and $\mu \geq 0$.

We show "$C \subseteq \text{cone}(R)$." Consider $\bar{x} \notin \text{cone}(R)$. It suffices to show that $\bar{x} \notin C$. Since $\bar{x} \notin \text{cone}(R)$, the system $R\mu = \bar{x}$, $\mu \geq 0$ in the variables μ is infeasible. By Farkas lemma (Theorem 3.5), there exists $y \in \mathbb{R}^n$ such that $yR \leq 0$ and $y\bar{x} > 0$. Since $yR \leq 0$, we have that $y \in C^*$, therefore there exists $\nu \geq 0$ such that $y = \nu A$. Since $y\bar{x} > 0$, we have that $\nu A\bar{x} > 0$. The latter and the fact that $\nu \geq 0$ imply that at least one component of $A\bar{x}$ must be positive. This shows that $\bar{x} \notin C$. □

Proposition 3.12. *Given a rational matrix $A \in \mathbb{R}^{m \times n}$, there exist rational vectors $r^1, \ldots, r^k \in \mathbb{R}^n$ such that $\{x : Ax \leq 0\} = \text{cone}(r^1, \ldots, r^k)$.*

Conversely, given rational vectors $r^1, \ldots, r^k \in \mathbb{R}^n$, there exists a rational matrix $A \in \mathbb{R}^{m \times n}$ such that $\text{cone}(r^1, \ldots, r^k) = \{x : Ax \leq 0\}$.

Proof. The statement follows from the proof of Theorem 3.11 and Remark 3.2(ii). □

3.5.2 Minkowski–Weyl Theorem for Polyhedra

Polyhedra are isomorphic to sections of polyhedral cones. Indeed, given $P := \{x \in \mathbb{R}^n : Ax \leq b\}$, consider the cone $C_P := \{(x, y) \in \mathbb{R}^n \times \mathbb{R} : Ax - by \leq 0, y \geq 0\}$. Then $P = \{x \in \mathbb{R}^n : (x, 1) \in C_P\}$. That is, any polyhedron can be seen as the intersection of a polyhedral cone (in a space whose dimension is increased by 1) with the hyperplane of equation $y = 1$. See Fig. 3.1.

A subset Q of $\mathbb{R}^n$ is a *polytope* if Q is the convex hull of a finite set of vectors in $\mathbb{R}^n$.

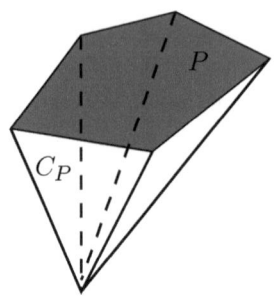

Figure 3.1: Polyhedron as the intersection of a polyhedral cone with a hyperplane

Given subsets V, R of $\mathbb{R}^n$, the *Minkowski sum* of V, R is the set
$$V + R := \{x \in \mathbb{R}^n : \text{there exist } v \in V,\ r \in R \text{ such that } x = v + r\}.$$
If one of V, R is empty, the Minkowski sum of V, R is empty. The next theorem shows that a polyhedron can be expressed as the Minkowski sum of a polytope and a finitely generated cone (Fig. 3.2).

Theorem 3.13 (Minkowski–Weyl Theorem [279, 348]). *A subset P of $\mathbb{R}^n$ is a polyhedron if and only if $P = Q + C$ for some polytope $Q \subset \mathbb{R}^n$ and finitely generated cone $C \subseteq \mathbb{R}^n$.*

Proof. Let P be a subset of $\mathbb{R}^n$. We need to show that the following two conditions are equivalent.

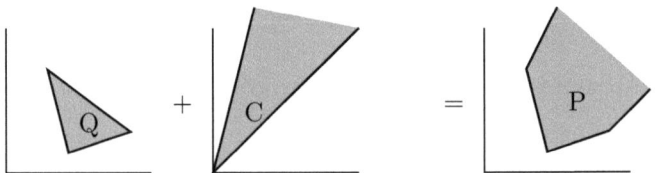

Figure 3.2: Illustration of the Minkowski–Weyl theorem for polyhedra

1. There exist a matrix A and a vector b such that $P = \{x \in \mathbb{R}^n : Ax \leq b\}$.

2. There exist $v^1, \ldots, v^p \in \mathbb{R}^n$ and $r^1, \ldots, r^q \in \mathbb{R}^n$ such that
$$P = \text{conv}(v^1, \ldots, v^p) + \text{cone}(r^1, \ldots, r^q).$$

We show that (1) implies (2). Assume that (1) holds, and consider the polyhedral cone $C_P := \{(x,y) \in \mathbb{R}^{n+1} : Ax - by \le 0, \ y \ge 0\}$. By Theorem 3.11, the cone C_P is finitely generated. Since $y \ge 0$ for every vector $(x,y) \in C_P$, the generators of C_P can be normalized so that their $(n+1)$th component is either 0 or 1. That is, there exist $v^1, \ldots, v^p \in \mathbb{R}^n$ and $r^1, \ldots, r^q \in \mathbb{R}^n$ such that

$$C_P = \operatorname{cone}\left\{ \begin{pmatrix} v^1 \\ 1 \end{pmatrix}, \ldots, \begin{pmatrix} v^p \\ 1 \end{pmatrix}, \begin{pmatrix} r^1 \\ 0 \end{pmatrix}, \ldots, \begin{pmatrix} r^q \\ 0 \end{pmatrix} \right\}. \quad (3.5)$$

Since $P = \{x : (x,1) \in C_P\}$, this implies that $P = \operatorname{conv}\{v^1, \ldots, v^p\} + \operatorname{cone}\{r^1, \ldots, r^q\}$.

We show that (2) implies (1). Assume that (2) holds, and let $C_P \in \mathbb{R}^{n+1}$ be the finitely generated cone defined by (3.5). Note that, by definition, $P = \{x : (x,1) \in C_P\}$. By Theorem 3.11, C_P is a polyhedral cone, therefore there exists a matrix (A,b) such that $C_P := \{(x,y) \in \mathbb{R}^{n+1} : Ax - by \le 0\}$. It follows that $P = \{x \in \mathbb{R}^n : Ax \le b\}$. $\square$

Corollary 3.14. *(Minkowski–Weyl Theorem for Polytopes) A set $Q \subseteq \mathbb{R}^n$ is a polytope if and only if Q is a bounded polyhedron.*

3.6 Lineality Space and Recession Cone

Given a nonempty polyhedron P, the *recession cone* of P is the set

$$\operatorname{rec}(P) := \{r \in \mathbb{R}^n : x + \lambda r \in P \text{ for all } x \in P \text{ and } \lambda \in \mathbb{R}_+\}.$$

It follows from the definition that $\operatorname{rec}(P)$ is indeed a cone. We will refer to the rays of $\operatorname{rec}(P)$ as the *rays of the polyhedron P*.

The *lineality space* of P is the set

$$\operatorname{lin}(P) := \{r \in \mathbb{R}^n : x + \lambda r \in P \text{ for all } x \in P \text{ and } \lambda \in \mathbb{R}\}.$$

Note that $\operatorname{lin}(P) = \operatorname{rec}(P) \cap -\operatorname{rec}(P)$. When $\operatorname{lin}(P) = \{0\}$, we say that the polyhedron P is *pointed*. In other words, a nonempty polyhedron is pointed when it does not contain any line.

Proposition 3.15. *Let $P := \{x \in \mathbb{R}^n : Ax \le b\} = \operatorname{conv}(v^1, \ldots, v^p) + \operatorname{cone}(r^1, \ldots, r^q)$ be a nonempty polyhedron. Then*

$$\operatorname{rec}(P) = \{r \in \mathbb{R}^n : Ar \le 0\} = \operatorname{cone}(r^1, \ldots, r^q)$$

and $\operatorname{lin}(P) = \{r \in \mathbb{R}^n : Ar = 0\}$.

Proof. If $\bar{r}$ satisfies $Ar \leq 0$, then $A(x + \lambda \bar{r}) \leq b + \lambda A\bar{r} \leq b$, for every $x \in P$ and $\lambda \in \mathbb{R}_+$, so $x + \lambda \bar{r} \in P$. It follows that $\mathrm{rec}(P) \supseteq \{r \in \mathbb{R}^n : Ar \leq 0\}$. For the reverse inclusion, if $\bar{r} \in \mathbb{R}^n$ does not satisfy $Ar \leq 0$, then for any $x \in P$ there is a $\bar{\lambda} > 0$ such that $x + \bar{\lambda}\bar{r}$ does not satisfy $Ax \leq b$, therefore $\bar{r} \notin \mathrm{rec}(P)$. This shows that $\mathrm{rec}(P) = \{r \in \mathbb{R}^n : Ar \leq 0\}$.

If $\bar{r}$ is in $\mathrm{cone}(r^1, \ldots, r^q)$, then, because $P = \mathrm{conv}(v^1, \ldots, v^p) + \mathrm{cone}(r^1, \ldots, r^q)$, it follows that $x + \lambda \bar{r} \in P$ for all $x \in P$ and $\lambda \in \mathbb{R}_+$. Thus $\mathrm{rec}(P) \supseteq \mathrm{cone}(r^1, \ldots, r^q)$. For the reverse inclusion, let $\bar{r} \in \mathrm{rec}(P)$. Then, given $x \in \mathrm{conv}(v^1, \ldots, v^p)$, it follows that $x + \lambda \bar{r} \in P$ for all $\lambda \in \mathbb{R}_+$. Since $\mathrm{conv}(v^1, \ldots, v^p)$ is a bounded set, it follows that $\bar{r} \in \mathrm{cone}(r^1, \ldots, r^q)$. Therefore $\mathrm{rec}(P) = \mathrm{cone}(r^1, \ldots, r^q)$.

Finally, since $\mathrm{lin}(P) = \mathrm{rec}(P) \cap -\mathrm{rec}(P)$ and $\mathrm{rec}(P) = \{r \in \mathbb{R}^n : Ar \leq 0\}$, we have that $\mathrm{lin}(P) = \{r \in \mathbb{R}^n : Ar = 0\}$. □

3.7 Implicit Equalities, Affine Hull, and Dimension

Given a system of linear inequalities $Ax \leq b$, we denote by $a^i x \leq b_i$, $i \in M$, the inequalities in the system. We say that $a^i x \leq b_i$ is an *implicit equality* of $Ax \leq b$ if $a^i x = b_i$ is satisfied by every solution of $Ax \leq b$. Equivalently, $a^i x \leq b_i$ is an implicit equality of $Ax \leq b$ if the polyhedron $P := \{x \in \mathbb{R}^n : Ax \leq b\}$ is contained in the hyperplane $\{x \in \mathbb{R}^n : a^i x = b_i\}$.

In the remainder of this chapter, whenever we have a linear system $Ax \leq b$, we denote by $A^= x \leq b^=$ the system comprising all implicit equalities of $Ax \leq b$, and by $A^< x \leq b^<$ the system comprising all the remaining inequalities of $Ax \leq b$. Thus, $P = \{x \in \mathbb{R}^n : A^= x \leq b^=, A^< x \leq b^<\} = \{x \in \mathbb{R}^n : A^= x = b^=, A^< x \leq b^<\}$.

We will let $I^= \subseteq M$ be the set of indices of the implicit equalities, and $I^< \subseteq M$ the set of indices of the remaining inequalities. For every $i \in I^<$ there exists $\bar{x} \in P$ such that $a^i \bar{x} < b_i$. Note that in particular $I^< = \emptyset$ whenever $P = \emptyset$.

Remark 3.16. *If $P \neq \emptyset$, then P contains a point $\bar{x}$ such that $A^< \bar{x} < b^<$.*

Proof. By definition of implicit equality, for every $i \in I^<$, there is a point $x^i \in P$ such that $a^i x^i < b_i$. The statement is therefore satisfied by the point $\bar{x} := \frac{1}{|I^<|} \sum_{i \in I^<} x^i$. □

3.7. IMPLICIT EQUALITIES, AFFINE HULL, AND DIMENSION

Theorem 3.17. *Let* $P := \{x \in \mathbb{R}^n : Ax \leq b\}$ *be a nonempty polyhedron. Then*

$$\mathrm{aff}(P) = \{x \in \mathbb{R}^n : A^=x = b^=\} = \{x \in \mathbb{R}^n : A^=x \leq b^=\}.$$

Furthermore $\dim(P) = n - \mathrm{rank}(A^=)$.

Proof. Since P is contained in the affine space $\{x \in \mathbb{R}^n : A^=x = b^=\}$, then $\mathrm{aff}(P) \subseteq \{x \in \mathbb{R}^n : A^=x = b^=\}$. Also, trivially $\{x \in \mathbb{R}^n : A^=x = b^=\} \subseteq \{x \in \mathbb{R}^n : A^=x \leq b^=\}$.

We now prove $\{x \in \mathbb{R}^n : A^=x \leq b^=\} \subseteq \mathrm{aff}(P)$. Let $\hat{x} \in \mathbb{R}^n$ be such that $A^=\hat{x} \leq b^=$. We need to show that $\hat{x} \in \mathrm{aff}(P)$. By Remark 3.16, P contains a point $\bar{x}$ such that $A^<\bar{x} < b^<$. For some $\epsilon > 0$ small enough, the point $\tilde{x} := \bar{x} + \epsilon(\hat{x} - \bar{x})$ satisfies $A^<\tilde{x} \leq b^<$. Since $A^=\bar{x} = b^=$ and $A^=\hat{x} \leq b^=$, it follows that $A^=\tilde{x} \leq b^=$, therefore $\tilde{x} \in P$. Since $\bar{x}, \tilde{x} \in P$, it follows that $\mathrm{aff}(P)$ contains the whole line L passing through $\bar{x}$ and $\tilde{x}$. Note that $\hat{x} \in L$, therefore $\hat{x} \in \mathrm{aff}(P)$.

The last part of the statement follows from the fact that $\mathrm{aff}(P) = \{x \in \mathbb{R}^n : A^=x = b^=\}$, thus, by standard linear algebra, $\dim(\mathrm{aff}(P)) = n - \mathrm{rank}(A^=)$. □

A polyhedron $P \subseteq \mathbb{R}^n$ is *full-dimensional* if $\dim(P) = n$. By Theorem 3.17, this is equivalent to saying that any system $Ax \leq b$ that defines P has no implicit equality, except possibly for the trivial equality $0 = 0$.

Example 3.18. The *assignment polytope* is the following

$$P := \left\{ x \in \mathbb{R}^{n^2} : \begin{array}{rl} \sum_{j=1}^n x_{ij} &= 1, \quad i = 1, \ldots n \\ \sum_{i=1}^n x_{ij} &= 1, \quad j = 1, \ldots n \\ x_{ij} &\geq 0, \quad i, j = 1, \ldots n \end{array} \right\}.$$

We show that $\dim(P) = n^2 - 2n + 1$.

Let $Ax = \mathbf{1}$ be the system comprising the $2n$ equations in the definition of P, where $\mathbf{1}$ denotes the vector of all ones. Thus A is a $2n \times n^2$ matrix. We first show that $\mathrm{rank}(A) = 2n - 1$. Note that taking the sum the rows of A relative to the equations $\sum_{j=1}^n x_{ij} = 1$, $i = 1, \ldots n$ minus the some of the rows of A relative to the equations $\sum_{i=1}^n x_{ij} = 1$, $j = 1, \ldots n$, we obtain the 0 vector. Thus $\mathrm{rank}(A) \leq 2n - 1$. To prove equality, consider the columns of A corresponding to the $2n - 1$ variables x_{ii}, $i = 1, \ldots, n$, and $x_{i,i+1}$, $i = 1, \ldots, n-1$. One can easily show that these $2n-1$ columns of A are linearly independent, hence $\mathrm{rank}(A) = 2n - 1$.

This shows that $\dim(P) \leq n^2 - (2n - 1)$. On the other hand, it is immediate to see that $x_{ij} \geq 0$ is not an implicit equality for P, $1 \leq i, j \leq n$. Therefore, by Theorem 3.17, $\dim(P) = n^2 - \text{rank}(A) = n^2 - 2n + 1$. ∎

In the previous example, the polyhedron P was given explicitly. In integer programming, P is usually given as the convex hull of a set of integer points. We illustrate in the next three examples how to compute the dimension of P in such a situation.

Example 3.19. The 0,1 *knapsack polytope* is the convex hull of the 0,1 knapsack set defined in Sect. 2.1. That is, the polyhedron

$$P := \text{conv}(\{x \in \{0,1\}^n : ax \leq b\})$$

where $a \in \mathbb{R}^n_+$ and $b > 0$.

Let $J \subseteq \{1, \ldots, n\}$ be the set of indices j such that $a_j > b$. Note that $P \subseteq \{x \in \mathbb{R}^n : x_j = 0, j \in J\}$. This shows that $\dim(P) \leq n - |J|$. On the other hand, for every $j \notin J$, the jth unit vector e^j is a point of P. Also the origin is in P, thus 0, e^j, $j \notin J$, are $n - |J| + 1$ affinely independent points in P. This shows that $\dim(P) = n - |J|$. ∎

The above example illustrates a way to prove that a given nonempty set $S \subset \mathbb{R}^n$ has a certain dimension k:

(a) Find a system $Ax = b$ of equations such that $S \subseteq \{x \in \mathbb{R}^n : Ax = b\}$ and $\text{rank}(A) = n - k$;

(b) Exhibit $k + 1$ affinely independent points in S.

Another way is to find a system $Ax = b$ as in (a), and then prove that every equation $\alpha x = \beta$ satisfied by all $x \in S$ is a linear combination of the equations $Ax = b$, namely there exists a vector u such that $\alpha = uA$ and $\beta = ub$. Then the system $Ax = b$ is the affine hull of S and therefore S has dimension k. We give two examples.

Example 3.20. The *permutahedron* $\Pi_n \subset \mathbb{R}^n$ is the convex hull of the set S_n of the $n!$ vectors that can be obtained by permuting the entries of the vector $(1, 2, \ldots, n-1, n)$. We will show that $\dim(\Pi_n) = n - 1$.

Note that, for every vector in S_n, the sum of the components is $1 + 2 + \cdots + n = \binom{n+1}{2}$, thus Π_n is contained in the hyperplane $\{x \in \mathbb{R}^n : \sum_{i=1}^n x_i = \binom{n+1}{2}\}$. Therefore $\dim(\Pi_n) \leq n - 1$. To prove equality, we show that any equation $\alpha x = \beta$ satisfied by every $x \in S_n$ must be a multiple of $\sum_{i=1}^n x_i = \binom{n+1}{2}$. We may of course assume that $\alpha \neq 0$. Note that, by definition of S_n, in order for $\alpha x = \beta$ to hold for every $x \in S_n$, we must have

$\alpha_i = \alpha_j$ for all $i, j \in \{1, \ldots, n\}$. Hence we may assume, up to multiplying α by a nonzero scalar, that $\alpha_i = 1$, $i = 1, \ldots, n$. By substituting the point $(1, \ldots, n)$ in $\alpha x = \beta$, we obtain $\beta = \binom{n+1}{2}$. ∎

Example 3.21. The *Hamiltonian-path polytope* of a graph $G = (V, E)$ is the convex hull of incidence vectors of the Hamiltonian paths of G. Recall that a Hamiltonian path in G is a path that goes exactly once through each node. We show the following.

The dimension of the Hamiltonian-path polytope of the complete graph on n nodes is $\binom{n}{2} - 1$.

Let $P \subset \mathbb{R}^{\binom{n}{2}}$ be the Hamiltonian-path polytope of the complete graph G on n nodes. Note that, since all Hamiltonian paths on a graph with n nodes have $n - 1$ edges, the Hamiltonian-path polytope is contained in the hyperplane of equation $\sum_{e \in E} x_e = n - 1$. Therefore $\dim(P) \leq \binom{n}{2} - 1$. To prove equality, we show that, given any equation $\alpha x = \beta$ satisfied by all points in P, $\alpha \neq 0$, the equations $\alpha x = \beta$ and $\sum_{e \in E} x_e = n - 1$ are identical up to scalar multiplication. It suffices to show that $\alpha_e = \alpha_{e'}$ for every $e, e' \in E$. Given $e, e' \in E$, let H be a Hamiltonian tour in G containing both e and e'. Note that $Q = H \setminus \{e\}$ and $Q' = H \setminus \{e'\}$ are Hamiltonian paths. Thus, if $\bar{x}$ and $\bar{x}'$ are the incidence vectors of Q and Q', respectively, we have $0 = \alpha \bar{x} - \alpha \bar{x}' = \alpha_{e'} - \alpha_e$. ∎

3.8 Faces

An inequality $cx \leq \delta$ is *valid* for the set $P \subseteq \mathbb{R}^n$ if $cx \leq \delta$ is satisfied by every point in P. Note that we allow $c = 0$ in our definition, in which case the inequality $0 \leq \delta$ is valid for every set P if $\delta \geq 0$, and it is valid only for the empty set if $\delta < 0$.

Theorem 3.22. *Let $P := \{x \in \mathbb{R}^n : Ax \leq b\}$ be a nonempty polyhedron. An inequality $cx \leq \delta$ is valid for P if and only if there exists $u \geq 0$ such that $uA = c$ and $ub \leq \delta$.*

Proof. Let $c \in \mathbb{R}^n$ and $\delta \in \mathbb{R}$. Assume $uA = c$, $ub \leq \delta$, $u \geq 0$ is feasible. Then, for all $x \in P$, we have $cx = uAx \leq ub \leq \delta$. This shows that $cx \leq \delta$ is valid for P.

Conversely, assume that the inequality $cx \leq \delta$ is valid for P. Consider the linear program $\max\{cx : x \in P\}$. Since $P \neq \emptyset$ and $cx \leq \delta$ is a valid inequality for P, the above program admits a finite optimum and its value is $\delta' \leq \delta$. By Proposition 3.9, the set $D = \{u : uA = c, u \geq 0\}$ is nonempty. By Theorem 3.7 the dual $\min\{ub : u \in D\}$ has value δ', and there exists $u \in D$ such that $ub = \delta'$. Thus $uA = c, ub \leq \delta, u \geq 0$. □

Theorem 3.22 can be used to show the following variant (Exercise 3.17).

Corollary 3.23. *Let $P := \{x : Ax \leq b, Cx = d\}$ be a nonempty polyhedron. An inequality $cx \leq \delta$ is valid for P if and only if the system $uA+vC=c$, $ub + vd \leq \delta,\ u \geq 0$ is feasible.*

A *face* of a polyhedron P is a set of the form
$$F := P \cap \{x \in \mathbb{R}^n : cx = \delta\}$$
where $cx \leq \delta$ is a valid inequality for P. We say that the inequality $cx \leq \delta$ *defines* the face F. If a valid inequality $cx \leq \delta$ with $c \neq 0$ defines a nonempty face of P, the hyperplane $\{x \in \mathbb{R}^n : cx = \delta\}$ is called a *supporting hyperplane* of P. A face is itself a polyhedron since it is the intersection of the polyhedron P with another polyhedron (the hyperplane $cx = \delta$ when $c \neq 0$). Note that $\emptyset$ and P are always faces of P, since they are the faces defined by the valid inequalities $0 \leq -1$ and $0 \leq 0$, respectively. A face of P is said to be *proper* if it is nonempty and properly contained in P.

Inclusionwise maximal proper faces of P are called *facets*. Thus any face distinct from P is contained in a facet. Any valid inequality for P that defines a facet is called a *facet-defining inequality* (Fig. 3.3).

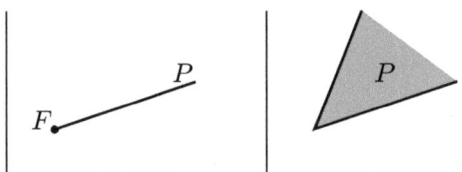

Figure 3.3: The polyhedron on the *left* (a half-line) has one proper face F, which is a facet; the unbounded polyhedron on the *right* has three proper faces, two of which have dimension one and are facets, while the third proper face has dimension zero

Theorem 3.24 (Characterization of the Faces). *Let $P := \{x \in \mathbb{R}^n : a^i x \leq b_i, i \in M\}$ be a nonempty polyhedron. For any $I \subseteq M$, the set*
$$F_I := \{x \in \mathbb{R}^n : a^i x = b_i, i \in I,\ a^i x \leq b_i, i \in M \setminus I\}$$
is a face of P. Conversely, if F is a nonempty face of P, then $F = F_I$ for some $I \subseteq M$.

3.8. FACES

Proof. For the first part of the statement, let $c := \sum_{i \in I} a^i$, $\delta := \sum_{i \in I} b_i$. Then $cx \leq \delta$ is a valid inequality. Furthermore, given $x \in P$, x satisfies $cx = \delta$ if and only if it satisfies $a^i x = b_i$, $i \in I$. Thus $F_I = P \cap \{x \in \mathbb{R}^n : cx = \delta\}$, so F_I is a face.

Conversely, let $F := \{x \in P : cx = \delta\}$ be a nonempty face of P defined by the valid inequality $cx \leq \delta$. Then F is the set of optimal solutions of the linear program $\max\{cx : x \in P\}$. Let $\bar{u}$ be an optimal solution to the dual problem $\min\{ub : uA = c, u \geq 0\}$, and $I = \{i \in M : \bar{u}_i > 0\}$. By Theorem 3.8, $F = F_I$. □

Proposition 3.25. *Given a polyhedron P, the following hold.*

(i) *The number of faces of P is finite.*

(ii) *For every nonempty face F of P, $\text{lin}(F) = \text{lin}(P)$.*

(iii) *Given faces F and F' of P, $F \cap F'$ is a face of P and there is a unique minimal face of P containing $F \cup F'$.*

(iv) *Two faces F and F' of P are distinct if and only if $\text{aff}(F) \neq \text{aff}(F')$.*

(v) *If F and F' are faces of P and $F \subset F'$, then $\dim(F) \leq \dim(F') - 1$.*

(vi) *Given a face F of P, the faces of F are exactly the faces of P contained in F.*

Proof. Let $P := \{x \in \mathbb{R}^n : a^i x \leq b_i, i \in M\}$.

(i) By Theorem 3.24, the number of faces of P does not exceed $2^{|M|}$.

(ii) It follows from Proposition 3.15 and Theorem 3.24 that $\text{lin}(F) = \{x \in \mathbb{R}^n : a^i x = 0, i \in M\} = \text{lin}(P)$.

(iii) Let I be the set of all $i \in M$ such that $a^i x = b_i$ is satisfied by all $x \in F$ (in particular $I = M$ if $F = \emptyset$), and J be the set of all $i \in M$ such that $a^i x = b_i$ is satisfied by all $x \in F'$. By Theorem 3.24, $F = F_I$ and $F' = F_J$. It follows that $F \cap F' = F_{I \cup J}$. Furthermore, $F_{I \cap J}$ is the unique minimal face of P containing $F \cup F'$.

(iv) The "if" direction is obvious. We prove the "only if" statement. Let F and F' be distinct faces of P. We assume F and F' are both nonempty, otherwise the statement holds. By Theorem 3.24, these exist distinct subsets $I, J \subseteq M$ such that $F = F_I$ and $F' = F_J$. Since $F \neq F'$, we may assume that there is a point $\bar{x} \in F \setminus F'$. Thus, there must exist some index $j \in J$ such that $a^j \bar{x} < b_j$. Since $j \in J$, we have that $\mathrm{aff}(F') \subseteq \{x \in \mathbb{R}^n : a^j x = b_j\}$ and since $\bar{x} \in F$, we have that $\bar{x}$ belongs to $\mathrm{aff}(F) \setminus \mathrm{aff}(F')$. Hence $\mathrm{aff}(F) \neq \mathrm{aff}(F')$.

(v) If $F \subset F'$, then $\mathrm{aff}(F) \subseteq \mathrm{aff}(F')$. By (iv), $\mathrm{aff}(F) \neq \mathrm{aff}(F')$, therefore $\dim(F) \leq \dim(F') - 1$.

(vi) is obvious.

□

3.9 Minimal Representation and Facets

Given a polyhedron $P := \{x \in \mathbb{R}^n : Ax \leq b\}$, we partition the rows of A into the implicit equalities and the rest. Namely $P = \{x \in \mathbb{R}^n : a^i x = b_i, i \in I^=, a^i x \leq b_i, i \in I^<\}$ where, for every $i \in I^<$, there exists $\bar{x} \in P$ such that $a^i \bar{x} < b_i$.

For any $j \in I^= \cup I^<$, we say that the jth constraint of the system is *redundant* if $P = \{x \in \mathbb{R}^n : a^i x = b_i, i \in I^= \setminus \{j\}, a^i x \leq b_i, i \in I^< \setminus \{j\}\}$. In other words, a constraint is redundant if removing it from the system does not change the feasible region. Constraints that are not redundant are called *irredundant*. Note that all constraints in the system defining P may be redundant; for example, if every inequality in the system is repeated twice.

Lemma 3.26. *Let $P := \{x \in \mathbb{R}^n : Ax \leq b\}$ be a nonempty polyhedron. Given $j \in I^<$ such that the inequality $a^j x \leq b_j$ is irredundant, let $F := \{x \in P : a^j x = b_j\}$. The following hold.*

(i) *The face F contains a point $\hat{x}$ such that $a^i \hat{x} < b_i$ for all $i \in I^< \setminus \{j\}$.*

(ii) $\mathrm{aff}(F) = \{x \in \mathbb{R}^n : a^i x = b_i, i \in I^= \cup \{j\}\}$.

(iii) $\dim(F) = \dim(P) - 1$.

Proof. (i) By Theorem 3.24, F is a face of P. By Remark 3.16, there exists $\bar{x} \in P$ satisfying $a^i \bar{x} < b_i$ for all $i \in I^<$. Furthermore, since $a^j x \leq b_j$ is not redundant, there exists a point $\tilde{x}$ satisfying

$$a^i \tilde{x} = b_i \ \forall i \in I^=, \quad a^i \tilde{x} \leq b_i \ \forall i \in I^< \setminus \{j\}, \quad a^j \tilde{x} > b_j.$$

Let $\hat{x}$ be the unique point in the intersection of F with the line segment joining $\bar{x}$ and $\tilde{x}$. Then $a^i \hat{x} < b_i$ for all $i \in I^< \setminus \{j\}$.

3.9. MINIMAL REPRESENTATION AND FACETS

(ii) By (i), the implicit equalities of the system $a^i x = b_i$, $i \in I^= \cup \{j\}$, $a^i x \le b_i$, $i \in I^< \setminus \{j\}$, are precisely the equations $a^i x = b_i$, $i \in I^= \cup \{j\}$. Thus, by Theorem 3.17, aff$(F) = \{x \in \mathbb{R}^n : a^i x = b_i, i \in I^= \cup \{j\}\}$.

(iii) Since $a^j x = b_j$ is not an implicit equation of P, it follows that F is strictly contained in P, therefore by Proposition 3.25 $\dim(F) \le \dim(P) - 1$. By (ii) and Theorem 3.17, $\dim(F) \ge \dim(P) - 1$. Thus $\dim(F) = \dim(P) - 1$.

$\square$

The system $a^i x = b_i$, $i \in I^=$, $a^i x \le b_i$, $i \in I^<$, defining P is said to be a *minimal representation* for P if all its constraints are irredundant. Note that every polyhedron P admits a minimal representation, since this can be obtained, starting from any system defining P, by iteratively removing a redundant constraint as long as one exists.

Theorem 3.27 (Characterization of the Facets). *Let $P := \{x \in \mathbb{R}^n : Ax \le b\}$ be a nonempty polyhedron, and let f be the number of its facets.*

(i) *For each facet F of P, there exists $j \in I^<$ such that the inequality $a^j x \le b_j$ defines F.*

(ii) *If $a^i x = b_i$, $i \in I^=$, $a^i x \le b_i$, $i \in I^<$ is a minimal representation for P, then $|I^<| = f$ and the facets of P are $F_i := \{x \in P : a^i x = b_i\}$, for $i \in I^<$.*

(iii) *A face F of P is a facet if and only if F is nonempty and $\dim(F) = \dim(P) - 1$.*

Proof. (i) Given a facet F of P, by Theorem 3.24 there exists $I \subseteq I^<$ such that $F = \{x \in P : a^i x = b_i, i \in I\}$. Since $F \ne P$, it follows that $I \ne \emptyset$. Choose $j \in I$ and let $F' := \{x \in P : a^j x = b_j\}$. Since $j \in I$, $F \subseteq F' \subset P$. Since F is a facet, it follows by maximality that $F = F'$. Thus F is defined by the inequality $a^j x \le b_j$.

(ii) Assume $a^i x = b_i$, $i \in I^=$, $a^i x \le b_i$, $i \in I^<$ is a minimal representation for P. Since $a^i x \le b_i$ is irredundant for every $i \in I^<$, it follows from Lemma 3.26 that F_i is a facet for every $i \in I^<$. Conversely, from (i) we have that every facet of P is of the form F_i for some $i \in I^<$. Therefore we only need to show that, for $j, k \in I^<$, $j \ne k$, the facets F_j and F_k are distinct. Indeed, by Lemma 3.26, there exists $\hat{x} \in F_j$ such that $a^i \hat{x} < b_i$ for all $i \in I^< \setminus \{j\}$, therefore $\hat{x} \in F_j \setminus F_k$.

(iii) follows from (ii) and from Lemma 3.26(iii).

$\square$

Example 3.28. The stable set problem has been introduced in Chap. 2. Given a graph $G = (V, E)$ the *stable set polytope* $\text{STAB}(G)$ is the convex hull of the characteristic vectors of all the stable sets of G. $\text{STAB}(G)$ is a full-dimensional polytope in $\mathbb{R}^V$, since it contains the origin and the unit vectors e^v, $v \in V$. Let $K \subseteq V$ be a maximal clique of G. The clique inequality $\sum_{v \in K} x_v \leq 1$ is valid for $\text{STAB}(G)$ (see Sect. 2.4.3). We show that it is facet-defining. Since K is a maximal clique of G, every node in $V \setminus K$ is contained in a stable set of size 2 containing one node of K. Consider the $|V|$ stable sets consisting of the $|V \setminus K|$ stable sets just defined, together with $|K|$ stable sets each consisting of a single node $v \in K$. Since the characteristic vectors of these $|V|$ stable sets are linearly independent, they are affinely independent as well. Therefore the inequality $\sum_{v \in K} x_v \leq 1$ defines a facet of $\text{STAB}(G)$. ∎

Example 3.29. Consider the permutahedron Π_n and the set S_n of permutation vectors defined in Example 3.20. Note that, for any $K \subset \{1, \ldots, n\}$, letting $k = |K|$, the inequality

$$\sum_{i \in K} x_i \geq \binom{k+1}{2} \tag{3.6}$$

is valid for Π_n. Indeed, given any permutation vector $\bar{x} \in S_n$, $\sum_{i \in K} \bar{x}_i \geq 1 + 2 + \cdots + k = \binom{k+1}{2}$, where equality holds if and only if $\{\bar{x}_i : i \in K\} = \{1, \ldots, k\}$. We show that (3.6) is facet-defining. Let F be the face of Π_n defined by (3.6). Since it is a proper face, it is contained in some facet, thus there exists a facet-defining inequality $\alpha x \geq \beta$ for Π_n such that $\alpha x = \beta$ for all $x \in F$. We will show that inequality (3.6) defines the same face as $\alpha x \geq \beta$, which implies that F is a facet.

We first show that $\alpha_i = \alpha_j$ if $i, j \in K$ or if $i, j \in \{1, \ldots, n\} \setminus K$. Indeed, given $i, j \in K$, $i \neq j$, and any vector $\bar{x} \in F \cap S_n$, the vector $\tilde{x}$ obtained from $\bar{x}$ by swapping the entries $\bar{x}_i$ and $\bar{x}_j$ belongs to F as well. Thus $0 = \alpha \bar{x} - \alpha \tilde{x} = \alpha_i(\bar{x}_i - \tilde{x}_i) + \alpha_j(\bar{x}_j - \tilde{x}_j) = (\alpha_i - \alpha_j)(\bar{x}_i - \bar{x}_j)$. Since $\bar{x}_i - \bar{x}_j \neq 0$, it follows that $\alpha_i = \alpha_j$. The proof that $\alpha_i = \alpha_j$ for all $i, j \in \{1, \ldots, n\} \setminus K$ is identical. Thus the inequality $\alpha x \geq \beta$ is of the form $\lambda \sum_{i \in K} x_i + \lambda' \sum_{i \notin K} x_i \geq \beta$. We further notice that $\lambda \geq \lambda'$. Indeed, given $\bar{x} \in F \cap S_n$, the vector $\tilde{x}$ obtained by swapping an entry $\bar{x}_i$, $i \in K$, with an entry $\bar{x}_j$, $j \notin K$, is also an element of S_n, and thus it satisfies $\alpha \tilde{x} \geq \beta$. It follows that $0 \leq \alpha \tilde{x} - \alpha \bar{x} = (\lambda - \lambda')(\bar{x}_j - \bar{x}_i)$. Since $\bar{x} \in F \cap S_n$, $\bar{x}_j > \bar{x}_i$, thus $\lambda \geq \lambda'$. It follows that $\alpha x \geq \beta$ is obtained by taking the sum of the equation $\sum_{i=1}^n x_i = \binom{n+1}{2}$ multiplied by λ' and the inequality (3.6) multiplied by $\lambda - \lambda'$. This shows that (3.6) and $\alpha x \geq \beta$ define the same face. ∎

3.9. MINIMAL REPRESENTATION AND FACETS

The next theorem shows that minimal representations of a polyhedron are essentially unique.

Theorem 3.30 (Uniqueness of the Minimal Representation)**.**
Let $P \subseteq \mathbb{R}^n$ be a nonempty polyhedron. Let $k := \dim(P)$ and f be the number of facets of P. Let $A^= x = b^=$, $A^< x \leq b^<$ and $C^= x = d^=$, $C^< x \leq d^<$ be two minimal representations for P. The following hold.

(i) The systems $A^= x = b^=$ and $C^= x = d^=$ have $n - k$ equations and every equation of $C^= x = d^=$ is of the form $(uA^=)x = ub^=$ for some $u \in \mathbb{R}^{n-k}$.

(ii) The systems $A^< x \leq b^<$ and $C^< x \leq d^<$ are comprised of f inequalities, say $a^i x \leq b_i$, $i = 1, \ldots, f$, and $c^i x \leq d_i$, $i = 1, \ldots, f$, respectively. Up to permuting the indices of the inequalities, for $j = 1, \ldots, f$, there exists $\lambda > 0$ and $u \in \mathbb{R}^{n-k}$ such that
$$c^j = \lambda a^j + uA^=, \quad d_j = \lambda b_j + ub^=.$$

Proof. (i) Both matrices $A^=$, $C^=$ have full row-rank, otherwise $A^= x = b^=$ or $C^= x = d^=$ would contain some redundant equation. By Theorem 3.17, $\{x \in \mathbb{R}^n : A^= x = b^=\} = \{x \in \mathbb{R}^n : C^= x = d^=\} = \mathrm{aff}(P)$. Thus, by standard linear algebra $\mathrm{rank}(A^=) = \mathrm{rank}(C^=) = n - k$, and every equation of $C^= x = d^=$ is a linear combination of equations of $A^= x = b^=$.

(ii) By Theorem 3.27(ii), $A^< x \leq b^<$ and $C^< x \leq d^<$ have f inequalities, all facet-defining, say $a^i x \leq b_i$, $i = 1, \ldots, f$, and $c^i x \leq d_i$, $i = 1, \ldots, f$, respectively. Up to permuting the indices, we may assume that $a^i x \leq b_i$ and $c^i x \leq d_i$ define the same facet F_i.

Let $j \in \{1, \ldots, f\}$. By Lemma 3.26, $\mathrm{aff}(F_j) = \mathrm{aff}(P) \cap \{x : a^j x = b_j\} = \mathrm{aff}(P) \cap \{x : c^j x = d_j\}$. It follows that $c^j x = d_j$ is a linear combination of the equations $A^= x = b^=$, $a^j x = b_j$, that is, there exists $\lambda \in \mathbb{R}$, $u \in \mathbb{R}^{n-k}$ such that $c^j = \lambda a^j + uA^=$ and $d_j = \lambda b_j + ub^=$. To show $\lambda > 0$, consider a point $\tilde{x} \in P \setminus F_j$. Then $0 < d_j - c^j \tilde{x} = \lambda(b_j - a^j \tilde{x}) + u(b^= - A^= \tilde{x}) = \lambda(b_j - a^j \tilde{x})$, which implies $\lambda > 0$ because $b_j - a^j \tilde{x} > 0$. □

For full-dimensional polyhedra the above theorem has a simpler form.

Corollary 3.31. *Let P be a full-dimensional polyhedron and let $Ax \leq b$ be a minimal representation of P. Then $Ax \leq b$ is uniquely defined up to multiplying inequalities by a positive scalar.*

Theorems 3.27 and 3.30 imply the following characterization of minimal representations.

Corollary 3.32. *Let $P = \{x \in A^=x = b^=, A^<x \leq b^<\}$ be a nonempty polyhedron. The system $A^=x = b^=$, $A^<x \leq b^<$ is a minimal representation for P if and only if $A^=$ has full row rank, $\operatorname{rank}(A^=) = n - \dim(P)$, and $A^<x \leq b^<$ has as many inequalities as the number of facets of P.*

3.10 Minimal Faces

Let P be a nonempty polyhedron in $\mathbb{R}^n$. A set $F \subseteq \mathbb{R}^n$ is a *minimal face* of P if F is a nonempty face of P that contains no proper face. The following characterization of minimal faces is due to Hoffman and Kruskal [204].

Theorem 3.33. *Let $P := \{x \in \mathbb{R}^n : Ax \leq b\}$ be a nonempty polyhedron.*

(i) A nonempty face F of P is minimal if and only if $F = \{x \in \mathbb{R}^n : A'x = b'\}$ for some system $A'x \leq b'$ of inequalities from $Ax \leq b$ such that $\operatorname{rank}(A') = \operatorname{rank}(A)$. Hence F is a translate of $\operatorname{lin}(P)$.

(ii) The dimensions of the nonempty faces of P take all values between $\dim(\operatorname{lin}(P))$ and $\dim(P)$.

Proof. We prove (i). Let F be a nonempty face of P. We first show the "if" part of the statement. Suppose $F = \{x : A'x = b'\}$ where $A'x \leq b'$ comprises some of the inequalities from $Ax \leq b$. Then F is an affine space, and therefore it has no proper face, implying that F is a minimal face of P. For the "only if" part, assume F is a minimal face. It follows that F has no facet. Let $A'x \leq b'$ be the system comprising all inequalities of $Ax \leq b$ that are satisfied at equality by every point in F, and let $A''x \leq b''$ be the system comprising all remaining inequalities of $Ax \leq b$. By Theorem 3.24, $F = \{x \in \mathbb{R}^n : A'x = b', A''x \leq b''\}$. Let $C^=x = d^=$, $C^<x \leq d^<$ be a minimal representation of F comprising constraints from $A'x = b'$, $A''x \leq b''$. Since F has no facet, it follows from Theorem 3.27 that $F = \{x : C^=x = d^=\}$. This implies that $F = \{x \in \mathbb{R}^n : A'x = b'\}$. It follows that F is an affine space, therefore $F = \{v\} + \operatorname{lin}(F)$ for any $v \in F$. By Proposition 3.25(ii), $\operatorname{lin}(F) = \operatorname{lin}(P)$, therefore $\dim(F) = \dim(\operatorname{lin}(P))$. It follows that $\operatorname{rank}(A') = n - \dim(F) = n - \dim(\operatorname{lin}(P)) = \operatorname{rank}(A)$.

(ii) follows from Theorem 3.27(iii), since this result implies that any face F of P of dimension greater than $\dim(\operatorname{lin}(P))$ contains a face of dimension $\dim(F) - 1$. □

Note that any polyhedral cone has a unique minimal face, namely, its lineality space.

Vertices

A face of dimension 0 is a *vertex* of P. By Theorem 3.33, P has a vertex if and only if $\text{lin}(P) = \{0\}$, that is: *P has a vertex if and only if P is pointed*.

Theorem 3.34. *Let $P := \{x \in \mathbb{R}^n : Ax \leq b\}$ be a pointed polyhedron, and let $\bar{x} \in P$. The following statements are equivalent.*

(i) $\bar{x}$ is a vertex.

(ii) $\bar{x}$ satisfies at equality n linearly independent inequalities of $Ax \leq b$.

(iii) $\bar{x}$ is not a proper convex combination of two distinct points in P (i.e., no distinct points $x', x'' \in P$ exist such that $\bar{x} = \lambda x' + (1 - \lambda)x''$ for some $0 < \lambda < 1$).

Proof. By Theorem 3.33, (i) and (ii) are equivalent. We show next that (i) implies (iii). Indeed, suppose $\bar{x}$ is a vertex, and let $cx \leq \delta$ be a valid inequality for P such that $\{\bar{x}\} = P \cap \{x : cx = \delta\}$. Given $x', x'' \in P$ and $0 < \lambda < 1$, such that $\bar{x} = \lambda x' + (1 - \lambda)x''$, it follows that $\delta = c\bar{x} = \lambda cx' + (1 - \lambda)cx'' \leq \delta$, therefore $cx' = cx'' = \delta$, hence $x' = x'' = \bar{x}$.

It remains to show that (iii) implies (ii). Let $A'x \leq b'$ be the system comprising all inequalities of $Ax \leq b$ satisfied at equality by $\bar{x}$. We will show that, if $\text{rank}(A') < n$, then $\bar{x}$ is a proper convex combination of two points in P. If $\text{rank}(A') < n$, then there exists a vector $\bar{y} \neq 0$ such that $A'\bar{y} = 0$. Let $A''x \leq b''$ be the system of inequalities of $Ax \leq b$ not in $A'x \leq b'$. By definition, $A''\bar{x} < b''$. Thus, for $\varepsilon > 0$ sufficiently small, the points $x' = \bar{x} + \varepsilon\bar{y}$ and $x'' = \bar{x} - \varepsilon\bar{y}$ are both in P. It is now clear that $\bar{x} \neq x', x''$ and $\bar{x} = \frac{1}{2}x' + \frac{1}{2}x''$. □

3.11 Edges and Extreme Rays

Let P be a nonempty polyhedron in $\mathbb{R}^n$. A face of dimension 1 is an *edge* of P. Note that an edge has at most two vertices, and it is bounded if and only if it has two vertices. Furthermore, if P is pointed, any edge has at least one vertex. If an edge F has two vertices x' and x'', then F is the line segment joining x' and x''. If an edge F of P has precisely one vertex, say $\bar{x}$, then F is a half-line starting from $\bar{x}$, that is, $F = \{\bar{x} + \lambda r : \lambda \geq 0\}$ for some ray r of $\text{rec}(P)$.

In particular, the edges of a pointed polyhedral cone are half-lines starting at the origin. An *extreme ray* of a pointed polyhedral cone C is an edge of C. The extreme rays of a pointed polyhedron P are the extreme rays of the recession cone of P.

Theorem 3.35. *Let $C := \{x \in \mathbb{R}^n : Ax \leq 0\}$ be a pointed cone, and let $\bar{r}$ be a ray of C. The following are equivalent.*

(i) $\bar{r}$ is an extreme ray of C.

(ii) $\bar{r}$ satisfies at equality $n-1$ linearly independent inequalities of $Ax \leq 0$.

(iii) $\bar{r}$ is not a proper conic combination of two distinct rays in C (i.e., no distinct rays $r', r'' \in C$ exist such that $\bar{r} = \mu' r' + \mu'' r''$ for some $\mu' > 0$, $\mu'' > 0$).

Proof. Consider $\bar{r} \in C \setminus \{0\}$. Let $\alpha \in \mathbb{R}^n$ be the sum of the rows of A and let $P := \{x \in C : \alpha x \geq -1\}$. Since C is pointed, $\text{rank}(A) = n$ (Proposition 3.15). Therefore, $\alpha x < 0$ for every $x \in C \setminus \{0\}$. In particular P is a polytope, because $\text{rec}(P) = \{x \in C : \alpha x \geq 0\} = \{0\}$. Up to multiplying $\bar{r}$ by a positive scalar, we may assume $\alpha \bar{r} = -1$. The theorem now follows by Theorem 3.34. Indeed, $\bar{r}$ satisfies $n-1$ linearly independent inequalities of $Ax \leq 0$ at equality if and only if $\bar{r}$ is a vertex of P, and this is the case if and only if $\bar{r}$ is not a proper convex combination of two distinct points in P. One can readily check that the latter condition is equivalent to (iii). □

The *skeleton* of a polytope P is the graph $G(P)$ whose nodes are the vertices of P and whose edges are the edges of P; that is, two nodes are adjacent in $G(P)$ if and only if there is an edge of P containing the two corresponding vertices.

Example 3.36. Let $G = (V, E)$ be the complete graph on n nodes. The *cut polytope* $P_n^{\text{cut}} \subseteq \mathbb{R}^{\binom{n}{2}}$ is the convex hull of the set S_n of characteristic vectors of the cuts of G. Note that $|S_n| = 2^{n-1}$ since $T \subseteq V$ and $V \setminus T$ define the same cut. We show that the cut polytope is a *neighborly polytope*, that is, any pair of vertices of P_n^{cut} is contained in an edge of P_n^{cut} (in other words, the skeleton of P_n^{cut} is a clique on 2^{n-1} nodes).

The vertices of P_n^{cut} are precisely the characteristic vectors of the cuts of G (this follows from Exercise 3.25).

Let $S, T \subseteq V$ such that $\delta(S) \neq \delta(T)$. We show that the characteristic vectors x^S of $\delta(S)$ and x^T of $\delta(T)$ are adjacent in P_n^{cut}. For this, we need to show that there exists an edge F of P_n^{cut} that contains x^S and x^T. We will describe a valid inequality $\alpha x \le \beta$ for P_n^{cut} such that $\alpha x^S = \alpha x^T = \beta$ and $\alpha x < \beta$ for all $x \in S_n \setminus \{x^S, x^T\}$. This will imply that $F := P_n^{\mathrm{cut}} \cap \{x : \alpha x = \beta\}$ is an edge of P_n^{cut} whose vertices are x^S and x^T.

We define α as follows. For all $uv \in E$, let

$$\alpha_{uv} := \begin{cases} 1 & \text{if } u \in S \cap T \text{ and } v \in V \setminus (S \cup T); \\ 1 & \text{if } u \in S \setminus T \text{ and } v \in T \setminus S; \\ -1 & \text{if } u,v \in S \setminus T, \text{ or } u,v \in T \setminus S, \text{ or } u,v \in S \cap T, \text{ or } u,v \in V \setminus (S \cup T); \\ 0 & \text{otherwise.} \end{cases}$$

Let β be the number of edges uv such that $\alpha_{uv} = 1$. By construction, $\alpha x^S = \alpha x^T = \beta$, while $\alpha x \le \beta$ for every $x \in S_n$. On the other hand, given $W \subseteq V$, the characteristic vector x^W satisfies $\alpha x = \beta$ if and only if $\delta(W)$ does not contain any edge uv such that $\alpha_{uv} = -1$ and $\delta(W)$ contains all the edges uv such that $\alpha_{uv} = 1$. One can verify that the only way this can happen is if W is one of the set $S, V \setminus S, T$, or $V \setminus T$. Since $\delta(S) = \delta(V \setminus S)$ and $\delta(T) = \delta(V \setminus T)$, it follows that x^W must be either x^S or x^T. ∎

3.12 Decomposition Theorem for Polyhedra

Theorem 3.37 (Decomposition Theorem for Polyhedra). *Let $P \subseteq \mathbb{R}^n$ be a nonempty polyhedron, and let $t := \dim(\mathrm{lin}(P))$. Let $F_1, \ldots, F_p$ be the family of minimal faces of P, and $R_1, \ldots, R_q$ the family of $(t+1)$-dimensional faces of $\mathrm{rec}(P)$. Let $v^i \in F_i$, $i = 1, \ldots, p$, and $r^i \in R_i \setminus \mathrm{lin}(P)$, $i = 1, \ldots, q$. Then*

$$P = \mathrm{conv}(v^1, \ldots, v^p) + \mathrm{cone}(r^1, \ldots, r^q) + \mathrm{lin}(P).$$

Furthermore if $P = \mathrm{conv}(X) + \mathrm{cone}(Y) + \mathrm{lin}(P)$, where $\mathrm{cone}(Y)$ is a pointed cone, then X contains a point in every t-dimensional face of P and Y contains a ray in each $(t+1)$-dimensional face of $\mathrm{rec}(P)$.

Proof. Assume first that P is pointed, i.e., $\dim(\mathrm{lin}(P)) = 0$. Then minimal faces are the vertices of P and one-dimensional faces of $\mathrm{rec}(P)$ are the extreme rays of $\mathrm{rec}(P)$. In this case the above theorem reduces to:

Let $v^1, \ldots, v^p$ be the vertices of P, and $r^1, \ldots, r^q$ the extreme rays of $\mathrm{rec}(P)$. Then $P = \mathrm{conv}(v^1, \ldots, v^p) + \mathrm{cone}(r^1, \ldots, r^q)$. Furthermore if $P = \mathrm{conv}(X) + \mathrm{cone}(Y)$, then $\{v^1, \ldots, v^p\} \subseteq X$ and $\{r^1, \ldots, r^q\} \subseteq Y$.

We prove the above statement. Assume $P = \mathrm{conv}(X) + \mathrm{cone}(Y)$ is a pointed polyhedron. By Theorem 3.34 all the vertices of P belong to X and if $x^j \in X$ is not a vertex of P, then $P = \mathrm{conv}(X \setminus \{x^j\}) + \mathrm{cone}(Y)$.

By Proposition 3.15, $\text{rec}(P) = \text{cone}(Y)$. Therefore by Theorem 3.35, Y contains all the extreme rays of $\text{rec}(P)$ and if $y^j \in Y$ is not an extreme ray of $\text{rec}(P)$, then $P = \text{conv}(X) + \text{cone}(Y \setminus \{y^j\})$.

Assume now that P is not pointed, i.e., $t := \dim(\text{lin}(P)) \geq 1$. Let $a^1, \ldots, a^t$ be a basis of $\text{lin}(P)$ and consider the polyhedron $Q := \{x \in P : a^i x = 0, i = 1, \ldots, t\}$. Note that Q is a pointed polyhedron and $P = Q + \text{lin}(P)$. Let $v^1, \ldots, v^p$ be the vertices of Q, and $r^1, \ldots, r^q$ be the extreme rays of $\text{rec}(Q)$. The statement above shows that $Q = \text{conv}(v^1, \ldots, v^p) + \text{cone}(r^1, \ldots, r^q)$. Hence $P = \text{conv}(v^1, \ldots, v^p) + \text{cone}(r^1, \ldots, r^q) + \text{lin}(P)$ and this is a representation of P where $\text{cone}(r^1, \ldots, r^q)$ is pointed and p and q are smallest.

By Theorem 3.33, the minimal faces of P are $F_i = v^i + \text{lin}(P)$, $i = 1, \ldots, p$. Furthermore the $(t+1)$-dimensional faces of $\text{rec}(P)$ are $R_i = \text{cone}(r^i) + \text{lin}(P)$, $i = 1, \ldots, p$. Hence in the representation $P = \text{conv}(v^1, \ldots, v^p) + \text{cone}(r^1, \ldots, r^q) + \text{lin}(P)$, any point in F_i can be substituted for v^i, $i = 1, \ldots, p$ and any ray in the cone R_i can be substituted for r^i, $i = 1, \ldots, q$. □

3.13 Encoding Size of Vertices, Extreme Rays, and Facets

Theorem 3.38. *Let $P := \{x \in \mathbb{R}^n : Ax \leq b\}$ be a pointed polyhedron where A and b have rational entries. Let $v^1, \ldots, v^p \in \mathbb{Q}^n$ be the vertices of P, and $r^1, \ldots, r^q \in \mathbb{Q}^n$ its extreme rays. If the encoding size of each coefficient of (A, b) is at most L, then each of the vectors $v^1, \ldots, v^p$ and $r^1, \ldots, r^q$ can be written so that their encoding size is polynomially bounded by n and L.*

Proof. By Theorem 3.34, a point $\bar{x} \in P$ is a vertex if and only if there exists a system $\bar{A}x \leq \bar{b}$ comprising n linearly independent inequalities from $Ax \leq b$ such that $\bar{x}$ is the unique solution of $\bar{A}x = \bar{b}$. By Proposition 1.2, the encoding size of $\bar{x}$ is polynomially bounded by n and L.

By Theorem 3.35, a ray r of P is extreme if and only if there exists a system $\bar{A}x \leq 0$ comprising $n-1$ linearly independent inequalities from $Ax \leq 0$ such that r is a solution of $\bar{A}x = 0$. Since $r \neq 0$ and $\text{rank}(A) = n$, there exists some inequality $\alpha x \leq 0$ of $Ax \leq 0$ such that $\alpha r < 0$. We may therefore assume that $\alpha r = -1$, thus, by Proposition 1.2, the encoding size of r is polynomially bounded by n and L. □

The above theorem can be easily extended to non-pointed polyhedra, showing that P can be written as $P = \text{conv}(v^1, \ldots, v^p) + \text{cone}(r^1, \ldots, r^q)$

where the encoding sizes of $v^1, \ldots, v^p$ and $r^1, \ldots, r^q$ are polynomially bounded by n and L, but for simplicity we only prove it in the pointed case.

Theorem 3.39. *Let $P = \mathrm{conv}(v^1, \ldots, v^p) + \mathrm{cone}(r^1, \ldots, r^q)$ be a polyhedron, where $v^1, \ldots, v^p$ and $r^1, \ldots, r^q$ are given rational vectors in $\mathbb{R}^n$. If the encoding size of each vector $v^1, \ldots, v^p$ and $r^1, \ldots, r^q$ is at most L, then there exist a rational matrix A and vector b such that $P = \{x : Ax \leq b\}$ and the encoding size of each entry of (A, b) is polynomially bounded by n and L.*

Proof. By Theorem 3.27, it suffices to show that, given any facet F of P, there exists a valid inequality $\alpha x \leq \beta$ for P defining F whose encoding size is polynomially bounded by n and L. Let $d = \dim(P)$. Let $q^1, \ldots, q^d$ be d affinely independent points in F and let $q^0 \in P \setminus F$. We may choose $q^0, \ldots, q^d$ to be elements of $\{v^1, \ldots, v^p\} + \{0, r^1, \ldots, r^q\}$, therefore the encoding sizes of $q^0, \ldots, q^d$ are polynomially bounded by n and L. Consider the system of equations in $n+1$ variables $\alpha_1, \ldots, \alpha_n, \beta$ and $d+1$ constraints

$$\begin{aligned} \alpha q^0 - \beta &= -1 \\ \alpha q^i - \beta &= 0 \quad i = 1, \ldots, d. \end{aligned} \tag{3.7}$$

Since $q^0, \ldots, q^d$ are affinely independent, the constraint matrix of (3.7) has rank $d+1$, thus the system admits a solution. By Lemma 1.2, any basic solution (α, β) of (3.7) has encoding size that is polynomially bounded by n and L. It follows by construction that $\alpha x \leq \beta$ is valid for P and defines F. □

3.14 Carathéodory's Theorem

Theorem 3.40 (Carathéodory). *If a vector $v \in \mathbb{R}^n$ is a conic combination of vectors in some set $X \subseteq \mathbb{R}^n$, then it is a conic combination of at most $\dim(X)$ linearly independent vectors in X.*

Proof. We can assume that X is finite, say $X = \{v^1, \ldots, v^k\}$. Since $v \in \mathrm{cone}(X)$, it follows that the polyhedron $P := \{\lambda \in \mathbb{R}_+^k : \sum_{i=1}^k \lambda_i v^i = v\}$ is nonempty. Polyhedron P is pointed because it is contained in $\mathbb{R}_+^k$, therefore it has a vertex $\bar{\lambda}$. By Theorem 3.34 $\bar{\lambda}$ satisfies at equality k linearly independent constraints of the system $\sum_{i=1}^k \lambda_i v^i = v$, $\lambda \geq 0$. It follows that the vectors in $\{v^i : \bar{\lambda}_i > 0\}$ are linearly independent. □

A direct proof of Carathéodory's theorem. Let S be an inclusionwise minimal subset of X such that $v \in \mathrm{cone}(S)$. Clearly S is finite, say $S = \{v^1, \ldots, v^k\}$, and there exists $\lambda \in \mathbb{R}_+^k$ such that $\sum_{i=1}^k \lambda_i v^i = v$. It suffices to show that the vectors in S are linearly independent. Suppose not, and let $\mu \in \mathbb{R}^k$, $\mu \neq 0$, such $\sum_{i=1}^k \mu_i v^i = 0$. Assuming without loss of generality that μ has a positive component, let $\theta := \min_{i\,:\,\mu_i>0} \frac{\lambda_i}{\mu_i}$, and let h be an index for which this minimum is attained. Defining $\lambda' := \lambda - \theta\mu$, we have that $\lambda' \geq 0$, $\sum_{i=1}^k \lambda'_i v^i = v$, and $\lambda'_h = 0$. It follows that $v \in \text{cone}(S \setminus \{v^h\})$, contradicting the minimality of S. □

The following is another form of Carathéodory's theorem.

Corollary 3.41. *If a point $v \in \mathbb{R}^n$ is a convex combination of points in some set $X \subseteq \mathbb{R}^n$, then it is a convex combination of at most $\dim(X) + 1$ affinely independent points in X.*

Proof. If $v \in \text{conv}(X)$, then $\binom{v}{1} \in \mathbb{R}^{n+1}$ is a conic combination of points in $X \times \{1\}$. By Theorem 3.40, there exist $v^1, \ldots, v^k \in X$ and $\lambda \in \mathbb{R}_+^k$ such that $\binom{v^1}{1}, \ldots, \binom{v^k}{1}$ are linearly independent, $\sum_{i=1}^k \lambda_i v^i = v$, and $\sum_{i=1}^k \lambda_i = 1$. In particular $v^1, \ldots, v^k$ are affinely independent and the statement follows. □

Finally, Carathéodory's theorem has the following implication for pointed polyhedra.

Corollary 3.42. *Let P be a pointed polyhedron. Every point of P is the sum of a convex combination of at most $\dim(P) + 1$ affinely independent vertices of P and of a conic combination of at most $\dim(\text{rec}(P))$ linearly independent extreme rays of P.*

Here are two more theorems on convexity.

Theorem 3.43 (Radon). *Let S be a subset of $\mathbb{R}^d$ with at least $d+2$ points. Then S can be partitioned into two sets S_1 and S_2 so that $\text{conv}(S_1) \cap \text{conv}(S_2) \neq \emptyset$.*

Theorem 3.44 (Helly). *Let $C_1, C_2, \ldots, C_h$ be convex sets in $\mathbb{R}^d$ such that $C_1 \cap C_2 \cap \cdots \cap C_h = \emptyset$, where $h \geq d+1$. Then there exist $d+1$ sets among $C_1, C_2, \ldots, C_h$ whose intersection is empty.*

The proofs are left as an exercise, see Exercises 3.30, 3.31.

Systems in Standard Equality Form and Basic Feasible Solutions

Let $A \in \mathbb{R}^{m \times n}$ and $b \in \mathbb{R}^m$. A system of the form

$$Ax = b \qquad (3.8)$$
$$x \geq 0$$

is said to be in *standard equality form*.

A solution $\bar{x}$ of $Ax = b$ is said to be *basic* if the columns of A corresponding to positive entries of $\bar{x}$ are linearly independent. A basic solution that satisfies (3.8) is a *basic feasible solution*. It follows from Carathéodory theorem (Theorem 3.40) that (3.8) has basic feasible solutions whenever it is feasible.

Note that, if (3.8) is feasible, then we may assume without loss of generality that the matrix A has full row rank. Given a basic solution $\bar{x}$, there exists a set $B \subseteq \{1, \ldots, n\}$, such that $|B| = m$, $\bar{x}_j = 0$ for all $j \notin B$, and the columns of A indexed by B form a square nonsingular matrix A_B. Such a set B is called a *basis*.

Assuming, possibly by permuting the columns of A, that $B = \{1, \ldots, m\}$, the matrix A is of the form $A = (A_B, A_N)$. Splitting the entries of $\bar{x}$ accordingly, it then follows that $A\bar{x} = A_B \bar{x}_B = b$. Since A_B is square and nonsingular, $\bar{x}$ is defined as follows

$$\bar{x}_B = A_B^{-1} b \qquad (3.9)$$
$$\bar{x}_N = 0.$$

For a linear program $\max\{cx : Ax = b, x \geq 0\}$, we say that a basis B is *primal feasible* if the solution $\bar{x}$ in (3.9) is feasible (i.e., if $A_B^{-1} b \geq 0$). We say that B is *dual feasible* if the vector $\bar{y} = c_B A_B^{-1}$ is feasible for the dual $\min\{yb : yA \geq c\}$. The *reduced costs* associated with B are the coefficients of the vector $\bar{c} = c - \bar{y}A$. Note that, by construction, $\bar{c}_j = 0$ for all $j \in B$, and $\bar{y}$ is feasible for the dual if and only if $\bar{c}_j \leq 0$ for all $j \in N$. If B is both primal and dual feasible, then by linear programming duality (Theorem 3.7) $\bar{x}$ is an optimal solution for the primal and $\bar{y}$ is an optimal solution for the dual, since $c\bar{x} = c_B A_B^{-1} b = \bar{y}b$. Therefore we say that B is an *optimal basis* in this case. It can be shown that there exists an optimal basis whenever the linear program $\max\{cx : Ax = b, x \geq 0\}$ admits a finite optimum. Indeed, the simplex method terminates with an optimal basis in this case.

3.15 Projections

The orthogonal projection of a set $S \subset \mathbb{R}^{n+p}$ onto the linear subspace $\mathbb{R}^n \times \{0\}^p$ is

$$\text{proj}_x(S) := \{x \in \mathbb{R}^n : \exists z \in \mathbb{R}^p \text{ s.t. } (x, z) \in S\}.$$

Figure 3.4 illustrates this definition.

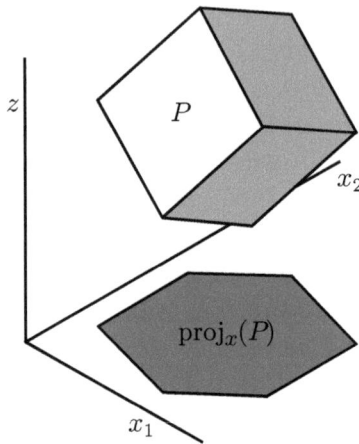

Figure 3.4: Projection $\text{proj}_x(P)$ of three-dimensional cube onto a two-dimensional space

In this book, our interest in projections comes from the fact that the two programs

$$\max\{f(x) : x \in \text{proj}_x(S)\} \quad \text{and} \quad \max\{f(x) + \mathbf{0}z : (x, z) \in S\}$$

have the same optimal solution x^*, but one of these two programs might be easier to solve than the other. We elaborate below.

Given a polyhedron $P := \{(x, z) \in \mathbb{R}^n \times \mathbb{R}^p : Ax + Gz \leq b\}$, Fourier's method applied to the system $Ax + Gz \leq b$ to eliminate all components of z, produces a system of inequalities $A^*x \leq b^*$ such that $\{x \in \mathbb{R}^n : A^*x \leq b^*\} = \{x \in \mathbb{R}^n : \exists z \in \mathbb{R}^p \text{ s.t. } (x, z) \in P\}$. By definition, this polyhedron is $\text{proj}_x(P)$.

The size of an irredundant system describing $\text{proj}_x(P)$ may be exponential in the size of the system describing P. To construct such an example, we take a polytope (in the x-space) with few vertices and an exponential number of facets. Its formulation as the convex hull of its vertices is small, but involves new variables z to express the convex combination; its projection onto the x-space has an exponential number of inequalities (Fig. 3.5).

3.15. PROJECTIONS

Example 3.45. Given a positive integer n, the *octahedron* is the following polytope:

$$\text{oct}_n := \{x \in \mathbb{R}^n : \sum_{i \in S} x_i - \sum_{i \notin S} x_i \leq 1 \quad S \subseteq \{1, \ldots, n\}\}.$$

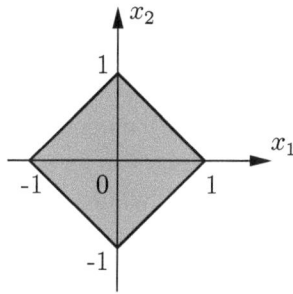

Figure 3.5: The set oct_2

We leave it as an exercise (Exercise 3.33) to show that:

(i) All the 2^n inequalities $\sum_{i \in S} x_i - \sum_{i \notin S} x_i \leq 1$, $S \subseteq \{1, \ldots, n\}$ define facets of oct_n.

(ii) oct_n has $2n$ vertices, namely the unit vectors e^i and their negatives $-e^i$, $i = 1, \ldots, n$.

Therefore if we define

$$P := \{(x, z) \in \mathbb{R}^n \times \mathbb{R}^{2n} : x = \sum_{i=1}^{n} z_i e^i - \sum_{i=1}^{n} z_{n+i} e^i, \sum_{i=1}^{2n} z_i = 1, z_i \geq 0\}$$

we have that $\text{proj}_x(P) = \text{oct}_n$. The reader can verify that $\dim(P) = 2n - 1$ and that P has exactly $2n$ facets, defined by the inequalities $z_i \geq 0$ for $i = 1, \ldots, 2n$. ∎

The above example shows that a polyhedron $Q \subset \mathbb{R}^n$ with a very large number of facets can sometimes be obtained as the projection of a polyhedron P in a higher-dimensional space that has a much smaller number of facets. A description of P by a system of linear inequalities is called an *extended formulation* of Q. When this system only has a number of variables and constraints that is polynomial in the size of the input used to describe Q, the extended formulation is said to be *compact*.

The Projection Cone

Given a polyhedron $P := \{(x,z) \in \mathbb{R}^n \times \mathbb{R}^p : Ax + Bz \leq b\}$, consider a valid inequality $ax \leq b$ for P that does not involve the variables z. It follows from the definition of $\operatorname{proj}_x(P)$ that $ax \leq b$ is also a valid inequality for $\operatorname{proj}_x(P)$. Since, for every vector u satisfying $u \geq 0, uB = 0$, the inequality $uAx \leq ub$ is valid for P, this observation shows that $uAx \leq ub$ is also valid for $\operatorname{proj}_x(P)$. The next theorem states that the converse also holds.

Define the *projection cone* of a polyhedron $P := \{(x,z) \in \mathbb{R}^n \times \mathbb{R}^p : Ax + Bz \leq b\}$ as the polyhedral cone $C_P := \{u \in \mathbb{R}^m : uB = 0, u \geq 0\}$. The reader can verify that $\operatorname{proj}_x(P) = \mathbb{R}^n$ when $C_P = \{0\}$ (Exercise 3.35). So we assume now that $C_P \neq \{0\}$. Since C_P is a pointed polyhedral cone, by Theorem 3.11 $C_P = \operatorname{cone}(r^1, \ldots, r^q)$, where $r^1, \ldots, r^q$ are the extreme rays of C_P.

Theorem 3.46. *Consider a polyhedron $P := \{(x,z) \in \mathbb{R}^n \times \mathbb{R}^p : Ax + Bz \leq b\}$, and let $r^1, \ldots, r^q$ be the extreme rays of C_P. Then $\operatorname{proj}_x(P) = \{x \in \mathbb{R}^n : r^t Ax \leq r^t b \text{ for } t = 1, \ldots, q\}$.*

Proof. It suffices to show that, for any $\bar{x} \in \mathbb{R}^n$, $\bar{x} \notin \operatorname{proj}_x(P)$ if and only if $r^t A\bar{x} > r^t b$ for some $t = 1, \ldots, q$. By definition, $\bar{x} \notin \operatorname{proj}_x(P)$ if and only if the system $Bz \leq b - A\bar{x}$ is infeasible. By Farkas' lemma (Theorem 3.4), the latter system is infeasible if and only if there exists a vector $u \in C_P$ such that $uA\bar{x} > ub$. Since $C_P = \operatorname{cone}(r^1, \ldots, r^q)$, such a vector u exists if and only if $r^t A\bar{x} > r^t b$ for some $t = 1, \ldots, q$. □

The projection cone has different descriptions depending on the form in which the polyhedron P is given. As an example, we state the result when the system defining P has equality constraints and the variables to be projected are nonnegative.

Corollary 3.47. *Given a polyhedron $P := \{(x,z) \in \mathbb{R}^n \times \mathbb{R}^p : Ax + Bz = b, z \geq 0\}$, its projection is $\operatorname{proj}_x(P) = \{x \in \mathbb{R}^n : uAx \leq ub \text{ for all } u \text{ s.t. } uB \geq 0\}$.*

Remark 3.48. *Theorem 3.46 shows that the inequalities $r^t Ax \leq r^t b$, $1 \leq t \leq q$, are sufficient to describe $\operatorname{proj}_x(P)$. However, some of these inequalities might be redundant even though $r^1, \ldots, r^q$ are extreme rays.*

We will give an example of projection using extreme rays of the projection cone in Theorem 4.11.

3.16 Polarity

Given a set $S \subseteq \mathbb{R}^n$, the *polar* of S is the set $S^* := \{y \in \mathbb{R}^n : yx \leq 1 \text{ for all } x \in S\}$. Note that S^* is a convex set that contains the origin. The polar of a polyhedron containing the origin has a simple expression.

Theorem 3.49. *Given $a^1, \ldots, a^m \in \mathbb{R}^n$ and $0 \leq k \leq m$, let $P := \{x \in \mathbb{R}^n : a^i x \leq 1, i=1 \ldots, k; a^i x \leq 0, i=k+1 \ldots, m\}$ and $Q := \mathrm{conv}(0, a^1, \ldots, a^k) + \mathrm{cone}(a^{k+1}, \ldots, a^m)$. Then $P^* = Q$ and $Q^* = P$.*

Proof. We first show that $Q \subseteq P^*$ and $P \subseteq Q^*$. It is sufficient to prove that, given $x \in P$ and $y \in Q$, $yx \leq 1$. If $y \in Q$ then there exists $\nu \in \mathbb{R}^m$ such that $y = \sum_{i=1}^m \nu_i a^i$, where $\nu \geq 0$, and $\sum_{i=1}^k \nu_i \leq 1$. Thus every $x \in P$ satisfies $yx = \sum_{i=1}^m \nu_i a^i x \leq \sum_{i=1}^k \nu_i \leq 1$.

To show $P^* \subseteq Q$, consider $\bar{y} \in P^*$. By the definition of polar, $\bar{y}x \leq 1$ is a valid inequality for P. By Theorem 3.22, there exists $\nu \geq 0$ such that $\bar{y} = \sum_{i=1}^m \nu_i a^i$ and $\sum_{i=1}^k \nu_i \leq 1$. Thus $\bar{y} \in Q$.

We now prove that $Q^* \subseteq P$. Let $\bar{x} \in Q^*$. By the definition of polar $a^i \bar{x} \leq 1$ for $i = 1, \ldots, k$. Since $0 \in Q$, $\lambda a^i \in Q$ for all $\lambda > 0$, $i = k+1, \ldots, m$. Since $\bar{x} \in Q^*$, $\lambda a^i \bar{x} \leq 1$ for all $\lambda \geq 0$, $i = k+1, \ldots, m$, hence $a^i \bar{x} \leq 0$. This shows $\bar{x} \in P$ and thus $Q^* \subseteq P$. □

This result, together with Minkowski–Weyl's theorem (Theorem 3.13), shows that, if P is a polyhedron containing the origin, then its polar P^* is also a polyhedron containing the origin. Furthermore, if C is a polyhedral cone, then $C^* = \{y \in \mathbb{R}^n : xy \leq 0 \text{ for all } x \in C\}$ is also a polyhedral cone.

Given a linear subspace L of $\mathbb{R}^n$, the *orthogonal complement* of L is the space $L^\perp := \{x \in \mathbb{R}^n : yx = 0 \text{ for all } y \in L\}$. It is a basic fact in linear algebra that, given a basis $\ell^1, \ldots, \ell^t$ of L, we have $L^\perp = \{x \in \mathbb{R}^n : \ell^i x = 0, i = 1, \ldots, t\}$. In particular $\dim(L) + \dim(L^\perp) = n$. We remark that $L^\perp$ is the polar of L.

Corollary 3.50. *Let $P \subseteq \mathbb{R}^n$ be a polyhedron containing the origin. The following hold.*

(i) $P^{**} = P$.

(ii) P^* *is bounded if and only if P contains the origin in its interior.*

(iii) $\mathrm{aff}(P^*)$ *is the orthogonal complement of* $\mathrm{lin}(P)$. *In particular* $\dim(P^*) = n - \dim(\mathrm{lin}(P))$.

Proof. Let $A \in \mathbb{R}^{m \times n}$ and $b \in \mathbb{R}^m$ such that $P = \{x \in \mathbb{R}^n : Ax \le b\}$. Since $0 \in P$, it follows that $b \ge 0$. Thus we can assume that $P = \{x \in \mathbb{R}^n : a^i x \le 1, i = 1 \ldots, k; \ a^i x \le 0, \ i = k+1 \ldots, m\}$. Without loss of generality we assume that, if A contains a row $a^i = 0$, then $i \le k$. Note that 0 is in the interior of P if and only if $k = m$, i.e., $P = \{x : Ax \le 1\}$.

(i) By applying Theorem 3.49 to P and to P^*, we get $P^{**} = P$.

(ii) By Theorem 3.49, $P^* = \text{conv}(0, a^1, \ldots, a^k) + \text{cone}(a^{k+1}, \ldots, a^m)$. Hence P^* is bounded if and only if $k = m$. That is P^* is bounded if and only if $P = \{x : Ax \le 1\}$. This shows ii).

(iii) Since $P^* = \text{conv}(0, a^1, \ldots, a^k) + \text{cone}(a^{k+1}, \ldots, a^m)$, it follows that $\dim(P^*)$ is equal to the number of linearly independent vectors in $a^1, \ldots, a^m$. That is, $\dim(P^*) = \text{rank}(A)$. Since $\text{lin}(P) = \{x \in \mathbb{R}^n : Ax = 0\}$, it follows that $\dim(\text{lin}(P)) = n - \text{rank}(A)$. Thus $\dim(P^*) = n - \dim(\text{lin}(P))$. Furthermore, $\text{aff}(P^*) = \langle a^1, \ldots, a^m \rangle$, thus $\text{aff}(P^*)$ is orthogonal to $\text{lin}(P)$.

□

3.17 Further Readings

There are many general references on polyhedral theory. Grünbaum [190] and Ziegler [357] are two excellent ones, with extensive treatments of the combinatorics and geometry of polytopes. Schrijver [325] gives a detailed account of topics related to optimization. Fukuda [153] provides a clear and accessible introduction.

As remarked in Sect. 3.1, Fourier's elimination procedure is not a polynomial algorithm for the problem of checking the feasibility of a system of linear inequalities. One of the most widely used algorithms to solve linear programs is the simplex method, which also has exponential-time worst-case complexity for most known pivoting rules (see, e.g., Klee and Minty [240]). However linear programs can be solved in polynomial time. The first polynomial-time algorithm to be discovered was the ellipsoid algorithm of Khachiyan [235], introduced earlier in the context of convex programming by Yudin and Nemirovski [356] and by Shor [332]. Maurras [272, 273] had previously described a polynomial algorithm for linear programs with totally unimodular constraint matrices. While the ellipsoid method is not efficient in practice, interior point algorithms provide both polynomial-time worst-case complexity and excellent performance. The first interior point method was introduced by Karmarkar [229]. We refer the readers to the book "Geometric Algorithms and Combinatorial Optimization," by Grötschel, Lovász, Schrijver [188], for a treatment of ellipsoid methods, and to "Linear Programming: Foundations and Extensions" by Vanderbei [343] for interior point methods.

Free software packages are available for generating the vertices and extreme rays of a polyhedron described by inequalities or, vice versa, for generating the inequalities when the vertices and extreme rays are given. Porta and cdd are two such packages based on the double description method. Another approach, the reverse search method of Avis and Fukuda [20], is the basis of the package lrs. In general cdd and Porta tend to be efficient for highly degenerate inputs whereas lrs tends to be efficient for nondegenerate or slightly degenerate problems. We discuss briefly the double description method.

The Double Description Method

We say that a pair of matrices (A, R), where A is an $m \times n$ matrix and R an $n \times k$ matrix, is an MW-pair if $\{x \in \mathbb{R}^n : Ax \leq 0\} = \text{cone}(R)$.

The *double description method* [280] is an algorithm that, given an $m \times n$ matrix A, constructs a matrix R such that (A, R) is an MW-pair, thus providing a constructive proof of Theorem 3.11. It has a similar flavor to the Fourier elimination procedure, which should not come as a surprise due to the correspondence through polarity. We give a description of the double description method.

Let A_i be the submatrix of A containing the first i rows. Suppose R_i is a matrix such that (A_i, R_i) is an MW-pair and let $a^{i+1}x \leq 0$ be the inequality associated with the $(i+1)$st row of A. The double description method constructs a matrix R_{i+1} such that (A_{i+1}, R_{i+1}) is an MW-pair as follows.

Let r^j, $j \in J$, denote the columns of R_i, and consider the following partition of J:
$$J^+ = \{j \in J : a^{i+1}r^j > 0\}$$
$$J^0 = \{j \in J : a^{i+1}r^j = 0\}$$
$$J^- = \{j \in J : a^{i+1}r^j < 0\}$$

Note that r^j is in $\{x : A_{i+1}x \leq 0\}$ if and only if $j \in J^- \cup J^0$. Furthermore, for every $j \in J^-$, $k \in J^+$, the vector
$$r^{jk} = (a^{i+1}r^k)r^j - (a^{i+1}r^j)r^k \tag{3.10}$$
is the unique ray in $\text{cone}(r^j, r^k) \cap \{x : a^{i+1}x = 0\}$.

Let R_{i+1} be the matrix whose columns are the vectors r^j, $j \in J^0 \cup J^-$, and r^{jk}, $j \in J^-, k \in J^+$, defined in (3.10). Lemma 3.51 states that (A_{i+1}, R_{i+1}) is an MW-pair.

The double description method iterates this procedure until the MW-pair $A = (A_n, R_n)$ is constructed.

The algorithm must start with an MW-pair (A_1, R_1). Note that the matrix R_1 can be constructed by choosing the vector $-a^1$ together with a basis for the subspace $\{x : a^1 x = 0\}$.

Lemma 3.51. *Let (A_i, R_i) be an MW-pair and let A_{i+1} be obtained from A_i by adding row a^{i+1}. Let R_{i+1} be the matrix constructed by the double description method. Then (A_{i+1}, R_{i+1}) is an MW-pair.*

Proof. Let $C := \{x : A_{i+1}x \le 0\}$ and $Q := \text{cone}(R_{i+1})$. By construction, every column r of R_{i+1} satisfies $A_{i+1}r \le 0$, therefore $Q \subseteq C$.

To prove $Q \supseteq C$, consider $\bar{x} \in C$. We need to prove that $\bar{x} \in \text{cone}(R_{i+1})$, that is, there exists $\mu \ge 0$ such that $R_{i+1}\mu = \bar{x}$.

Since $A_i \bar{x} \le 0$ and (A_i, R_i) is an MW-pair, it follows that $\bar{x} \in \text{cone}(R_i)$, thus there exists $\bar{\mu} \ge 0$ such that $R_i \bar{\mu} = \bar{x}$. If $\bar{\mu}_k = 0$ for every $k \in J^+$ then $\bar{x} \in \text{cone}(R_{i+1})$ and we are done. Otherwise, let $k \in J^+$ such that $\bar{\mu}_k > 0$. Since $a^{i+1}\bar{x} \le 0$, there must exist an index $j \in J^-$ such that $\bar{\mu}_j > 0$. By construction, the vector r^{jk} defined in (3.10) is a column of R_{i+1}. Let $\tilde{\mu}$ be obtained from $\bar{\mu}$ by setting $\tilde{\mu}_j := \bar{\mu}_j - \alpha(a^{i+1}r^k)$ and $\tilde{\mu}_k := \bar{\mu}_k + \alpha(a^{i+1}r^j)$, where α is the largest number such that $\tilde{\mu}_j, \tilde{\mu}_k \ge 0$. It follows from (3.10) that $\bar{x} = R_i \tilde{\mu} + \alpha r^{jk}$ and that at least one among $\tilde{\mu}_j$ and $\tilde{\mu}_k$ is 0. That is, we obtain an expression of $\bar{x}$ as a conic combination of the vectors in R_i and of r^{jk}, except that one less vector in R_i appears in the combination. Repeat this process until no vector r^k, $k \in J^+$, appears in the combination. □

A constructive proof of Theorem 3.11. We have already shown in Sect. 3.5.1 how to use Fourier elimination to obtain from a matrix R a matrix A such that (A, R) in an MW-pair. The double description method, whose correctness has been proven in Lemma 3.51, shows constructively how to obtain from a matrix A a matrix R such that (A, R) in an MW-pair. □

Observe that at each iteration of the double description method, if R_i has q columns then the matrix R_{i+1} might have up to $(q/2)^2$ columns. So this method produces a matrix R_n with a finite set of columns, but potentially very large. In general not all the newly added columns are extreme rays of $\{x : A_{i+1}x \le 0\}$ and to control the process it may be desirable to eliminate

3.17. FURTHER READINGS

the ones that are not extreme. In general, testing if a ray is extreme is equivalent to testing feasibility of a system of inequalities and this can be done using linear programming. Even if at the end of each iteration only the extreme rays are kept, it is possible that the number of extreme rays that are computed by the algorithm is exponentially large, in terms of the size of the matrix A and the final number of extreme rays.

Vertex Enumeration, Polytope Verification

We describe fundamental complexity issues associated with the Minkowski–Weyl theorem. For simplicity, we state them for full-dimensional polytopes. We distinguish polytopes by their representation. A V-polytope is the convex hull of a finite set of points and an H-polytope is the intersection of a finite number of half spaces.

The *vertex enumeration* problem asks to enumerate all vertices of an H-polytope P_H. It is not known whether there is an algorithm to solve this problem whose complexity is polynomial in the size of the input plus the size of the output. (That is, the number of facets and vertices of P_H). Avis and Fukuda [20] give such an algorithm when P_H is nondegenerate, i.e., every vertex of P_H belongs to exactly n facets. Vertex enumeration is obviously equivalent to *facet enumeration*: Compute all the facets of a V-polytope. The method of Fourier or the double description method solve these problems, but they are not polynomial algorithms, as discussed.

Given a V-polytope P_V and an H-polytope P_H, *polytope verification* asks whether $P_V = P_H$. Checking if every vertex of P_V satisfies the inequality description of P_H settles the question $P_V \subseteq P_H$. Freund and Orlin [150] show that the question $P_H \subseteq P_V$ is co-NP-complete. (The problem is in co-NP because the "no" answer has a short certificate, namely a point v in P_H and a valid inequality for P_V that strictly separates v from P_V.) However the complexity status of the question $P_H \subseteq P_V$, assuming $P_V \subseteq P_H$, is not known. Hence the complexity status of polytope verification is also open. Is can be easily shown that if polytope verification can be solved in polynomial time, then vertex enumeration can also be solved in polynomial time. These questions and other computational issues for polyhedra are surveyed by Fukuda [152].

Khachiyan et al. [236] show that the following decision problem is NP-hard: "Given an H polyhedron P_H and a set V of vertices, does there exist a set R of rays such that $P_H = \text{conv}(V) + \text{cone}(R)$?"

3.18 Exercises

Exercise 3.1. Prove Theorem 3.6 using Theorem 3.4.

Exercise 3.2. Given an $m \times n$ matrix, show that the system $Ax < 0$ is feasible if and only if $uA = 0, u \geq 0, u\mathbf{1} = 1$ is infeasible.

Exercise 3.3. The *fractional knapsack problem* is the linear relaxation of the 0, 1 knapsack problem, i.e., $\max\{\sum_{j=1}^{n} c_j x_j : \sum_{j=1}^{n} a_j x_j \leq b, 0 \leq x \leq 1\}$ where $a_j > 0$, $j = 1, \ldots, n$, and $c \in \mathbb{R}_+^n$. Using complementary slackness, show that an optimal solution x^* of the fractional knapsack problem is the following: assuming w.l.o.g. that $\frac{c_1}{a_1} \geq \frac{c_2}{a_2} \geq \ldots \geq \frac{c_n}{a_n}$, let h be the largest index such that $\sum_{j=1}^{h} a_j \leq b$, set $x_j^* = 1$ for $j = 1, \ldots, h$, $x_{h+1}^* = \frac{b - \sum_{j=1}^{h} a_j}{a_{h+1}}$, $x_j^* = 0$ for $j = h+2, \ldots, n$.

Exercise 3.4. Let V be the set of vertices of a polytope P, let S be a subset of V with the property that, for every $v \in V \setminus S$, the set S contains all the vertices adjacent to v.

(i) Show that the following algorithm solves the problem $\min\{cx : x \in S\}$.

   Compute a vertex v of P minimizing $\{cx : x \in P\}$.

   If $v \in S$, return v.

   Otherwise return a vertex v' of P minimizing $\{cx : x \text{ is adjacent to } v\}$.

   [Hint: Use the simplex algorithm.]

(ii) Give a polynomial algorithm to optimize a linear function over the set of 0,1 vectors in $\mathbb{R}^n$ with an even number of 1's.

Exercise 3.5. Show that all bases of a linear space have the same cardinality.

Exercise 3.6. Using the definition of linear space given in Sect. 3.4, prove that:

(i) A subset $\mathcal{L}$ of $\mathbb{R}^n$ is a linear space if and only if $\mathcal{L} = \{x \in \mathbb{R}^n : Ax = 0\}$ for some matrix A.

(ii) If $\mathcal{L} = \{x \in \mathbb{R}^n : Ax = 0\}$, then the dimension of $\mathcal{L}$ is $n - \text{rank}(A)$.

Exercise 3.7. Let $x^0, x^1, \ldots, x^q \in \mathbb{R}^n$. Show that the following statements are equivalent.

(i) $x^0, x^1, \ldots, x^q$ are affinely independent;

(ii) $x^1 - x^0, \ldots, x^q - x^0$ are linearly independent;

(iii) The vectors $\binom{x^0}{1}, \ldots, \binom{x^q}{1} \in \mathbb{R}^{n+1}$ are linearly independent.

Exercise 3.8. Using the definition of affine space given in Sect. 3.4, prove that

(i) A set $\mathcal{A} \subseteq \mathbb{R}^n$ is an affine space if and only if there exists a linear space $\mathcal{L}$ such that for any $x^* \in \mathcal{A}$, $\mathcal{A} = \{x^*\} + \mathcal{L}$;

(ii) A set $\mathcal{A} \subseteq \mathbb{R}^n$ is an affine space if and only if there is a matrix A and a vector b such that $\mathcal{A} = \{x : Ax = b\}$. Furthermore $\mathcal{L}$ defined in (i) satisfies $\mathcal{L} := \{x \in \mathbb{R}^n : Ax = 0\}$ and $\dim(\mathcal{A}) = \dim(\mathcal{L})$

Exercise 3.9. The intersection of any family of convex sets is a convex set.

Exercise 3.10. Let $V, R \subseteq \mathbb{R}^n$. Show that $\text{conv}(V + R) = \text{conv}(V) + \text{conv}(R)$, i.e., the convex hull of a Minkowski sum is the Minkowski sum of the convex hulls.

Exercise 3.11. Given a polyhedron $P := \{x \in \mathbb{R}^n : a^i x \leq b^i, i \in I\}$, let $I' \subseteq I$ be the set of indices i such that $a^i r = 0$ for every $r \in \text{rec}(P)$. Show that the polyhedron $P + \text{rec}(-P)$ is described by the system $a^i x \leq b^i$, $i \in I'$.

Exercise 3.12. Given a nonempty polyhedron $P = \{x \in \mathbb{R}^n : Ax \leq b\}$, consider the polyhedron $Q = \{y \in \mathbb{R}^n : \exists u \geq 0, v \leq 0 : y = uA = vA, ub = vb\}$. Show that $\dim(P) + \dim(Q) = n$.

Exercise 3.13. Does Theorem 3.22 hold without the assumption $P \neq \emptyset$? Prove or give a counterexample.

Exercise 3.14. Let $P := \{x \in \mathbb{R}^n : Ax \leq b\}$ be a polyhedron such that $\dim(P) > \dim(\text{rec}(P))$. Show that an inequality $cx \leq \delta$ is valid for P if and only if there exists $u \geq 0$ such that $uA = c$ and $ub = \delta$.

Does the statement still hold if we replace the hypotheses that $\dim(P) > \dim(\text{rec}(P))$ by the hypothesis $P \neq \emptyset$?

Exercise 3.15. Let $P := \{x \in \mathbb{R}^n : Ax \geq b\}$ be a polyhedron. Let $j \in \{1, \ldots, n\}$, and suppose that $\Pi := P \cap \{x \in \mathbb{R}^n : x_j = 0\} \neq \emptyset$. Show that if $\alpha x \geq \beta$ is a valid inequality for Π, there exists a $\lambda \in \mathbb{R}$ such that $\alpha x + \lambda x_j \geq \beta$ is valid for P.

Exercise 3.16. Let $a^1 x \leq b_1$, $a^2 x \leq b_2$ be redundant inequalities for $Ax \leq b$. Let $A^2 x \leq b^2$ be the system obtained from $Ax \leq b$ by removing the second inequality. Is $a^1 x \leq b_1$ always a redundant inequality for $A^2 x \leq b^2$?

Exercise 3.17. Prove Corollary 3.23.

Exercise 3.18. Let P be a nonempty affine space, and $cx \leq \delta$ be a valid inequality for P. Show that either $cx = \delta$ for every $x \in P$, or $cx < \delta$ for every $x \in P$.

Exercise 3.19. Let $n \geq 2$ be an integer. Find the dimension of the polytope $\{x \in [0,1]^n : \sum_{j \in J} x_j + \sum_{j \notin J}(1 - x_j) \geq \frac{n}{2}, \text{ for all } J \subseteq \{1, 2, \cdots, n\}\}$.

Exercise 3.20. Show that the cut polytope, defined in Example 3.36, is full dimensional.

Exercise 3.21. List all the facets of the assignment polytope
$$\{x \in \mathbb{R}^{n^2} : \begin{array}{ll} \sum_{j=1}^n x_{ij} = 1 & \text{for all } i = 1, \ldots n \\ \sum_{i=1}^n x_{ij} = 1 & \text{for all } j = 1, \ldots n \\ x_{ij} \geq 0 & \text{for all } i, j = 1, \ldots n\}. \end{array}$$
(Hint: Distinguish the cases $n = 1$, $n = 2$ and $n \geq 3$.)

Exercise 3.22. Let $n \geq 3$ and $m \geq 2$ be two integers. The simple plant location polytope is the convex hull of the points $(x, y) \in \{0, 1\}^{m \times n} \times \{0, 1\}^n$ satisfying

$$\sum_{j=1}^n x_{ij} = 1 \quad i = 1, \ldots, m$$
$$0 \leq x_{ij} \leq y_j \leq 1 \quad i = 1, \ldots, m, \ j = 1, \ldots, n$$

(i) Find the dimension of the simple plant location polytope.

(ii) Show that $x_{ij} \geq 0$ defines a facet of the simple plant location polytope for all $i = 1, \ldots, m$ and $j = 1, \ldots, n$.

(iii) Show that $y_j \leq 1$ defines a facet of the simple plant location polytope for all $j = 1, \ldots, n$.

(iv) Show that $x_{ij} \leq y_j$ defines a facet of the simple plant location polytope for all $i = 1, \ldots, m$ and $j = 1, \ldots, n$.

Exercise 3.23. A nonempty polytope P is *simple* if every vertex of P belongs to exactly $\dim(P)$ facets of P. Prove that the permutahedron Π_n is a simple polytope.

Exercise 3.24. Using a polynomial-time algorithm for linear programming as subroutine, give a polynomial-time algorithm to solve the following problem. Given two systems of linear inequalities $Ax \leq b$, $Cx \leq d$ in n variables, decide whether $\{x \in \mathbb{R}^n : Ax \leq b\} = \{x \in \mathbb{R}^n : Cx \leq d\}$.

3.18. EXERCISES

Exercise 3.25. Let $S \subseteq \{0,1\}^n$, and let $P = \text{conv}(S)$. Show that S is the set of vertices of P.

Exercise 3.26. Show that a closed convex set $S \subseteq \mathbb{R}^n$ is bounded if and only if S contains no ray.

Exercise 3.27. Prove that, given a pointed polyhedron P and extreme ray r of P, there exists an edge of P of the form $\{\bar{x} + \lambda r : \lambda \geq 0\}$ for some vertex $\bar{x}$ of P.

Exercise 3.28. Prove that two distinct vertices v, w of a polyhedron P are adjacent if and only if the midpoint $\frac{1}{2}(v+w)$ is not a convex combination of two points of P outside the line segment joining v and w.

Exercise 3.29. Consider the polyhedron $P = \{(x_0, \ldots, x_n) \in \mathbb{R}^{n+1} : x_0 + x_t \geq b_t,\ t = 1, \ldots, n,\ x_0 \geq 0\}$. Show that P is pointed, $(0, b_1, \ldots, b_n)$ is the unique vertex of P, and that the extreme rays of P are precisely the $n+1$ vectors $r^0, \ldots, r^n$ defined by:

$$r_i^0 = \begin{cases} 1 & \text{for } i = 0 \\ -1 & \text{otherwise} \end{cases} \qquad r_i^t = \begin{cases} 1 & \text{for } i = t \\ 0 & \text{otherwise} \end{cases} \quad t = 1, \ldots, n.$$

Exercise 3.30. Prove Theorem 3.43. Hint: Since $S \subseteq \mathbb{R}^d$ and $|S| \geq d+2$, the points in S are affinely dependent. Use this fact to find the required partition into S_1 and S_2.

Exercise 3.31. Prove Theorem 3.44. Hint: Use induction on h. If $h = d+1$, the theorem is trivial. Assume that the theorem is true for $h < k (\geq d+2)$ and prove that the theorem holds for $h = k$ as follows: Assume by contradiction that for all $j = 1, \ldots h$, $S_j := \bigcap_{i \neq j} C_i \neq \emptyset$. Use the fact that $h \geq d+2$ and apply Radon's Theorem 3.43.

Exercise 3.32. Show that the convexity assumption in Theorem 3.44 is necessary.

Exercise 3.33. Consider the octahedron oct_n defined in Example 3.45. Show that:

(i) All the 2^n inequalities $\sum_{i \in S} x_i - \sum_{i \notin S} x_i \leq 1$, $S \subseteq \{1, \ldots, n\}$, define facets of oct_n.

(ii) oct_n has $2n$ vertices, namely the unit vectors e^i and their negatives $-e^i$, $i = 1, \ldots, n$.

Define $P = \{(x, z) \in \mathbb{R}^n \times \mathbb{R}^{2n} : x = \sum_{i=1}^n z_i e^i - \sum_{i=1}^n z_{n+i} e^i, \sum_{i=1}^{2n} z_i = 1, z_i \geq 0\}$. Show that

(iii) $\text{proj}_x(P) = \text{oct}_n$.

(iv) $\dim(P) = 2n - 1$.

(v) P has exactly $2n$ facets, defined by the inequalities $z_i \geq 0$ for $i = 1, \ldots, 2n$.

Exercise 3.34. Show that the operations of taking the projection and taking the convex hull commute. Specifically, for $S \subseteq \mathbb{R}^{n+p}$, define as usual $\text{proj}_x(S) := \{x \in \mathbb{R}^n : \exists y \in \mathbb{R}^p \ (x, y) \in S\}$. Show that $\text{proj}_x(\text{conv}(S)) = \text{conv}(\text{proj}_x(S))$.

Exercise 3.35. Consider a polyhedron $P := \{(x, z) \in \mathbb{R}^n \times \mathbb{R}^p : Ax + Bz \leq b\}$. Show that if its projection cone C_P is equal to $\{0\}$, then $\text{proj}_x(P) = \mathbb{R}^n$.

Exercise 3.36. Prove Corollary 3.47.

Exercise 3.37. Let $P \in \mathbb{R}^n$ be a full-dimensional pointed polyhedron, and let $v^1, \ldots, v^p$ and $r^1, \ldots, r^q$ be its vertices and extreme rays, respectively. Then P is the orthogonal projection onto the space of the x variables of the polyhedron $Q = \{(x, \lambda, \mu) \in \mathbb{R}^m \times \mathbb{R}_+^p \times \mathbb{R}_+^q : x - V\lambda - R\mu = 0, \mathbf{1}\lambda = 1\}$, where V is the matrix with columns $v^1, \ldots, v^p$ and R is the matrix with columns $r^1, \ldots, r^q$, while $\mathbf{1}$ is the vector of all ones in $\mathbb{R}^p$. Show that an inequality $\alpha x \leq \beta$ is facet-defining for P if and only if (α, β) is an extreme ray of the pointed cone $C = \{(\alpha, \beta) \in \mathbb{R}^{n+1} : \alpha V - \mathbf{1}\beta \leq 0, \alpha R \leq 0\}$.

Chapter 4

Perfect Formulations

A *perfect formulation* of a set $S \subseteq \mathbb{R}^n$ is a linear system of inequalities $Ax \leq b$ such that $\text{conv}(S) = \{x \in \mathbb{R}^n : Ax \leq b\}$. For example, Proposition 1.5 gives a perfect formulation of a 2-variable mixed integer linear set. When a perfect formulation is available for a mixed integer linear set, the corresponding integer program can be solved as a linear program. In this chapter, we present several classes of integer programming problems for which a perfect formulation is known. For pure integer linear sets, a classical case is when the constraint matrix is totally unimodular. Important combinatorial problems on directed or undirected graphs such as network flows and matchings in bipartite graphs have a totally unimodular constraint matrix. We also give a perfect formulation for nonbipartite matchings and spanning trees. It is often the case that perfect formulations go together with polynomial-time algorithms to solve the associated integer programs. In particular, we will describe polynomial algorithms for maximum flow, matchings and minimum cost spanning trees. Perfect formulations can be obtained from linear systems of inequalities that are total dual integral. We give the example of the submodular polyhedron. We show a fundamental result of Meyer, which states that every mixed integer linear set has a perfect formulation when the data are rational. We then prove a theorem of Balas on the union of polyhedra. For selected mixed integer linear sets, this theorem can lead to perfect formulations, albeit ones that use additional variables. More generally, we discuss extended formulations and a theorem of Yannakakis on the smallest size of an extended formulation.

4.1 Properties of Integral Polyhedra

A convex set $P \subseteq \mathbb{R}^n$ is *integral* if $P = \text{conv}(P \cap \mathbb{Z}^n)$.

Theorem 4.1. *Let P be a rational polyhedron. The following conditions are equivalent.*

(i) *P is an integral polyhedron.*

(ii) *Every minimal face of P contains an integral point.*

(iii) *$\max\{cx : x \in P\}$ is attained by an integral vector x for each $c \in \mathbb{R}^n$ for which the maximum is finite.*

(iv) *$\max\{cx : x \in P\}$ is integer for each integral c for which the maximum is finite.*

Proof. (i)$\Rightarrow$(ii) Since P is integral, every $x \in P$ is a convex combination of integral points in P. Given a minimal face F of P, let $x \in F$. Then $x = \sum_{j=1}^{k} \lambda_j x^j$, where $x^1, \ldots, x^k \in P \cap \mathbb{Z}^n$, $\lambda_j > 0$, $j = 1, \ldots, k$, $\sum_{j=1}^{k} \lambda_j = 1$. This implies that $x^1, \ldots, x^k$ are integral points in F.
The implication (ii)$\Rightarrow$(iii) follows from the definition of a face. The implication (iii)$\Rightarrow$(iv) is immediate.

(iv)$\Rightarrow$(ii) We prove the result by contradiction. Let $P := \{x \in \mathbb{R}^n : Ax \leq b\}$ where $A \in \mathbb{Z}^{m \times n}$ and $b \in \mathbb{Z}^m$. Suppose that a minimal face F does not contain an integral point. By Theorem 3.33, there is a subsystem $A^F x \leq b^F$ of $Ax \leq b$ such that $F = \{x \in \mathbb{R}^n : A^F x = b^F\}$. Since the system $A^F x = b^F$ has no integral solution, it follows from Theorem 1.20 that there exists $u \in \mathbb{R}^m$ such that $c := uA$ is integral, $z := ub$ is fractional, and $u_i = 0$ for those rows of $Ax \leq b$ that are not in $A^F x \leq b^F$. Since A, b are integral, we may assume that $u \geq 0$ (possibly by replacing u with $u - \lfloor u \rfloor$).

For $x \in P$, $cx = uAx \leq ub = z$ where the inequality holds as an equality for $x \in F$. Therefore $\max\{cx : x \in P\} = z$, which is fractional. This contradicts (iv).

(ii)$\Rightarrow$(i) Let $F_1, \ldots, F_p$ be the minimal faces of P, and let v^i be an integral point in F_i, $i = 1, \ldots, p$. By Theorem 3.37, $P = \text{conv}(v^1, \ldots, v^p) + \text{cone}(r^1, \ldots, r^q)$, where $r^1, \ldots, r^q$ are generators of $\text{rec}(P)$. Since $\text{rec}(P)$ is a rational cone, by Proposition 3.12 we can choose $r^1, \ldots, r^q$ to be rational, and so in particular we can assume that $r^1, \ldots, r^q$ are integral. Therefore $\text{conv}(P \cap \mathbb{Z}^n) \supseteq \text{conv}(v^1, \ldots, v^p) + \text{cone}(r^1, \ldots, r^q) = P$. □

Corollary 4.2. *A rational polyhedron P is integral if and only if every rational supporting hyperplane for P contains an integral point.*

Proof. "$\Longrightarrow$" Assume P is an integral polyhedron and let H be a supporting hyperplane. Then $H \cap P$ contains a minimal face F of P. Since P is integral, by Theorem 4.1(ii) F contains an integral point.
"$\Longleftarrow$" Assume that $P \subseteq \mathbb{R}^n$ is a rational polyhedron and every rational supporting hyperplane of P contains an integral point. Then $\max\{cx : x \in P\}$ is an integer for every $c \in \mathbb{Z}^n$ for which the maximum is finite. By Theorem 4.1, P is an integral polyhedron. □

Theorem 4.1 can be extended, in part, to the mixed integer case.

Theorem 4.3. *Let $P \subseteq \mathbb{R}^n \times \mathbb{R}^p$ be a rational polyhedron, and let $S := P \cap (\mathbb{Z}^n \times \mathbb{R}^p)$. The following are equivalent.*

(i) $P = \mathrm{conv}(S)$.

(ii) Every minimal face of P contains a point in $\mathbb{Z}^n \times \mathbb{R}^p$.

(iii) $\max\{cx + hy : (x, y) \in P\}$ is attained by a point in $\mathbb{Z}^n \times \mathbb{R}^p$ for each $(c, h) \in \mathbb{R}^n \times \mathbb{R}^p$ for which the maximum is finite.

The proof of this theorem is similar to the proof of Theorem 4.1 (we leave it to the reader to check this). Corollary 4.2 does not extend to the mixed integer case (see Exercise 4.3).

4.2 Total Unimodularity

In this section we address the following question:

Which integral matrices A have the property that the polyhedron $\{x : Ax \leq b, x \geq 0\}$ is integral for every integral vector b?

A matrix A is *totally unimodular* if every square submatrix has determinant $0, \pm 1$. It follows from the definition that a totally unimodular matrix has all entries equal to $0, \pm 1$. For example, the matrix

$$\begin{pmatrix} -1 & 1 & 0 \\ 1 & 0 & 1 \\ 0 & 1 & 1 \end{pmatrix}$$

is totally unimodular because its determinant is 0 and all its proper square submatrices are triangular after permutation of rows and columns, and thus they have determinant equal to $0, \pm 1$.

Theorem 4.4 (Hoffman and Kruskal). *Let A be an $m \times n$ integral matrix. The polyhedron $\{x : Ax \leq b, x \geq 0\}$ is integral for every $b \in \mathbb{Z}^m$ if and only if A is totally unimodular.*

Proof. For all $b \in \mathbb{Z}^m$, let $P(b) := \{x : Ax \leq b, x \geq 0\}$. Since $P(b)$ is a rational pointed polyhedron for all $b \in \mathbb{Z}^m$, by Theorem 4.1 $P(b)$ is an integral polyhedron if and only if all its vertices are integral.

We first prove the "if" part of the theorem. Assume that A is a totally unimodular matrix, let $b \in \mathbb{Z}^m$ and $\bar{x}$ be a vertex of $P(b)$. We need to show that $\bar{x}$ is integral. By Theorem 3.34, $\bar{x}$ satisfies at equality n linearly independent inequalities of $Ax \leq b$, $x \geq 0$. Let $Bx = d$ be a system comprised of n such equations. Then $\bar{x} = B^{-1}d$. Up to permuting columns, we may assume that B is a matrix of the form

$$B = \begin{pmatrix} C & D \\ 0 & I \end{pmatrix}. \tag{4.1}$$

In particular C is a square nonsingular submatrix of A and $\det(B) = \det(C)$. Since A is totally unimodular, $\det(B) = \det(C) = \pm 1$. By Cramer's rule, B^{-1} is equal to the adjugate matrix of B divided by $\det(B)$. Since B is an integral matrix, its adjugate is integral as well, therefore B^{-1} is also integral. It follows that $\bar{x}$ is integral because d is an integral vector.

We show the "only if" part of the theorem. Suppose A is not totally unimodular and let C be a square submatrix of A with $\det(C) \neq 0, \pm 1$. Up to permuting rows and columns, we may assume that C is indexed by the first k rows and columns of A. Since $\det(C^{-1}) = 1/\det(C)$ is not an integer, C^{-1} is not an integral matrix. Let γ be a column of C^{-1} with a fractional entry. Define $\bar{x} \in \mathbb{R}^n$ by $\bar{x}_j = \lceil \gamma_j \rceil - \gamma_j$ for $j = 1, \ldots, k$, $\bar{x}_j = 0$ for $j = k+1, \ldots, n$. Let $b := \lceil A\bar{x} \rceil$. Clearly b is an integral vector and $\bar{x} \in P(b)$. Observe that $\bar{x}$ satisfies at equality the first k inequalities of $Ax \leq b$, $x \geq 0$. Indeed, if we let (A', b') be the matrix comprised of the first k rows of (A, b), by construction $A'\bar{x} = C\lceil \gamma \rceil - C\gamma = b'$, where the last equality follows from the fact that $C\gamma$ is integral because it is a unit vector, and from the fact that $b' = \lceil A'\bar{x} \rceil$. The point $\bar{x}$ satisfies at equality also the $n - k$ nonnegativity constraints $x_j \geq 0$, $j = k+1, \ldots, n$. The constraint matrix B relative to these n constraints has the form (4.1), therefore B is nonsingular because $\det(B) = \det(C)$. In particular $\bar{x}$ is a vertex of $P(b)$ but $\bar{x} \notin \mathbb{Z}^n$. This shows that $P(b)$ is not an integral polyhedron. $\square$

Exercise 4.5 shows that the Hoffman–Kruskal theorem can be restated in the following form.

4.2. TOTAL UNIMODULARITY

Theorem 4.5. *Let A be an $m \times n$ integral matrix. The polyhedron $Q := \{x : c \leq Ax \leq d, \, l \leq x \leq u\}$ is integral for all integral vectors c, d, l, u if and only if A is totally unimodular.*

Ghouila-Houri [165] gives a useful characterization of total unimodularity. An *equitable bicoloring* of a matrix A is a partition of its columns into two sets (one of which may possibly be empty), say "red" and "blue" columns, such that the sum of the red columns minus the sum of the blue columns is a vector whose entries are $0, \pm 1$.

Theorem 4.6. *A matrix A is totally unimodular if and only if every column submatrix of A admits an equitable bicoloring.*

Proof. For the "only if" part, let A be a totally unimodular matrix and let B be a column submatrix of A. Let P_B be the polytope described by the system

$$\lfloor \tfrac{1}{2} B \mathbf{1} \rfloor \leq Bx \leq \lceil \tfrac{1}{2} B \mathbf{1} \rceil$$
$$0 \leq x \leq 1$$

where $\mathbf{1}$ denotes the vector of all ones. The polytope P_B is nonempty, since it contains the point $\tfrac{1}{2}\mathbf{1}$, and by Theorem 4.5, P_B is integral. Therefore P_B contains a $0, 1$ vector $\bar{x}$. An equitable bicoloring of B is obtained by painting red the columns j where $\bar{x}_j = 1$ and blue the columns j where $\bar{x}_j = 0$.

For the "if" part, let us assume that every column submatrix of A admits an equitable bicoloring. We prove that the determinant of every $k \times k$ submatrix of A is $0, \pm 1$. We proceed by induction on k. The fact that every matrix consisting of only one column of A admits an equitable bicoloring implies that the entries of A are $0, \pm 1$; this establishes the base case $k = 1$. We assume by induction that the determinant of every $k \times k$ submatrix of A is $0, \pm 1$. Let B be a $(k+1) \times (k+1)$ submatrix of A, and let $\delta := \det(B)$. We need to show that $\delta = 0, \pm 1$. We assume that $\delta \neq 0$, otherwise we are done. Since the determinant of every $k \times k$ submatrix of B is $0, \pm 1$, by Cramer's rule the entries of the matrix δB^{-1} are $0, \pm 1$. Let d be the first column of δB^{-1}. Then $Bd = \delta e^1$. Let B^* be the submatrix of B containing the columns j for which $d_j = \pm 1$. Since $Bd = \delta e^1$, the sum of the elements in rows $2, \ldots, k+1$ of B^* is even. Assume first that δ is even. Then the sum of the elements in every row of B^* is even. Since B^* is bicolorable, the columns of B^* are linearly dependent because the 0-vector can be obtained by summing the red columns and subtracting the sum of the blue columns of B^*. Hence B is a singular matrix and $\delta = 0$. Assume now that δ is odd.

The bicoloring assumption applied to B^* shows that there is a $0, \pm 1$ vector $\bar{x}$ (which has the same support as d) such that $B\bar{x} = e^1$. Since $Bd = \delta e^1$ and B is nonsingular, we have that $\delta \bar{x} = d$. Since d and $\bar{x}$ are $0, \pm 1$-vectors, this shows that $\delta = \pm 1$. □

We will also use the above theorem in its transposed form. An *equitable row-bicoloring* of a matrix A is a partition of its rows into two sets, red and blue, such that the sum of the red rows minus the sum of the blue rows is a vector whose entries are $0, \pm 1$.

Corollary 4.7. *A matrix A is totally unimodular if and only if every row submatrix of A admits an equitable row-bicoloring.*

When A has at most two nonzero entries per column, Corollary 4.7 can be simplified to the following result of Heller and Tompkins [199].

Corollary 4.8. *A $0, \pm 1$ matrix A with at most two nonzero elements in each column is totally unimodular if and only if A admits an equitable row-bicoloring.*

This is because, when A has at most two nonzero entries per column, an equitable row-bicoloring of A trivially induces one for any row submatrix.

4.3 Networks

In this section we relate total unimodularity to several well-known combinatorial optimization problems on graphs.

Given a digraph $D = (V, A)$, an arc e of D has a *tail* u and a *head* v with $u \neq v$, and it is denoted by uv. The *incidence matrix* A_D of a digraph $D = (V, A)$ is the $|V| \times |A|$ matrix whose rows correspond to the nodes of D, whose columns correspond to the arcs of D, and whose entries are, for every node w and arc $e = uv$,

$$a_{we} = \begin{cases} -1 & \text{if } w = u, \\ 1 & \text{if } w = v, \\ 0 & \text{if } w \neq u, v. \end{cases}$$

Note that the incidence matrices of digraphs are those $0, \pm 1$ matrices with exactly one $+1$ and one -1 in each column. In particular, the rows of the incidence matrix of a digraph are not linearly independent, since they sum to the zero vector (see Exercise 4.12 for a precise characterization of the rank of incidence matrices).

4.3. NETWORKS

Theorem 4.9. *Incidence matrices of digraphs are totally unimodular.*

Proof. Since the incidence matrix of a digraph has two nonzero entries in each column, the statement follows from Corollary 4.8 by coloring all rows the same color. □

This theorem implies that several classical network problems can be formulated as linear programs. We will give examples in the remainder of this section. In each case there is a faster "ad-hoc" algorithm than solving the straightforward linear program.

4.3.1 Circulations

Let $D = (V, A)$ be a digraph. A vector $x \in \mathbb{R}^{|A|}$ that satisfies $A_D x = 0$, $x \geq 0$ is called a *circulation*. The set of circulations of D is a polyhedral cone, called the *circulation cone*.

A *circuit* C in D is a set of arcs such that the digraph induced by C is connected and, for every node $v \in V$, the number of arcs of C entering v equals the number of arc of C leaving v. A circuit C is *simple* if every node of the digraph induced by C has exactly one arc of C entering and one leaving. Note that every circuit is the disjoint union of simple circuits. It is easy to show that the characteristic vector of a simple circuit is an extreme ray of the circulation cone (Exercise 4.13). We prove that the converse is also true.

Lemma 4.10. *Let D be a digraph. The extreme rays of the circulation cone are the characteristic vectors of the simple circuits of D.*

Proof. Let $\bar{x}$ be an extreme ray of $\{x \in \mathbb{R}^A : A_D x = 0, \ x \geq 0\}$. Possibly by multiplying $\bar{x}$ by a positive scalar, we may assume that $\bar{x} \leq 1$ and that $\bar{x}$ is a vertex of the polytope $Q := \{x \in \mathbb{R}^A : A_D x = 0, \ 0 \leq x \leq 1\}$. Since A_D is totally unimodular by Theorem 4.9, it follows from Theorem 4.5 that $\bar{x}$ is a $0, 1$ vector. Let $C := \{e \in A : \bar{x}_e = 1\}$. Then $\bar{x}$ is the characteristic vector of C. Since $A_D \bar{x} = 0$, it follows that the number of arcs in C entering any node of D is equal to the number of arcs leaving the node. Therefore the connected components of C are circuits. In particular, C is the disjoint union of simple circuits $C_1, \ldots, C_k$, hence $\bar{x}$ is the sum of the characteristic vectors of $C_1, \ldots, C_k$. Since $\bar{x}$ is an extreme ray of the circulation cone, it follows that C is a simple circuit. □

Application: Comparing Formulations for the Traveling Salesman Problem

In Sect. 2.7 we described several formulations for the traveling salesman problem; the one proposed by Dantzig et al. [103] in 1954 is based on subtour elimination constraints; the one proposed by Miller et al. [278] in 1960 is more compact but involves additional variables. How do these two formulations compare? To answer this question, we will apply what we learned about circulations. Since the Miller–Tucker–Zemlin formulation involves more variables, in order to compare it with the subtour formulation we will need to project out these extra variables using Theorem 3.46.

Consider a digraph $D = (V, A)$ with costs c_a, $a \in A$, on the arcs. Assume $V = \{1, \ldots, n\}$. Recall from Sect. 2.7 that both formulations have binary variables x_{ij} for every arc $ij \in A$, that must satisfy the constraints

$$\sum_{a \in \delta^+(i)} x_a = 1 \quad \text{for all } i \in V$$
$$\sum_{a \in \delta^-(i)} x_a = 1 \quad \text{for all } i \in V \quad (4.2)$$
$$x_a \geq 0 \quad \text{for all } a \in A.$$

These constraints ensure that every node has exactly one arc of the tour entering and one arc leaving, but additional constraints are needed to exclude subtours. In the Dantzig–Fulkerson–Johnson formulation this is achieved using exponentially many constraints

$$\sum_{a \in \delta^+(S)} x_a \geq 1 \quad \text{for } \emptyset \subset S \subset V. \quad (4.3)$$

Summing the equations $\sum_{a \in \delta^+(i)} x_a = 1$ from (4.2) for all $i \in S$ and subtracting (4.3), we get an equivalent form of (4.3):

$$\sum_{i,j \in S \,:\, ij \in A} x_{ij} \leq |S| - 1 \quad \text{for } \emptyset \subset S \subset V. \quad (4.4)$$

In the Miller–Tucker–Zemlin formulation, instead of (4.4), we have extra variables u_i that represent the position of node $i \geq 2$ in the tour, and we have the following constraints

$$u_i - u_j + 1 \leq n(1 - x_{ij}) \quad \text{for all } ij \in A \text{ such that } i, j \neq 1. \quad (4.5)$$

In order to compare the two formulations, define

$$P_{MTZ} := \{(x, u) \in \mathbb{R}^A \times \mathbb{R}^{V \setminus \{1\}} : (x, u) \text{ satisfies } (4.2), (4.5)\},$$
$$P_{subtour} := \{x \in \mathbb{R}^A : \text{ satisfies } (4.2), (4.4)\}.$$

We project P_{MTZ} onto the x-space to be able to compare it with $P_{subtour}$.

4.3. NETWORKS

Theorem 4.11. $P_{subtour} \subset \text{proj}_x(P_{MTZ})$.

Proof. By Theorem 3.46 we have that

$$\text{proj}_x(P_{MTZ}) = \{x \in \mathbb{R}^A : x \text{ satisfies } (4.2) \text{ and } v((n-1)\mathbf{1} - nx) \geq 0 \text{ for all } v \in Q\}$$

where $Q := \{v \geq 0 : \sum_{j \neq 1: ij \in A} v_{ij} - \sum_{j \neq 1: ji \in A} v_{ji} = 0 \text{ for all } i \in V \setminus \{1\}\}$ is the projection cone.

Observe that the polyhedral cone Q is the set of circulations in $D \setminus \{1\}$. By Lemma 4.10, the extreme rays of Q are the characteristic vectors of the simple circuits of $D \setminus \{1\}$. Then, if we denote by $\mathcal{C}$ the set of simple circuits of $D \setminus \{1\}$,

$$\text{proj}_x(P_{MTZ}) = \{x \in \mathbb{R}^A : x \text{ satisfies } (4.2) \text{ and } (n-1)|C| - n\sum_{ij \in C} x_{ij} \geq 0 \text{ for all } C \in \mathcal{C}\}.$$

The inequality $(n-1)|C| - n\sum_{ij \in C} x_{ij} \geq 0$ can be rewritten as

$$\sum_{ij \in C} x_{ij} \leq |C| - \frac{|C|}{n}.$$

This inequality is much weaker than the inequality (4.4) for the set S of nodes in the simple circuit C, since the latter has more terms on the left-hand side as well as a smaller right-hand side. □

It is therefore not surprising that the Miller–Tucker–Zemlin formulation is effective only for small values of n. For large instances, tighter formulations such as $P_{subtour}$ are needed. We will come back to the solution of traveling salesman problems in Chap. 7.

4.3.2 Shortest Paths

Consider a digraph $D = (V, A)$ with lengths ℓ_e on its arcs $e \in A$, and two distinct nodes $s, t \in V$. A *directed s,t-path* is a minimal set of arcs of the form $sv_1, v_1v_2, \ldots, v_k t$. In this section we will simply write "s,t-path" for "directed s,t-path." The *length* of an s,t path is $\ell(P) := \sum_{e \in P} \ell_e$. The *shortest path problem* consists in finding an s,t-path P of minimum length $\ell(P)$.

Let us assume for simplicity that there exists at least one s,t-path in D. The problem can be formulated as follows. We introduce a binary variable x_e for each $e \in A$, where $x_e = 1$ if e is in the path P, 0 otherwise. For every node $v \neq s, t$, the number of arcs of P entering v (none or one) equals the number of arcs of P leaving v. Furthermore, there must be exactly one arc leaving s and one arc entering t in P. These conditions can be expressed by the equations

$$\sum_{e \in \delta^-(v)} x_e - \sum_{e \in \delta^+(v)} x_e = 0 \quad v \in V \setminus \{s,t\}$$
$$\sum_{e \in \delta^-(s)} x_e - \sum_{e \in \delta^+(s)} x_e = -1 \qquad (4.6)$$
$$\sum_{e \in \delta^-(t)} x_e - \sum_{e \in \delta^+(t)} x_e = 1.$$

The above system can be written as $A_D x = b$ where $b \in \mathbb{R}^V$ is the vector with zero in all components except $b_s = -1$, $b_t = 1$. We formulate the shortest path problem as follows.

$$\begin{aligned} \min \quad & \ell x \\ & A_D x = b \\ & x \geq 0. \end{aligned} \qquad (4.7)$$

Since A_D is totally unimodular, the polyhedron $Q := \{x \in \mathbb{R}^A : A_D x = b,\ x \geq 0\}$ is integral. The recession cone of Q is $\operatorname{rec}(Q) = \{x \in \mathbb{R}^A : A_D x = 0,\ x \geq 0\}$, which is the set of circulations of D. By Proposition 4.10, the extreme rays of Q are the characteristic vectors of the simple circuits of D. Since Q is an integral polyhedron and the extreme rays of Q are the characteristic vectors of the simple circuits of D, one can readily verify that the vertices of Q are the characteristic vectors of the s,t-paths. We say that a circuit C of D is a *negative-length circuit* if $\ell(C) < 0$.

Remark 4.12. *If D contains a negative-length circuit, then (4.7) is unbounded. If D does not contain a negative-length circuit, then (4.7) has an optimal solution, and any basic optimal solution is the characteristic vector of a shortest s,t-path.*

Thus, when D has no negative-length circuit, we can determine in polynomial time a shortest s,t-path by solving (4.7). On the other hand, for general objective functions the shortest s,t-path problem is NP-hard (see [158]). There are several specialized algorithms to solve the shortest path problem in graphs without negative-length circuits.

Here we explain the *algorithm of Bellman–Ford*, which computes the lengths of the shortest paths from s to all other nodes in V. The algorithm is based on the following simple observation.

Observation 4.13. *Given $k \in \{1, \ldots, n-1\}$ and $v \in V$, let P be an s,v-path such that $|P| \leq k$ and $\ell(P)$ is minimum among all s,v-paths of cardinality at most k. Let uv be the last arc in P and let P' be the s,u-path contained in P. If D has no negative-length circuit, then P' has minimum length $\ell(P')$ among all s,u-paths of cardinality at most $k-1$.*

4.3. NETWORKS

Proof. Suppose that there exists an s,u-path R such that $|R| \leq k-1$ and $\ell(R) < \ell(P')$. If v is not in R, then $R' = R \cup uv$ is an s,v-path, $|R| \leq k$ and $\ell(R') < \ell(P)$, a contradiction. Hence v must be in R. Let T and T' be, respectively, the s,v-path and the v,u-path contained in R. Note that $C := T' \cup uv$ is a simple circuit. Since T is an s,v-path and $|T| \leq k$, it follows by the choice of P that $\ell(P) \leq \ell(T)$. Hence $\ell(C) = \ell(T') + \ell_{uv} = \ell(R) - \ell(T) + \ell_{uv} < \ell(P') - \ell(P) + \ell_{uv} = 0$. Thus C is a negative-length circuit. □

Assume D does not contain any negative-length circuit. For $k = 0, \ldots, n-1$ and every $v \in V$, let $d_k(v)$ be the minimum length of an s,v-path of cardinality at most k (where $d_k(v) = \infty$ if there is no such path). Clearly $d_0(s) = 0$ and $d_0(v) = \infty$ for all $v \in V \setminus \{s\}$. By Observation 4.13, for all $v \in V$ and for $k = 1, \ldots, n-1$ we have the following recursion:

$$d_k(v) = \min\{d_{k-1}(v), \min_{uv \in A}(d_{k-1}(u) + \ell_{uv})\}.$$

Since any path in D has cardinality at most $n-1$, the length of a shortest s,v-path, $v \in V$, is $d_{n-1}(v)$. By the above recursion, these numbers can be computed in time $O(|V||A|)$.

We conclude by showing the connection of Bellman–Ford's algorithm to linear programming duality. Let $y_v := d_{n-1}(v)$ for all $v \in V$. By Observation 4.13, the vector $y \in \mathbb{R}^V$ satisfies the conditions

$$y_v - y_u \leq \ell_{uv}, \quad \text{for all } uv \in A.$$

These conditions are precisely the constraints of the dual of (4.7). Furthermore, the value of the dual objective function is $y_t - y_s = d_{n-1}(t)$. Since $d_{n-1}(t)$ is the length of a shortest s,t-path P, the characteristic vector x^P of P and the vector y are a primal and a dual solution to (4.7), respectively, with the same objective value $d_{n-1}(t)$. Hence x^P and y are primal and dual optimal solutions.

4.3.3 Maximum Flow and Minimum Cut

Given a digraph $D = (V, A)$ and two distinct nodes $s, t \in V$, an s,t-*flow* in D is a nonnegative vector $x \in \mathbb{R}^A$ – where the quantity x_e, $e \in A$, is called the *flow on arc* e—such that the total amount of flow entering any node $v \neq s, t$ equals the total amount of flow leaving v. That is,

$$\sum_{e \in \delta^-(v)} x_e - \sum_{e \in \delta^+(v)} x_e = 0, \quad v \in V \setminus \{s, t\}.$$

It can be verified that the total amount of flow leaving node s equals the total amount of flow entering node t. Such quantity is called the *value of the s,t-flow x*, denoted by

$$\mathrm{val}(x) := \sum_{e \in \delta^+(s)} x_e - \sum_{e \in \delta^-(s)} x_e.$$

Given nonnegative capacities c_e, $e \in A$, on the arcs, a *feasible s,t-flow* is an s,t-flow which does not exceed the arc capacities, that is $x_e \leq c_e$, $e \in A$. An s,t-flow x is *integral* if it is an integral vector.

The *maximum flow problem* consists in finding a feasible s,t-flow of maximum value. See Fig. 4.1 for an example. The maximum flow problem can be formulated as the following linear program

$$\begin{array}{rl} \max \nu & \\ & \sum_{e \in \delta^-(v)} x_e - \sum_{e \in \delta^+(v)} x_e = 0 \qquad v \in V \setminus \{s,t\} \\ \nu + & \sum_{e \in \delta^-(s)} x_e - \sum_{e \in \delta^+(s)} x_e = 0 \\ -\nu + & \sum_{e \in \delta^-(t)} x_e - \sum_{e \in \delta^+(t)} x_e = 0 \\ & 0 \leq x_e \leq c_e \qquad\qquad\qquad\qquad e \in A. \end{array} \qquad (4.8)$$

Note that the constraint matrix of the $|V|$ equations in the above linear program is the incidence matrix of the digraph D' obtained by adding to D a new arc from t to s, corresponding to the variable ν. Thus the constraint matrix of the above problem is totally unimodular. Whenever the capacities c_e, $e \in A$, are all integer numbers, Theorem 4.5 implies that the system of linear constraints (4.8) describes an integral polyhedron. In particular, finding a maximum integral s,t-flow amounts to solving a linear program. For a thorough account on algorithms for solving max-flow problems we refer the reader to [17].

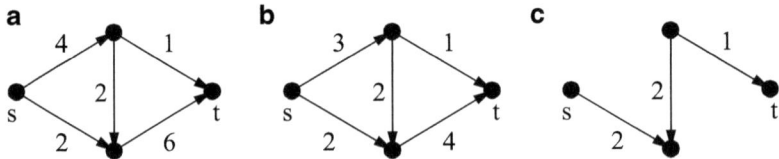

Figure 4.1: (**a**) Network with capacities, (**b**) Maximum flow, (**c**) Minimum cut

A closely related problem is the min-cut problem. An *s,t-cut* is a set of arcs of the form $\Gamma := \delta^+(S)$ where $s \in S \subseteq V \setminus \{t\}$. The *capacity* of the cut Γ is $c(\Gamma) := \sum_{e \in \delta^+(S)} c_e$. The *min-cut problem* consists in finding an s,t-cut of minimum capacity.

4.3. NETWORKS

Lemma 4.14. *Given a digraph D, two distinct nodes $s, t \in V$, nonnegative capacities c_e, $e \in A$, a feasible s, t-flow x and an s, t-cut $\Gamma = \delta^+(S)$, we have that $\mathrm{val}(x) \leq c(\Gamma)$. Furthermore $\mathrm{val}(x) = c(\Gamma)$ if and only if $x_e = c_e$, $e \in \delta^+(S)$ and $x_e = 0$, $e \in \delta^-(S)$. In this case x is a maximum s, t-flow and Γ is an s, t-cut of minimum capacity.*

Proof. Since $\Gamma := \delta^+(S)$ is an s, t-cut, $s \in S$ and $t \notin S$. Therefore adding all the equations (4.8) for the nodes in S one obtains

$$\mathrm{val}(x) = \sum_{e \in \delta^+(S)} x_e - \sum_{e \in \delta^-(S)} x_e \leq c(\Gamma)$$

where the inequality follows from $x_e \leq c_e$, $e \in \delta^+(S)$ and $-x_e \leq 0$, $e \in \delta^-(S)$. Therefore $\mathrm{val}(x) = c(\Gamma)$ if and only if x satisfies all these inequalities at equality. □

A classical theorem of Ford and Fulkerson states that the maximum value of an s, t-flow and the minimum capacity of an s, t-cut coincide. We give two proofs of this fact. The first one uses linear programming duality and total unimodularity, while the second is based on an algorithm that explicitly computes a feasible s, t-flow x and an s, t-cut Γ such that $\mathrm{val}(x) = c(\Gamma)$.

The following is the dual of the linear program (4.8), where the dual variable y_v corresponds to the equation relative to node v, and the dual variable z_e to the constraint $x_e \leq c_e$, $e \in A$.

$$\begin{aligned} \min \; & \sum_{e \in A} c_e z_e \\ & y_v - y_u + z_{uv} \geq 0 \quad uv \in A \\ & y_s - y_t = 1 \\ & z \geq 0. \end{aligned} \quad (4.9)$$

Theorem 4.15 (Max-Flow Min-Cut Theorem). *Given a digraph D, two distinct nodes $s, t \in V$, and nonnegative capacities c_e, $e \in A$,*

$$\max\{\mathrm{val}(x) \,:\, x \text{ is a feasible } s, t\text{-flow}\} = \min\{c(\Gamma) \,:\, \Gamma \text{ is an } s, t\text{-cut}\}. \quad (4.10)$$

Proof. Let x^* be a feasible s, t-flow of maximum value. By Lemma 4.14 it suffices to show that there exists an s, t-cut Γ such that $\mathrm{val}(x^*) \geq c(\Gamma)$. Let (y^*, z^*) be an optimal solution for (4.9). Since the constraint matrix of (4.9) is totally unimodular and the right-hand side is integral, we may assume that (y^*, z^*) is an integral vector. By linear programming duality (Theorem 3.7), $\mathrm{val}(x^*) = \sum_{e \in A} c_e z_e^*$. Let $\bar{A} = \{e \in A \,:\, z_e^* > 0\}$. We will show that $\bar{A}$

contains some s,t-cut Γ. It will then follow that $\mathrm{val}(x^*) = \sum_{e \in A} c_e z_e^* \geq \sum_{e \in \bar{A}} \bar{c}_e \geq c(\Gamma)$, where the first inequality follows from the fact that z^* is a nonnegative integral vector, and the second from the fact that $\Gamma \subseteq \bar{A}$.

Consider an s,t-path $P = v_0 v_1, v_1 v_2, \ldots, v_{k-1} v_k$ (thus $v_0 = s$, $v_k = t$). We have $\sum_{e \in P} z_e^* = \sum_{i=1}^{k} z_{v_{i-1} v_i}^* \geq \sum_{i=1}^{k} (y_{v_{i-1}}^* - y_{v_i}^*) = y_s^* - y_t^* = 1$. Thus, every s,t-path P in D has at least one arc in $\bar{A}$, that is, there is no s,t-path in $A \setminus \bar{A}$. Let S be the set of all nodes $v \in V$ such that there exists an s,v-path in $A \setminus \bar{A}$ (in particular $s \in S$), and let $\Gamma := \delta^+(S)$. Since there is no s,t-path in $A \setminus \bar{A}$, it follows that $t \notin S$ and $\Gamma \subseteq \bar{A}$. □

Given a digraph $D = (V, A)$, nodes s, t in V, nonnegative capacities c_e, $e \in A$ and a feasible s,t-flow x, the *residual digraph* (relative to x) $D_x := (V, A_x)$ has the same node set as D and $A_x := \overrightarrow{A}_x \cup \overleftarrow{A}_x$ (Fig. 4.2), where

$$\overrightarrow{A}_x := \{uv : uv \in A,\ x_{uv} < c_{uv}\} \text{ and } \overleftarrow{A}_x := \{vu : uv \in A,\ x_{uv} > 0\}.$$

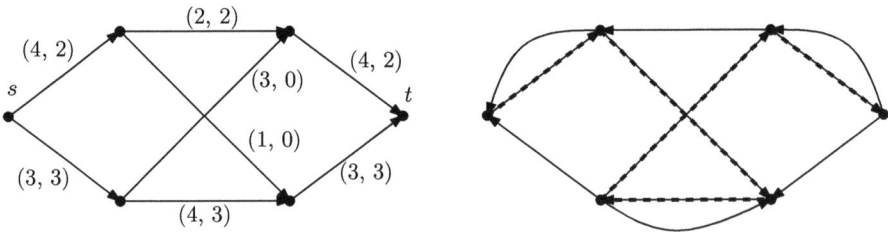

Figure 4.2: Figure to the *left*: labels on the arcs are pairs of numbers, where the first component represents the arc capacity and the second represents the flow on the arc. Figure to the *right*: residual graph. *Dashed arcs* represent an augmenting path

Theorem 4.16. *Given a digraph D, two distinct nodes $s, t \in V$, and nonnegative capacities c_e, $e \in A$, a feasible flow x is maximum if and only if the residual digraph $D_x = (V, A_x)$ contains no s,t-path.*

Proof. Assume first $D_x = (V, A_x)$ contains an s,t-path P. Let

$$\varepsilon := \min \left\{ \min_{uv \in \overrightarrow{A}_x \cap P} c_{uv} - x_{uv},\ \min_{vu \in \overleftarrow{A}_x \cap P} x_{uv} \right\}.$$

4.3. NETWORKS

Since P is a path in D_x, $\varepsilon > 0$. Let x' be defined as follows:

$$x'_{uv} = \begin{cases} x_{uv} + \varepsilon & uv \in \vec{A}_x \cap P \\ x_{uv} - \varepsilon & vu \in \overleftarrow{A}_x \cap P \\ x_{uv} & \text{otherwise.} \end{cases} \quad (4.11)$$

It is routine to check that x' is a feasible flow and $\text{val}(x') = \text{val}(x) + \varepsilon$. Thus x is not a maximum flow.

For the converse, assume that D_x contains no s,t-path. Let S be the subset of nodes of V that can be reached by a directed path in D_x originating at s. Then $s \in S$ and $t \notin S$, so $\Gamma := \delta^+(S)$ is an s,t-cut. Since no arc of A_x leaves S, it follows by construction of D_x that $x_{uv} = c_{uv}$ for every arc $uv \in \delta^+(S)$ and $x_{uv} = 0$ for every arc of $uv \in \delta^-(S)$. Thus, by Lemma 4.14, x is a maximum s,t-flow and Γ is a minimum s,t-cut. □

Given a feasible s,t-flow x, an s,t-path P in D_x is called an *x-augmenting path*, and the new flow x' defined in (4.11) is said to be obtained by *augmenting x along P*. The above proof gives an algorithm to solve the max-flow problem.

Augmenting Paths Algorithm. *Start with the initial flow $x := 0$. While there exists an s,t-path P in the residual digraph D_x, augment the current s,t-flow x along P.*

Upon termination of the algorithm, x is a maximum s,t-flow. Moreover, if we let S be the set of nodes reachable from s by a directed path in D_x at termination of the algorithm, the proof of Theorem 4.16 shows that $\delta^+(S)$ is a minimum s,t-cut whose capacity equals the value of the optimum flow.

If the capacities c_e are rational, the algorithm terminates because in each iteration the flow augmentation ε is bounded away from 0 by $\frac{1}{D}$ where D is the least common denominator of the positive capacities. If the capacities c_e are integer, then ε is an integer whenever the current s,t-flow is integral, and therefore the s,t-flow obtained by augmenting along P is also integral. Since we start with the feasible integral s,t-flow $x := 0$, the algorithm terminates with an s,t-flow of maximum value which is integral.

The augmenting paths algorithm does not guarantee polynomial-time convergence in general (see Exercise 4.16 for an example). However, it can be easily modified to run in polynomial time as follows. At each iteration, given the current feasible flow x, if x is not maximal then augment x along a *shortest* s,t-path P in D_x (here the length of a path is its number of arcs).

This algorithm is referred to as the *shortest augmenting path algorithm*. The following theorem was discovered independently by Dinic [120] and Edmonds-Karp [129].

Theorem 4.17. *Given a digraph $D = (V, A)$, distinct nodes $s, t \in V$, and nonnegative capacities c_e, $e \in A$, the shortest augmenting path algorithm terminates in at most $|V||A|$ iterations.*

Proof. Given a feasible s,t-flow x, denote by $d(u,v)$ the distance from $u \in V$ to $v \in V$ in the residual digraph $D_x = (V, A_x)$. Let us say that an arc of A_x is *relevant* in D_x if it is contained in a shortest s,t-path in D_x. If there is no s,t-path, then no arc is relevant in D_x. Note that for every $uv \in A_x$ we have $d(s,v) \leq d(s,u) + 1$, and if uv is relevant in D_x we have $d(s,v) = d(s,u) + 1$.

Let x' be the s,t-flow obtained by augmenting x along a shortest s,t-path P in D_x. Then

$$A_x \setminus A_{x'} = \{uv \in \overrightarrow{A}_x \cap P : c_{uv} - x_{uv} = \varepsilon\} \cup \{uv \in \overleftarrow{A}_x \cap P : x_{vu} = \varepsilon\} \neq \emptyset; \quad (4.12)$$

$$A_{x'} \setminus A_x = \{uv : vu \in \overrightarrow{A}_x \cap P, x_{vu} = 0\} \cup \{uv : vu \in \overleftarrow{A}_x \cap P : x_{uv} = c_{uv}\}. \quad (4.13)$$

Denote by $d'(u,v)$ the distance from $u \in V$ to $v \in V$ in $D_{x'}$. We prove next that, for every $v \in V$, $d(s,v) \leq d'(s,v)$ and $d(v,t) \leq d'(v,t)$. If not, there must exist at least one arc $uw \in A_{x'}$ such that $d(s,w) > d(s,u) + 1$. It follows that $uw \notin A_x$, so $wu \in P$ by (4.13), but we should have $d(s,u) = d(s,w) + 1$ because P is a shortest path, a contradiction.

We now show that, if $d'(s,t) = d(s,t)$, then the number of relevant arcs in $D_{x'}$ is strictly less than in D_x. Note that (4.12) implies that at least one relevant arc of D_x is not present in $A_{x'}$. Thus it suffices to show that, for any relevant arc uv of $D_{x'}$, uv is also relevant in D_x. Suppose $uv \notin A_x$. By (4.13) $vu \in P$, thus $d'(s,u) \geq d(s,u) = d(s,v) + 1$ and $d'(v,t) \geq d(v,t) = d(s,t) - d(s,v)$. It follows that any s,t-path in $D_{x'}$ that uses uv has length at least $d(s,t) + 2$, contradicting the fact that uv is relevant in $D_{x'}$. Thus $uv \in A_x$. We have that $d(s,u) + d(v,t) \leq d'(s,u) + d'(v,t) = d(s,t) - 1$, therefore uv is relevant in D_x.

Thus, at each iteration, the distance from s to t in the residual graph does not increase, and when it does not decrease the number of relevant arcs decreases. The distance from s to t is never more than $|V| - 1$, and the number of relevant arcs is never more than $|A|$. It follows that the shortest augmenting path algorithm terminates after at most $|V||A|$ iterations. □

4.4 Matchings in Graphs

A *matching* in an undirected graph $G = (V, E)$ is a set $M \subseteq E$ of pairwise disjoint edges, where an edge is viewed here as a set of two distinct nodes.

A matching is *perfect* if it *covers* every node of the graph (that is, every node is contained in exactly one edge of the matching).

A basic problem in combinatorial optimization is the *maximum cardinality matching* problem, that is, finding a matching of G of maximum cardinality. Obviously a perfect matching of G has maximum cardinality, but in general a perfect matching might not exist. The problem can be formulated as an integer program with binary variables x_e, $e \in E$, where $x_e = 1$ if and only if e belongs to the matching M. Since each node $v \in V$ can be covered by at most one edge of M, it follows that x must satisfy the *degree constraints* $\sum_{e \in \delta(v)} x_e \leq 1$, $v \in V$. Hence the maximum cardinality problem can be formulated as

$$\max \sum_{e \in E} x_e$$
$$\sum_{e \in \delta(v)} x_e \leq 1, \quad v \in V \qquad (4.14)$$
$$x \in \{0, 1\}^E.$$

The *incidence matrix* A_G of a graph $G = (V, E)$ is the $|V| \times |E|$ matrix whose rows correspond to the nodes of G and columns correspond to the edges of G, and whose entries are defined as follows: For every edge $e = uv$, $a_{ue} = a_{ve} = 1$ and $a_{we} = 0$ for all $w \in V \setminus \{u, v\}$. Note that any $0, 1$ matrix A_G with exactly two nonzero entries in each column is the incidence matrix of some graph G.

Formulation (4.14) can be written in terms of the incidence matrix A_G

$$\max \sum_{e \in E} x_e$$
$$A_G x \leq 1$$
$$x \in \{0, 1\}^E,$$

where $\mathbf{1}$ is the vector of all ones in $\mathbb{R}^V$. G has a perfect matching if $A_G x = \mathbf{1}$ admits a solution $x \in \{0, 1\}^E$.

Theorem 4.18. *Let A_G be the incidence matrix of a graph G. Then A_G is totally unimodular if and only if G is bipartite.*

Proof. Since A_G has two nonzero entries in each column, by Corollary 4.8 A_G is totally unimodular if and only if it has an equitable row-bicoloring.

Note that an equitable row-bicoloring of A_G corresponds to a bipartition of the nodes of G such that each edge has an endnode in each side of the bipartition. Such a bicoloring exists if and only if G is bipartite. □

The *matching polytope* of G is the convex hull of all characteristic vectors of matchings of G, and the *perfect matching polytope* of G is the convex hull of characteristic vectors of perfect matchings in G. Note that the perfect matching polytope of G is the face of the matching polytope of G defined by setting the degree constraints at equality. By Theorems 4.4 and 4.18, the polytope $\{x \in \mathbb{R}^E : A_G x \leq 1, x \geq 0\}$ is integral when G is bipartite, thus we have the following.

Corollary 4.19. *If G is a bipartite graph, then the matching polytope of G is the set $\{x \in \mathbb{R}^E : A_G x \leq 1, x \geq 0\}$, and the perfect matching polytope of G is the set $\{x \in \mathbb{R}^E : A_G x = 1, x \geq 0\}$.*

This shows that in the bipartite case finding the maximum cardinality matching amounts to solving the linear relaxation of (4.14). We will see in Sect. 4.4.4 that a set of inequalities describing the matching polytope can be given even for general graphs, but it is considerably more involved.

4.4.1 Augmenting Paths

There are specialized polynomial-time algorithms to solve the maximum cardinality matching problem. These algorithms rely on the concept of *augmenting path*.

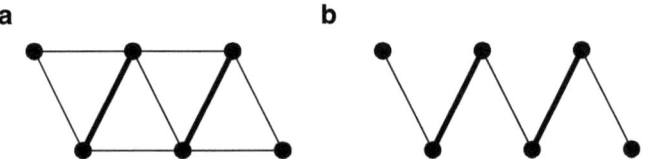

Figure 4.3: (a) A graph and a matching M in bold, (b) an M-augmenting path

Given a matching M, a path P in $G = (V, E)$ is said to be M-*alternating* if each node in P, except the two endnodes, belongs to one edge in M and one edge in $E \setminus M$. In other words, the edges alternate along P between edges in M and edges in $E \setminus M$. An M-alternating path P is said to be M-*augmenting* if the two endnodes of P are not covered by any edge of M. See Fig. 4.3 for an example. If P is an M-augmenting path, then $|P \setminus M| = |P \cap M| + 1$. By definition of M-augmenting path, the set

$$N := M \triangle P = (M \setminus P) \cup (P \setminus M)$$

is a matching, and $|N| = |M| + |P \setminus M| - |P \cap M| = |M| + 1$. The matching N is said to be obtained by *augmenting M along P*. In particular, if there is an M-augmenting path, then M is not optimal and we can find a larger matching. A classic result of Petersen [307], later rediscovered by Berge [50], states that the converse also holds.

Theorem 4.20. *A matching M has maximum cardinality if and only if there is no M-augmenting path.*

Proof. We already proved the "only if" part of the statement. We now show that, if M is not of maximum cardinality, then there exists an M-augmenting path. Let N be a matching in G such that $|N| > |M|$, and consider the graph $G' = (V, E')$, where $E' = M \triangle N$. Note that, since every edge in E' belongs to M or N, but not both, and since every node of V belongs to at most one element of M and at most one element of N, every node in G' has degree at most two. In particular, E' is the node-disjoint union of cycles $C^1, \ldots, C^p$ and paths $P^1, \ldots, P^q$. Note that every node in a cycle C^i, $i = 1, \ldots, p$, belongs to one edge in $C^i \cap M$ and one edge in $C^i \cap N$, thus $|C^i \cap M| = |C^i \cap N|$. Since $|N| > |M|$, there must therefore exist a path P^i, $i \in \{1, \ldots, q\}$, such that $|P^i \cap N| > |P^i \cap M|$. Since every node in P^i, except possibly the endnodes, belongs to an edge in $P^i \cap M$ and one edge in $P^i \cap N$, it follows that the two endnodes of P^i are covered by N but not by M. Therefore P^i is an M-augmenting path. □

The above theorem suggests the following algorithm for the maximum cardinality matching: start with any matching M of G and look for an M-augmenting path. If one exists, then augment M and repeat, else M is optimal. The issue at this point is of course how to find an M-augmenting path. In the general case it is far from obvious how to find one in polynomial time, and this problem was settled by Edmonds [123]. In the bipartite case the situation is much easier.

4.4.2 Cardinality Bipartite Matchings

We now describe an algorithm for finding a maximum cardinality matching in a bipartite graph $G = (V, E)$. Let the two sides of the bipartition be the sets U and W. Let M be a matching in G. We construct the following *auxiliary digraph* (with respect to M). Let $D_M = (V, A_M)$ be the digraph with the same nodeset as G, where the arcs in A_M are obtained by orienting the edges in E from U to W if they do not belong to M, and from W to U otherwise. That is, $A_M := \overrightarrow{A}_M \cup \overleftarrow{A}_M$, where

$$\overrightarrow{A}_M := \{uv : u \in U, v \in W, uv \in E \setminus M\}, \quad \overleftarrow{A}_M := \{vu : u \in U, v \in W, uv \in M\}. \tag{4.15}$$

To simplify notation, we identify any arc in A_M with the corresponding edge in E.

Let S denote the set of nodes of U that are not covered by M, and let T be the set of nodes of W not covered by M. Observe that, if P is an M-augmenting path in G, then P has odd length, thus one endnode of P is in S and the other in T. Also, since the edges of P alternate between edges in M and edges in $E \setminus M$, it follows from the construction of D_M that P is a directed path in D_M from a node in S to a node in T. Conversely, if P is a directed path in D_M from a node in S to a node in T, it follows similarly that P is an M-augmenting path. Thus, the problem of finding an M-augmenting path in G amounts to determining if there exists a directed path in D_M from a node in S to a node in T, which can be achieved by any algorithm to explore a digraph (such as, for example, breadth-first search or depth-first search).

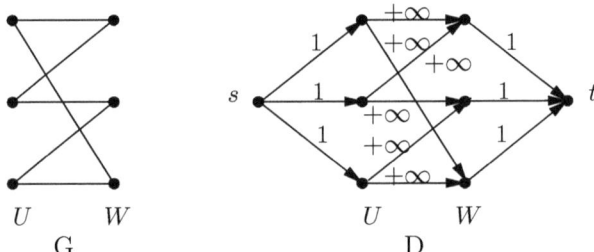

Figure 4.4: Transformation of bipartite matching into maximum flow

One can also solve the maximum cardinality matching problem in a bipartite graph by transforming it into a maximum flow problem (Fig. 4.4). Let $G = (V, E)$ be a bipartite graph and let U and W be the two sides of the bipartition. Define the digraph $D = (V', A)$ as follows:

$$V' := V \cup \{s, t\}, \quad A := \{su : u \in U\} \cup \{uw : u \in U, w \in W, uw \in E\} \cup \{wt : w \in W\}.$$

For $e \in A$, define capacity $c_e = 1$ if s or t belongs to e and $c_e = \infty$ otherwise. Since $\delta^-(u) = \{su\}$ for every $u \in U$, $\delta^+(w) = \{wt\}$ for every

4.4. MATCHINGS IN GRAPHS 149

$w \in W$ and $c_e = 1$ if s or t belongs to e, it follows that an integral s, t-flow x is a $0, 1$ vector, the set $M := \{e \in E : x_e = 1\}$ is a matching of G, and $\text{val}(x) = |M|$.

4.4.3 Minimum Weight Perfect Matchings in Bipartite Graphs

Let $G = (V, E)$ be a graph with weights w_e, $e \in E$ on the edges. The *minimum weight perfect matching problem* is to determine a perfect matching M of minimum total weight $w(M) := \sum_{e \in M} w_e$. In this section, we assume that G is bipartite and we let U and W be the sides of the bipartition. Clearly, in order for a perfect matching to exist, U and W must have the same cardinality, say n. A matching M in G is perfect if and only if it has cardinality n. The minimum weight perfect matching problem in a bipartite graph is also known under the name of *assignment problem* (recall (1.4)).

By Corollary 4.19, the minimum weight perfect matching can be determined by solving the linear program $\min\{\sum_{e \in E} w_e x_e : A_G x = \mathbf{1}, x \geq \mathbf{0}\}$. Next we present an "ad-hoc" algorithm to solve the minimum weight perfect matching problem, which is known in the literature as the *Hungarian method* [245].

The Hungarian method consists of at most n iterations. At the kth iterations, $0 \leq k \leq n - 1$, we have a matching M of cardinality k such that $w(M)$ is minimum among all matchings of cardinality k. We want to compute a minimum weight matching of cardinality $k + 1$. We reduce this problem to a shortest path problem.

Construct the auxiliary digraph $D_M = (V, A_M)$ relative to M, where $A_M = \overrightarrow{A}_M \cup \overleftarrow{A}_M$, and $\overrightarrow{A}_M, \overleftarrow{A}_M$ are defined in (4.15). Define a length ℓ_e for each arc $e \in A_M$ by

$$\ell_e = w_e \text{ for all } e \in \overrightarrow{A}_M, \quad \ell_e = -w_e \text{ for all } e \in \overleftarrow{A}_M.$$

Observation 4.21. D_M *does not contain any negative-length circuit.*

Proof. By contradiction, suppose there exists a circuit C of D_M with $\ell(C) < 0$. Since C is a circuit, the arcs in C alternate between arcs in $\overrightarrow{A}_M$ and arcs in $\overleftarrow{A}_M$. That is, the edges in C alternate between edges in M and edges in $E \setminus M$. Thus, the set $M' = M \triangle C$ is a matching of cardinality k. We have that $w(M') = w(M) - w(M \cap C) + w(C \setminus M) = w(M) + \ell(C) < w(M)$, contradicting the fact that M has minimum weight among all matchings of cardinality k. □

Let S denote the set of nodes of U that are not covered by M, and let T be the set of nodes of W not covered by M. By Theorem 4.20, G has a matching of cardinality $k+1$ if and only if G has an M-augmenting path. As seen in the previous section, this is the case if and only if there exists a directed path in D_M between a node of S and a node of T. If no such path exists, then we conclude that G has no perfect matching, otherwise we choose a shortest such path P with respect to the length ℓ.

Such a path can be computed in polynomial time with Bellman–Ford's algorithm (Sect. 4.3.2), because there are no negative-length circuits in D_M. Let $N := M \triangle P$. Then N is a matching of cardinality $k+1$ and $w(N) = w(M) + w(P \setminus M) - w(P \cap M) = w(M) + \ell(P)$.

Observation 4.22. *N has minimum weight among all matchings of G of cardinality $k+1$.*

Proof. Consider a matching N' of G with $|N'| = k+1$. We need to show that $w(N') \geq w(N)$. Since $|N'| > |M|$, it follows that $M \triangle N'$ contains an M-augmenting path P'. By the choice of P, we have $\ell(P) \leq \ell(P')$. Let $M' := N' \triangle P'$. Since P' is an M-augmenting path contained in $M \triangle N'$, it follows that P' is N'-alternating and that its endnodes are covered by N', hence M' is a matching of cardinality k. Since M has minimum weight among all matchings of cardinality k, we have $w(M) \leq w(M')$. Finally, observe that $w(N') = w(M') + \ell(P') \geq w(M) + \ell(P) = w(N)$, where the inequality follows from the facts that $w(M) \leq w(M')$ and $\ell(P) \leq \ell(P')$. □

4.4.4 The Matching Polytope

Let $G = (V, E)$ be a graph. Corollary 4.19 states that, whenever G is bipartite, its matching polytope is described by the nonnegativity constraints $x \geq 0$ and the degree constraints

$$\sum_{e \in \delta(v)} x_e \leq 1, \quad v \in V. \tag{4.16}$$

This statement does not carry through in the general case. For example, suppose G is just a triangle, with nodes $\{1, 2, 3\}$ and edges $12, 13, 23$. The system formed by the degree and nonnegativity constraints is

$$\begin{aligned}
x_{12} + x_{13} &\leq 1 \\
x_{12} \phantom{{}+ x_{13}} + x_{23} &\leq 1 \\
x_{13} + x_{23} &\leq 1 \\
x &\geq 0.
\end{aligned}$$

Point $(1/2, 1/2, 1/2)$ is a vertex of the polytope described by the above constraints, which shows that these are not sufficient to describe the matching polytope of the triangle. What are the conditions not captured by the degree constraints in this example?

For a graph $G = (V, E)$ and a set of nodes $U \subseteq V$, let $E[U] := \{uv \in E : u, v \in U\}$. Given a matching M of G, every edge of $M \cap E[U]$ covers two nodes of U. Since each node of U is covered by at most one edge of M, it follows that $|M \cap E[U]| \leq |U|/2$. If U has odd cardinality, this means that $|M \cap E[U]| \leq (|U|-1)/2$. Thus the characteristic vector x of any matching must satisfy the *blossom inequality*

$$\sum_{e \in E[U]} x_e \leq \frac{|U|-1}{2}, \quad U \subseteq V,\ |U|\ \text{odd}. \tag{4.17}$$

In the above example, the blossom inequality relative to the odd set $\{1, 2, 3\}$ would be $x_{12} + x_{13} + x_{23} \leq 1$. Adding this inequality cuts off the point $(1/2, 1/2, 1/2)$.

Edmonds [124] showed that, in fact, adding the blossom inequalities to the nonnegativity and degree constraints is always sufficient to describe the matching polytope (Theorem 4.24).

Recall that the perfect matching polytope of G is the convex hull of all characteristic vectors of perfect matchings. It is the face of the matching polytope of G obtained by setting the degree constraints at equality

$$\sum_{e \in \delta(v)} x_e = 1, \quad v \in V. \tag{4.18}$$

For the perfect matching polytope, the blossom inequalities (4.17) can be written in an equivalent form. Given an odd set $U \subseteq V$, summing (4.18) over all $v \in U$ gives $|U| = \sum_{v \in U} \sum_{e \in \delta(v)} x_e = \sum_{e \in \delta(U)} x_e + 2\sum_{e \in E[U]} x_e$. Subtracting the blossom inequality relative to U multiplied by 2, we obtain the so-called *odd cut inequality*

$$\sum_{e \in \delta(U)} x_e \geq 1, \quad U \subseteq V,\ |U|\ \text{odd}. \tag{4.19}$$

The odd cut inequalities define the same faces of the perfect matching polytope as the blossom inequalities. They express the fact that, for every odd set U, some node in U is matched to some node in $V \setminus U$ in every perfect matching. When U or $V \setminus U$ is a singleton, the odd cut inequality relative to U is implied by one of the degree constraints (4.18).

Theorem 4.23 (Perfect Matching Polytope Theorem). *The perfect matching polytope of a graph $G = (V, E)$ is the set $\{x \in \mathbb{R}^E : x \geq 0,$ x satisfies (4.18)(4.19)$\}$.*

Proof. Let $P(G) := \{x \in \mathbb{R}^E : x \geq 0, x$ satisfies (4.18)(4.19)$\}$. Clearly $P(G) \cap \mathbb{Z}^E$ is the set of characteristic vectors of the perfect matchings of G, therefore it suffices to show that $P(G)$ is an integral polytope. Assume not and let $G = (V, E)$ be a graph with $|V| + |E|$ smallest possible such that $P(G)$ is not an integral polytope. Let $\bar{x}$ be a fractional vertex of $P(G)$. By the minimality of G, G is connected and $0 < \bar{x}_e < 1$ for every edge $e \in E$. By Corollary 4.19 G is not bipartite. If $|V|$ is odd then $P(G) = \emptyset$ because the odd cut inequality relative to V implies $1 \leq \sum_{e \in \delta(V)} x_e = 0$, therefore $|V|$ must be even.

We show next that $\sum_{e \in \delta(U)} \bar{x}_e = 1$ for some $U \subseteq V$ such that $|U|$ odd, $|U| \geq 3$, $|V \setminus U| \geq 3$. Suppose not. Since $\bar{x}_e > 0$ for every $e \in E$, $\bar{x}$ is the unique solution of the $|V|$ equalities (4.18). It follows that $|E| \leq |V|$. Then G must be a tree plus possibly an edge, because G is connected. Since $\bar{x}_e < 1$ for every $e \in E$ and $\bar{x}$ satisfies (4.18), it follows that every node of G has degree at least 2, so G is a cycle. Since $|V|$ is even, G is an even cycle, contradicting the fact that G is not bipartite.

Let $U \subseteq V$ be an odd set such that $|U| \geq 3$, $|V \setminus U| \geq 3$, and $\sum_{e \in \delta(U)} \bar{x}_e = 1$. Let $G' = (V', E')$ be the graph obtained from G by shrinking U into a single node v' (we identify the edges in $\delta_{G'}(v')$ with the original ones in $\delta_G(U)$, and we may create parallel edges). Analogously, let $G'' = (V'', E'')$ be the graph obtained by shrinking $V \setminus U$ into a single node v''. Since $|U| \geq 3$ and $|V \setminus U| \geq 3$, both G', G'' are smaller than G. By the minimality of G, both polytopes $P(G')$, $P(G'')$ are integral. Let $\bar{x}'$ and $\bar{x}''$ be the restrictions of $\bar{x}$ to the edges of G', G'' respectively. Next we show that $\bar{x}' \in P(G')$ and $\bar{x}'' \in P(G'')$. By symmetry, it suffices to show that $\bar{x}' \in P(G')$. First observe that $\sum_{e \in \delta_{G'}(v')} \bar{x}'_e = \sum_{e \in \delta(U)} \bar{x}_e = 1$. Furthermore, consider $W' \subseteq V'$ such that $|W'|$ is odd. Let $W := W'$ if $v' \notin W'$, and $W := W' \setminus \{v'\} \cup U$ if $v' \in W'$. Then $\sum_{e \in \delta_{G'}(W')} \bar{x}'_e = \sum_{e \in \delta(W)} \bar{x}_e \geq 1$.

It follows that $\bar{x}' \in P(G')$ and $\bar{x}'' \in P(G'')$, thus $\bar{x}'$ and $\bar{x}''$ can be written as convex combinations of incidence vectors of perfect matchings of G' and G''. Equivalently, there is a positive integer k such that $k\bar{x}'$ and $k\bar{x}''$ are the sum of k incidence vectors of perfect matchings of G' and G'', say $M'_1, \ldots, M'_k$ and $M''_1, \ldots, M''_k$, respectively. Note that every edge $e \in \delta(U)$ is contained in exactly $k\bar{x}_e$ matchings among $M'_1, \ldots, M'_k$ and $k\bar{x}_e$ matchings among $M''_1, \ldots, M''_k$. Thus we may relabel the indices so that $M'_i \cap \delta(U) = M''_i \cap \delta(U)$ for $i = 1, \ldots, k$. If we define $M_i := M'_i \cup M''_i$,

4.5. SPANNING TREES

$i = 1, \ldots, k$, it follows that $M_1, \ldots, M_k$ are perfect matchings of G and $k\bar{x}$ is the sum of their incidence vectors. Hence $\bar{x}$ is a convex combination of incidence vectors of matchings of G, contradicting the assumption that $\bar{x}$ is a fractional vertex of $P(G)$. □

Theorem 4.24 (Matching Polytope Theorem). *The matching polytope of a graph $G = (V, E)$ is the set $\{x \in \mathbb{R}^E : x \geq 0, \ x \text{ satisfies } (4.16)(4.17)\}$.*

Proof. Let $\tilde{G} = (\tilde{V}, \tilde{E})$ be the graph defined by creating two disjoint copies of G and connecting every pair of corresponding nodes. Formally, $\tilde{V} := V \cup V'$, where V' contains a copy v' of each node $v \in V$, and $\tilde{E} := E \cup E' \cup \{vv' : v \in V\}$, where $E' := \{u'v' : uv \in E\}$. The restriction to G of a perfect matching of $\tilde{G}$ is a matching of G and, conversely, any matching of G can be extended to a perfect matching of $\tilde{G}$. Therefore the matching polytope of G is the projection onto $\mathbb{R}^E$ of the perfect matching polytope $P(\tilde{G})$ of $\tilde{G}$.

Thus, it suffices to show that, given $\bar{x} \in \mathbb{R}^E_+$, satisfying (4.16)(4.17), there exists $\tilde{x} \in P(\tilde{G})$ such that $\bar{x}_e = \tilde{x}_e$ for all $e \in E$. We define $\tilde{x} \in \mathbb{R}^{\tilde{E}}$ by $\tilde{x}_{uv} := \tilde{x}_{u'v'} := \bar{x}_{uv}$ for every $uv \in E$, and $\tilde{x}_{vv'} := 1 - \sum_{e \in \delta_G(v)} \bar{x}_e$ for every $v \in V$. By construction, $\tilde{x} \geq 0$ and $\sum_{e \in \delta_{\tilde{G}}(w)} \tilde{x}_e = 1$ for all $w \in \tilde{V}$. Let $\tilde{U} \subseteq \tilde{V}$ be an odd set. Let $U := \{v \in V : |\{v, v'\} \cap \tilde{U}| = 1\}$ and $T := \{v \in V : v, v' \in \tilde{U}\}$. Since $|\tilde{U}| = |U| + 2|T|$, $|U|$ is odd. Denoting $(U : T) := \{uv \in E : u \in U, v \in T\}$, one can verify that

$$\sum_{e \in \tilde{E}[\tilde{U}]} \tilde{x}_e \leq \sum_{e \in E[U]} \bar{x}_e + 2 \sum_{e \in E[T]} \bar{x}_e + \sum_{e \in (U:T)} \bar{x}_e + \sum_{v \in T} \tilde{x}_{vv'} \leq \frac{|U|-1}{2} + |T| = \frac{|\tilde{U}|-1}{2},$$

where the second inequality follows from the fact that $\sum_{v \in T} \tilde{x}_{vv'} = |T| - 2\sum_{e \in E[T]} \bar{x}_e - \sum_{e \in \delta(T)} \bar{x}_e$. Now Theorem 4.23 implies that $\tilde{x} \in P(\tilde{G})$. □

4.5 Spanning Trees

In a graph $G = (V, E)$, a *spanning tree* is a set $T \subseteq E$ of edges such that the graph (V, T) is acyclic and connected. See Fig. 4.5 for an example. The convex hull of the incidence vectors of all spanning trees of G is the *spanning tree polytope of G*. It is well known that G has a spanning tree if and only if G is connected, and that, if T is a spanning tree, then $|T| = |V| - 1$ and $|T \cap E[S]| \leq |S| - 1$ for every $\emptyset \neq S \subset V$.

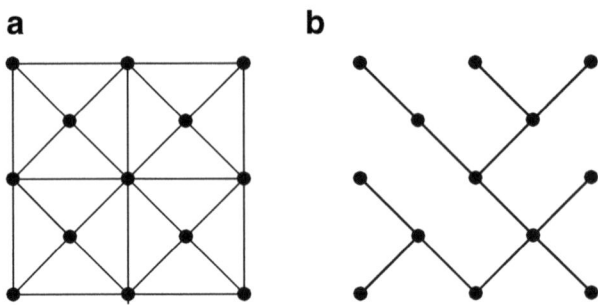

Figure 4.5: (a) A graph, (b) a spanning tree

Theorem 4.25. *Let $G = (V, E)$ be a graph. The spanning tree polytope of G is described by the following inequalities*

$$\begin{aligned}
\sum_{e \in E[S]} x_e &\leq |S| - 1 & \emptyset \neq S \subset V \\
\sum_{e \in E} x_e &= |V| - 1 & \\
x_e &\geq 0 & e \in E.
\end{aligned} \quad (4.20)$$

Proof. The statement is true if G is not connected, since in that case the spanning tree polytope is empty and the system (4.20) is infeasible. Indeed, given a connected component $S \neq \emptyset$, we have that $\sum_{e \in E} x_e = \sum_{e \in E[S]} x_e + \sum_{e \in E[V \setminus S]} x_e \leq (|S| - 1) + (|V \setminus S| - 1) = |V| - 2$.

Hence we assume that G is connected. It suffices to show that, for every $c \in \mathbb{R}^E$, the linear programming problem $\max\{cx : x \text{ satisfies } (4.20)\}$ has an optimal solution that is the incidence vector of a spanning tree. In order to show this, we present an algorithm that constructs such a tree. We refer to c_e as the *cost* of edge $e \in E$.

Let $F_0 := \emptyset$. For $i = 1, \ldots, n-1$, let e_i be an edge of maximum cost among all edges with endnodes in distinct connected components of (V, F_{i-1}), and let $F_i := F_{i-1} \cup \{e_i\}$.

Clearly, F_{n-1} is a spanning tree, and therefore its incidence vector x^* is a feasible solution to (4.20). To show that it is an optimal solution to the linear program, we give a feasible dual solution y^* such that x^*, y^* satisfy complementary slackness (Theorem 3.8). The dual is

$$\begin{aligned}
\min \sum_{\emptyset \neq S \subseteq V} (|S| - 1) y_S & & \\
\sum_{S : e \in E[S]} y_S &\geq c_e & e \in E \\
y_S &\geq 0 & \emptyset \neq S \subset V.
\end{aligned} \quad (4.21)$$

Note that the variable y_V is unrestricted in sign.

For $i = 1, \ldots, n-1$, let S_i be the connected component of (V, F_i) containing both endnodes of e_i. Note that $S_{n-1} = V$. For $i = 1, \ldots, n-2$, we define $y^*_{S_i} := c_{e_i} - c_{e_j}$, where j is the smallest index in $\{i+1, \ldots, n-1\}$ such that $S_i \subset S_j$. Let $y^*_V = c_{e_{n-1}}$ and $y^*_S = 0$ for all $S \subset V$ such that $S \neq S_1, \ldots, S_{n-1}$. Note that $c_{e_1} \geq c_{e_2} \geq \ldots \geq c_{e_{n-1}}$, therefore y^*_S is nonnegative for every $S \subset V$.

We show that y^* is a feasible dual solution. Consider $e \in E$ and let $I := \{i : e \in E[S_i]\}$. Assume $I = \{j_1, j_2, \ldots, j_k\}$ where $j_1 \leq j_2 \leq \ldots \leq j_k$. Clearly $j_k = n-1$. By definition both edges e and e_{j_1} have endnodes in distinct components of (V, F_{j_1-1}), hence by the algorithm described above $c_{e_{j_1}} \geq c_e$. Therefore

$$\sum_{S : e \in E[S]} y^*_S = \sum_{i=1}^k y^*_{S_{j_i}} = \sum_{i=1}^{k-1}(c_{e_{j_i}} - c_{e_{j_{i+1}}}) + c_{e_{n-1}} = c_{e_{j_1}} \geq c_e.$$

This shows that y^* is feasible and that, for $i = 1, \ldots, n-1$, the dual constraint corresponding to edge e_i is satisfied at equality. Thus, in order to prove that x^* and y^* satisfy complementary slackness, we only need to observe that, for $i = 1, \ldots, n-1$, $|F_i \cap E[S_i]| = |S_i| - 1$, which implies that $\sum_{e \in E[S_i]} x^*_e = |S_i| - 1$. □

The algorithm presented in the proof of Theorem 4.25 solves the problem of computing a maximum cost spanning tree in a graph with costs on the edges. The algorithm is due to Kruskal [244]. It can be modified in a straightforward way to compute also a minimum cost spanning tree.

4.6 Total Dual Integrality

A rational system $Ax \leq b$ is *totally dual integral* (TDI, for short) if, for every integral vector c for which the the linear program $\max\{cx : Ax \leq b\}$ admits a finite optimum, its dual $\min\{yb : yA = c, y \geq 0\}$ admits an integral optimal solution.

For example, the system (4.20) in the previous section is TDI, since the dual solution constructed in the proof of Theorem 4.25 is integral whenever the vector c is integral.

Theorem 4.26. *Let $Ax \leq b$ be a totally dual integral system and b be an integral vector. Then $P := \{x : Ax \leq b\}$ is an integral polyhedron.*

Proof. For every c for which the value $z_c := \max\{cx : Ax \leq b\}$ is finite, we have that z_c is integer since $z_c = by$ for some integral optimal solution y to the dual $\min\{yb : yA = c, y \geq 0\}$. By Theorem 4.1(iv), P is an integral polyhedron. □

In combinatorial optimization, total dual integrality plays a central role in the derivation of min–max theorems (see Schrijver [327] for an extensive treatment).

The property of being totally dual integral pertains to the system $Ax \leq b$ and is not invariant under scaling of the system (see Exercise 4.23).

It must be emphasized that there are integral polyhedra $\{x : Ax \leq b\}$ where A, b are integral but the system $Ax \leq b$ is not TDI (see Exercise 4.24 for example). However, any integral polyhedron can always be represented by a TDI system whose coefficients are all integer.

Theorem 4.27. *If P is a rational polyhedron then there exists a TDI system $Ax \leq b$ with an integral matrix A such that $P = \{x : Ax \leq b\}$. Furthermore, if P is an integral polyhedron, then the vector b can be chosen to be integral.*

Proof. If $P = \emptyset$, then the theorem holds by writing P as $P = \{x : 0x \leq -1\}$. Thus we assume $P \neq \emptyset$. Since P is a rational polyhedron, we may assume that $P := \{x : Mx \leq d\}$ where M is an $m \times n$ integral matrix. Let

$$C := \{c \in \mathbb{Z}^m : c = \lambda M,\ 0 \leq \lambda \leq 1\}.$$

Note that the set C is finite, since it is the set of integral vectors in a polytope. For every $c \in C$, let $\delta_c := \max\{cx : x \in P\}$ (since $P \neq \emptyset$ this program admits a finite optimum bounded above by λd). Let $Ax \leq b$ be the system comprising the inequalities $cx \leq \delta_c$, for all $c \in C$. Clearly, $P \subseteq \{x : Ax \leq b\}$. Since every row of M is in A, we actually have $P = \{x : Mx \leq d\} = \{x : Ax \leq b\}$. Furthermore, by Theorem 4.1, b is an integral vector whenever P is an integral polyhedron. We show next that $Ax \leq b$ is TDI.

Let c be an integral vector such that $\max\{cx : Ax \leq b\}$ is finite. We will show how to construct an integral optimal solution to the dual problem $\min\{zb : zA = c,\ z \geq 0\}$. We have $\max\{cx : Ax \leq b\} = \max\{cx : x \in P\}$. Let y^* be an optimal solution of the dual problem $\min\{yd : yM = c,\ y \geq 0\}$. Let $\lambda = y^* - \lfloor y^* \rfloor$ and let $c' = \lambda M$, $c'' = \lfloor y^* \rfloor M$. Note that, since $c = c' + c''$ and y^* is optimal for $\min\{yd : yM = c,\ y \geq 0\}$, it follows that λ is an optimal solution for the problem $\min\{yd : yM = c',\ y \geq 0\}$ and $\lfloor y^* \rfloor$ is an optimal solution for $\min\{yd : yM = c'',\ y \geq 0\}$.

Also, $c' = c - c''$ is an integral vector because c and c'' are integral vectors, thus $c' \in C$ because $0 \leq \lambda \leq 1$. In particular, $c'x \leq \delta_{c'}$ is an inequality in the system $Ax \leq b$.

Let us assume without loss of generality that (M, d) is the matrix defined by the first m rows of (A, b). Let k be the number of rows of A and assume that $c'x \leq \delta_{c'}$ is the hth inequality of $Ax \leq b$. Note that the hth unit vector e^h of $\mathbb{R}^k$ is an optimal solution to the problem $\min\{zb : zA = c', z \geq 0\}$ since $e^h b = \delta_{c'} = \max\{c'x : x \in P\}$, and that the vector $\bar{z}$, defined by $\bar{z}_i = \lfloor y_i^* \rfloor$ for $i = 1, \ldots, m$, $\bar{z}_i = 0$ for $i = m+1, \ldots, k$, is an optimal solution to $\min\{zb : zA = c'', z \geq 0\}$. It follows that the vector $z^* = \bar{z} + e^h$ is an integral optimal solution to the program $\min\{zb : zA = c, z \geq 0\}$. □

4.7 Submodular Polyhedra

In this section, we introduce an integral polyhedron defined by a TDI system of inequalities. Given a finite set $N := \{1, \ldots, n\}$, a set function $f : 2^N \to \mathbb{R}$ is *submodular* if

$$f(S) + f(T) \geq f(S \cap T) + f(S \cup T) \quad \text{for all } S, T \subseteq N. \tag{4.22}$$

Example 4.28. Given a graph $G = (V, E)$, the *cut function* of G is the function $f : 2^V \to \mathbb{R}$ defined by $f(S) = |\delta(S)|$ for all $S \subseteq V$. An easy counting argument shows that, for any $S, T \subseteq V$ the following holds:

$$|\delta(S)| + |\delta(T)| = |\delta(S \cap T)| + |\delta(S \cup T)| + 2|(T \setminus S : S \setminus T)|$$

where $(T \setminus S : S \setminus T)$ is the set of edges with one endnode in $S \setminus T$ and the other in $T \setminus S$. Therefore the cut function is submodular. ■

Given a set $N := \{1, \ldots, n\}$ and a submodular function $f : 2^N \to \mathbb{R}$, the *submodular polyhedron* is the set

$$P := \{x \in \mathbb{R}^n : \sum_{j \in S} x_j \leq f(S) \text{ for all } S \subseteq N\}. \tag{4.23}$$

Theorem 4.29. *Let $N := \{1, \ldots, n\}$ and let $f : 2^N \to \mathbb{Z}$ be an integer-valued submodular function such that $f(\emptyset) = 0$. Then the submodular polyhedron (4.23) is integral. Furthermore, the system $\sum_{j \in S} x_j \leq f(S)$, $S \subseteq N$, is TDI.*

Proof. We show that both the linear program

$$\max\{cx : \sum_{j \in S} x_j \leq f(S), S \subseteq N\} \tag{4.24}$$

and its dual have optimal solutions that are integral for each $c \in \mathbb{Z}^n$ for which the maximum in (4.24) is finite. Observe that the maximum is finite if and only if c is a nonnegative vector.

We may assume $c_1 \geq c_2 \geq \ldots \geq c_n \geq 0$. Let $S_0 := \emptyset$ and $S_j := \{1, \ldots, j\}$ for $j \in N$. For $j \in N$, let $\bar{x}_j := f(S_j) - f(S_{j-1})$. Since f is integer valued, $\bar{x}$ is an integral vector. We claim that $\bar{x}$ is an optimal solution.

We first show that $\bar{x} \in P$. We need to show that, for every $T \subseteq N$, $\bar{x}(T) \leq f(T)$ (Here we use the notation $x(T) := \sum_{j \in T} x_j$). The proof is by induction on $|T|$, the case $T = \emptyset$ being trivial. Let $k := \max\{j \in T\}$. Note that $T \subseteq S_k$. By induction, $\bar{x}(T \setminus \{k\}) \leq f(T \setminus \{k\})$. Therefore

$$\bar{x}(T) = \bar{x}(T\setminus\{k\}) + \bar{x}_k \leq f(T\setminus\{k\}) + \bar{x}_k = f(T\setminus\{k\}) + f(S_k) - f(S_{k-1}) \leq f(T)$$

where the last inequality follows from the submodularity of f, since $T \cap S_{k-1} = T \setminus \{k\}$ and $T \cup S_{k-1} = S_k$.

To show that $\bar{x}$ is optimal, we give an integral feasible solution for the dual of (4.24) whose objective value equals $c\bar{x}$. The dual is the following.

$$\begin{aligned} \min \ & \sum_{S \subseteq N} f(S) y_S \\ & \sum_{S \ni j} y_S = c_j \quad j \in N \\ & y_S \geq 0 \quad S \subseteq N. \end{aligned}$$

We define a dual solution $\bar{y}$ as follows.

$$\bar{y}_S := \begin{cases} c_j - c_{j+1} & \text{for } S = S_j, \ j = 1, \ldots, n-1 \\ c_n & \text{for } S = N \\ 0 & \text{otherwise.} \end{cases}$$

Since $c \in \mathbb{Z}^n$, $\bar{y}$ is integral. It is immediate to verify that $\bar{y}$ is dual feasible. Furthermore

$$\sum_{S \subseteq N} f(S) \bar{y}_S = \sum_{j=1}^{n-1} f(S_j)(c_j - c_{j+1}) + f(N) c_n = \sum_{j=1}^{n}(f(S_j) - f(S_{j-1})) c_j = \sum_{j=1}^{n} c_j \bar{x}_j,$$

where the second equation follows from $f(\emptyset) = 0$. Therefore $\bar{x}$ is optimal for the primal and $\bar{y}$ is optimal for the dual. □

The proof of the above theorem gives an algorithm to compute the optimum of (4.24). Namely, order the variables so that $c_1 \geq c_2 \geq \ldots \geq c_n \geq 0$, and let $S_0 := \emptyset$ and $S_j := \{1, \ldots, j\}$ for $j \in N$. An optimal solution $\bar{x}$ is defined by $\bar{x}_j := f(S_j) - f(S_{j-1})$ for $j \in N$. This algorithm is known as the *greedy algorithm*.

4.8 The Fundamental Theorem of Integer Programming

Let $S := \{(x,y) \in \mathbb{Z}^n \times \mathbb{R}^p : Ax + Gy \le b\}$ be a mixed integer linear set, where matrices A, G and vector b have rational entries. We prove in this section that $\text{conv}(S)$ admits a perfect formulation that is a rational polyhedron.

Note first that, if S contains finitely many vectors (for instance this happens when S is the set of integral points in a polytope), the above result follows from Corollary 3.14.

Theorem 4.30 (Meyer [276]). *Given rational matrices A, G and a rational vector b, let $P := \{(x,y) : Ax + Gy \le b\}$ and let $S := \{(x,y) \in P : x \text{ integral}\}$.*

1. *There exist rational matrices A', G' and a rational vector b' such that $\text{conv}(S) = \{(x,y) : A'x + G'y \le b'\}$.*

2. *If S is nonempty, the recession cones of $\text{conv}(S)$ and P coincide.*

Fig. 4.6 illustrates Meyer's theorem and its proof.

Proof. The theorem is obvious if S is empty, so we assume that S is nonempty. By Theorem 3.13, there exist $v^1, \ldots, v^t$ and $r^1, \ldots, r^q$ such that $P = \text{conv}(v^1, \ldots, v^t) + \text{cone}(r^1, \ldots, r^q)$. Since P is a rational polyhedron, by Proposition 3.12 we can assume that the vectors $v^1, \ldots, v^t$ are rational and that $r^1, \ldots, r^q$ are integral. Consider the following truncation of P

$$T := \{(x,y) : (x,y) = \sum_{i=1}^{t} \lambda_i v^i + \sum_{j=1}^{q} \mu_j r^j,\ \sum_{i=1}^{t} \lambda_i = 1, \lambda \ge 0,\ 0 \le \mu \le 1\}. \tag{4.25}$$

T is a polytope whose vertices are of the form $v^i + \sum_{j \in Q} r^j$ for some $i \in \{1, \ldots, t\}$ and some $Q \subseteq \{1, \ldots, q\}$. Since $v^1, \ldots, v^t$ and $r^1, \ldots, r^q$ are rational, T is a rational polytope. Let $T_I := \{(x,y) \in T : x \text{ integral}\}$, and let $R_I := \{\sum_{i=1}^{q} \mu_i r^i : \mu \in \mathbb{Z}_+^q\}$. We show next that

$$S = T_I + R_I. \tag{4.26}$$

Clearly $S \supseteq T_I + R_I$. For the reverse inclusion, let $(\bar{x}, \bar{y})$ be a point in S. Then $\bar{x}$ is integral and there exist multipliers $\lambda \ge 0$, $\sum_{i=1}^{t} \lambda_i = 1$, and $\mu \ge 0$ such that

$$(\bar{x}, \bar{y}) = \sum_{i=1}^{t} \lambda_i v^i + \sum_{j=1}^{q} \mu_j r^j.$$

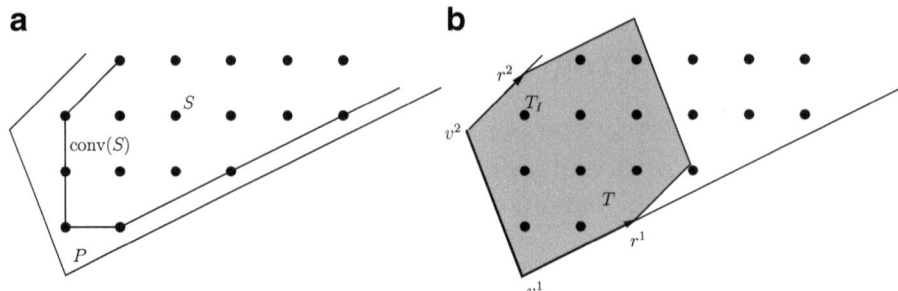

Figure 4.6: (a) Illustration of Meyer's theorem, (b) illustration of the proof

Let $(x', y') := \sum_{i=1}^{t} \lambda_i v^i + \sum_{j=1}^{q}(\mu_j - \lfloor \mu_j \rfloor)r^j$ and $r := \sum_{j=1}^{q} \lfloor \mu_j \rfloor r^j$. Then $(\bar{x}, \bar{y}) = (x', y') + r$. Note that r is integral and thus also x' is integral. Since $0 \le \mu_j - \lfloor \mu_j \rfloor \le 1$, the point (x', y') is in T_I, while by definition $r \in R_I$. It follows that $(\bar{x}, \bar{y}) \in T_I + R_I$.

Claim. $\text{conv}(T_I)$ *is a rational polytope.*

Since T is a polytope, the set $X := \{x : \exists y \text{ s.t. } (x, y) \in T_I\}$ is finite. For fixed $\bar{x} \in X$, the set $T_{\bar{x}} := \{(\bar{x}, y) : (\bar{x}, y) \in T_I\}$ is a rational polytope, thus it is the convex hull of the set $V_{\bar{x}}$ of its vertices. Since the set X is finite, also the set $V := \bigcup_{\bar{x} \in X} V_{\bar{x}}$ is finite. Since $\text{conv}(T_I) = \text{conv}(V)$, we have that $\text{conv}(T_I)$ is a rational polytope and this proves the claim.

Recall that the convex hull of a Minkowski sum is the Minkowski sum of the convex hulls (Exercise 3.10). Therefore (4.26) implies $\text{conv}(S) = \text{conv}(T_I) + \text{conv}(R_I)$. Since $\text{conv}(R_I) = \text{cone}(r^1, \ldots, r^q)$, we have $\text{conv}(S) = \text{conv}(T_I) + \text{cone}(r^1, \ldots, r^q)$. By the above claim, $\text{conv}(T_I)$ is a rational polytope. Thus $\text{conv}(S)$ is a rational polyhedron having the same recession cone as P, namely $\text{cone}(r^1, \ldots, r^q)$. □

Given $R \subseteq \mathbb{R}^n$, we denote by $\text{intcone}(R)$ the set of nonnegative integer combinations of vectors in R, i.e., $\text{intcone}(R) := \{\sum_{i=1}^{t} \mu_i r^i : r^i \in R, \mu_i \in \mathbb{Z}_+, i = 1, \ldots, t\}$.

Corollary 4.31. *Let $P \subseteq \mathbb{R}^{n+p}$ be a rational polyhedron and $S := P \cap (\mathbb{Z}^n \times \mathbb{R}^p)$. There exist finitely many rational polytopes $P_1, \ldots, P_k \subseteq \mathbb{R}^{n+p}$ and vectors $r^1, \ldots, r^q \in \mathbb{Z}^{n+p}$ such that*

$$S = \bigcup_{i=1}^{k} P_i + \text{intcone}\{r^1, \ldots, r^q\}.$$

4.8. THE FUNDAMENTAL THEOREM OF INTEGER...

Proof. Following the proof of Theorem 4.30, by (4.26) $S = T_I + R_I$, where $R_I = \text{intcone}\{r^1, \ldots, r^q\}$ and T_I is the union of finitely many rational polytopes. □

Remark 4.32. *In Theorem 4.30:*

- *If matrices A, G are not rational, then $\text{conv}(S)$ may not be a polyhedron, see Exercise 4.30.*

- *If A, G are rational matrices but b is not rational, then $\text{conv}(S)$ is a polyhedron that has the same recession cone as P. However $\text{conv}(S)$ is not always a rational polyhedron. (This can be inferred from the above proof).*

Next we give an explicit construction of $\text{conv}(S)$ for the case of the "mixing set" (recall Sect. 2.9).

4.8.1 An Example: The Mixing Set

We consider the mixed integer linear set, known as the *mixing set*:

$$MIX := \{(x_0, \ldots, x_n) \in \mathbb{R} \times \mathbb{Z}^n : x_0 + x_t \geq b_t, \ t = 1, \ldots, n, \ x_0 \geq 0\},$$

where $b_1, \ldots, b_n \in \mathbb{Q}$. Let $f_t = b_t - \lfloor b_t \rfloor$. We assume that $0 \leq f_1 \leq \cdots \leq f_n$ and we define $f_0 := 0$. Note that MIX is always nonempty.

Let $P^{\text{mix}} := \text{conv}(MIX)$. Since P^{mix} is nonempty, it follows by Theorem 4.30 that P^{mix} is a rational polyhedron, and that its recession cone is equal to the recession cone of the polyhedron $P := \{x \in \mathbb{R}^{n+1} : x_0 + x_t \geq b_t, \ t = 1, \ldots, n, \ x_0 \geq 0\}$. Note that P is the polyhedron considered in Exercise 3.29 and the extreme rays of P^{mix} are the $n+1$ vectors $r^0, \ldots, r^n$ defined by:

$$r_i^0 = \begin{cases} 1 & \text{for } i = 0 \\ -1 & \text{otherwise} \end{cases} \qquad r_i^t = \begin{cases} 1 & \text{for } i = t \\ 0 & \text{otherwise} \end{cases} \qquad t = 1, \ldots, n.$$

Proposition 4.33. *The number of vertices of P^{mix} is the number of distinct values in the sequence $f_0, f_1, \ldots, f_n$. Furthermore, the vertices of P^{mix} are among the $n+1$ points $v^0, \ldots, v^n$ defined by*

$$v_i^t = \begin{cases} f_t & \text{if } i = 0 \\ \lfloor b_i \rfloor & \text{if } 1 \leq i \leq t \\ \lceil b_i \rceil & \text{if } t+1 \leq i \leq n \end{cases} \qquad t = 0, \ldots, n.$$

Proof. Let $\bar{x}$ be a vertex of P^{mix}. We first show $\bar{x}_0 < 1$. Suppose not. Then P^{mix} contains both points $\bar{x}+r^0$ and $\bar{x}-r^0$. Since $\bar{x} = \frac{1}{2}((\bar{x}+r^0)+(\bar{x}-r^0))$, $\bar{x}$ is not a vertex, a contradiction.

We next show that either $\bar{x}_0 = 0$ or $\bar{x}_0 = f_t$ for some $t = 1,\ldots,n$. Suppose that $\bar{x}_0 \neq 0$ and $\bar{x}_0 \neq f_t, 1 \leq t \leq n$. Since $\bar{x}_1\ldots,\bar{x}_n$ are integer, $\bar{x}$ does not satisfy at equality any of the inequalities $x_0 + x_t \geq b_t$, $1 \leq t \leq n$, $x_0 \geq 0$. Therefore $(\bar{x}_0 \pm \epsilon, \bar{x}_1, \ldots, \bar{x}_n) \in P^{\mathrm{mix}}$ for $\epsilon > 0$ sufficiently small, thus $\bar{x}$ is not a vertex.

A similar argument shows that $\bar{x}_i = \lceil b_i - \bar{x}_0 \rceil$, $i = 1, \ldots, n$. Hence $\bar{x} = v^t$ for some $t \in \{0, \ldots, n\}$, and the number of vertices of P^{mix} is the number of distinct values in the sequence $f_0, f_1, \ldots, f_n$. □

From the above proposition, $P^{\mathrm{mix}} = \mathrm{conv}(v^0, \ldots, v^n) + \mathrm{cone}(r^0, \ldots, r^n)$. Therefore $P^{\mathrm{mix}} = \{x \in \mathbb{R}^{n+1} : \exists (\lambda, \mu) \in \mathbb{R}^{n+1}_+ \times \mathbb{R}^{n+1}_+ \text{ s.t. } (x, \lambda, \mu) \in Q\}$, where Q is

$$\{(x, \lambda, \mu) \in \mathbb{R}^{n+1} \times \mathbb{R}^{n+1}_+ \times \mathbb{R}^{n+1}_+ : x - \sum_{t=0}^n v^t \lambda_t - \sum_{t=0}^n r^t \mu_t = 0, \sum_{t=0}^n \lambda_t = 1\}. \quad (4.27)$$

We now obtain P^{mix} by projecting Q onto the space of x variables. By Theorem 3.46, every facet-defining inequality for P^{mix} is of the form $\sum_{i=0}^n \alpha_i x_i \geq \beta$, where $(\alpha_0, \ldots, \alpha_n, \beta)$ is an extreme ray of the projection cone, which is described by the following inequalities

$$\begin{aligned} -\alpha v^t + \beta &\leq 0 \quad t = 0, \ldots, n \\ -\alpha r^t &\leq 0 \quad t = 0, \ldots, n. \end{aligned}$$

By the change of variables $x_t := x_t - \lfloor b_t \rfloor$, $t = 1, \ldots, n$, we may assume that $b_t = f_t$. Under this transformation, the points v^t become

$$v^t_i = \begin{cases} f_t & \text{if } i = 0 \\ 0 & \text{if } 1 \leq i \leq t \\ 1 & \text{if } t+1 \leq i \leq n \end{cases} \quad t = 0, \ldots, n.$$

Therefore the above system can be written as

$$\begin{aligned} f_t \alpha_0 + \sum_{i=t+1}^n \alpha_i &\geq \beta \quad t = 0, \ldots, n-1 \\ f_n \alpha_0 &\geq \beta \\ \alpha_0 - \sum_{i=1}^n \alpha_i &\geq 0 \\ \alpha_1, \ldots, \alpha_n &\geq 0. \end{aligned} \quad (4.28)$$

Let $(\bar{\alpha}, \bar{\beta})$ be an extreme ray of the projection cone. Note that the inequalities $\alpha_0 - \sum_{i=1}^n \alpha_i \geq 0$ and $\alpha_1, \ldots, \alpha_n \geq 0$ imply that $\bar{\alpha}_0 > 0$. We may therefore assume that $\bar{\alpha}_0 = 1$. Hence if $\bar{\alpha}_1, \ldots, \bar{\alpha}_n = 0$, since $f_0 = 0$, we must

4.8. THE FUNDAMENTAL THEOREM OF INTEGER...

also have $\bar{\beta} = 0$, and the inequality corresponding to this solution is $x_0 \geq 0$. We now assume that the set $I := \{i \in \{1, \ldots, n\} : \bar{\alpha}_i > 0\}$ is nonempty and suppose $I = \{i_1, \ldots, i_m\}$, where $i_1 < \ldots < i_m$. Then $(1, \bar{\alpha}_1, \ldots, \bar{\alpha}_n, \bar{\beta})$ is an extreme ray of (4.28) if and only if $(\bar{\alpha}_{i_1}, \ldots, \bar{\alpha}_{i_m}, \bar{\beta})$ satisfies at equality $m + 1$ linearly independent inequalities among the following $m + 2$

$$\begin{array}{rcl} \sum_{h=1}^{m} \alpha_{i_h} & \geq & \beta \\ f_{i_k} + \sum_{h=k+1}^{m} \alpha_{i_h} & \geq & \beta \quad k = 1, \ldots, m-1 \\ f_{i_m} & \geq & \beta \\ \sum_{h=1}^{m} \alpha_{i_h} & \leq & 1. \end{array}$$

Since $\sum_{h=1}^{m} \alpha_{i_h} \geq \beta$ and $\sum_{h=1}^{m} \alpha_{i_h} \leq 1$ are linearly dependent inequalities, all remaining m inequalities must be satisfied at equality and we have two possibilities.

If $\sum_{h=1}^{m} \bar{\alpha}_{i_h} = \bar{\beta}$, then it can be readily verified that the solution is $\bar{\beta} = f_{i_m}$, $\bar{\alpha}_{i_1} = f_{i_1}$, $\bar{\alpha}_{i_k} = f_{i_k} - f_{i_{k-1}}$ ($k = 2, \ldots, m$), which corresponds to the inequality

$$x_0 + f_{i_1}(x_{i_1} - \lfloor b_{i_1} \rfloor) + \sum_{k=2}^{m}(f_{i_k} - f_{i_{k-1}})(x_{i_k} - \lfloor b_{i_k} \rfloor) \geq f_{i_m}. \quad (4.29)$$

If $\sum_{h=1}^{m} \bar{\alpha}_{i_h} = 1$, then it can be readily verified that the solution is $\bar{\beta} = f_{i_m}$, $\bar{\alpha}_{i_1} = 1 + f_{i_1} - f_{i_m}$, $\bar{\alpha}_{i_k} = f_{i_k} - f_{i_{k-1}}$ ($k = 2, \ldots, m$), which corresponds to the inequality

$$x_0 + (1 + f_{i_1} - f_{i_m})(x_{i_1} - \lfloor b_{i_1} \rfloor) + \sum_{k=2}^{m}(f_{i_k} - f_{i_{k-1}})(x_{i_k} - \lfloor b_{i_k} \rfloor) \geq f_{i_m}. \quad (4.30)$$

Observe that, in both cases, we must have $0 < f_{i_1} < \ldots < f_{i_m}$ because $\bar{\alpha}_j > 0$ for all $j \in I$. Inequalities (4.29), (4.30) are called the *mixing inequalities*. Note that inequalities (4.30) comprise $x_0 + x_t \geq b_t$, $t = 1, \ldots, n$ (obtained when $m = 1$ and $i_1 = t$ in (4.30)). We have just proved the following result due to Günlük and Pochet [195].

Theorem 4.34. P^{mix} *is described by the inequality $x_0 \geq 0$ and the mixing inequalities (4.29),(4.30), for all sequences $1 \leq i_1 < \ldots < i_m \leq n$ such that $0 < f_{i_1} < \ldots < f_{i_m}$.*

4.8.2 Mixed Integer Linear Programming is in NP

We conclude this section by proving an important consequence of Theorem 4.30, namely, that the problem of deciding whether a mixed integer linear set is nonempty is in NP. Since we can always express a free variable as the difference of two nonnegative variables, we can assume that all variables in the mixed integer linear set are nonnegative.

Lemma 4.35. *Given $A \in \mathbb{Q}^{m \times n}$, $G \in \mathbb{Q}^{m \times p}$, and $b \in \mathbb{Q}^m$, let $S := \{(x, y) \in \mathbb{Z}^n \times \mathbb{R}^p : Ax + Gy \leq b, \ x \geq 0, \ y \geq 0\}$. If the encoding size of every coefficient of (A, G, b) is at most L, then every vertex of $\mathrm{conv}(S)$ has an encoding whose size is polynomially bounded by $n + p$ and L.*

Proof. Let $P := \{(x, y) : Ax + Gy \leq b, \ x \geq 0, \ y \geq 0\}$. Let $v^1, \ldots, v^t$ be the vertices of P and $r^1, \ldots, r^q$ be its extreme rays. By Theorem 3.38, $v^1, \ldots, v^t$ and $r^1, \ldots, r^q$ can be written as rational vectors whose encoding size is polynomially bounded by $n + p$ and L. By Remark 1.1, we may assume that $r^1, \ldots, r^q$ are integral vectors.

Let $(\bar{x}, \bar{y})$ be a vertex of $\mathrm{conv}(S)$. Then $\bar{x}$ is integral and there exist multipliers $\lambda \geq 0$, $\sum_{i=1}^{t} \lambda_i = 1$ and $\mu \geq 0$ such that $(\bar{x}, \bar{y}) = \sum_{i=1}^{t} \lambda_i v^i + \sum_{j=1}^{q} \mu_j r^j$.

By Carathéodory's theorem (Theorem 3.42), we may assume that at most $n+p+1$ components of (λ, μ) are positive. Furthermore, it follows from the proof of Meyer's theorem (Theorem 4.30) that $\mu_i \leq 1$ for $i = 1, \ldots, q$. If θ is the largest absolute value of an entry in the vectors $v^1, \ldots, v^t, r^1, \ldots, r^q$, it follows that $|\bar{x}_i| \leq (n+p+1)\theta$, $i = 1, \ldots, n$. Since $\bar{x}$ is integral, it follows that the encoding size of $\bar{x}$ is polynomially bounded by $n + p$ and L.

Furthermore, since $(\bar{x}, \bar{y})$ is a vertex of $\mathrm{conv}(S)$, it follows that $\bar{y}$ is a vertex of the polyhedron $\{y \in \mathbb{R}^p : Gy \leq b - A\bar{x}, \ y \geq 0\}$. Since the encoding size of $\bar{x}$ is polynomial, Theorem 3.38 implies that the encoding size of $\bar{y}$ is also polynomially bounded by $n + p$ and L. □

Note that the polynomial bound in the above lemma does not depend on the number m of constraints.

The *MILP feasibility problem* is the following: "Given $A \in \mathbb{Q}^{m \times n}$, $G \in \mathbb{Q}^{m \times p}$, and $b \in \mathbb{Q}^m$, is the mixed integer linear set $S := \{(x, y) \in \mathbb{Z}^n \times \mathbb{R}^p : Ax + Gy \leq b, \ x \geq 0, \ y \geq 0\}$ nonempty?."

Theorem 4.36. *The MILP feasibility problem is in NP.*

Proof. We need to show that, if an instance of the MILP feasibility problem has a "yes" answer, then there exists a certificate that can be checked in polynomial time. Since $S \neq \emptyset$, $\mathrm{conv}(S)$ has a vertex $(\bar{x}, \bar{y})$. By Lemma 4.35, $(\bar{x}, \bar{y})$ provides such a certificate, since one can verify in polynomial time that $\bar{x}$ is integral and $A\bar{x} + G\bar{y} \leq b$. □

4.8. THE FUNDAMENTAL THEOREM OF INTEGER...

The following fact is a consequence of Lemma 4.35, and will be used in Chap. 9 when discussing Lenstra's polynomial-time algorithm for pure integer linear programming in fixed dimension.

Corollary 4.37. *Given $A \in \mathbb{Q}^{m \times n}$ and $b \in \mathbb{Q}^m$, let L be the maximum encoding size of the coefficients of (A, b). If the system $Ax \leq b$ has an integral solution, then it has an integral solution $\bar{x}$ whose encoding size is polynomially bounded by n and L.*

Proof. The only difficulty arises from the fact that $P := \{x \in \mathbb{R}^n : Ax \leq b\}$ might not be pointed. However, $Ax \leq b$ has an integral solution if and only if the system $Ax^+ - Ax^- \leq b$, $x^+, x^- \geq 0$ has an integral solution. By Lemma 4.35, if the latter has an integral solution, then it has one, say $(\bar{x}^+, \bar{x}^-)$, whose encoding size is polynomially bounded by n and L. It follows that $\bar{x} := \bar{x}^+ - \bar{x}^-$ is an integral solution of $Ax \leq b$ whose encoding size is polynomially bounded by n and L. □

It is important to note that, in all these results, *explicit* polynomial bounds could always be constructed by working carefully through the proofs. This, however, is quite laborious. We refer to [325], Sect. 17.1.

4.8.3 Polynomial Encoding of the Facets of the Integer Hull

Lemma 4.38. *Given $A \in \mathbb{Q}^{m \times n}$, $G \in \mathbb{Q}^{m \times p}$, and $b \in \mathbb{Q}^m$, let $S := \{(x, y) \in \mathbb{Z}^n \times \mathbb{R}^p : Ax + Gy \leq b, x \geq 0, y \geq 0\}$. If the encoding size of every coefficient of (A, G, b) is at most L, then for every facet F of $\mathrm{conv}(S)$ there exists an inequality $\alpha z \leq \beta$ defining F such that the encoding size of the vector (α, β) is polynomially bounded by $n + p$ and L.*

Proof. By Lemma 4.35 the encoding size of the vertices of $\mathrm{conv}(S)$ is polynomially bounded by $n + p$ and L. By Theorem 4.30, the recession cone of $\mathrm{conv}(S)$ is $\{(x, y) : Ax + Gy \leq 0, x \geq 0, y \geq 0\}$, therefore by Theorem 3.38 the extreme rays of $\mathrm{conv}(S)$ can be written as rational vectors whose size is polynomially bounded by $n + p$ and L. The statement now follows from Theorem 3.39. □

Explicit bounds for the length of the encoding of a facet of $\mathrm{conv}(S)$ are given in [325], Sect. 17.4.

4.9 Union of Polyhedra

In this section, we present a result of Balas [24, 26] about the union of k polyhedra in $\mathbb{R}^n$. See Fig. 4.7 for an example. Recall that we treated the special case of bounded polyhedra in Sect. 2.11. We now present the general case.

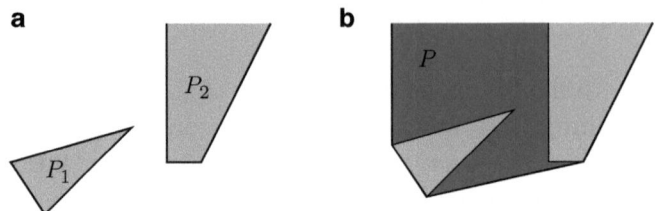

Figure 4.7: (a) Polyhedra P_1 and P_2 (note that P_2 is unbounded) (b) $P := \overline{\text{conv}}(P_1 \cup P_2)$

Theorem 4.39 (Balas [24, 26]). *Given k polyhedra $P_i := \{x \in \mathbb{R}^n : A_i x \leq b^i\}$, $i = 1, \ldots k$, let $C_i := \{x : A_i x \leq 0\}$, and let $R^i \subset \mathbb{R}^n$ be a finite set such that $C_i = \text{cone}(R^i)$. For every $i \in \{1, \ldots, k\}$ such that $P_i \neq \emptyset$, let $V^i \subset \mathbb{R}^n$ be a finite set such that $P_i = \text{conv}(V^i) + \text{cone}(R^i)$.*

Consider the polyhedron

$$P := \text{conv}(\bigcup_{i:P_i \neq \emptyset} V^i) + \text{cone}(\bigcup_{i=1}^k R^i)$$

and let $Y \subseteq \mathbb{R}^n \times (\mathbb{R}^n)^k \times \mathbb{R}^k$ be the polyhedron described by the following system

$$\begin{array}{rcll} A_i x^i & \leq & \delta_i b^i & i = 1, \ldots, k \\ \sum_{i=1}^k x^i & = & x & \\ \sum_{i=1}^k \delta_i & = & 1 & \\ \delta_i & \geq & 0 & i = 1, \ldots, k. \end{array} \quad (4.31)$$

Then $P = \text{proj}_x(Y) := \{x \in \mathbb{R}^n : \exists (x^1, \ldots, x^k, \delta) \in (\mathbb{R}^n)^k \times \mathbb{R}^k \text{ s.t. } (x, x^1, \ldots, x^k, \delta) \in Y\}$.

Proof. Assume first that $P = \emptyset$. This implies $P_i = \emptyset$ for all $i = 1, \ldots, k$. Since the system describing Y, includes $\delta_i \geq 0$, $i = 1, \ldots, k$, and $\sum_{i=1}^k \delta_i = 1$, then at least one of the variables δ_i is positive. Since the system $A_i x^i \leq b^i$ is infeasible, the system describing Y is also infeasible. This shows that the theorem holds in this case.

4.9. UNION OF POLYHEDRA

We now assume that P is nonempty, i.e., the index set $K := \{1, \ldots, k\}$ can be partitioned into a nonempty set $K^N := \{i : P_i \neq \emptyset\}$ and $K^E := K \setminus K^N$.

Let $x \in P$. Then there exist points $v^i \in \text{conv}(V^i)$ and scalars $\delta_i \geq 0$ for $i \in K^N$, and vectors $r^i \in C^i$ for $i \in K$, such that

$$x = \sum_{i \in K^N} \delta_i v^i + \sum_{i \in K^N} r^i + \sum_{i \in K^E} r^i, \quad \sum_{i \in K^N} \delta_i = 1.$$

For $i \in K^N$ define $x^i := \delta_i v^i + r^i$, and for $i \in K^E$ define $x^i := r^i$ and $\delta_i := 0$. By construction, $x = \sum_{i \in K} x^i$. Since $r^i \in C_i$, it follows that $A_i x^i \leq \delta_i b^i$ for all $i \in K$. Thus $(x, x^1, \ldots, x^k, \delta) \in Y$. This shows $P \subseteq \text{proj}_x(Y)$.

Let $(x, x^1, \ldots, x^k, \delta)$ be a vector in Y. Let $K^P := \{i \in K : \delta^i > 0\}$ and let $z^i := \frac{x^i}{\delta_i}$, $i \in K^P$. Then $A_i z^i \leq b^i$. So $P_i \neq \emptyset$ in this case and $z^i \in \text{conv}(V^i) + \text{cone}(R^i)$.

For $i \in K \setminus K^P$, $A_i x^i \leq 0$, and therefore $x^i \in \text{cone}(R^i)$. Since $x = \sum_{i \in K^P} \delta_i z^i + \sum_{i \in K \setminus K^P} x^i$ and $\sum_{i \in K^P} \delta_i = 1, \delta_i \geq 0$, it follows that $x \in \text{conv}(\bigcup_{i \in K^P} V^i) + \text{cone}(\bigcup_{i \in K} R^i)$ and therefore $x \in P$. This shows that $\text{proj}_x(Y) \subseteq P$. □

Remark 4.40. *Theorem 4.39 shows that the system of inequalities describing Y gives an extended formulation of the polyhedron P that uses $nk+n+k$ variables and the size of this formulation is approximately the sum of the sizes of the formulations that describe the polyhedra P_i.*

The polyhedron P defined in Theorem 4.39 contains the convex hull of $\bigcup_{i=1}^{k} P_i$ but in general this inclusion is strict. Indeed, the recession cone of P contains cone $C_i, i = 1, \ldots, k$, even if P_i is empty. Furthermore, even if the polyhedra P_i are all nonempty but have different recession cones, the set $\text{conv}(\bigcup_{i=1}^{k} P_i)$ may not be closed, and therefore it may not be a polyhedron. For example, in $\mathbb{R}^2$, the convex hull of a line L and a point not in L is not a closed set. The next lemma shows that, when the polyhedra P_i are all nonempty, the polyhedron P defined in Theorem 4.39 satisfies $P = \overline{\text{conv}}(\bigcup_{i=1}^{k} P_i)$ (where for any set $X \subseteq \mathbb{R}^n$ we denote by $\overline{\text{conv}}(X)$ the topological closure of $\text{conv}(X)$).

Lemma 4.41. *Let $P_1, \ldots, P_k \subseteq \mathbb{R}^n$ be nonempty polyhedra. For $i = 1, \ldots, k$, let $V^i, R^i \subset \mathbb{R}^n$ be finite sets such that $P_i = \text{conv}(V^i) + \text{cone}(R^i)$. Then*

$$\overline{\text{conv}}(\cup_{i=1}^{k} P_i) = \text{conv}(\cup_{i=1}^{k} V^i) + \text{cone}(\cup_{i=1}^{k} R^i).$$

Proof. Let $Q := \text{conv}(\cup_{i=1}^{k} V^i)$ and $C := \text{cone}(\cup_{i=1}^{k} R^i)$.

We first show $\overline{\text{conv}}(\cup_{i=1}^{k} P_i) \subseteq Q + C$. Note that we just need to show $\text{conv}(\cup_{i=1}^{k} P_i) \subseteq Q + C$ because $Q + C$ is a polyhedron, and so it is closed. Let $x \in \text{conv}(\cup_{i=1}^{k} P_i)$. Then x is a convex combination of a finite number of points in $\cup_{i=1}^{k} P_i$. Since P_i is convex, we can write x as a convex combination of points $x^i \in P_i$, say $x = \sum_{i=1}^{k} \lambda_i x^i$ where $\lambda_i \geq 0$ for $i = 1, \ldots, k$ and $\sum_{i=1}^{k} \lambda_i = 1$. Since $P_i = \text{conv}(V^i) + \text{cone}(R^i)$, $x^i = v^i + r^i$ where $v^i \in \text{conv}(V^i)$, $r^i \in \text{cone}(R^i)$, thus $x = \sum_{i=1}^{k} \lambda_i v^i + \sum_{i=1}^{k} \lambda_i r^i$, so $x \in Q + C$ since $\sum_{i=1}^{k} \lambda_i v^i \in Q$ and $\sum_{i=1}^{k} \lambda_i r^i \in C$.

We now show $Q + C \subseteq \overline{\text{conv}}(\cup_{i=1}^{k} P_i)$. Let $x \in Q + C$. Then $x = \sum_{i=1}^{k} \lambda_i v^i + \sum_{i=1}^{k} r^i$ where $v^i \in \text{conv}(V^i)$, $r^i \in \text{cone}(R^i)$, $\lambda_i \geq 0$ for $i = 1, \ldots, k$, and $\sum_{i=1}^{k} \lambda_i = 1$. We need to show that there exist points in $\text{conv}(\cup_{i=1}^{k} P_i)$ that are arbitrarily close to x.

Let $I := \{i : \lambda_i > 0\}$. For all $\epsilon > 0$ small enough so that $\lambda_i - \frac{k}{|I|}\epsilon \geq 0$ for all $i \in I$, define the point

$$x^\epsilon := \sum_{i \in I}(\lambda_i - \frac{k}{|I|}\epsilon)v^i + \sum_{i=1}^{k} \epsilon(v^i + \frac{1}{\epsilon}r^i).$$

Observe that $x^\epsilon \in \text{conv}(\cup_{i=1}^{k} P_i)$ because $\sum_{i \in I}(\lambda_i - \frac{k}{|I|}\epsilon) + \sum_{i=1}^{k} \epsilon = 1$ and $v^i + \frac{1}{\epsilon}r^i \in P_i$. Since $\lim_{\epsilon \to 0^+} x^\epsilon = x$, it follows that $x \in \overline{\text{conv}}(\cup_{i=1}^{k} P_i)$. □

Theorem 4.42 (Balas [24, 26]). *Let $P_i := \{x \in \mathbb{R}^n : A_i x \leq b^i\}$ be k polyhedra such that $\cup_{i=1}^{k} P_i \neq \emptyset$, and let Y be the polyhedron defined in Theorem 4.39. Let $C_i := \{x : A_i x \leq 0\}$ and let $R^i \subset \mathbb{R}^n$ be a finite set such that $C_i = \text{cone}(R^i)$, $i = 1, \ldots, k$. Then $\overline{\text{conv}}(\cup_{i=1}^{k} P_i)$ is the projection of Y onto the x-space if and only if $C_j \subseteq \text{cone}(\cup_{i : P_i \neq \emptyset} R^i)$ for every $j = 1, \ldots, k$.*

Proof. For every $i \in \{1, \ldots, k\}$ such that $P_i \neq \emptyset$, let $V^i \subset \mathbb{R}^n$ be a finite set such that $P_i = \text{conv}(V^i) + C_i$. Let $P := \text{conv}(\cup_{i : P_i \neq \emptyset} V^i) + \text{cone}(\cup_{i=1}^{k} R^i)$ and let $P' := \text{conv}(\cup_{i : P_i \neq \emptyset} V^i) + \text{cone}(\cup_{i : P_i \neq \emptyset} R^i)$. By Theorem 4.39 $P = \text{proj}_x(Y)$, whereas by Lemma 4.41 $P' = \overline{\text{conv}}(\cup_{i=1}^{k} P_i)$.

By definition, $P = P'$ if and only if $\text{cone}(\cup_{i=1}^{k} R^i) = \text{cone}(\cup_{i : P_i \neq \emptyset} R^i)$. In turn, this occurs if and only if $C_j \subseteq \text{cone}(\cup_{i : P_i \neq \emptyset} R^i)$ for every $j = 1, \ldots, k$. □

Remark 4.43. *Note that if the k polyhedra $P_i := \{x \in \mathbb{R}^n : A_i x \leq b^i\}$ are all nonempty, then Theorem 4.42 implies that $\overline{\text{conv}}(\cup_{i=1}^{k} P_i)$ is the projection of Y onto the x-space.*

4.9. UNION OF POLYHEDRA

Corollary 4.44. *If $P_1, \ldots, P_k$ are nonempty polyhedra with identical recession cones, then $\mathrm{conv}(\cup_{i=1}^k P_i)$ is a polyhedron.*

Proof. We leave it as an exercise for the reader to check how the last part of the proof of Lemma 4.41 simplifies to show $Q + C \subseteq \mathrm{conv}(\cup_{i=1}^k P_i)$. □

4.9.1 Example: Modeling Disjunctions

Suppose we are given a system of linear constraints in n variables $Ax \leq b$, and we want to further impose the following disjunctive constraint

$$cx \leq d_1 \quad \text{or} \quad cx \geq d_2,$$

where $c \in \mathbb{R}^n$ and $d_1 < d_2$.

If we define $P := \{x \in \mathbb{R}^n : Ax \leq b\}$, $P_1 := \{x \in P : cx \leq d_1\}$, $P_2 := \{x \in P : cx \geq d_2\}$, the set of feasible solutions is $P_1 \cup P_2$. The next lemma shows that $\mathrm{conv}(P_1 \cup P_2)$ is a polyhedron. Note that, to prove this, we cannot apply Corollary 4.44 because P_1 and P_2 may have different recession cones.

Lemma 4.45. *The set $\mathrm{conv}(P_1 \cup P_2)$ is a polyhedron. Furthermore, $\mathrm{conv}(P_1 \cup P_2)$ is the projection onto the space of x variables of the polyhedron Q described by*

$$\begin{aligned} Ax^1 &\leq \lambda b \\ cx^1 &\leq \lambda d_1 \\ Ax^2 &\leq (1-\lambda)b \\ cx^2 &\geq (1-\lambda)d_2 \\ x^1 + x^2 &= x \\ 0 \leq \lambda &\leq 1. \end{aligned}$$

Proof. The lemma holds when $P_1 = P_2 = \emptyset$ by Theorem 4.39. We assume in the remainder that $P_1 \cup P_2 \neq \emptyset$. Let $C_1 = \{r : Ar \leq 0, cr \leq 0\}$ and $C_2 = \{r : Ar \leq 0, cr \geq 0\}$. Note that $\mathrm{rec}(P) = C_1 \cup C_2$.

We observe that, given a point $\bar{x} \in P_1$ and a vector $r \in C_2 \setminus C_1$, (resp. $\bar{x} \in P_2$, $r \in C_1 \setminus C_2$), we have $\bar{x} + r \in \mathrm{conv}(P_1 \cup P_2)$ and $P_2 \neq \emptyset$ (resp. $P_1 \neq \emptyset$). Indeed, given $\bar{x} \in P_1$ and $r \in C_2 \setminus C_1$, it follows that $cr > 0$ and, if we let $\lambda := \max(1, \frac{d_2 - c\bar{x}}{cr})$, the point $\bar{x} + \lambda r$ is in P_2 and $\bar{x} + r$ is in the line segment joining $\bar{x}$ and $\bar{x} + \lambda r$.

The above observation shows that, if $P_2 = \emptyset$ (resp. $P_1 = \emptyset$), then $C_2 \subseteq C_1$ (resp. $C_1 \subseteq C_2$), therefore the cone condition of Theorem 4.42 holds in this case. It also trivially holds when both $P_1, P_2 \neq \emptyset$. Thus, in all cases, Theorem 4.42 implies that $\overline{\mathrm{conv}}(P_1 \cup P_2)$ is the projection onto the space of x variables of the polyhedron Q defined in the statement of the lemma.

Therefore, to prove the lemma, we only need to show $\overline{\text{conv}}(P_1 \cup P_2) = \text{conv}(P_1 \cup P_2)$. We assume $P_1, P_2 \neq \emptyset$ otherwise the statement is obvious. Let $Q_1, Q_2 \subset \mathbb{R}^n$ be two polytopes such that $P_1 = Q_1 + C_1$ and $P_2 = Q_2 + C_2$. By Lemma 4.41, and because $\text{rec}(P) = C_1 \cup C_2$, $\overline{\text{conv}}(P_1 \cup P_2) = \text{conv}(Q_1 \cup Q_2) + \text{rec}(P)$, thus we only need to show that $\text{conv}(Q_1 \cup Q_2) + \text{rec}(P) \subseteq \text{conv}(P_1 \cup P_2)$. Let $\bar{x} \in \text{conv}(Q_1 \cup Q_2) + \text{rec}(P)$. Then there exist $x^1 \in Q_1$, $x^2 \in Q_2$, $0 \leq \lambda \leq 1$, $r \in \text{rec}(P)$, such that $\bar{x} = \lambda x^1 + (1-\lambda)x^2 + r$. By symmetry we may assume $\lambda > 0$. By the initial observation, $x^1 + \frac{r}{\lambda} \in \text{conv}(P_1 \cup P_2)$, thus $\bar{x} = \lambda(x^1 + \frac{r}{\lambda}) + (1-\lambda)x^2 \in \text{conv}(P_1 \cup P_2)$. □

4.9.2 Example: All the Even Subsets of a Set

Consider the "all even" set $S_n^{\text{even}} := \{x \in \{0,1\}^n : x \text{ has an even number of ones}\}$. The following theorem, due to Jeroslow [212], characterizes the facet-defining inequalities of $\text{conv}(S_n^{\text{even}})$. We do not give the proof, but we give some further details in Exercise 4.31.

Theorem 4.46. *Let S be the family of subsets of $N = \{1, \ldots, n\}$ having odd cardinality. Then*

$$\text{conv}(S_n^{\text{even}}) = \left\{ x \in \mathbb{R}^n : \begin{array}{ll} \sum_{i \in S} x_i - \sum_{i \in N \setminus S} x_i \leq |S| - 1, & S \in \mathcal{S} \\ 0 \leq x_i \leq 1, & i \in N \end{array} \right\}.$$

The formulation of $\text{conv}(S_n^{\text{even}})$ given in Theorem 4.46 has exponentially many constraints. However, Theorem 4.39 gives us the means of obtaining a compact extended formulation. We present it next.

Let $S_n^k := \{x \in \{0,1\}^n : x \text{ has } k \text{ ones}\}$. It is easy to show that $\text{conv}(S_n^k) = \{x \in \mathbb{R}^n : 0 \leq x_i \leq 1, i \in N; \sum_{i \in N} x_i = k\}$ (for example, this follows from the total unimodularity of the constraint matrix).

Let $N^{\text{even}} := \{k : 0 \leq k \leq n, k \text{ even}\}$. Consider the polytope Q described by the following system:

$$x_i - \sum_{k \in N^{\text{even}}} x_i^k = 0 \quad i \in N$$

$$\sum_{i \in N} x_i^k = k\lambda_k \quad k \in N^{\text{even}}$$

$$\sum_{k \in N^{\text{even}}} \lambda_k = 1$$

$$x_i^k \leq \lambda_k \quad i \in N, k \in N^{\text{even}}$$

$$x_i^k \geq 0 \quad i \in N, k \in N^{\text{even}}$$

$$\lambda_k \geq 0 \quad k \in N^{\text{even}}.$$

4.9. UNION OF POLYHEDRA

Since $S_n^{\text{even}} = \bigcup_{k \in N_{\text{even}}} S_n^k$, by Theorem 4.39, we have that $\text{conv}(S_n^{\text{even}}) = \text{proj}_x(Q)$.

4.9.3 Mixed Integer Linear Representability

Here we present a result of Jeroslow and Lowe [213]. Given a set $S \subseteq \mathbb{R}^n$, we are interested in understanding whether S can be described as the set of feasible solutions to a mixed integer linear set, possibly using additional variables. Clearly, optimizing a linear function over S and over $\text{conv}(S)$ is equivalent, but if the objective function is nonlinear, then this is not the same, so the problem arises. We say that S is *mixed integer linear representable* if there exist *rational* matrices A, B, C and a *rational* vector d such that S is the set of points x for which there exist y and z satisfying

$$Ax + By + Cz \leq d \tag{4.32}$$
$$x \in \mathbb{R}^n,\ y \in \mathbb{R}^p,\ z \in \mathbb{Z}^q. \tag{4.33}$$

In other words, S is mixed integer linear representable if it is the projection onto the x-space of the region described by (4.32)–(4.33). Deciding whether a set S has this property is an interesting problem already for $n = 1$.

As a first example, consider the set $S_1 := \{0\} \cup [2, 3] \cup [4, 5] \cup [6, 7] \cup \ldots$ (first set in Fig. 4.8). This set can be represented as the following mixed integer linear set:

$$x = y + 2z$$
$$0 \leq y \leq 1,\ y \leq z,\ z \geq 0$$
$$x \in \mathbb{R},\ y \in \mathbb{R},\ z \in \mathbb{Z}.$$

Similarly, the set $S_2 := \{0\} \cup [3/2, +\infty)$ (second set in Fig. 4.8) has the following mixed integer linear representation:

$$x = y + (3/2)z$$
$$0 \leq y \leq 3/2,\ y \leq (3/2)z,\ z \geq 0$$
$$x \in \mathbb{R},\ y \in \mathbb{R},\ z \in \mathbb{Z}.$$

Now consider the set $S_3 := [0, 1] \cup \{2, 3, 4, \ldots\}$ (third set in Fig. 4.8), which can be seen as a counterpart of S_1 with the roles of segments and points switched. We will see that, unlike S_1, set S_3 is not mixed integer linear representable. The same holds for the sets $S_4 := \{x : x = 2^r,\ r \in \mathbb{N}\}$ and $S_5 := (-\infty, 0] \cup [1, +\infty)$ (last two sets in Fig. 4.8).

Theorem 4.47. *A set $S \subseteq \mathbb{R}^n$ is mixed integer linear representable if and only if there exist rational polytopes $P_1, \ldots, P_k \subseteq \mathbb{R}^n$ and vectors $r^1, \ldots, r^t \in \mathbb{Z}^n$ such that*

$$S = \bigcup_{i=1}^{k} P_i + \text{intcone}\left\{r^1, \ldots, r^t\right\}. \tag{4.34}$$

Proof. We first prove the "only if" part. Let $S \subseteq \mathbb{R}^n$ be a mixed integer linear representable set. Then S is the projection onto the x-space of a set $\tilde{S}$ of the form (4.32)–(4.33), where A, B, C are rational matrices and d is a rational vector. Let Q be the rational polyhedron described by system (4.32), and let $\tilde{S} := \{(x, y, z) \in Q : z \in \mathbb{Z}^q\}$. By Corollary 4.31, there exist finitely many rational polytopes $Q_1, \ldots, Q_k$ and integral vectors $s^1, \ldots, s^t$ such that $\tilde{S} = \cup_{i=1}^{k} Q_i + \text{intcone}\left\{s^1, \ldots, s^t\right\}$.

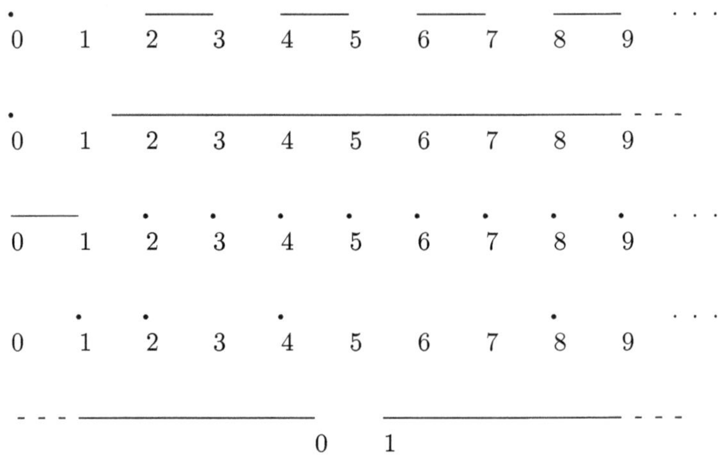

Figure 4.8: Only the first two of these five subsets of $\mathbb{R}$ are mixed integer linear representable

Defining $P_i := \text{proj}_x Q_i$, $i = 1, \ldots, k$, and letting r^j be the vector obtained from s^j by dropping the y and z-components, $j = 1, \ldots, t$, we have $S = \cup_{i=1}^{k} P_i + \text{intcone}\left\{r^1, \ldots, r^t\right\}$. Since each set P_i is a rational polytope, the conclusion follows.

For the converse, assume that S has the form (4.34). For $i = 1, \ldots, k$, let $A_i x \leq b^i$ be a rational linear system describing P_i. By introducing, for each $i = 1, \ldots, k$, an indicator variable δ_i whose value is 1 if and only if we select a point in P_i, we can write S as the set of vectors $x \in \mathbb{R}^n$ that satisfy the following conditions:

4.9. UNION OF POLYHEDRA

$$x = \sum_{i=1}^{k} x^i + \sum_{j=1}^{t} \mu_j r^j$$

$$A_i x^i \leq \delta_i b^i, \qquad i = 1, \ldots, k$$

$$\sum_{i=1}^{k} \delta_i = 1 \tag{4.35}$$

$$\delta, \mu \geq 0$$

$$\delta \in \mathbb{Z}^k, \ \mu \in \mathbb{Z}^t. \tag{4.36}$$

(Note that since each P_i is a polytope, when $\delta_i = 0$ we have $x_i = 0$.) Thus S is mixed integer linear representable. □

We remark that, in the statement of Theorem 4.47, the vectors $r^1, \ldots, r^t$ can equivalently be required to be rational instead of integral. We also note that the family of sets that can be written in the form (4.34) contains all rational polyhedra: indeed, it is known (and shown in the above proof) that a rational polyhedron Q can always be written as $Q = P + \text{intcone}\{r^1, \ldots, r^t\}$ for some polytope P and some integral vectors $r^1, \ldots, r^t$.

We remark that the sets S_3, S_4 and S_5 are not mixed integer linear representable, as they cannot be written in the form (4.34). Consider for instance the set S_3 and assume by contradiction that it can be written in the form (4.34). Since S_3 is not bounded from above, the set $\{r^1, \ldots, r^t\}$ contains at least one positive integer; and since S_3 contains the segments $[0, 1]$, at least one of the polytopes $P_1, \ldots, P_k$, say P_1, is a segment. But then $P_1 + \text{intcone}\{r^1, \ldots, r^t\}$ would necessarily contain some point not in S_3, a contradiction.

Finally, we observe that if a set is mixed integer linear representable, then its convex hull is the projection of the polyhedron described by the linear relaxation of the representation (4.35) given in the proof of Theorem 4.47.

Corollary 4.48. *Let S be a set of the form (4.34). For $i = 1, \ldots, k$, let $A_i x \leq b^i$ be a rational linear system describing P_i. Let Q be the polyhedron defined by the inequalities in (4.35). Then $\text{conv}(S) = \text{proj}_x(Q)$.*

Proof. If (4.34) holds, then clearly

$$S \subseteq \text{conv}(P_1 \cup \cdots \cup P_k) + \text{cone}\{r^1, \ldots, r^t\} \subseteq \text{conv}(S).$$

Because the set in the middle is a polyhedron and $\mathrm{conv}(S)$ is the smallest convex set containing S, we have indeed $\mathrm{conv}(S) = \mathrm{conv}(P_1 \cup \cdots \cup P_k) + \mathrm{cone}\{r^1, \ldots, r^t\}$. Since $P_1, \ldots, P_k$ are polytopes, a description of $\mathrm{conv}(P_1 \cup \cdots \cup P_k)$ by means of linear inequalities is given by Theorem 4.39:

$$x = \sum_{i=1}^{k} x^i$$
$$A_i x^i \leq \delta_i b^i, \quad i = 1, \ldots, k$$
$$\sum_{i=1}^{k} \delta_i = 1$$
$$\delta_i \geq 0, \quad i = 1, \ldots, k.$$

Now, in order to obtain a linear formulation for $\mathrm{conv}(S)$, we have to take into account the rays $r^1, \ldots, r^t$. This yields precisely the formulation given by the inequalities in (4.35). □

4.10 The Size of a Smallest Perfect Formulation

Many integral polyhedra of interest in combinatorial optimization have an exponential number of facet-defining inequalities (the matching polytope, the traveling salesman polytope, the spanning tree polytope and many others). Let $P := \{x \in \mathbb{R}^n : Ax \leq b\}$ be such a polyhedron. By introducing a polynomial number of new variables, it is sometimes possible to obtain an extended formulation with polynomially many constraints whose projection onto the original space $\mathbb{R}^n$ is the polyhedron P itself. For example, the convex hull $P^{\mathrm{mix}} \subseteq \mathbb{R}^{n+1}$ of the mixing set (discussed in Sect. 4.8.1) has an exponential number of facet-defining inequalities, but it can be represented as the projection of the higher dimensional polyhedron Q defined in (4.27), where Q can be described by a system of inequalities in $3n + 3$ variables and $3n + 4$ constraints. Martin [270] shows that the spanning tree polytope (recall Theorem 4.25) is the projection of a polytope defined by $O(n^3)$ inequalities, where n is the number of nodes in the graph.

Given a polyhedron $P \subseteq \mathbb{R}^n$, what is the smallest number of variables and constraints needed in an extended formulation? Yannakakis [355] gave an elegant answer to this question.

Let S be an $m \times n$ nonnegative matrix. The *nonnegative rank* of S is the smallest integer t such that $S = FW$ where F and W are nonnegative matrices of size $m \times t$ and $t \times p$.

4.10. THE SIZE OF A SMALLEST PERFECT FORMULATION

Given a polytope $P \subseteq \mathbb{R}^n$, let $v^1, \ldots, v^p$ be its vertices and let $Ax \leq b$ be a system of linear inequalities such that $P = \{x \in \mathbb{R}^n : Ax \leq b\}$. Let $a^i x \leq b_i$, $i = 1, \ldots, m$, denote the inequalities in $Ax \leq b$. Define the *slack matrix* of $Ax \leq b$ to be the $m \times p$ matrix S where the entry

$$s_{ij} := b_i - a^i v^j$$

is the slack in the ith inequality when computed at vertex v^j. Note that S is a nonnegative matrix.

Lemma 4.49. *Let $Ax \leq b$ and $Cx \leq d$ be systems describing the same polytope P and let S, S' be the corresponding slack matrices. Then the nonnegative ranks of S and S' coincide.*

In particular, adding redundant inequalities to a linear system $Ax \leq b$ does not modify the nonnegative rank of the slack matrix. To prove Lemma 4.49, we will need the following result on polytopes.

Lemma 4.50. *Let $P := \{x \in \mathbb{R}^n : Ax \leq b\}$ be a polytope of dimension at least one. An inequality $cx \leq \delta$ is valid for P if and only if there exists $u \geq 0$ such that $uA = c$ and $ub = \delta$.*

Proof. Let $cx \leq \delta$ be valid for P. By Theorem 3.22 there exists $u \geq 0$ such that $uA = c$ and $ub \leq \delta$. Since P is a polytope and $\dim(P) \geq 1$, there exists $r \in \mathbb{R}^n$ such that both $\delta_1 := \min\{rx : x \in P\}$ and $\delta_2 := \max\{rx : x \in P\}$ are finite, and such that $\delta_1 < \delta_2$. Possibly by multiplying r by some positive number we may assume $\delta_2 - \delta_1 = 1$. By strong duality (Theorem 3.7), there exist $u^1, u^2 \geq 0$ such that $u^1 A = -r$, $u^1 b = -\delta_1$, $u^2 A = r$, $u^2 b = \delta_2$. Define $\bar{u} := u + (\delta - ub)(u^2 + u^1)$. It follows that $\bar{u} \geq 0$, $\bar{u}A = c$ and $\bar{u}b = \delta$. □

Proof of Lemma 4.49. The statement is obvious if P consists of only one point, thus we assume $\dim(P) \geq 1$. By symmetry it suffices to show that the nonnegative rank of S' is at most the nonnegative rank of S. By Lemma 4.50 there exists a nonnegative matrix L such that $C = LA$ and $d = Lb$. It follows that $S' = LS$. Given any factorization $S = FW$ where F, W are nonnegative matrices, if we let $F' = LF$, we have that F' is a nonnegative matrix and $S' = F'W$. □

By Lemma 4.49, we may define the *nonnegative rank of the polytope P* as the nonnegative rank of the slack matrix S of any system describing P.

Theorem 4.51 (Yannakakis). *Let $P \subset \mathbb{R}^n$ be a polytope, and let t be its nonnegative rank. Every extended formulation of P has at least t constraints, and there exists an extended formulation of P with at most $t+n$ constraints and $t+n$ variables.*

Proof. The statement is obvious if P consists of only one point, thus we may assume that $\dim(P) \geq 1$. Let $Ax \leq b$ be a system describing P, and let S be the slack matrix of $Ax \leq b$.

We first prove that every extended formulation has at least t constraints. Indeed, let
$$Q = \{(x,z) \in \mathbb{R}^n \times \mathbb{R}^k : Rx + Tz \leq d\}$$
be a polyhedron such that $P = \text{proj}_x(Q)$, where (R,T) is an $h \times (n+k)$ matrix. By Lemma 4.50, there exists a nonnegative vector $f^i \in \mathbb{R}^h$ such that $f^i T = 0$, $f^i R = a^i$, $f^i d = b_i$, $i = 1, \ldots, m$. Also, for any vertex v^j of P there exists $u^j \in \mathbb{R}^k$ such that $(v^j, u^j) \in Q$. Let $w^j := d - Rv^j - Tu^j$, $j = 1, \ldots, p$. Note that w^j is nonnegative. For $i = 1, \ldots, m$ and $j = 1, \ldots, p$, we have $f^i w^j = f^i(d - Rv^j - Tu^j) = b_i - a^i v^j$. If we define the $m \times h$ matrix F with rows f^i, $i = 1, \ldots, m$, and the $h \times p$ matrix W with columns w^j, $j = 1, \ldots, p$, we have $S = FW$. Since F and W are nonnegative matrices and F has h columns, it follows that $t \leq h$.

Next we show that there exists an extended formulation with at most $t+n$ constraints and $t+n$ variables. Indeed, let F and W be nonnegative matrices of size $m \times t$ and $t \times p$ such that $S = FW$. Let $f^1, \ldots, f^m \in \mathbb{R}^t$ be the rows of F and $w^1, \ldots, w^p \in \mathbb{R}^t$ be the columns of W. Define
$$Q := \{(x,z) \in \mathbb{R}^n \times \mathbb{R}^t : Ax + Fz = b,\ z \geq 0\}.$$

We will show that $P = \text{proj}_x(Q)$. To prove the inclusion $\text{proj}_x(Q) \subseteq P$, note that for any $(x,z) \in Q$ we have that $Ax = b - Fz \leq b$, because $F \geq 0$ and $z \geq 0$, and therefore $x \in P$. For the inclusion $P \subseteq \text{proj}_x(Q)$, it suffices to show that every vertex v^j of P is contained in $\text{proj}_x(Q)$. This is the case since, for any vertex v^j of P, the point (v^j, w^j) is in Q because $w^j \geq 0$ and $f^i w^j = b_i - a^i v^j$ for $i = 1, \ldots, m$.

We have shown that $P = \text{proj}_x(Q)$. Since $Q \subseteq \mathbb{R}^n \times \mathbb{R}^t$, at most $n+t$ of the equations $Ax + Fz = b$ are linearly independent. Expressing basic variables in terms of nonbasic variables and substituting any basic z_ℓ variable into the inequality $z_\ell \geq 0$, we are left with at most n equations and t inequalities. Thus Q can be described with at most $t+n$ constraints and $t+n$ variables. □

4.10. THE SIZE OF A SMALLEST PERFECT FORMULATION

Note that expressing the polytope P by a linear system in the original variables corresponds to the factorization $S = IS$ with the identity matrix I.

Given a factorization of S as FW, the proof of Yannakakis' theorem is constructive for obtaining an extended formulation. The catch is that factorizing S is a difficult task in general. Vavasis [344] shows that the problem of testing whether the rank of a given matrix coincides with its nonnegative rank is NP-complete. On the other hand, Yannakakis' theorem is a powerful tool for proving lower bounds on the size of extended formulation. Indeed, the following problem posed by Yannakakis [355] in 1991,

> "prove that the matching and traveling salesman polytopes cannot be expressed by polynomial size linear programs",

was solved by Fiorini et al. [136] in 2012 for the case of the traveling salesman polytope, and by Rothvoß [318] in 2014 for the case of the matching polytope, by providing exponential lower-bounds of the nonnegative rank of the slack matrices. We give more details on these proofs in the next three sections. We first state the following easy lemma.

Lemma 4.52. *Let S be an $m \times n$ nonnegative matrix with at least one positive entry. The nonnegative rank of S is the smallest number t such that S is the sum of t nonnegative rank-1 matrices.*

Proof. Let $\phi^1, \ldots, \phi^t \in \mathbb{R}^m$ be nonnegative column vectors, and $\omega^1, \ldots, \omega^t \in \mathbb{R}^n$ be nonnegative row vectors. Let $F \in \mathbb{R}^{m \times t}$ be the matrix with columns $\phi^1, \ldots, \phi^t$, $W \in \mathbb{R}^{t \times n}$ be the matrix with rows $\omega^1, \ldots, \omega^t$, and let $T^h \in \mathbb{R}^{m \times n}$ be the nonnegative rank-1 matrix defined by $T^h = \phi^h \omega^h$, $h = 1, \ldots, t$. The statement now follows from the simple observation that $S = FW$ if and only if $S = \sum_{h=1}^{t} T^h$. □

4.10.1 Rectangle Covering Bound

In this section, we define the "rectangle covering number" of a matrix and we show that, when applied to the slack matrix of any system of linear inequalities describing a polytope, it provides a lower bound on the nonnegative rank of the polytope. We then give a family of matrices U^n whose rectangle covering number grows exponentially with n. The matrix U^n will appear in Sect. 4.10.2 and its rectangle covering number will be a lower bound on the nonnegative rank of the cut polytope. By Yannakakis' theorem (Theorem 4.51), this will give a lower bound on the size of any extended formulation of the cut polytope.

An $m \times n$ matrix is a *rectangle matrix* if it is a 0, 1 matrix whose support is of the form $I \times J$ for some nonempty sets $I \subseteq \{1, \ldots, m\}$ and $J \subseteq \{1, \ldots, n\}$. Note that rectangle matrices are precisely the 0, 1 matrices of rank 1. Given a matrix $S \in \mathbb{R}^{m \times n}$, a *rectangle cover* of S is a family of rectangle matrices $R^1, \ldots, R^k$ such that the support of S (the set $\{(i, j) \in \{1, \ldots, m\} \times \{1, \ldots, n\} : s_{ij} \neq 0\}$) is equal to the support of $R^1 + \cdots + R^k$. The *rectangle covering number* of S is the smallest cardinality k of a rectangle cover of S.

Lemma 4.53. *Let $P \subset \mathbb{R}^n$ be a polytope. Every extended formulation of P has a number of constraints at least equal to the rectangle covering number of the slack matrix of any system of linear inequalities describing P.*

Proof. Let $S \in \mathbb{R}^{m \times n}$ be the slack matrix of a system describing P, and let t be its nonnegative rank. The lemma holds when $S = 0$, so we assume $S \neq 0$. By Lemma 4.52, $S = \sum_{h=1}^{t} T^h$ where $T^1, \ldots, T^t$ are $m \times n$ nonnegative matrices with rank 1. For $h = 1, \ldots, t$, define the $m \times n$ matrix R^h whose entries have value 1 if they belong to the support of T^h, and 0 otherwise. One can readily verify that $R^1, \ldots, R^t$ is a rectangle cover of S, therefore the nonnegative rank of S is at least the rectangle covering number of S. The statement now follows from Theorem 4.51. □

Rectangle Covering Number of the Unique Disjointness Matrix. The *unique disjointness matrix* U^n is the $2^n \times 2^n$ 0, 1 matrix whose rows and columns are indexed by the 2^n subsets of the set $\{1, \ldots, n\}$, defined by

$$U_{ab}^n = \begin{cases} 1 & \text{if } |a \cap b| \neq 1, \\ 0 & \text{if } |a \cap b| = 1, \end{cases} \quad \text{for all } a, b \subseteq \{1, \ldots, n\}.$$

We show next that the rectangle covering number of U^n grows exponentially with n. This result, due to Razborov [313], will be used in the next section to prove exponential bounds on the extended formulations of certain families of polytopes. The proof given here is due to Kaibel and Weltge [223].

Theorem 4.54. *The rectangle covering number of U^n is at least 1.5^n.*

Proof. Let $\mathcal{D} := \{(a, b) : a, b \subseteq \{1, \ldots, n\} \text{ and } a \cap b = \emptyset\}$. Note that $|\mathcal{D}| = 3^n$, since for each $j \in \{1, \ldots, n\}$ we have three possibilities for a, b, namely (i) $a \ni j$, $b \not\ni j$, (ii) $a \not\ni j$, $b \ni j$, and (iii) $a \not\ni j$, $b \not\ni j$.

Given $D \subseteq \mathcal{D}$, we define $A_D := \{a : (a, b) \in D \text{ for some } b\}$ and $B_D := \{b : (a, b) \in D \text{ for some } a\}$. We say that $D \subseteq \mathcal{D}$ is a *valid family for n* if $|a \cap b| \neq 1$ for every $(a, b) \in A_D \times B_D$. That is, D is valid if and only if

4.10. THE SIZE OF A SMALLEST PERFECT FORMULATION

the submatrix of U^n whose rows are indexed by A_D and whose columns are indexed by B_D has all entries equal to 1. We denote by $\varrho(n)$ the largest cardinality of a valid family for n.

We now compute a lower bound on the rectangle covering number of U^n using $\mathcal{D}$ and $\varrho(n)$. Note that, for every $2^n \times 2^n$ rectangle matrix R whose support is contained in the support of U^n, the family $\mathcal{D}_R := \{(a,b) \in \mathcal{D} : R_{ab} = 1\}$ is valid. It follows that any rectangle covering of U^n contains at least $\frac{|\mathcal{D}|}{\varrho(n)} = \frac{3^n}{\varrho(n)}$ elements. We will show next that $\varrho(n) = 2^n$. This implies that $\frac{|\mathcal{D}|}{\varrho(n)} = 1.5^n$, and the main statement follows.

First, $\varrho(n) \geq 2^n$ because $\{(a, \emptyset) : a \subseteq \{1, \ldots, n\}\}$ is a valid family. We show $\varrho(n) \leq 2^n$ by induction on n. The case $n = 0$ holds because $\varrho(0) = 1$ ($\{(\emptyset, \emptyset)\}$ is the only valid family in this case). Let D be a valid family for $n \geq 1$. Define

$$D'_1 := \{(a,b) \in D : n \in a\} \cup \{(a,b) \in D : n \notin b \text{ and } (a \cup \{n\}, b) \notin D\},$$
$$D'_2 := \{(a,b) \in D : n \in b\} \cup \{(a,b) \in D : n \notin a \text{ and } (a, b \cup \{n\}) \notin D\}.$$

We show next that $D \subseteq D'_1 \cup D'_2$. Let $(a,b) \in D$. Then $(a,b) \in D'_1 \cup D'_2$ when $n \in a$ or $n \in b$. Assume that $n \notin a$ and $n \notin b$. Since $|(a \cup \{n\}) \cap (b \cup \{n\})| = 1$ and D is a valid family, we have that at least one between $(a \cup \{n\}, b)$ and $(a, b \cup \{n\})$ is not in D. Therefore $(a,b) \in D'_1 \cup D'_2$ and this proves that $D \subseteq D'_1 \cup D'_2$.

Let $D_1 := \{(a \setminus \{n\}, b) : (a,b) \in D'_1\}$ and $D_2 := \{(a, b \setminus \{n\}) : (a,b) \in D'_2\}$. By the definition of D'_1, D'_2, we have that $|D_1| = |D'_1|$ and $|D_2| = |D'_2|$. Note that $a \subseteq \{1, \ldots, n-1\}$ for every (a,b) in D'_2 and $b \subseteq \{1, \ldots, n-1\}$ for every (a,b) in D'_1. It follows that D_1, D_2 are valid families for $n-1$. By the inductive hypothesis and the fact that $D \subseteq D'_1 \cup D'_2$, we conclude that $|D| \leq |D'_1| + |D'_2| = |D_1| + |D_2| \leq 2\varrho(n-1) = 2^n$. □

4.10.2 An Exponential Lower-Bound for the Cut Polytope

Fiorini et al. [136] showed that the traveling salesman, cut and stable set polytopes cannot have a polynomial-size extended formulation. Here we present the proof for the cut polytope.

Let P_n^{cut} be the *cut polytope* of order n, that is, the convex hull of the characteristic vectors of all the cuts of the complete graph $G_n = (V_n, E_n)$ on n nodes (including $\delta(V_n) = \emptyset$).

The proof of the lower bound is not done directly on the cut polytope, but on a polytope that is linearly isomorphic to it. Two polytopes $P \subseteq \mathbb{R}^p$ and $Q \subseteq \mathbb{R}^q$ are *linearly isomorphic* if there exists a linear function $f : \mathbb{R}^p \to \mathbb{R}^q$

such that $f(P) = Q$ and for all $y \in Q$ there is a unique $x \in P$ for which $f(x) = y$. Exercise 4.37 shows that this notion is symmetric (P, Q are linearly isomorphic if and only if Q, P are linearly isomorphic).

The *correlation polytope* P_n^{corr} of order n is the convex hull of the 2^n $n \times n$ matrices of the form bb^T, for all column vectors $b \in \{0,1\}^n$.

Lemma 4.55. *For all n, P_n^{corr} and P_{n+1}^{cut} are linearly isomorphic.*

Proof. Let $f : \mathbb{R}^{n \times n} \to \mathbb{R}^{E_{n+1}}$ be the linear function that maps each $x \in \mathbb{R}^{n \times n}$ to the element $y \in \mathbb{R}^{E_{n+1}}$ defined by

$$y_{ij} = \begin{cases} x_{ii} & \text{if } 1 \leq i \leq n,\ j = n+1 \\ x_{ii} + x_{jj} - 2x_{ij} & \text{if } 1 \leq i < j \leq n. \end{cases}$$

One can verify that $f(P_n^{\text{corr}}) = P_{n+1}^{\text{cut}}$ and, for every $y \in P_{n+1}^{\text{cut}}$, the point $x \in \mathbb{R}^{n \times n}$ defined by

$$x_{ij} = \begin{cases} y_{i,n+1} & \text{if } i = j \\ \frac{1}{2}(y_{i,n+1} + y_{j,n+1} - y_{ij}) & \text{if } i \neq j \end{cases} \qquad i,j = 1,\ldots,n.$$

is the only point in P_n^{corr} such that $f(x) = y$. $\square$

Given a vector $a \in \mathbb{R}^n$, we denote by $\text{Diag}(a)$ the $n \times n$ matrix with a on the diagonal and 0 everywhere else. We denote by $\langle \cdot, \cdot \rangle$ the *Frobenius product* of two matrices having the same dimensions, that is $\langle A, B \rangle := \sum_i \sum_j a_{ij} b_{ij}$. We recall that the *trace* of a square matrix C, denoted by $\text{tr}(C)$, is the sum of its diagonal entries. It is easy to verify that

$$\langle A, B \rangle = \text{tr}(A^T B) = \text{tr}(BA^T). \tag{4.37}$$

Lemma 4.56. *For all column vectors $a \in \{0,1\}^n$, the inequality*

$$\langle 2\text{Diag}(a) - aa^T, Y \rangle \leq 1 \tag{4.38}$$

is valid for P_n^{corr}. Furthermore, the slack of each vertex $Y = bb^T$ with $b \in \{0,1\}^n$, is $(1 - a^T b)^2$.

Proof. For all column vectors $a, b \in \{0,1\}^n$, we have

$$1 - \langle 2\text{Diag}(a) - aa^T, bb^T \rangle = 1 - 2a^T b + \text{tr}(aa^T bb^T) = 1 - 2a^T b + \text{tr}(b^T aa^T b) = (1 - a^T b)^2$$

where the first equality follows from the fact that $b \in \{0,1\}^n$, the second follows from (4.37), and the third because $a^T b$ is a scalar. Since $(1 - a^T b)^2 \geq 0$, (4.38) is a valid inequality for P_n^{corr}. $\square$

4.10. THE SIZE OF A SMALLEST PERFECT FORMULATION

Theorem 4.57. *Every extended formulation for P_n^{cut} has at least 1.5^n constraints.*

Proof. Let $Ay \leq b$ be a system of inequalities such that $P_n^{\text{corr}} = \{y \in \mathbb{R}^{n \times n} : Ay \leq b\}$ and such that all the inequalities (4.38) are included in $Ay \leq b$. Define S to be the slack matrix of $Ay \leq b$ and let M be the $2^n \times 2^n$ matrix whose rows and columns are indexed by the vectors in $\{0,1\}^n$, defined by $M_{ab} = (1 - a^T b)^2$ for all $a, b \in \{0,1\}^n$. Then M is the row-submatrix of the slack matrix S relative to the 2^n constraints (4.38). By Lemma 4.56, $M_{ab} = 0$ if and only if $|a \cap b| = 1$, therefore the rectangle covering number of M and of the unique disjointness matrix U^n coincide. It follows by Theorem 4.51 and Lemma 4.53 that every extended formulation for P_n^{corr} has at least 1.5^n constraints. By Lemma 4.55, the same holds for P_{n+1}^{cut}. □

Recall that, for a graph G, the *stable set polytope* STAB(G) is the convex hull of the characteristic vectors of all the stable sets in G. Fiorini et al. [136] show that, for every n, there exists a graph H with $O(n^2)$ nodes such that STAB(H) has a face that is an extended formulation for P_n^{cut}. This implies that there exists an infinite family of graphs $\{H_n\}_{n \in \mathbb{N}}$ such that every extended formulation for STAB(H_n) has at least $1.5^{\sqrt{n}}$ constraints.

4.10.3 An Exponential Lower-Bound for the Matching Polytope

Yannakakis' problem of showing that the matching polytope does not admit compact extended formulations was solved by Rothvoß [318]. He showed that *"Every extended formulation for the perfect matching polytope on a complete graph with n vertices (obviously n even) needs at least 2^{cn} constraints for some constant $c > 0$."*

This is the first problem that is polynomially solvable for which it is proven that no compact extended formulation exists.

Since the slack matrix of the perfect matching polytope (4.23) on a complete graph with n vertices can be covered with n^2 rectangles (see Exercise 4.36), the rectangle covering bound used to prove that the cut polytope does not admit a compact extended formulation is too weak to prove exponential lower bounds for the perfect matching polytope.

Rothvoß uses the following known but unpublished "hyperplane separation bound," which he learned from Fiorini. Intuitively, the hyperplane separation bound says that, given an $m \times n$ nonnegative matrix S, if a matrix W exists such that $\langle W, S \rangle$ is large but $\langle W, R \rangle$ is small for every rectangle R of S, then the nonnegative rank of S is large.

Theorem 4.58. *Let S be an $m \times n$ nonnegative matrix and let t be its nonnegative rank. Then, for every $m \times n$ matrix W,*

$$t \geq \frac{\langle W, S \rangle}{\alpha \cdot \|S\|_\infty}$$

where $\alpha := \max\{\langle W, R \rangle : R \text{ is an } m \times n \text{ rectangle matrix}\}$.

Proof. Let $\alpha^* = \max\{\langle W, R \rangle : R$ is a rank-1 matrix in $[0,1]^{m \times n}\}$ be the linear relaxation of α. Since any rank-1 matrix $R \in [0,1]^{m \times n}$ can be written as $R = xy^T$, where $x \in [0,1]^m$ and $y \in [0,1]^n$, we have that $\alpha^* = \max\{\sum_{i=1}^m \sum_{j=1}^n w_{ij} x_i y_j : x_i \in [0,1], y_j \in [0,1]\}$. Since for fixed x the objective function is linear in y and vice versa, there exist $x \in \{0,1\}^m$ and $y \in \{0,1\}^n$ that achieve this maximum. This shows $\alpha^* = \alpha$.

By Lemma 4.52, there exist rank-1 nonnegative matrices $R_1, \ldots, R_t$ such that $S = \sum_{i=1}^t R_i$. We obtain

$$\langle W, S \rangle = \sum_{i=1}^t \|R_i\|_\infty \langle W, \frac{R_i}{\|R_i\|_\infty} \rangle \leq \alpha \sum_{i=1}^t \|R_i\|_\infty \leq \alpha t \|S\|_\infty$$

where the first inequality follows from the fact that $\langle W, \frac{R_i}{\|R_i\|_\infty} \rangle \leq \alpha^* = \alpha$ and the second inequality from the fact that $\|R_i\|_\infty \leq \|S\|_\infty$. □

When S is the slack matrix of the perfect matching polytope, Rothvoß constructs a matrix W for which the hyperplane separation bound is exponential. However his beautiful construction is too involved to be presented here.

Given a graph G, the *traveling salesman polytope* is the convex hull of the characteristic vectors of all the edge sets that induce a Hamiltonian tour in G. Yannakakis [355] showed that the perfect matching polytope on n nodes is a face of the traveling salesman polytope on $3n$ nodes. Using this reduction, together with the exponential bound on the matching polytope, Rothvoß proves that every extended formulation for the traveling salesman polytope needs a number of inequalities that is exponential in the number of nodes, thus improving the bound $2^{\sqrt{n}}$ of Fiorini et al. [136].

4.11 Further Readings

Combinatorial Optimization

Following the pioneering work of Fulkerson, Edmonds, Hoffman (among others), there has been a tremendous amount of research in polyhedral combinatorics and combinatorial optimization. This is treated in the monumental monograph of Schrijver [327], which is (and will remain for many years) the main reference for researchers in the area.

A first, important book that explores the link between polynomial-time algorithms and perfect formulations for combinatorial problems is "Combinatorial Optimization: Networks and Matroids" by Lawler [252]. The books by Cook et al. [87] and Korte and Vygen [243] are modern and well-written accounts.

Integral Polyhedra

Papadimitriou and Yannakakis [311] showed that the following problem is co-NP-complete: *Given a rational system of inequalities $Ax \leq b$, is the polyhedron $Q := \{x : Ax \leq b\}$ integral?*

Total Unimodularity

Connections with matroid theory are presented in the books of Truemper [338] and Oxley [296]. Camion and Gomory give other characterizations of totally unimodular matrices, see Theorem 19.3 in [325]. In particular, Camion shows that

Theorem 4.59. *A $0, \pm 1$ matrix A is totally unimodular if and only if A does not contain a square submatrix with an even number of nonzero entries in each row and column whose sum of the entries divided by 2 is odd.*

Tutte [339] characterized the $0, 1$ matrices that can be signed to be totally unimodular (i.e., some $+1$ entries can be turned to -1 so that the resulting matrix is totally unimodular), in terms of "forbidden minors." Gerards [164] gave a simple proof of Tutte's theorem. Seymour [329] gave a structure theorem for totally unimodular matrices which essentially says that every totally unimodular matrix can be constructed by piecing together totally unimodular matrices belonging to an infinite class (the incidence matrices of digraphs, described below, and their transpose) and copies of specific 5×5 totally unimodular matrices. The piecing of two matrices is done so that the gluing part that ties the two matrices has rank at most two. This result has the following important computational consequence:

There exist polynomial algorithms that solve the following problems:

– *Is a given matrix totally unimodular?*
– *Find an optimal solution for $\max\{cx : Ax \leq b, x \text{ integer}\}$, where A is totally unimodular.*

When A is totally unimodular and b is not an integral vector, solving the integer program $\max\{cx \,:\, Ax \leq b, x \in \mathbb{Z}_+^n\}$ amounts to solving the linear program $\max\{cx \,:\, Ax \leq \lfloor b \rfloor, x \geq 0\}$, see Exercise 4.6. Solving the mixed integer linear program $\max\{cx \,:\, Ax \leq b,\ x \in \mathbb{Z}_+^p \times \mathbb{R}_+^{n-p}\}$ with $0 < p < n$ is NP-Hard even when A is totally unimodular, see [81].

Perfect and Ideal Matrices

A $0,1$ matrix A is *perfect* if the set packing polytope $\{x \in \mathbb{R}_+^n \,:\, Ax \leq \mathbf{1}\}$ is integral and A is *ideal* if the set covering polyhedron $\{x \in \mathbb{R}_+^n \,:\, Ax \geq \mathbf{1}\}$ is integral. A $0,1$ totally unimodular matrix is obviously perfect and ideal. The theory of perfect and ideal matrices is full of deep, elegant and difficult results. These are surveyed in Cornuéjols' monograph [92] and in Schrijver's book [327]. In 1960, Berge made two beautiful conjectures about perfection, one stronger than the other. The first was solved by Lovász [259] in 1972 and is known as the "perfect graph theorem." The second, now called the "strong perfect graph theorem," was solved by Chudnovsky et al. [72] in 2006. Polynomial recognition of perfect matrices was solved in [71]. The situation for ideal matrices is still largely open.

Network Flows and Matchings

The minimum cut problem in a graph can be solved in polynomial time when the edge weights are nonnegative (Nagamochi and Ibaraki [282], Stoer and Wagner [333], Karger [230]). Formulation (2.4) is not perfect, but there exist compact extended formulations (see Carr et al. [69] and [79] for instance).

The book of Ford and Fulkerson [147] provides a short and elegant account of network flows and their connection with combinatorial theory. A modern and comprehensive book focusing on algorithms is that by Ahuja et al. [17].

The characterization of the matching polytope is due to Edmonds [124]. References for matching theory and flows are the book of Lovász and Plummer [262] and Volume A of Schrijver's monograph [327].

Total Dual Integrality

An extensive treatment of total dual integrality can be found in Chap. 22 of [325].

Theorem 4.26 is due to Edmonds and Giles [128] and extends previous results of Fulkerson [154] and Hoffman [203]. Theorem 4.27 is due to Giles and Pulleyblank [166] and Schrijver [324]. Total dual integrality provides an elegant framework for some beautiful and important min–max relations of graphs, see e.g., the book of Frank [148].

Vectors $a_1,\ldots a_k$ form a *Hilbert basis* if every integral vector in cone$(a_1,\ldots a_k)$ can be written as a conic combination of $a_1,\ldots a_k$ with integer coefficients. Every rational cone admits a Hilbert basis. However the smallest such basis may be exponentially large with respect to the set of generators of the cone. (see Exercise 4.39)

Giles and Pulleyblank [166] show the following.

Theorem 4.60. *The rational system $Ax \leq b$ is TDI if and only if for each face F of $\{x \in \mathbb{R}^n : Ax \leq b\}$, the set of active rows forms a Hilbert basis.*

Cook et al. [89] show the following theorem of Carathéodory type.

Theorem 4.61. *Let $a_1,\ldots a_k \in \mathbb{Z}^n$ form a Hilbert basis and let* cone $(a_1,\ldots a_k)$ *be pointed. Then every integral vector in* cone$(a_1,\ldots a_k)$ *can be written as conic combination with integral coefficients of at most $2n-1$ vectors among $a_1,\ldots a_k$.*

Pap [304] proved that testing whether a given system of inequalities is TDI is NP-hard.

Submodular Functions

Submodular functions are linked to the theory of matroid and polymatroids and the greedy algorithm, as nondecreasing submodular functions are rank functions of polymatroids. This was pioneered by Edmonds [126]. There are combinatorial algorithms to minimize a submodular function [209, 326].

Submodular functions are also linked to connectivity problems on graphs. This is covered in the recent book by Frank [148].

Mixed Integer Linear Representability

Section 4.9.3 is taken from a manuscript of Conforti, Di Summa, Faenza, and Zambelli. Jeroslow [211] establishes that a subset $S \subseteq \mathbb{R}^n$ can be represented as a mixed integer linear set with bounded (or, equivalently, binary) integer variables if and only if S is the union of a finite number of polyhedra having the same recession cone (Exercise 4.32). Vielma [347] has a comprehensive survey that treats several issues on mixed integer linear representability.

Extended Formulations

The survey of Vanderbeck and Wolsey [342] focuses on extended formulations in integer programming, while the surveys of Conforti et al. [79] and Kaibel [219] treat extended formulations in combinatorial optimization.

An early example of an extended formulation was given by Balas and Pulleyblank for the "perfectly matchable subgraph polytope" [33]. The book of Pochet and Wolsey [308] gives several examples of mixed integer linear sets, mostly arising in production planning, that admit compact extended formulations. Uncapacitated lot-sizing provides a nice illustration [37].

Martin [269] and Martin et al. [271] show that a large class of dynamic programming recursions yield perfect extended formulations. Martin [270] shows that given polyhedron P, separation for P can be expressed with a compact linear program if and only if P admits a compact extended formulation.

Balas's extended formulation for the union of polyhedra (Theorem 4.39) has been used by many authors to construct extended formulations for various mixed integer linear sets, several of which arising in production planning (see, for instance, Van Vyve [341], Atamtürk [15], and Conforti and Wolsey [82]).

Kaibel and Pashkovich [220] show that reflection polytopes (that is, polytopes of the form $\text{conv}(P \cup P')$ where P' is the reflection of polytope $P = \{x : Ax \leq b\}$ with respect to a hyperplane H such that P lies on only one side of H) admit an extended formulation that uses only one more variable and two more constraints than $Ax \leq b$. Using this result they were able to provide compact extended formulations for families of polytopes whose description in the original space is currently not known.

Goemans [170] gives an $O(n \log n)$ extended formulation for the permutahedron, and shows that this is asymptotically the smallest possible.

Rothvoß [317] proves that there are matroid polytopes that do not have polynomial-size extended formulations.

The fundamental paper of Yannakakis [355] was inspired by a series of (incorrect) papers giving compact systems of inequalities that describe a polytope that is nonempty if and only if a given graph is Hamiltonian. He proves that under certain symmetry assumptions, the traveling salesman polytope does not admit a compact extended formulation. Kaibel et al. [221] show that there are polytopes that admit a compact extended formulation, but no compact symmetric extended formulation.

The characterization of the smallest size of an extended formulation in terms of nonnegative rank of the slack matrix (Theorem 4.51) has been a source of inspiration for a number of bounds, culminating in the exponential lower bounds of Fiorini et al. [136] and of Rothvoß [318]. This result has been generalized to other representations of a polytope in an extended space (such as semi-definite representations) by Gouveia et al. [182].

An exponential lower bound on the rectangle covering number for the unique disjointness matrix was given by De Wolf [111], based on a result of Razborov [313]. The improved 1.5^n bound given in Theorem 4.54, as well as the self-contained proof we presented, are due to Kaibel and Weltge [223]. Rectangle covering bounds have been studied in the context of communication complexity [137].

4.12 Exercises

Exercise 4.1. Find a perfect formulation for the set:
$$x_1 \neq x_2$$
$$x_1 \neq x_3$$
$$x_2 \neq x_3$$
$$1 \leq x_i \leq 3, \text{integer}.$$

Exercise 4.2. Let V_n be the set of vertices of the n-hypercube H_n and $V = \{v_1, \ldots, v_k\}$ be a subset of nonadjacent vertices in H_n. Let $N := \{1, \ldots, n\}$. Given vertex v_i of H_n let $S_i \subseteq N$ be its support. Show that

$$\text{conv}(V_n \setminus V) = \{x \in [0,1]^n : \sum_{j \in S_i} x_j - \sum_{j \in N \setminus S_i} x_j \leq |S_i| - 1, \forall i : v_i \in V\}$$

Exercise 4.3. Let $P := \{(x,y) \in \mathbb{R}^n \times \mathbb{R}^p : Ax + Gy \leq b\}$ be a rational polyhedron, and let $S := P \cap (\mathbb{Z}^n \times \mathbb{R}^p)$. Give a counter-example to the following statement. $P = \text{conv}(S)$ if and only if every rational supporting hyperplane of P contains a point in $\mathbb{Z}^n \times \mathbb{R}^p$.

Exercise 4.4. Show that the following matrix A is not totally unimodular but that the polyhedron $\{x \in \mathbb{R}^3 : Ax = b\}$ is integral for every $b \in \mathbb{Z}^3$.

$$A := \begin{pmatrix} 1 & 1 & 1 \\ 0 & -1 & 1 \\ 0 & 0 & 1 \end{pmatrix}$$

Exercise 4.5. Let A be an integral $m \times n$ matrix. Show that the following are all equivalent.

1. A is totally unimodular.

2. A^T is totally unimodular.

3. The $m \times 2n$ matrix $(A \mid -A)$ is totally unimodular.

4. The $m \times (n+m)$ matrix $(A \mid I)$ is unimodular (recall the definition of a unimodular matrix given in Chap. 1).

5. Any matrix A' obtained from A by changing the signs of all entries in some rows is totally unimodular.

Exercise 4.6. Given the integer set $S := \{x \in \mathbb{Z}_+^n \ : \ Ax \leq b\}$, where A is a totally unimodular matrix and b is a vector that may have fractional components, show that $\text{conv}(S) = \{x \in \mathbb{R}_+^n \ : \ Ax \leq \lfloor b \rfloor\}$.

Exercise 4.7. A $0,1$ matrix can be *signed to be totally unimodular* if some $+1$ entries can be turned into -1 so that the resulting matrix is totally unimodular. Consider the two following 0,1 matrices. Show that the matrix on the left cannot be signed to be totally unimodular, whereas the matrix on the right can be signed to be totally unimodular.

$$\begin{pmatrix} 1 & 1 & 1 & 0 \\ 1 & 0 & 1 & 1 \\ 1 & 1 & 0 & 1 \end{pmatrix} \qquad \begin{pmatrix} 1 & 1 & 0 & 0 & 1 \\ 1 & 1 & 1 & 0 & 0 \\ 0 & 1 & 1 & 1 & 0 \\ 0 & 0 & 1 & 1 & 1 \\ 1 & 0 & 0 & 1 & 1 \end{pmatrix}$$

Exercise 4.8. A $0,1$-matrix A has the *consecutive 1's property* if $a_{ij} = a_{i\ell} = 1$, $j < \ell$ implies $a_{ik} = 1$, $j \leq k \leq \ell$.

1. Give a polynomial-time algorithm to test if a given $0,1$-matrix has the consecutive 1's property.

2. Prove that a $0,1$-matrix with the consecutive 1's property is totally unimodular.

Exercise 4.9. Let $\mathcal{S}$ be a family of subsets of a nonempty finite set V, and let $A_{\mathcal{S}}$ denote the $|\mathcal{S}| \times |V|$ incidence matrix of $\mathcal{S}$. The family $\mathcal{S}$ is *laminar* if, for all $S, T \in \mathcal{S}$ such that $S \cap T \neq \emptyset$, either $S \subseteq T$ or $T \subseteq S$. Show the following.

1. If $\mathcal{S}$ is a laminar family, then $|\mathcal{S}| \leq 2|V|$.

2. If $\mathcal{S}$ is a laminar family, then $A_{\mathcal{S}}$ is totally unimodular. (Hint: Apply Corollary 4.7).

4.12. EXERCISES

Exercise 4.10. Given integer $k \geq 2$, an *equitable k-coloring* of matrix A is a partition of its columns into k sets or "colors" such that every pair of colors I and J induces an equitable bicoloring of the submatrix A_{IJ} formed by the columns of A in $I \cup J$. Prove that every totally unimodular matrix admits an equitable k-coloring, for every $k \geq 2$.

Exercise 4.11. Given a rational polyhedron $P = \{x \in \mathbb{R}_+^n : Ax \leq b\}$ and integer $k \geq 1$, let $kP := \{kx : x \in P\} = \{x \in \mathbb{R}_+^n : Ax \leq kb\}$. We say that P has the *integer decomposition property* if, for every integer $k \geq 1$, every integral vector in kP is the sum of k integral vectors in P. Show the following.

1. If P has the integer decomposition property, then P is an integral polyhedron.

2. If A is totally unimodular and b is integral, then P has the integer decomposition property. (Hint: There are several possible proofs. One idea is to use Exercise 4.10. Let $\bar{x}$ be an integral vector in kP. Let A' be the matrix containing $\bar{x}_j$ copies of column j of A. Consider an equitable k-coloring of A').

Exercise 4.12. Let $D = (V, A)$ be a digraph. A *cycle* of D is a set of arcs that forms a cycle in the underlying undirected graph (ignoring the orientation of the arcs).

1. Let $F \subseteq A$ be a set of arcs. Show that the columns of the incidence matrix A_D indexed by the elements of F are linearly independent if and only if F does not contain any cycle.

2. Show that, if D has k connected components, then $\operatorname{rank}(A_D) = |V| - k$.

Exercise 4.13. Show that the characteristic vector of a simple circuit is an extreme ray of the circulation cone.

Exercise 4.14. Let $D = (V, A)$ be a digraph. For every $a \in A$, let $\ell_a, u_a \in \mathbb{R}$ be given such that $\ell_a \leq u_a$. Show that the set of circulations $\{x \in \mathbb{R}^{|A|} : A_D x = 0, \ell \leq x \leq u\}$ is nonempty if and only if

$$\sum_{a \in \delta^-(X)} \ell_a \leq \sum_{a \in \delta^+(X)} u_a \text{ for all } X \subseteq V.$$

Exercise 4.15. Let G be a graph whose edge set C is a cycle. Show that $|\det(A_G)| = 2$ if C is an odd cycle and $\det(A_G) = 0$ if C is an even cycle.

Exercise 4.16. Consider the instances of the maximum s,t-flow problem on the digraph in the figure below, where labels on the arcs represent capacities and u is a positive integer.

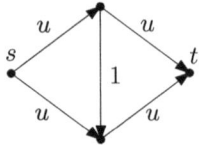

Show that, if at every iteration of the augmenting paths algorithm we choose the *longest* s,t-augmenting path (here the *length* of a path is its number of arcs), then the number of iterations is exponential in input size.

Exercise 4.17. Let G be a graph and A_G be its incidence matrix. Let $\mathcal{F}$ be the family of sets $F \subseteq E$ such that every connected component of the graph induced by F is a tree plus possibly an additional edge, which closes an odd cycle.

1. Given $F \subseteq E$, let A^F be the submatrix of A_G whose columns are indexed by the elements of F. Show that the columns of A^F are linearly independent if and only if $F \in \mathcal{F}$. Furthermore, if $F \in \mathcal{F}$ is not a tree, then A^F is square and $\det(A^F) = \pm 2$.

2. Give a formula that involves parameters of G (number of nodes, number of components, etc.) that characterizes $\text{rank}(A_G)$.

Exercise 4.18. Let A_G be the incidence matrix of a graph G and b be a vector such that $\frac{b}{2}$ is integral. Show that $\{x : A_G x \leq b\}$ is an integral polyhedron.

Exercise 4.19. Let $G = (V, E)$ be an (undirected) graph, with lengths ℓ_e, $e \in E$. Given nodes $s, t \in V$, the *undirected shortest s,t-path* problem consists in finding an (undirected) path with endnodes s, t of minimum length. Show that, if $\ell_e \geq 0$ for all $e \in E$, then the undirected shortest s,t-path problem in G can be reduced to solving the directed shortest s, t-path problem on an appropriate digraph. Explain why the construction cannot be applied if some of the lengths are negative.

Exercise 4.20. A *vertex cover* of a graph $G = (V, E)$ is a set of nodes $U \subseteq V$ such that every edge in E contains at least an element of U.

1. Show that, for every matching M and every vertex cover U, $|M| \leq |U|$.

2. Using linear programming duality and total unimodularity, show the following statement: If G is bipartite, then $\max\{|M| : M$ is a matching$\} = \min\{|U| : U$ is a vertex cover$\}$.

3. Give another proof of the statement in 2) using the max-flow/min-cut theorem (Theorem 4.15).

4. Does the statement in 2) hold if G is not bipartite?

Exercise 4.21. An *edge cover* of a graph $G = (V, E)$ is a subset F of E such that each node of G is covered by at least one edge of E'. Show the following.

1. If S is a stable set of G and F is an edge cover, then $|S| \leq |F|$.

2. Let S^* and F^* be respectively a stable set of maximum cardinality and an edge cover of minimum cardinality. Show that if G is bipartite, $|S^*| = |F^*|$.

3. Show a graph for which $|S^*| < |F^*|$.

Exercise 4.22. Given a graph $G = (V, E)$, the *edge-chromatic number* of G is the minimum number of disjoint matchings that cover E. Let $\delta^{\min}$ and $\delta^{\max}$ be, respectively, the minimum and maximum among the degrees of the nodes of G. If G is bipartite, show that:

1. The edge-chromatic number of G equals $\delta^{\max}$.

2. The maximum number of disjoint edge-covers of G equals $\delta^{\min}$.

Exercise 4.23. Let $Ax \leq b$ be a rational system. Show that there exists a positive integer k such that the system $\frac{A}{k} x \leq \frac{b}{k}$ is TDI.

Exercise 4.24. Let

$$A = \begin{bmatrix} 1 & 1 & 0 & 1 & 0 & 0 \\ 1 & 0 & 1 & 0 & 1 & 0 \\ 0 & 1 & 1 & 0 & 0 & 1 \\ 0 & 0 & 0 & 1 & 1 & 1 \end{bmatrix}.$$

Show that

- The polyhedron $P := \{x \geq 0 : Ax \geq 1\}$ is integral.

- The system $x \geq 0$, $Ax \geq 1$ is not TDI.

Exercise 4.25. Let $N := \{1, \ldots, n\}$ and let $f : 2^N \to \mathbb{R}$ be a submodular integer-valued function such that $f(\emptyset) = 0$. Let $\bar{x}$ be a vertex of the polyhedron $P = \{x \in \mathbb{R}^n : \sum_{j \in S} x_j \leq f(S) \text{ for all } S \subseteq N\}$. Show that $\bar{x}$ satisfies at equality n linearly independent inequalities $\sum_{j \in S} x_j \leq f(S_i)$ $i = 1, \ldots, n$ such that the family $\{S_i, i = 1, \ldots, n\}$ is laminar.

Exercise 4.26. Show that the permutahedron $\Pi_n \subset \mathbb{R}^n$ (introduced in Example 3.20 and discussed in Chap. 3) is described by the following inequalities.

$$\sum_{i \in S} x_i \geq \binom{|S|+1}{2} \quad \emptyset \subset S \subset V$$
$$\sum_{i=1}^n x_i = \binom{n+1}{2}.$$

Hint: note that the above system is equivalent to

$$\sum_{i \in S} x_i \leq \binom{n+1}{2} - \binom{n-|S|+1}{2} \quad \emptyset \subset S \subset V$$
$$\sum_{i=1}^n x_i = \binom{n+1}{2}.$$

In Chap. 3 it is shown that the equation defines the affine hull of Π_n and the inequalities are facet-defining.

Exercise 4.27 (Adjacencies of the Matching Polytope). Let $G = (V, E)$ be an undirected graph, and denote by P its matching polytope. Show the following.

1. Given two matchings M and N in G, their incidence vectors are adjacent vertices of P if and only if $M \triangle N$ consists of a cycle or a path.

2. For $k = 1, \ldots, |V|$, the polytope $P \cap \{x \in \mathbb{R}^E : \sum_{e \in E} x_e = k\}$ is the convex hull of the characteristic vectors of matchings of cardinality k. (Hint: use 1.)

Exercise 4.28. Let $G = (V, E)$ be an undirected graph.

1. Show that the following system of inequalities describes the perfect matching polytope:

$$\begin{aligned} \sum_{e \in \delta(v)} x_e &= 1 & v \in V, \\ \sum_{e \in \delta(U)} x_e &\geq 1 & U \subseteq V, |U| \text{ odd}, \\ x_e &\geq 0 & e \in E. \end{aligned}$$

2. Let G be a bridgeless cubic graph. Using the previous part, show that for every edge e, G contains at least one perfect matching containing e and at least two distinct perfect matchings not containing e. (A graph is *bridgeless* if removing any edge does not increase the number of connected components, and it is *cubic* if every node has degree three.)

Exercise 4.29. Given a graph $G = (V, E)$, an *edge cover* C is a subset of E such that every node of G is contained in some element of C.

4.12. EXERCISES

1. Show that C is a minimum cardinality edge cover if and only if C is obtained from a maximum cardinality matching M by adding one edge that is incident to each node that is not covered by M.

2. Give a perfect formulation for the *edge-cover polytope* (i.e., the convex-hull of incidence vectors of edge covers).

Exercise 4.30. Consider the integer program
$$\min \sqrt{2}x_1 - x_2$$
$$1 \le x_2 \le \sqrt{2}x_1$$
$$x_1, x_2 \text{ integer.}$$

(i) Show that this integer program has no optimal solution.

(ii) Construct feasible solutions x^k such that $\lim_{k \to \infty} \sqrt{2}x_1^k - x_2^k = 0$.

(iii) Prove that $\text{conv}\{x \in \mathbb{Z}^2 : 1 \le x_2 \le \sqrt{2}x_1\}$ is not a polyhedron.

Exercise 4.31. The purpose of this exercise is to prove Theorem 4.46. For every $c \in \mathbb{R}^n$, consider the primal–dual pair of problems

$$\max \sum_{i \in N} c_i x_i \qquad\qquad \min \sum_{S \in \mathcal{S}}(|S|-1)y_S + \sum_{i \in N} z_i$$

$$\sum_{i \in S} x_i - \sum_{i \in N \setminus S} x_i \le |S| - 1 \quad S \in \mathcal{S} \qquad \sum_{S \in \mathcal{S}, S \ni i} y_S - \sum_{S \in \mathcal{S}, S \not\ni i} y_S + z_i \ge c_i \quad i \in N$$

$$x_i \le 1 \quad i \in N \qquad\qquad z_i \ge 0 \quad i \in N$$
$$x_i \ge 0 \quad i \in N \qquad\qquad y_S \ge 0 \quad S \in \mathcal{S}.$$

Characterize the optimal solution x^* of the problem $\max\{cx : x \in S_n^{\text{even}}\}$, and show that the dual problem (on the right) has a feasible solution (y^*, z^*) of value cx^*.

Exercise 4.32. We say that a set $S \subseteq \mathbb{Z}^n \times \mathbb{R}^p$ is *mixed binary linear representable* if there exists a system of constraints with rational coefficients

$$Ax + By + Cz \le d$$
$$x \in \mathbb{Z}^n \times \mathbb{R}^p, \ y \in \mathbb{R}^m, \ z \in \{0,1\}^q$$

such that S is the projection onto the space of x variables of the set of the solutions of the above conditions.

Recall that a set $S \subseteq \mathbb{Z}^n \times \mathbb{R}^p$ is a *mixed integer linear set* if $S = P \cap (\mathbb{Z}^n \times \mathbb{R}^p)$ for some rational polyhedron $P \subseteq \mathbb{R}^{n+p}$. By Theorem 4.30, if S is a mixed integer linear set, then all polyhedra P satisfying $S = P \cap (\mathbb{Z}^n \times \mathbb{R}^p)$ have the same recession cone, which we will refer to as the recession cone of S. Prove the following:

A nonempty set $S = \mathbb{Z}^n \times \mathbb{R}^p$ is mixed binary linear representable if and only if it is the union of finitely many mixed integer linear sets in $\mathbb{Z}^n \times \mathbb{R}^p$ having the same recession cone.

Exercise 4.33. Let $S \subseteq \mathbb{Z}^n \times \mathbb{R}^p$ be bounded. Show that S is mixed integer linear representable if and only if it is mixed binary linear representable (this notion is defined in Exercise 4.32).

Exercise 4.34. Let $P \subseteq \mathbb{R}^n_+$ be a polytope, and let k be a positive integer. Show that the set $\{x \in P : x \text{ has at least } k \text{ positive entries}\}$ is not mixed integer linear representable, while the set $\{x \in P : x \text{ has at most } k \text{ positive entries}\}$ is mixed integer linear representable.

Exercise 4.35. Show that the rank of the slack matrix of any polytope $P \subset \mathbb{R}^n$ is at most $n + 1$.

Exercise 4.36. Show that the rectangle covering number of the slack matrix of the perfect matching polytope on a complete graph on n nodes is at most n^4. (Hint: For every pair of edges consider the matchings and the odd cuts that contain both of them.)

Exercise 4.37. Let $P \subset \mathbb{R}^p$ and $Q \subset \mathbb{R}^q$ be two polytopes. Show that P, Q are linearly isomorphic if and only if Q, P are linearly isomorphic.

Exercise 4.38. Given polytopes P and Q such that $P = \text{proj}_x(Q)$, show that

1. Q has at least as many faces as P.

2. The number of facets of Q is at least the logarithm of the number of faces of P.

3. If $\text{proj}_x(Q)$ is the permutahedron $\Pi_n \subset \mathbb{R}^n$, then Q has at least $\frac{n}{2} \log \frac{n}{2}$ facets.

Exercise 4.39. Consider the cone $C \subset \mathbb{R}^2$ generated by $(0, 1)$ and $(k, 1)$. Show that the unique minimal Hilbert basis for C consists of vectors $(i, 1)$, $i = 0, \ldots, k$.

Chapter 5

Split and Gomory Inequalities

Chapter 4 dealt with perfect formulations. What can one do when one is handed a formulation that is not perfect? A possible option is to strengthen the formulation in an attempt to make it closer to being perfect. One of the most successful strengthening techniques in practice is the addition of Gomory's mixed integer cuts. These inequalities have a geometric interpretation, in the context of Balas' disjunctive programming. They are known as split inequalities in this context, and they are the topic of interest in this chapter. They are also related to the so-called mixed integer rounding inequalities. We show that the convex set defined by intersecting all split inequalities is a polyhedron. For pure integer problems and mixed 0,1 problems, iterating this process a finite number of times produces a perfect formulation. We study Chvátal inequalities and lift-and-project inequalities, which are important special cases of split inequalities. Finally, we discuss cutting planes algorithms based on Gomory inequalities and lift-and-project inequalities, and provide convergence results.

5.1 Split Inequalities

Let $P := \{x \in \mathbb{R}^n : Ax \leq b\}$ be a polyhedron, let $I \subseteq \{1, \ldots, n\}$, and let $S := \{x \in P : x_j \in \mathbb{Z}, j \in I\}$ be a mixed integer set, where I indexes the integer variables. We define $C := \{1, \ldots, n\} \setminus I$ to be the index set of the continuous variables. In this chapter, we study a general principle for generating valid inequalities for $\mathrm{conv}(S)$.

Given a vector $\pi \in \mathbb{Z}^n$ such that $\pi_j = 0$ for all $j \in C$, the scalar product πx is integer for all $x \in S$. Thus, for any $\pi_0 \in \mathbb{Z}$, it follows that every $x \in S$ satisfies exactly one of the terms of the disjunction $\pi x \leq \pi_0$ or $\pi x \geq \pi_0 + 1$. We refer to the latter as a *split disjunction*, and say that a vector $(\pi, \pi_0) \in \mathbb{Z}^n \times \mathbb{Z}$ such that $\pi_j = 0$ for all $j \in C$ is a *split*.

Given P and I, an inequality $\alpha x \leq \beta$ is a *split inequality* [90] if there exists a split (π, π_0) such that $\alpha x \leq \beta$ is valid for both sets

$$\begin{aligned}\Pi_1 &:= P \cap \{x : \pi x \leq \pi_0\} \\ \Pi_2 &:= P \cap \{x : \pi x \geq \pi_0 + 1\}.\end{aligned} \qquad (5.1)$$

It follows from the above discussion that $S \subseteq \Pi_1 \cup \Pi_2$, therefore split inequalities are valid for $\mathrm{conv}(S)$. We define $P^{(\pi,\pi_0)} := \mathrm{conv}(\Pi_1 \cup \Pi_2)$. Clearly $\mathrm{conv}(S) \subseteq P^{(\pi,\pi_0)}$ and an inequality is a split inequality if and only if it is valid for $P^{(\pi,\pi_0)}$ for some split (π, π_0) (see Fig. 5.1).

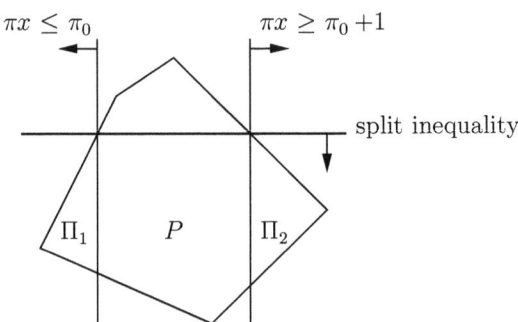

Figure 5.1: A split inequality

By Lemma 4.45, $P^{(\pi,\pi_0)}$ is a polyhedron. Furthermore $P^{(\pi,\pi_0)}$ is the projection in the x-space of the polyhedron defined by the system

$$\begin{aligned} Ax^1 &\leq \lambda b \\ \pi x^1 &\leq \lambda \pi_0 \\ Ax^2 &\leq (1-\lambda)b \\ \pi x^2 &\geq (1-\lambda)(\pi_0 + 1) \\ x^1 + x^2 &= x \\ 0 \leq \lambda &\leq 1. \end{aligned} \qquad (5.2)$$

The polyhedron $P^{(\pi,\pi_0)}$ could have a large number of facets compared to the number of constraints defining P (see Exercise 5.2). However, the extended formulation (5.2) has only $2m + n + 4$ constraints, where m is the number of constraints in the system $Ax \leq b$.

5.1. SPLIT INEQUALITIES

The *split closure* of P is the set defined by

$$P^{\text{split}} := \bigcap P^{(\pi,\pi_0)}, \tag{5.3}$$

where the intersection is taken over all splits (π, π_0). Clearly $\text{conv}(S) \subseteq P^{\text{split}} \subseteq P$. Although each of the sets $P^{(\pi,\pi_0)}$ is a polyhedron, it is not obvious that P^{split} itself is a polyhedron, since it is defined as the intersection of an infinite number of polyhedra. We will prove in Sect. 5.1.2 that, if P is a rational polyhedron, then P^{split} is also a rational polyhedron.

Remark 5.1. *Given a split (π, π_0), let g be the greatest common divisor of the entries of vector π, and let (π', π'_0) be the split defined by $\pi'_j = \frac{\pi_j}{g}$ for $j \in I$, and $\pi'_0 = \lfloor \frac{\pi_0}{g} \rfloor$. Since $\pi'_0 \leq \frac{\pi_0}{g}$ and $\pi'_0 + 1 \geq \frac{\pi_0+1}{g}$, it follows that $P^{(\pi',\pi'_0)} \subseteq P^{(\pi,\pi_0)}$. In particular P^{split} is the intersection of all polyhedra $P^{(\pi,\pi_0)}$ relative to splits (π, π_0) such that the entries of π are relatively prime.*

Proposition 5.2. *Assume that the polyhedron P is nonempty.*

(i) *Let (π, π_0) be a split. If $\Pi_1, \Pi_2 \neq \emptyset$ and $V_1, V_2 \subseteq \mathbb{R}^n$ are finite sets such that $\Pi_1 = \text{conv}(V_1) + \text{rec}(\Pi_1)$ and $\Pi_2 = \text{conv}(V_2) + \text{rec}(\Pi_2)$, then $P^{(\pi,\pi_0)} = \text{conv}(V_1 \cup V_2) + \text{rec}(P)$.*

(ii) *Let (π, π_0) be a split. There is strict inclusion $P^{(\pi,\pi_0)} \subset P$ if and only if some minimal face of P lies in the region defined by $\pi_0 < \pi x < \pi_0 + 1$.*

(iii) *If P is a rational polyhedron, $P^{\text{split}} \subset P$ if and only if $\text{conv}(S) \subset P$.*

Proof. By Lemma 4.45, $P^{(\pi,\pi_0)}$ is a polyhedron, i.e., $P^{(\pi,\pi_0)} := \text{conv}(\Pi_1 \cup \Pi_2) = \overline{\text{conv}}(\Pi_1 \cup \Pi_2)$. Therefore (i) follows from Lemma 4.41 and the fact that $\text{rec}(\Pi_1) \cup \text{rec}(\Pi_2) = \text{rec}(P)$. For the sake of simplicity, we prove (ii) and (iii) under the hypothesis that P is pointed.

(ii) Assuming P is pointed, let V be the set of vertices of P. By Theorem 3.37, $P = \text{conv}(V) + \text{rec}(P)$. If no point of V lies in the region defined by $\pi_0 < \pi x < \pi_0 + 1$, then $\text{conv}(V) \subseteq \text{conv}(\Pi_1 \cup \Pi_2)$, therefore $P \subseteq P^{(\pi,\pi_0)}$. Conversely, given $v \in V$ such that $\pi_0 < \pi v < \pi_0 + 1$, it follows from Theorem 3.34(iii) that $v \notin P^{(\pi,\pi_0)}$.

(iii) If $\text{conv}(S) = P$ then clearly $P = P^{\text{split}}$. Conversely, assume $\text{conv}(S) \subset P$. Since P is a rational pointed polyhedron, by Theorem 4.3 there exists a vertex v of P and an index $j \in I$ such that $v_j \notin \mathbb{Z}$. By (ii), since $\lfloor v_j \rfloor < v_j < \lfloor v_j \rfloor + 1$, v does not belong to $P^{(e^j, \lfloor v_j \rfloor)}$, where e^j is the jth unit vector in $\mathbb{R}^n$. Thus $v \notin P^{\text{split}}$. □

Let $(\pi, \pi_0) \in \mathbb{Z}^n \times \mathbb{Z}$ be a split (i.e., $\pi_j = 0$ for all $j \in C$). The next lemma, due to Bonami [58], gives a necessary and sufficient condition for a point to be in $P^{(\pi,\pi_0)}$.

Lemma 5.3. *Given $\bar{x} \in P$ such that $\pi_0 < \pi\bar{x} < \pi_0 + 1$, $\bar{x}$ belongs to $P^{(\pi,\pi_0)}$ if and only if there exists $\tilde{x} \in \Pi_2$ such that*

$$b - A\tilde{x} \leq \frac{b - A\bar{x}}{\pi\bar{x} - \pi_0}. \tag{5.4}$$

Proof. For the "if" direction, let $\tilde{x}$ be as in the statement and define $\lambda := \frac{\pi\bar{x} - \pi_0}{\pi_0 + 1 - \pi\bar{x}}$. Note that $\lambda > 0$. We show that the point $\hat{x} = \bar{x} + \lambda(\bar{x} - \tilde{x})$ belongs to Π_1, thus $\bar{x} \in P^{(\pi,\pi_0)}$ since it is a convex combination of $\tilde{x} \in \Pi_2$ and $\hat{x} \in \Pi_1$. Indeed,

$$\pi\hat{x} = \pi\bar{x} + \lambda(\pi\bar{x} - \pi\tilde{x}) \leq \pi\bar{x} + \lambda(\pi\bar{x} - (\pi_0 + 1)) = \pi_0,$$

$$A\hat{x} - b = (1+\lambda)A\bar{x} - \lambda A\tilde{x} - b = (1+\lambda)(A\bar{x} - b) + \lambda(b - A\tilde{x}) \leq 0,$$

where the last inequality follows from (5.4) and the fact that $\frac{\lambda}{1+\lambda} = \pi\bar{x} - \pi_0$. Thus $\hat{x} \in \Pi_1$ and $\bar{x} \in P^{(\pi,\pi_0)}$.

For the "only if" part, suppose there exist $x^1 \in \Pi_1$ and $x^2 \in \Pi_2$ such that $\bar{x} = (1-\lambda)x^1 + \lambda x^2$ where $0 < \lambda < 1$. We can choose x^1 and x^2 so that $\pi x^1 = \pi_0$ and $\pi x^2 = \pi_0 + 1$. It follows that $\pi\bar{x} = \pi_0 + \lambda$, thus $\lambda = \pi\bar{x} - \pi_0$. We show that $\tilde{x} := x^2$ satisfies (5.4). Indeed,

$$b - A\tilde{x} = b - \frac{A\bar{x} - (1-\lambda)Ax^1}{\lambda} = \frac{b - A\bar{x} - (1-\lambda)(b - Ax^1)}{\lambda} \leq \frac{b - A\bar{x}}{\pi\bar{x} - \pi_0},$$

where the last inequality follows from $\lambda < 1$ and $Ax^1 \leq b$. □

By Lemma 5.3, given $\bar{x} \in P$ such that $\pi_0 < \pi\bar{x} < \pi_0 + 1$, in order to decide whether $\bar{x}$ is in $P^{(\pi,\pi_0)}$ one can solve the following linear programming problem

$$\max \pi x$$
$$b - \frac{b - A\bar{x}}{\pi\bar{x} - \pi_0} \leq Ax \leq b. \tag{5.5}$$

If the optimal value is greater than or equal to $\pi_0 + 1$, then $\bar{x} \in P^{(\pi,\pi_0)}$, otherwise $\bar{x} \notin P^{(\pi,\pi_0)}$.

5.1.1 Inequality Description of the Split Closure

Let $P := \{x \in \mathbb{R}^n : Ax \leq b\}$ be a polyhedron, let $I \subseteq \{1,\ldots,n\}$ and $C := \{1,\ldots,n\} \setminus I$ index integer and continuous variables respectively, and let $S := \{x \in P : x_j \in \mathbb{Z}, j \in I\}$ be the corresponding mixed integer set.

The main goal of this section is to prove that all split inequalities that are necessary to describe $P^{(\pi,\pi_0)}$ can be written in the following form. Let m be the number of rows of A. For $u \in \mathbb{R}^m$, let u^+ be defined by $u_i^+ := \max\{0, u_i\}$, $i = 1,\ldots,m$, and let $u^- := (-u)^+$, so $u = u^+ - u^-$. Throughout this chapter, we denote by A_I and A_C the matrices comprising the columns of A with indices in I and C, respectively, and for any vector $\alpha \in \mathbb{R}^n$ we define α_I and α_C accordingly.

For $u \in \mathbb{R}^m$ such that uA_I is integral, $uA_C = 0$, and $ub \notin \mathbb{Z}$, consider the inequality

$$\frac{u^+(b - Ax)}{f} + \frac{u^-(b - Ax)}{1 - f} \geq 1, \tag{5.6}$$

where $f := ub - \lfloor ub \rfloor$.

Lemma 5.4. *Let $u \in \mathbb{R}^m$ such that uA_I is integral, $uA_C = 0$, and $ub \notin \mathbb{Z}$. Define $\pi := uA$ and $\pi_0 := \lfloor ub \rfloor$. The inequality (5.6) is valid for $P^{(\pi,\pi_0)}$, thus it is a split inequality for P.*

Proof. By definition, π_I is integral and $\pi_C = 0$, thus (π, π_0) is a split. It suffices to show that (5.6) is valid for $P^{(\pi,\pi_0)}$. We show that (5.6) is valid for Π_1, the argument for Π_2 being symmetric. Given $\bar{x} \in \Pi_1$, let $s^1 := u^+(b - A\bar{x})$ and $s^2 := u^-(b - A\bar{x})$. Observe that $s^1 - s^2 = ub - uA\bar{x} = ub - \pi\bar{x}$. Thus $(1-f)s^1 + fs^2 = (1-f)(s^1 - s^2) + s^2 = (1-f)(ub - \pi\bar{x}) + s^2 \geq (1-f)f$, where the last inequality follows from $\pi\bar{x} \leq \pi_0$ and from $s^2 \geq 0$. □

Let $B_\pi \subseteq \mathbb{R}^m$ denote the set of basic solutions to the system $uA = \pi$. (Recall that u is basic if the rows of A corresponding to the nonzero entries of u are linearly independent.)

Theorem 5.5. *Let $P := \{x \in \mathbb{R}^n : Ax \leq b\}$ be a polyhedron, let $I \subseteq \{1,\ldots,n\}$, and let $S := \{x \in P : x_j \in \mathbb{Z}, j \in I\}$. Given a split $(\pi, \pi_0) \in \mathbb{Z}^{n+1}$, $P^{(\pi,\pi_0)}$ is the set of all points in P satisfying the inequalities*

$$\frac{u^+(b - Ax)}{ub - \pi_0} + \frac{u^-(b - Ax)}{\pi_0 + 1 - ub} \geq 1, \quad \text{for all } u \in B_\pi \text{ s.t. } \pi_0 < ub < \pi_0 + 1. \tag{5.7}$$

Proof. It follows from Lemma 5.4 that the inequalities (5.7) are valid for $P^{(\pi,\pi_0)}$. Thus we only need to show that, given a point $\bar{x} \in P \setminus P^{(\pi,\pi_0)}$, there exists an inequality (5.7) violated by $\bar{x}$. Since $\bar{x} \in P \setminus P^{(\pi,\pi_0)}$, it satisfies $\pi_0 < \pi\bar{x} < \pi_0 + 1$, and it follows from Lemma 5.3 that the linear program (5.5) has optimal value less than $\pi_0 + 1$. The dual of (5.5) is

$$\begin{aligned} \min \quad & (u^1 - u^2)b + u^2 \frac{b - A\bar{x}}{\pi\bar{x} - \pi_0} \\ & (u^1 - u^2)A = \pi \\ & u^1, u^2 \geq 0. \end{aligned} \quad (5.8)$$

Let (u^1, u^2) be an optimal basic solution of (5.8), and let $u := u^1 - u^2$. Since (u^1, u^2) is basic, it follows that, for $i = 1, \ldots, m$, at most one among u_i^1 and u_i^2 is nonzero, therefore $u^+ = u^1$ and $u^- = u^2$. Furthermore, the rows of A corresponding to nonzero components of u are linearly independent, hence $u \in B_\pi$. By the linear programming duality theorem (Theorem 3.7), Problem (5.8) has optimal value less than $\pi_0 + 1$, thus

$$ub + u^- \frac{b - A\bar{x}}{\pi\bar{x} - \pi_0} < \pi_0 + 1. \quad (5.9)$$

Using the fact that $uA = \pi$ and $\pi\bar{x} - \pi_0 > 0$, (5.9) is equivalent to

$$u^-(b - A\bar{x}) < (\pi_0 + 1 - ub)(uA\bar{x} - \pi_0). \quad (5.10)$$

Since $uA\bar{x} - \pi_0 = ub - \pi_0 - u(b - A\bar{x})$, (5.10) can be expressed as

$$(\pi_0+1-ub)u^+(b-A\bar{x})+(ub-\pi_0)u^-(b-A\bar{x}) < (ub-\pi_0)(\pi_0+1-ub). \quad (5.11)$$

We prove that $\pi_0 < ub < \pi_0 + 1$. Since $u^- \frac{b-A\bar{x}}{\pi\bar{x}-\pi_0} \geq 0$, (5.9) implies that $ub < \pi_0 + 1$. Inequality (5.11) is equivalent to $u^+(b - A\bar{x}) < (ub - \pi_0)(\pi_0 + 1 - ub + u(b - A\bar{x}))$. Since $u^+(b - A\bar{x}) \geq 0$ and $\pi_0 + 1 - ub + u(b - A\bar{x}) = \pi_0 + 1 - \pi\bar{x} > 0$, it follows that $ub - \pi_0 > 0$.

Therefore $(ub - \pi_0)(\pi_0 + 1 - ub) > 0$, which implies that (5.11) can be rewritten as

$$\frac{u^+(b - A\bar{x})}{ub - \pi_0} + \frac{u^-(b - A\bar{x})}{\pi_0 + 1 - ub} < 1,$$

showing that $\bar{x}$ violates the inequality (5.7) relative to u. □

Theorem 5.5 implies the following characterization of P^{split}.

Corollary 5.6. P^{split} *is the set of all points in P satisfying the inequalities (5.6) for all $u \in \mathbb{R}^m$ such that uA_I is integral, $uA_C = 0$, $ub \notin \mathbb{Z}$, and the rows of A corresponding to nonzero entries of u are linearly independent.*

5.1. SPLIT INEQUALITIES

Another consequence of Theorem 5.5 is the following result of Andersen et al. [11]. Let $a^1, \ldots, a^m$ denote the rows of A. Let $k := \text{rank}(A)$, and denote by $\mathcal{B}$ the family of bases of $Ax \leq b$, that is, sets $B \subseteq \{1, \ldots, m\}$ such that $|B| = k$ and the vectors a^i, $i \in B$, are linearly independent. For every $B \in \mathcal{B}$, let $P_B := \{x : a^i x \leq b_i, i \in B\}$.

Corollary 5.7. $P^{\text{split}} = \bigcap_{B \in \mathcal{B}} P_B^{\text{split}}$.

Note that, given $B \in \mathcal{B}$, the polyhedron $P_B := \{x : a^i x \leq b_i, i \in B\}$ has a unique minimal face, namely $F_B := \{x : a^i x = b_i, i \in B\}$. In particular P_B is a translate of its recession cone, that is, $P_B = v + \text{rec}(P_B)$ for any $v \in F_B$. Theorem 5.5 implies that the polyhedron $P_B^{(\pi, \pi_0)}$ is defined by introducing only one split inequality.

Remark 5.8. Let (π, π_0) be a split. For all $B \in \mathcal{B}$ such that $P_B^{(\pi, \pi_0)} \neq P_B$,

$$P_B^{(\pi, \pi_0)} = P_B \cap \left\{ x : \frac{\bar{u}^+(b - Ax)}{\bar{u}b - \pi_0} + \frac{\bar{u}^-(b - Ax)}{\pi_0 + 1 - \bar{u}b} \geq 1 \right\},$$

where $\bar{u}$ is the unique vector such that $\bar{u}A = \pi$ and $\bar{u}_i = 0$ for all $i \notin B$.

Any basic solution u in B_π needed in Theorem 5.5 is defined by some $B \in \mathcal{B}$ by $uA = \pi$ and $u_i = 0$ for all $i \notin B$. Note that B needs not be a feasible basis of the system $Ax \leq b$ defining P. By this we mean that the minimal face F_B of P_B may not be a face of P, since it could be that $F_B \cap P = \emptyset$. Figure 5.2 illustrates the fact that the description of $P^{(\pi, \pi_0)}$ may require split inequalities generated from infeasible bases. Indeed, the

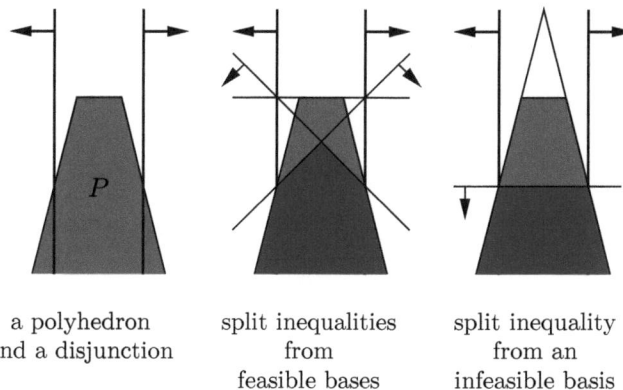

a polyhedron and a disjunction

split inequalities from feasible bases

split inequality from an infeasible basis

Figure 5.2: A split inequality from an infeasible basis can be stronger than split inequalities from feasible bases

polyhedron P on the left has two feasible bases and one infeasible one. Split inequalities from the two feasible bases give the dark polyhedron in the middle. The split inequality defined by the infeasible basis gives the dark polyhedron on the right; in this case this single split inequality is sufficient to define $P^{(\pi,\pi_0)}$.

5.1.2 Polyhedrality of the Split Closure

Let $P \subseteq \mathbb{R}^n$ be a polyhedron, let $I \subseteq \{1,\ldots,n\}$, let $C := \{1,\ldots,n\} \setminus I$, and let $S := P \cap (\mathbb{Z}^I \times \mathbb{R}^C)$. Cook et al. [90] showed that, if P is a rational polyhedron, then P^{split} is also a polyhedron. Next we present a simpler proof of this fact, due to Dash et al. [107]. The idea is to prove that a finite number of splits (π, π_0) are sufficient to generate P^{split} defined in (5.3). The result follows, since then P^{split} is the intersection of a finite number of polyhedra $P^{(\pi,\pi_0)}$ and therefore is a polyhedron.

Throughout the rest of this section, we will assume that P is rational. Therefore, we may assume without loss of generality that $P = \{x \in \mathbb{R}^n : Ax \leq b\}$ where A and b are integral. We say that a split inequality $\alpha x \leq \beta$ is *dominated* by another split inequality $\alpha' x \leq \beta'$ if $P \cap \{x : \alpha' x \leq \beta'\} \subset P \cap \{x : \alpha x \leq \beta\}$. If $\alpha x \leq \beta$ is not dominated by any split inequality we say that it is *undominated*.

Lemma 5.9. *Let Δ be the largest among 1 and the absolute values of all subdeterminants of A_C. Let $u \in \mathbb{R}^m$ be such that $uA_I \in \mathbb{Z}^I$, $uA_C = 0$, $ub \notin \mathbb{Z}$, and the inequality (5.6) is undominated. Then $|u_i| \leq m\Delta$, $i = 1, \ldots, m$.*

Proof. Let $u \in \mathbb{R}^m$ be as in the statement, and let $f := ub - \lfloor ub \rfloor$. Consider the set of indices $M^- := \{i \in [m] : u_i < 0\}$ and $M^+ := \{i \in [m] : u_i \geq 0\}$, and define $C_u := \{w \in \mathbb{R}^m : wA_C = 0; w_i \leq 0, i \in M^-; w_i \geq 0, i \in M^+\}$. Note that C_u is a pointed cone. We first observe that, if there exist $u^1, u^2 \in C_u$ such that $u = u^1 + u^2$ and $u^2 \in \mathbb{Z}^m \setminus \{0\}$, the split inequality (5.6) defined by u is dominated. Indeed, since $u^2 \in \mathbb{Z}^m$, we have that $u^1 A_I \in \mathbb{Z}^I$ and $u^1 b - \lfloor u^1 b \rfloor = f$, thus u^1 defines the split inequality

$$(u^1)^+ \frac{b - Ax}{f} + (u^1)^- \frac{b - Ax}{1 - f} \geq 1,$$

which dominates (5.6) because $(u^1)^+ \leq u^+$ and $(u^1)^- \leq u^-$, since $u^1, u^2 \in C_u$.

Suppose now that u does not satisfy $-m\Delta \mathbf{1} \leq u \leq m\Delta \mathbf{1}$. We will show that there exist $u^1, u^2 \in C_u$ such that $u = u^1 + u^2$ and $u^2 \in \mathbb{Z}^m \setminus \{0\}$, thus concluding the proof. Let $r^1, \ldots, r^q \in \mathbb{R}^m$ be the extreme rays of C_u. Since A_C is a rational matrix, we can choose $r^1, \ldots, r^q$ integral

5.1. SPLIT INEQUALITIES

(if $C = \emptyset$, $r^1, \ldots, r^q$ are chosen to be the unit vectors or their negatives), and by standard linear algebra we can choose them so that $-\Delta \mathbf{1} \leq r^t \leq \Delta \mathbf{1}$, $t = 1, \ldots, q$. Since $u \in C_u$, by Carathéodory's theorem $u = \sum_{t=1}^{q} \nu_t r^t$ where at most m of the ν_t are positive, while the others are 0. Let $u^1 := \sum_{t=1}^{q}(\nu_t - \lfloor \nu_t \rfloor) r^t$, $u^2 := \sum_{t=1}^{q} \lfloor \nu_t \rfloor r^t$. Clearly $u^1, u^2 \in C_u$ and $u = u^1 + u^2$. Since $r^1, \ldots, r^q$ are integral vectors, u^2 is integral. Finally, since at most m of the ν_t are positive and $-\Delta \mathbf{1} \leq r^t \leq \Delta \mathbf{1}$, $t = 1, \ldots, q$, it follows that $-m\Delta \mathbf{1} \leq u^1 \leq m\Delta \mathbf{1}$, thus $u^2 \neq 0$, as u does not satisfy $-m\Delta \mathbf{1} \leq u \leq m\Delta \mathbf{1}$. □

Theorem 5.10 (Cook et al. [90]). *Let $P \subseteq \mathbb{R}^n$ be a rational polyhedron and let $S := P \cap (\mathbb{Z}^I \times \mathbb{R}^C)$. Then P^{split} is a rational polyhedron.*

Proof. Let (A, b) be an integral matrix such that $P = \{x \in \mathbb{R}^n : Ax \leq b\}$. By Theorem 5.5 and Lemma 5.9, P^{split} is the intersection of all polyhedra $P^{(\pi, \pi_0)}$ where (π, π_0) is a split such that $\pi = uA$ and $\pi_0 = \lfloor ub \rfloor$ for some $u \in \mathbb{R}^m$ such that $|u_i| \leq m\Delta$ for $i = 1, \ldots, m$. Since (π, π_0) is integral, there is a finite number of such vectors (π, π_0). Hence P^{split} is the intersection of a finite number of rational polyhedra and hence is a rational polyhedron. □

A natural question is whether one can optimize a linear function over P^{split} in polynomial time. It turns out that this problem is NP-hard (Caprara and Letchford [68], Cornuéjols and Li [95]). The equivalence between optimization and separation (see Theorem 7.26 stated later in this book) implies that, given a positive integer n, a rational polyhedron $P \subset \mathbb{R}^n$, a set $I \subseteq \{1, \ldots, n\}$ of integer variables, and a rational point $\bar{x} \in P$, it is NP-hard to find a split inequality that cuts off $\bar{x}$ or show that none exists.

5.1.3 Split Rank

Let $P := \{x \in \mathbb{R}^n : Ax \leq b\}$ be a rational polyhedron, let $I \subseteq \{1, \ldots, n\}$, and let $S := \{x \in P : x_j \in \mathbb{Z}, j \in I\}$. Let us denote the split closure P^{split} of P by P^1 and, for $k \geq 2$, let P^k denote the split closure of P^{k-1}. We refer to P^k as the *kth split closure* relative to P. By Theorem 5.10, P^k is a polyhedron for all k. One may ask whether, by repeatedly applying the split closure operator, one eventually obtains $\text{conv}(S)$. The next example, due to Cook et al. [90], shows that this is not the case.

Example 5.11. Let $S := \{(x, y) \in \mathbb{Z}_+^2 \times \mathbb{R}_+ : x_1 \geq y, x_2 \geq y, x_1 + x_2 + 2y \leq 2\}$. Starting from $P := \{(x_1, x_2, y) \in \mathbb{R}_+^3 : x_1 \geq y, x_2 \geq y, x_1 + x_2 + 2y \leq 2\}$, we claim that there is no finite k such that $P^k = \text{conv}(S)$.

To see this, note that P is a simplex with vertices $O = (0, 0, 0)$, $A = (2, 0, 0)$, $B = (0, 2, 0)$ and $C = (\frac{1}{2}, \frac{1}{2}, \frac{1}{2})$ (see Fig. 5.3). S is contained in the plane $y = 0$. So $\text{conv}(S) = P \cap \{(x_1, x_2, y) : y \leq 0\}$. More generally, consider a simplex P with vertices O, A, B and $C = (\frac{1}{2}, \frac{1}{2}, t)$ with $t > 0$. Define $C_1 := C$, let C_2 be the point on the edge from C to A with coordinate $x_1 = 1$ and C_3 the point on the edge from C to B with coordinate $x_2 = 1$. Observe that no split inequality removes all three points C_1, C_2, C_3. Indeed, the projections of C_1, C_2, C_3 onto the plane $y = 0$ are inner points on the edges of the triangle T with vertices $(1, 0), (0, 1), (1, 1)$. Since these vertices are integral, any split leaves at least one edge of T entirely on one side of the disjunction. It follows that the corresponding C_i is not removed by the split inequality.

Let Q_i be the intersection of all split inequalities that do not cut off C_i. All split inequalities belong to at least one of these three sets, thus $P^1 = Q_1 \cap Q_2 \cap Q_3$. Let S_i be the simplex with vertices O, A, B, C_i. Clearly, $S_i \subseteq Q_i$. Thus $S_1 \cap S_2 \cap S_3 \subseteq P^1$. It is easy to verify that $(\frac{1}{2}, \frac{1}{2}, \frac{t}{3}) \in S_i$ for $i = 1, 2$ and 3. Thus $(\frac{1}{2}, \frac{1}{2}, \frac{t}{3}) \in P^1$. By induction, $(\frac{1}{2}, \frac{1}{2}, \frac{t}{3^k}) \in P^k$. Therefore $P^k \neq \text{conv}(S)$ for every positive integer k. ∎

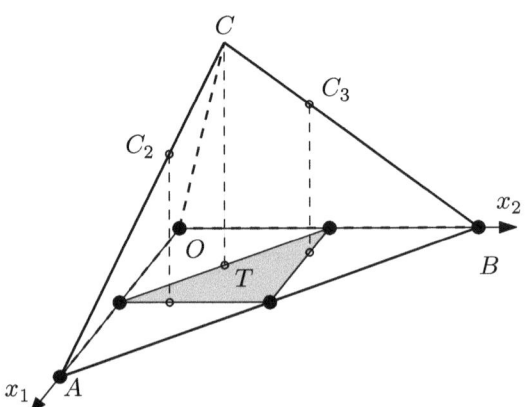

Figure 5.3: Example showing that the split rank can be unbounded, and representation of the proof

The smallest k such that $P^k = \text{conv}(S)$ is called the *split rank* of P, if such an integer k exists. The *split rank of a valid inequality* $\alpha x \leq \beta$ for $\text{conv}(S)$ is the smallest k such that $\alpha x \leq \beta$ is valid for P^k. In the above example the inequality $y \leq 0$ does not have finite split rank. By contrast, we will see that the split rank is always finite for pure integer programs (Theorem 5.18) and at most $|I|$ for mixed 0,1 linear programs (Theorem 5.22).

5.1.4 Gomory's Mixed Integer Inequalities

Gomory's mixed integer inequalities [176], introduced in 1960, were the first example of general-purpose valid inequalities for mixed integer linear programs. They can be interpreted as split inequalities for problems written in standard equality form. Let $P := \{x \in \mathbb{R}^n : Ax = b, x \geq 0\}$ be a polyhedron expressed in standard equality form. Let $I \subseteq \{1,\ldots,n\}$, $C := \{1,\ldots,n\} \setminus I$, and $S := \{x \in P : x_j \in \mathbb{Z}, j \in I\}$.

By Corollary 5.6 applied to $P = \{x \in \mathbb{R}^n : Ax \leq b, -Ax \leq -b, -x \leq 0\}$, any undominated split inequality is determined by a vector $(u,v) \in \mathbb{R}^m \times \mathbb{R}^n$ such that $uA_I - v_I \in \mathbb{Z}^I$, $uA_C - v_C = 0$, and $ub \notin \mathbb{Z}$, and can be written in the form

$$\frac{v^+ x}{f} + \frac{v^- x}{1-f} \geq 1, \tag{5.12}$$

where $f := ub - \lfloor ub \rfloor > 0$.

Let $\alpha := uA$, $\beta := ub$, and $f_j := \alpha_j - \lfloor \alpha_j \rfloor$, $j \in I$. Since the variables defining P are all nonnegative, for a given u, the choice of $v_j = v_j^+ - v_j^-$, $j \in I$ that gives the strongest inequality is the smallest possible value that satisfies the requirement $\alpha_j - (v_j^+ - v_j^-) \in \mathbb{Z}$. Therefore $v_j = f_j$ whenever $v_j = v_j^+$ and $v_j = f_j - 1$ whenever $v_j = -v_j^-$. This implies that $v_j = v_j^+$ if $\frac{f_j}{f} \leq \frac{1-f_j}{1-f}$ (i.e., if $f_j \leq f$), and $v_j = v_j^-$ otherwise. Finally observe that $uA_C - v_C = 0$ is equivalent to $v_j^+ - v_j^- = \alpha_j$, $j \in C$. Hence $v_j^+ = \alpha_j$ if $\alpha_j \geq 0$, otherwise $v_j^- = -\alpha_j$. It follows that the undominated split inequalities are of the form

$$\sum_{\substack{j \in I \\ f_j \leq f}} \frac{f_j}{f} x_j + \sum_{\substack{j \in I \\ f_j > f}} \frac{1-f_j}{1-f} x_j + \sum_{\substack{j \in C \\ \alpha_j \geq 0}} \frac{\alpha_j}{f} x_j - \sum_{\substack{j \in C \\ \alpha_j < 0}} \frac{\alpha_j}{1-f} x_j \geq 1. \tag{5.13}$$

This is *Gomory's mixed integer inequality* derived from the equation $\alpha x = \beta$ [176]. Note that (5.13) is a split inequality relative to the split (π, π_0) defined by $\pi_0 = \lfloor ub \rfloor$ and, for $j = 1, \ldots, n$,

$$\pi_j = \begin{cases} \lfloor \alpha_j \rfloor & \text{if } j \in I \text{ and } f_j \leq f_0 \\ \lceil \alpha_j \rceil & \text{if } j \in I \text{ and } f_j > f_0 \\ 0 & \text{if } j \in C. \end{cases} \tag{5.14}$$

In practice, Gomory's mixed integer inequalities have turned out to be effective cutting planes in branch-and-cut algorithms. Implementation of the mixed integer inequalities will be discussed in Sect. 5.3.

5.1.5 Mixed Integer Rounding Inequalities

Mixed integer rounding inequalities, introduced by Nemhauser and Wolsey [286], offer an alternative, yet equivalent, definition of the split inequalities.

As seen in Sect. 1.4.1, the convex hull of the two-dimensional mixed integer set $\{(\xi, v) \in \mathbb{Z} \times \mathbb{R}_+ : \xi - v \leq \beta\}$ is defined by the original inequalities $v \geq 0$, $\xi - v \leq \beta$ and the *simple rounding inequality*

$$\xi - \frac{1}{1-f} v \leq \lfloor \beta \rfloor,$$

where $f := \beta - \lfloor \beta \rfloor$. The simple rounding inequality is a split inequality, relative to the split disjunction $(\xi \leq \lfloor \beta \rfloor) \vee (\xi \geq \lceil \beta \rceil)$.

Mixed integer rounding inequalities for general mixed integer linear sets are inequalities that can be derived as a simple rounding inequality using variable aggregation. Formally, let $P := \{x \in \mathbb{R}^n : Ax \leq b\}$, $S := \{x \in P : x_j \in \mathbb{Z}, j \in I\}$, and $C := \{1, \ldots, n\} \setminus I$. Suppose that a given valid inequality for P can be written, rearranging the variables, in the form

$$\pi x - (\gamma - cx) \leq \beta \qquad (5.15)$$

such that π_I is integral, $\pi_C = 0$, and $cx \leq \gamma$ is a valid inequality for P. Clearly πx is an integer and $\gamma - cx \geq 0$ for all $x \in S$. Deriving the simple mixed integer rounding with the variable substitution $\xi = \pi x$ as an integer variable and $v = \gamma - cx$ as a nonnegative continuous variable, we obtain the inequality

$$\pi x - \frac{1}{1-f}(\gamma - cx) \leq \lfloor \beta \rfloor. \qquad (5.16)$$

Inequalities that can be obtained with the above derivation are the *mixed integer rounding inequalities*. Nemhauser–Wolsey [286] showed that mixed integer rounding and split inequalities are equivalent. More formally, define the *mixed integer rounding closure* P^{MIR} of P as the set of points in P satisfying all mixed integer rounding inequalities.

Theorem 5.12. $P^{\mathrm{MIR}} = P^{\mathrm{split}}$.

Proof. By construction, the mixed integer rounding inequality (5.16) is a split inequality relative to the split $(\pi, \lfloor \beta \rfloor)$, thus $P^{\mathrm{MIR}} \supseteq P^{\mathrm{split}}$.

To prove the converse, by Theorem 5.5 it suffices to show that any inequality of the form (5.6) is a mixed integer rounding inequality. Let $u \in \mathbb{R}^m$ such that uA_I is integral and $uA_C = 0$. The inequality $u^+ Ax \leq u^+ b$ is valid for P, and it can be written as

$$uAx - u^-(b - Ax) \leq ub.$$

Since $u^- Ax \leq u^- b$ is valid for P, we can derive the mixed integer rounding inequality

$$uAx - \frac{u^-}{1-f}(b - Ax) \leq \lfloor ub \rfloor \qquad (5.17)$$

where $f := ub - \lfloor ub \rfloor$. One can readily verify that (5.17) is equivalent to the split inequality (5.6). □

5.2 Chvátal Inequalities

Given a polyhedron $P := \{x \in \mathbb{R}^n : Ax \leq b\}$, let $I \subseteq \{1, \ldots, n\}$, $C := \{1, \ldots, n\} \setminus I$, and $S := \{x \in P : x_j \in \mathbb{Z}, j \in I\}$. Given a split (π, π_0), the inequality $\pi x \leq \pi_0$ is a *Chvátal inequality* if $P \cap \{x : \pi x \geq \pi_0 + 1\} = \emptyset$. Chvátal inequalities are valid for conv(S), since πx is an integer for every $x \in S$. Note that Chvátal inequalities are split inequalities, relative to the split (π, π_0).

For $\pi \in \mathbb{Z}^n$, setting $\delta := \max\{\pi x : x \in P\}$ and $\pi_0 = \lfloor \delta \rfloor$, the Chvátal inequality $\pi x \leq \pi_0$ cuts off a part of P if and only if $\delta \notin \mathbb{Z}$ (see Fig. 5.4). By Remark 5.1, the only relevant Chvátal inequalities are the ones for which the coefficients of π are relatively prime. When the coefficients of π are relatively prime, by Corollary 1.9 the equation $\pi x = \pi_0$ admits an integral solution, therefore, by Theorem 4.3, conv$\{x \in \mathbb{Z}^I \times \mathbb{R}^C : \pi x \leq \delta\} = \{x \in \mathbb{R}^n : \pi x \leq \pi_0\}$.

Assume that $P \neq \emptyset$ and let (π, π_0) be a split. By Theorem 3.22, $\pi x \leq \pi_0$ is a Chvátal inequality if and only if there exists a vector u such that

$$u \geq 0, \; uA = \pi, \; \pi_0 \geq \lfloor ub \rfloor. \qquad (5.18)$$

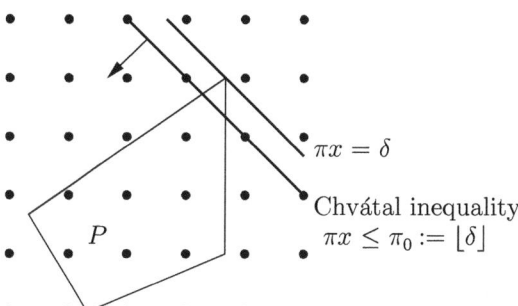

Figure 5.4: Chvátal inequality

The *Chvátal closure* P^{Ch} of P is the set of points in P satisfying all the Chvátal inequalities. Equivalently,

$$P^{Ch} := \{x \in P : (uA)x \leq \lfloor ub \rfloor \text{ for all } u \geq 0 \text{ s.t. } uA_I \in \mathbb{Z}^I, uA_C = 0\}. \tag{5.19}$$

Note that $S \subseteq P^{\text{split}} \subseteq P^{Ch} \subseteq P$.

5.2.1 The Chvátal Closure of a Pure Integer Linear Set

The above presentation of Chvátal inequalities is not the most standard: Chvátal's seminal paper [73] introduced the closure P^{Ch} for pure integer linear sets.

In this section, we focus on pure integer sets of the type $S := P \cap \mathbb{Z}^n$, where $P := \{x \in \mathbb{R}^n : Ax \leq b\}$ is a rational polyhedron. Thus we can assume that A, b have integer entries. A Chvátal inequality $\pi x \leq \pi_0$ is derived from a vector u satisfying

$$u \geq 0, \ uA = \pi \in \mathbb{Z}^n, \ \lfloor ub \rfloor = \pi_0. \tag{5.20}$$

As before, P^{Ch} is the set of points in P satisfying all the Chvátal inequalities. The next lemma shows that the vectors u with some of their components equal to 1 or larger are not needed in the description of P^{Ch}.

Lemma 5.13. *P^{Ch} is the set of all points in P satisfying the Chvátal inequalities $uAx \leq \lfloor ub \rfloor$ for all u such that $uA \in \mathbb{Z}^n$ and $0 \leq u \leq 1$.*

Proof. Given $u \geq 0$ such that uA is integral, let $u^1 := u - \lfloor u \rfloor$ and $u^2 = \lfloor u \rfloor$. Note that $u^2 \geq 0$, thus $(u^2 A)x \leq u^2 b$ is valid for P. Since u^2 is an integral vector and A, b are integral, $u^2 A$ is an integral vector and $u^2 b \in \mathbb{Z}$. Because uA is integral, it follows that $u^1 A = uA - u^2 A$ is integral as well. Furthermore, by definition $0 \leq u^1 \leq 1$. The Chvátal inequality $uAx \leq \lfloor ub \rfloor$ is the sum of $u^1 Ax \leq \lfloor u^1 b \rfloor$ and $u^2 Ax \leq u^2 b$. Since $u^2 Ax \leq u^2 b$ is valid for P, it follows that $P \cap \{x : u^1 Ax \leq \lfloor u^1 b \rfloor\} \subseteq P \cap \{x : uAx \leq \lfloor ub \rfloor\}$. □

Theorem 5.14 (Chvátal [73]). *P^{Ch} is a rational polyhedron.*

Proof. Since $\{uA \in \mathbb{R}^n : 0 \leq u < 1\}$ is a bounded set, the set $\{uA \in \mathbb{Z}^n : 0 \leq u < 1\}$ is finite. It follows from Lemma 5.13 that P^{Ch} is a polyhedron. □

As for the split closure, the *membership problem* for the Chvátal closure P^{Ch} "Given a rational polyhedron $P \subset \mathbb{R}^n$ and a rational point $\bar{x} \in P$, does $\bar{x} \in P^{Ch}$?" is NP-hard, as proved by Eisenbrand [130].

5.2.2 Chvátal Rank

In this section, we consider a pure integer set $S := P \cap \mathbb{Z}^n$ where $P := \{x \in \mathbb{R}^n : Ax \leq b\}$ is a rational polyhedron. We denote $\mathrm{conv}(S)$ by P_I. The Chvátal closure P^{Ch} of P will be denoted by $P^{(1)}$ in this section. We can iterate the closure process to obtain the Chvátal closure of $P^{(1)}$. We denote by $P^{(2)}$ this second Chvátal closure. Iteratively, we define the tth Chvátal closure $P^{(t)}$ of P to be the Chvátal closure of $P^{(t-1)}$, for $t \geq 2$ integer. An inequality that is valid for $P^{(t)}$ but not $P^{(t-1)}$ is said to have *Chvátal rank* t. Are there inequalities of arbitrary large Chvátal rank or is there a value t after which $P^{(t)} = P^{(t+1)}$? The main result of this section is that the second statement is the correct one. In fact, we will prove that there exists a finite t such that $P^{(t)} = P_I$. The smallest such t is called the *Chvátal rank of the polyhedron* P. Therefore, every valid inequality for $P_I := \mathrm{conv}(S)$ has a bounded Chvátal rank. This result for the pure integer case is in contrast to the situation for the mixed case, as we saw in Example 5.11.

Lemma 5.15. *Let $P \subseteq \mathbb{R}^n$ be a nonempty rational polyhedron such that $\mathrm{aff}(P) \cap \mathbb{Z}^n \neq \emptyset$. If $P_I = \emptyset$, then $\dim(\mathrm{rec}(P)) < \dim(P)$.*

Proof. Let $d := \dim(P) = \dim(\mathrm{aff}(P))$ and assume $P_I = \emptyset$. Suppose, by contradiction, that $\dim(\mathrm{rec}(P)) = d$. Then, since P is a rational polyhedron, there exist d linearly independent integral vectors $r^1, \ldots, r^d \in \mathrm{rec}(P)$. Given $z \in P$, the points $z, z + r^1, \ldots, z + r^d$ define a basis of $\mathrm{aff}(P)$. Since $\mathrm{aff}(P) \cap \mathbb{Z}^n \neq \emptyset$, it follows that there exist $\mu_1, \ldots, \mu_d \in \mathbb{R}$ such that $z + \sum_{i=1}^d \mu_i r^i \in \mathbb{Z}^n$. Thus $z + \sum_{i=1}^d (\mu_i - \lfloor \mu_i \rfloor) r^i$ is an integral point in P, contradicting the fact that $P_I = \emptyset$. □

A consequence of the above lemma is that every rational polyhedron having full-dimensional recession cone contains an integer point.

Lemma 5.16. *Let $P \subseteq \mathbb{R}^n$ be a nonempty rational polyhedron such that $\mathrm{aff}(P) \cap \mathbb{Z}^n \neq \emptyset$. Then $P_I = \{x : Ax \leq b\} \cap \mathrm{aff}(P)$ for some integral A and b such that, for every row a^i of A,*

1. *a^i is not orthogonal to $\mathrm{aff}(P)$;*

2. *there exists $d_i \in \mathbb{R}$ such that $a^i x \leq d_i$ is valid for P.*

Proof. Assume first $P_I \neq \emptyset$. Then by Meyer's theorem (Theorem 4.30) there exist an integral matrix A and an integral vector b such that $P_I = \{x : Ax \leq b\} \cap \mathrm{aff}(P)$ and no row of A is orthogonal to $\mathrm{aff}(P)$. We prove 2). Since $\mathrm{rec}(P_I) = \mathrm{rec}(P)$ by Theorem 4.30, for every row a^i of A, $d_i = \max\{a^i x : x \in P\}$ is finite, thus $a^i x \leq d_i$ is valid for P.

Assume now $P_I = \emptyset$. By standard linear algebra, aff$(P) = z + L$ where $z \in P$ and L is a rational linear subspace of $\mathbb{R}^n$ such that $\dim(L) = \dim(P)$. Notice that $\text{rec}(P) \subseteq L$. By Lemma 5.15, $\dim(\text{rec}(P)) < \dim(P)$, thus there exists an integral $a \in L$ such that a is orthogonal to $\text{rec}(P)$. Hence both $u = \max\{ax : x \in P\}$ and $l = \min\{ax : x \in P\}$ are finite. Therefore $P_I = \{x : ax \leq -1, -ax \leq 0\} = \emptyset$, $a, -a$ are not orthogonal to aff(P), and $ax \leq u$, $-ax \leq -l$ are valid for P. □

Lemma 5.17. *Let P be a rational polyhedron and F a nonempty face of P. Then $F^{(s)} = P^{(s)} \cap F$ for every positive integer s.*

Proof. Since P is a rational polyhedron, we may assume that $P := \{x \in \mathbb{R}^n : Ax \leq b\}$ where (A, b) is an integral matrix.

Since every Chvátal inequality for P is also a Chvátal inequality for F, it follows that $F^{(1)} \subseteq P^{(1)} \cap F$. To show that $F^{(1)} \supseteq P^{(1)} \cap F$ it suffices to prove the following statement.

Consider any valid inequality $cx \leq d$ for F where $c \in \mathbb{Z}^n$. Then there is a valid inequality $c^*x \leq d^*$ for P with $c^* \in \mathbb{Z}^n$ such that $F \cap \{x : c^*x \leq \lfloor d^* \rfloor\} \subseteq F \cap \{x : cx \leq \lfloor d \rfloor\}$.

By Theorem 3.24, we can partition the inequalities in $Ax \leq b$ into two systems $A'x \leq b'$ and $A''x \leq b''$ so that $F = \{x : A'x \leq b', A''x = b''\}$. Since $cx \leq d$ is valid for F, $d \geq \max\{cx : x \in F\}$, thus by the linear programming duality theorem (Theorem 3.7) there exist vectors y', y'' such that

$$y'A' + y''A'' = c, \quad y'b' + y''b'' \leq d, \quad y' \geq 0.$$

If we define c^* and d^* as

$$c^* := y'A' + (y'' - \lfloor y'' \rfloor)A'', \quad d^* := y'b' + (y'' - \lfloor y'' \rfloor)b'',$$

the inequality $c^*x \leq d^*$ is valid for P since y' and $y'' - \lfloor y'' \rfloor$ are nonnegative vectors. We have $c^* = c - \lfloor y'' \rfloor A''$ and $d^* \leq d - \lfloor y'' \rfloor b''$. Since A'' is an integral matrix and b'', c are integral vectors, it follows that c^* is integral and $\lfloor d^* \rfloor \leq \lfloor d \rfloor - \lfloor y'' \rfloor b''$. Therefore $F \cap \{x : c^*x \leq \lfloor d^* \rfloor\} = F \cap \{x : \lfloor y'' \rfloor A''x = \lfloor y'' \rfloor b'', c^*x \leq \lfloor d^* \rfloor\} \subseteq F \cap \{x : cx \leq \lfloor d \rfloor\}$. □

Theorem 5.18 (Chvátal [73], Schrijver [323]). *Let P be a rational polyhedron. Then there exists a positive integer t such that $P^{(t)} = P_I$.*

Proof. The proof is by induction on $d = \dim(P)$, the cases $d = -1$, $d = 0$ being trivial. If aff$(P) \cap \mathbb{Z}^n = \emptyset$, by Theorem 1.20 there exist an integral vector a and a scalar $d \notin \mathbb{Z}$ such that $P \subseteq \{x : ax = d\}$, hence

5.2. CHVÁTAL INEQUALITIES

$P_I = \emptyset = \{x : ax \leq \lfloor d \rfloor, -ax \leq -\lceil d \rceil\} = P^{(1)}$. Therefore we may assume $\text{aff}(P) \cap \mathbb{Z}^n \neq \emptyset$. By Lemma 5.16, $P_I = \{x : Ax \leq b\} \cap \text{aff}(P)$ for some integral A and b such that, for every row a^i of A, a^i is not orthogonal to $\text{aff}(P)$ and $a^i x \leq d_i$ is valid for P for some $d_i \in \mathbb{R}$.

We only need to show that, for any row a^i of A, there exists a positive integer t such that the inequality $a^i x \leq b_i$ is valid for $P^{(t)}$. Suppose not. Since $a^i x \leq d_i$ is valid for $P^{(0)} := P$, there exist integers $d > b_i$ and $r \geq 0$ such that, for every $s \geq r$, $a^i x \leq d$ is valid for $P^{(s)}$ but $a^i x \leq d-1$ is not valid for $P^{(s)}$. Then $F := P^{(r)} \cap \{x : a^i x = d\}$ is a face of $P^{(r)}$ and $F_I = \emptyset$ where $F_I := \text{conv}(F \cap \mathbb{Z}^n)$. Since a^i is not orthogonal to $\text{aff}(P)$, $\dim(F) < \dim(P)$, therefore, by induction, there exists h such that $F^{(h)} = \emptyset$. By Lemma 5.17, $F^{(h)} = P^{(r+h)} \cap F$, hence $a^i x < d$ for every $x \in P^{(r+h)}$, therefore $a^i x \leq d-1$ is valid for $P^{(r+h+1)}$, contradicting the choice of d and r. □

Eisenbrand and Schulz [132] prove that for any polytope P contained in the unit cube $[0,1]^n$, one can choose $t = O(n^2 \log n)$ in the above theorem. Rothvoß and Sanitá [319] prove that there is a polytope contained in the unit cube whose Chvátal rank has order n^2, thus showing that the above bound is tight, up to a logarithmic factor.

5.2.3 Chvátal Inequalities for Other Forms of the Linear System

Consider a polyhedron P in the form $P := \{x \in \mathbb{R}^n : Ax \leq b, x \geq 0\}$. Given $I \subseteq \{1,\ldots,n\}$ and $C := \{1,\ldots,n\} \setminus I$, let $S := P \cap \{x : x_j \in \mathbb{Z}, j \in I\}$. Any Chvátal inequality for the above system is of the form

$$(uA - v)x \leq \lfloor ub \rfloor,$$

where $u \in \mathbb{R}^m_+$, $v \in \mathbb{R}^n_+$, $uA_I - v_I \in \mathbb{Z}^I$ and $uA_C - v_C = 0$. Let $\alpha := uA$, $\beta := ub$. Clearly $v_j = \alpha_j \geq 0$ for all $j \in C$, and $v_j \geq \alpha_j - \lfloor \alpha_j \rfloor$ for all $j \in I$. It follows that the only relevant Chvátal inequalities are of the form

$$\sum_{j \in I} \lfloor uA \rfloor x_j \leq \lfloor ub \rfloor, \tag{5.21}$$

for all vectors $u \in \mathbb{R}^m_+$ such that $uA_C \geq 0$.

Consider now a polyhedron P in the form $P := \{x \in \mathbb{R}^n_+ : Ax = b\}$ and, given $I \subseteq \{1,\ldots,n\}$, let $S := P \cap \{x : x_j \in \mathbb{Z}, j \in I\}$. The same argument as before shows that any irredundant Chvátal inequality for P is of the form $\sum_{j \in I} \lfloor uA \rfloor x_j \leq \lfloor ub \rfloor$, where $u \in \mathbb{R}^m$ is such that $uA_C \geq 0$ and $ub \notin \mathbb{Z}$.

5.2.4 Gomory's Fractional Cuts

The first general-purpose cutting plane method was proposed by Gomory [175]. The method is based on the so-called fractional cuts, and it applies to pure integer linear programming problems. Gomory showed how to ensure finite convergence of this method [177].

Given a rational $m \times n$ matrix A and a rational vector $b \in \mathbb{R}^n$, let $P := \{x \in \mathbb{R}^n_+ : Ax = b\}$ and let $S := P \cap \mathbb{Z}^n$.

Let $c \in \mathbb{R}^n$, and consider the pure integer programming problem $\max\{cx : x \in S\}$. Let B be an optimal basis of the linear programming relaxation $\max\{cx : x \in P\}$, and let $N := \{1, \ldots, n\} \setminus B$. The tableau relative to B is of the form

$$x_i + \sum_{j \in N} \bar{a}_{ij} x_j = \bar{b}_i, \quad i \in B. \tag{5.22}$$

The corresponding optimal solution to the linear programming relaxation is $x_i^* = \bar{b}_i$, $i \in B$, $x_j^* = 0$, $j \in N$. If x^* is integral, then it is an optimal solution to the integer programming problem. Otherwise, there exists some $h \in B$ such that $\bar{b}_h \notin \mathbb{Z}$.

The Chvátal inequality relative to the hth row of the tableau is

$$x_h + \sum_{j \in N} \lfloor \bar{a}_{hj} \rfloor x_j \leq \lfloor \bar{b}_h \rfloor. \tag{5.23}$$

Note that (5.23) cuts off x^*, since $x_h^* = \bar{b}_h > \lfloor \bar{b}_h \rfloor$ and $x_j^* = 0$ for all $j \in N$. Introducing a nonnegative slack variables x_{n+1}, the above inequality becomes

$$x_h + \sum_{j \in N} \lfloor \bar{a}_{hj} \rfloor x_j + x_{n+1} = \lfloor \bar{b}_h \rfloor. \tag{5.24}$$

Note that, since all the coefficients of (5.23) are integer, x_{n+1} takes an integer value for every integer solution x. Therefore we may impose the constraint that x_{n+1} is also an integer variable. Finally, by subtracting the hth tableau row $x_h + \sum_{j \in N} \bar{a}_{hj} = \bar{b}_h$ from (5.24), we obtain the *Gomory fractional cut*

$$\sum_{j \in N} -f_j x_j + x_{n+1} = -f_0, \tag{5.25}$$

where $f_j := \bar{a}_{hj} - \lfloor \bar{a}_{hj} \rfloor$ and $f_0 := \bar{b}_h - \lfloor \bar{b}_h \rfloor$.

Juxtaposing the latter equation at the bottom of the tableau (5.22), we obtain the tableau with respect to the basis $B' := B \cup \{n+1\}$. The new tableau is not primal feasible, since in the primal solution associated with the tableau the variable x_{n+1} has value $-f_0 < 0$. However, the new tableau

5.2. CHVÁTAL INEQUALITIES

is dual feasible because the reduced costs are unchanged, and so they are all nonpositive. We may therefore apply the dual simplex method starting from the basis B' to solve the linear relaxation of the new problem, and repeat the argument starting from the new optimal solution.

An example of this method was given in Sect. 1.2.2. We remark that the method, as just described, does not guarantee finite convergence, but we will show in the next section how the above cutting-plane scheme can be turned into a finite algorithm by carefully choosing both the optimal basis of each linear relaxation and the tableau row from which the cut is generated.

5.2.5 Gomory's Lexicographic Method for Pure Integer Programs

In light of Theorem 5.18 one can in principle solve any pure integer linear programming problem by adding a finite number of Chvátal cuts. How to discover such cuts is however not at all obvious. A finite cutting plane algorithm for pure integer programming problems was described by Gomory [175, 177]. The algorithm is based on fractional cuts, and it guarantees finite convergence by a careful choice of the basis defining the optimal solution that we intend to cut off, and of the tableau row used to generate the cut (descriptions and proofs of the method can be also found in [159, 325]). A difference with the method described earlier in Sect. 5.2.4 is that here we also generate cuts from the tableau row corresponding to the objective function.

Given a rational $m \times n$ matrix A, a rational vector $b \in \mathbb{R}^n$, and an integral vector $c \in \mathbb{Z}^n$, we want to solve the pure integer program

$$\max\{x_0 : x_0 - cx = 0,\ Ax = b,\ x \geq 0,\ (x_0, x) \in \mathbb{Z}^{n+1}\}. \tag{5.26}$$

Because c is an integral vector, also $x_0 = cx$ is integer for every $x \in \mathbb{Z}^n$. We will assume that the feasible region of the linear relaxation of (5.26) is bounded. The latter assumption is without loss of generality, since by Corollary 4.37 there exists a function f of n and of the maximum encoding size L of the coefficients of (A, b) such that $P \cap \{x \in \mathbb{R}^n : x \leq f(n, L)\}$ contains an optimal solution to (5.26), if any exists.

At each iteration, the linear programming relaxation of the current problem is solved, and if the optimal solution is not integral a new cut is generated. To express the new inequality as an equality, a slack variable is added to the problem. Therefore, at each iteration t, the current linear programming relaxation of (5.26) is of the form

$$\max\{x_0 : x_0 - cx = 0,\ A^{(t)}x^{(t)} = b^{(t)},\ x^{(t)} \geq 0\}, \tag{5.27}$$

where $x^{(t)}$ denotes the vector with $n+t+1$ variables defined by the original variables $x_0, x_1 \ldots, x_n$ of (5.26) and the slack variables $x_{n+1}, \ldots, x_{n+t}$ of the t cuts added so far, and $A^{(t)}$ is an $(m+t) \times (n+t+1)$ matrix whose first column is the vector of all zeroes. In particular, $A^{(0)} := (0, A) \in \mathbb{R}^{m \times (n+1)}$, $b^{(0)} := b$.

Assuming that (5.27) has a finite optimum, let $B \subseteq \{1, \ldots, n+t\}$ be an optimal basis and $N := \{1, \ldots, n+t\} \setminus B$. The tableau relative to B is of the form

$$x_i + \sum_{j \in N} \bar{a}_{ij} x_j = \bar{a}_{i0}, \quad i \in B \cup \{0\}. \tag{5.28}$$

The corresponding basic optimal solution is $x_i = \bar{a}_{i0}$, $i \in B$, $x_j = 0$, $j \in N$, with objective value $x_0 = \bar{a}_{00}$. If the solution has some fractional component, say $\bar{a}_{h0} \notin \mathbb{Z}$ for some $h \in B \cup \{0\}$, we may generate the cut

$$x_h + \sum_{j \in N} \lfloor \bar{a}_{hj} \rfloor x_j + x_{n+t+1} = \lfloor \bar{a}_{h0} \rfloor, \tag{5.29}$$

where x_{n+t+1} is a nonnegative integer slack variable. The new cut is included in the formulation, and the argument repeated until an optimal integral solution is found.

Gomory [177] showed that, if one applies a certain lexicographic rule both in the pivots of the simplex method and in the selection of the source row of the cut, the algorithm converges in finite time. What we present next is a slight variation of Gomory's algorithm.

A vector $(\alpha_0, \ldots, \alpha_n)$ is *lexicographically larger* than a vector $(\beta_0, \ldots, \beta_n)$ if there exists $i \in \{0, \ldots, n\}$ such that $\alpha_i > \beta_i$ and $\alpha_k = \beta_k$ for all $k < i$. For example, the four vectors $(3, -1, -2)$, $(2, 5, 0)$, $(2, 5, -2)$, $(1, 12, 60)$ are sorted from the lexicographically largest to the lexicographically smallest. The algorithm requires, at each iteration t, to find the lexicographically largest feasible solution $(\bar{x}_0, \ldots, \bar{x}_{n+t})$ of (5.27). Note that such a solution is basic and it is optimal for (5.27), since $\bar{x}_0$ is largest possible.

Given a feasible basis $B \subseteq \{1, \ldots, n+t\}$ of (5.27), consider the tableau (5.28) with respect to B. We say that B is *lexicographically optimal* if, for every $j \in N$, the smallest index $i \in \{0\} \cup B$ such that $\bar{a}_{ij} \neq 0$ satisfies $i < j$ and $\bar{a}_{ij} > 0$.

One can verify that, if B is a lexicographically optimal basis, then the basic solution $(\bar{x}_0, \ldots, \bar{x}_{n+t})$ associated with B is the lexicographically largest feasible solution for (5.27). Since (5.27) is bounded, a lexicographically optimal basis exists whenever the problem is feasible, and it can be computed with the simplex algorithm (Exercise 5.11).

5.2. CHVÁTAL INEQUALITIES

Gomory's Lexicographic Cutting Plane Method

Start with $t := 0$;

1. If (5.27) has no solution, then (5.26) is infeasible. Otherwise, compute a lexicographically optimal basis for (5.27) and let $(\bar{x}_0, \ldots, \bar{x}_{n+t})$ be the corresponding basic solution.

2. If $(\bar{x}_1, \ldots, \bar{x}_n)$ is integral, then it is optimal for (5.26). Otherwise

3. Let $h \in \{0, \ldots, n\}$ be the smallest index such that $\bar{x}_h$ is fractional. Add the cut (5.29) generated from the tableau row relative to variable x_h. Let $t := t + 1$ and go to 1.

Theorem 5.19. *Gomory's lexicographic cutting plane method terminates in a finite number of iterations.*

Proof. At iteration t, let $\bar{x}^t$ denote the basic solution of (5.27) computed in Step 1. Observe that, by construction, the sequence of vectors $(\bar{x}_0^t, \ldots, \bar{x}_n^t)$ is lexicographically nonincreasing as t increases.

We need to show that, for $k = 0, \ldots, n$, after a finite number of iterations the value of $\bar{x}_k^t$ is integer and does not change in subsequent iterations. Suppose not, and let k be the smallest index contradicting the claim. In particular, for $i = 0, \ldots, k-1$, after a finite number T of iterations, the values of $\bar{x}_i^t$ are integer and do not change in subsequent iterations, i.e., $\bar{x}_i^t := \bar{x}_i \in \mathbb{Z}$ for all $t \geq T$, $i = 0, \ldots, k-1$.

By construction, the sequence $\{\bar{x}_k^t\}_{t>T}$ is nonincreasing. Note that the sequence is bounded from below because the feasible region of (5.27) is bounded. This implies that $\lim_{t \to +\infty} \bar{x}_k^t := \ell$ exists. Therefore there exists some iteration $\bar{t} > T$ such that $\bar{x}_k^{\bar{t}} = \lfloor \ell \rfloor + f$, where $\ell - \lfloor \ell \rfloor \leq f < 1$. Observe that $f > 0$, otherwise ℓ is integer and $\bar{x}_k^t = \ell$ for all $t \geq \bar{t}$, contradicting the choice of k. Since $\bar{x}_i^{\bar{t}} = \bar{x}_i$ is integer for $i = 0, \ldots, k-1$, while $\bar{x}_k^{\bar{t}} \notin \mathbb{Z}$, it follows that at Step 3 the procedure generates the cut (5.29) from the tableau row relative to x_k, namely

$$x_k + \sum_{j \in N} \lfloor \bar{a}_{kj} \rfloor x_j \leq \lfloor \ell \rfloor. \qquad (5.30)$$

If $k = 0$, then $\bar{a}_{kj} \geq 0$ for every $j \in N$, since B is lexicographically optimal. In particular (5.30) implies the inequality $x_0 \leq \lfloor \ell \rfloor$. It follows that, for every $t \geq \bar{t} + 1$, $x_0^t \leq \lfloor \ell \rfloor$, thus ℓ is integer and $x_0^t = \ell$ for all $t > \bar{t}$.

Assume $k \geq 1$. Since B is a lexicographically optimal basis, for every $j \in N$ such that $\bar{a}_{kj} < 0$ there exists $h \in \{0\} \cup B$, such that $h < k$ and $\bar{a}_{hj} > 0$. One can then easily determine numbers $\delta_i \geq 0$, $i \in \{0\} \cup (B \cap \{1, \ldots, k-1\})$, such that, for all $j \in N$,

$$\alpha_j := \sum_{\substack{i=0 \\ i \in B \cup \{0\}}}^{k-1} \delta_i \bar{a}_{ij} + \lfloor \bar{a}_{kj} \rfloor \geq 0.$$

If we let $\delta_i := 0$ for all $i \in N \cap \{1, \ldots, k-1\}$, (5.30) is equivalent to

$$\sum_{i=0}^{k-1} \delta_i x_i + x_k + \sum_{j \in N} \alpha_j x_j \leq \lfloor \ell \rfloor + \sum_{i=0}^{k-1} \delta_i \bar{x}_i.$$

Since $\bar{x}_i^t = \bar{x}_i$ for every $t \geq \bar{t}+1$, $i = 0, \ldots, k-1$, it follows from the above inequality and the fact that $\alpha_j \geq 0$ for all $j \in N$ that $\bar{x}_k^t \leq \lfloor \ell \rfloor$ for all $t \geq \bar{t}+1$. Thus ℓ must be integer and $\bar{x}_k^t = \ell$ for every $t \geq \bar{t}+1$. This contradicts the choice of k. □

One can infer from Example 5.11 that, for the mixed integer case, finite convergence cannot be established for any cutting plane method based on Gomory's mixed integer inequalities. However, if we modify the algorithm outlined in this section in a straightforward fashion by using the Gomory mixed integer cuts described in Sect. 5.1.4 (see also Sect. 5.3) instead of the Gomory fractional cuts, the same arguments as in the proof of Theorem 5.19 can be used to prove finite convergence if the objective function depends only on the integer variables, or more generally if we can assume that the objective function value x_0 is integer.

5.3 Gomory's Mixed Integer Cuts

Gomory generalized his fractional cuts to handle mixed integer linear programs [176]. These cuts have become known as the Gomory mixed integer cuts.

Consider a rational polyhedron $P := \{x \in \mathbb{R}_+^n : Ax = b\}$ defined by a system of linear equations in nonnegative variables. Given $I \subseteq \{1, \ldots, n\}$, let $S := P \cap \{x : x_j \in \mathbb{Z}, j \in I\}$ be a mixed integer set. As usual let $C = \{1, \ldots, n\} \setminus I$.

5.3. GOMORY'S MIXED INTEGER CUTS

Let B be a feasible basis of the system $Ax = b$, $x \geq 0$. The tableau associated with B is of the form

$$x_i + \sum_{j \in N} \bar{a}_{ij} x_j = \bar{b}_i, \quad i \in B.$$

The basic solution associated with such a tableau is $x_i^* = \bar{b}_i$, $i \in B$, $x_j^* = 0$, $j \in N$. This vector belongs to S if and only if $\bar{b}_i \in \mathbb{Z}$ for all $i \in B \cap I$. If not, consider an index $i \in B_I$ such that $f_0 := \bar{b}_i - \lfloor \bar{b}_i \rfloor > 0$, and let $f_j := \bar{a}_{ij} - \lfloor \bar{a}_{ij} \rfloor$ for all $j \in N$. The Gomory mixed integer inequality (5.13) derived from the tableau equation relative to x_i is

$$\sum_{\substack{j \in N \cap I \\ f_j \leq f_0}} \frac{f_j}{f_0} x_j + \sum_{\substack{j \in N \cap I \\ f_j > f_0}} \frac{1-f_j}{1-f_0} x_j + \sum_{\substack{j \in N \cap C \\ \bar{a}_{ij} \geq 0}} \frac{\bar{a}_{ij}}{f_0} x_j - \sum_{\substack{j \in N \cap C \\ \bar{a}_{ij} < 0}} \frac{\bar{a}_{ij}}{1-f_0} x_j \geq 1. \quad (5.31)$$

Clearly the above inequality cuts off the basic solution x^* defined by B, since $x_j^* = 0$ for all $j \in N$.

Remark 5.20. *For pure integer sets (i.e., $C = \emptyset$), the Gomory mixed integer cut (5.31) dominates the Gomory fractional cut (5.25), because $\frac{f_j}{f_0} > \frac{1-f_j}{1-f_0}$ whenever $f_j > f_0$.*

Example 5.21. Consider the following pure integer programming problem, which we solved in Sect. 1.2.2 using Gomory fractional cuts.

$$\begin{array}{rl}
\max z = & 5.5x_1 + 2.1x_2 \\
& -x_1 + x_2 \leq 2 \\
& 8x_1 + 2x_2 \leq 17 \\
& x_1, x_2 \geq 0 \\
& x_1, x_2 \text{ integer.}
\end{array}$$

We first add slack variables x_3 and x_4 to turn the inequality constraints into equalities. The problem becomes:

$$\begin{array}{rl}
z \quad -5.5x_1 \quad -2.1x_2 & = 0 \\
-x_1 \quad +x_2 \quad +x_3 & = 2 \\
8x_1 \quad +2x_2 \quad\quad +x_4 & = 17 \\
x_1, x_2, x_3, x_4 \geq 0 \text{ integer.} &
\end{array}$$

Solving the linear programming relaxation, we get the optimal tableau:

$$\begin{array}{rrrl} z & +0.58x_3 +0.76x_4 & = & 14.08 \\ x_2 & +0.8x_3\ \ +0.1x_4 & = & 3.3 \\ x_1 & -0.2x_3 +0.1x_4 & = & 1.3 \\ & x_1, x_2, x_3, x_4 \geq 0. & & \end{array}$$

The corresponding basic solution is $x_3^* = x_4^* = 0$, $x_1^* = 1.3$, $x_2^* = 3.3$ and $z^* = 14.08$. This solution is not integer. Let us generate the Gomory mixed integer cut corresponding to the equation

$$x_2 + 0.8x_3 + 0.1x_4 = 3.3$$

found in the above optimal tableau. We have $f_0 = 0.3$, $f_3 = 0.8$ and $f_4 = 0.1$. Applying formula (5.31), we get the Gomory mixed integer cut

$$\frac{1-0.8}{1-0.3}x_3 + \frac{0.1}{0.3}x_4 \geq 1, \quad \text{i.e.} \quad 6x_3 + 7x_4 \geq 21.$$

We could also generate a Gomory mixed integer cut from the other equation in the final tableau $x_1 - 0.2x_3 + 0.1x_4 = 1.3$. It turns out that, in this case, we get exactly the same Gomory mixed integer cut.

Since $x_3 = 2 + x_1 - x_2$ and $x_4 = 17 - 8x_1 - 2x_2$, we can express the above Gomory mixed integer cut in the space (x_1, x_2). This yields

$$5x_1 + 2x_2 \leq 11.$$

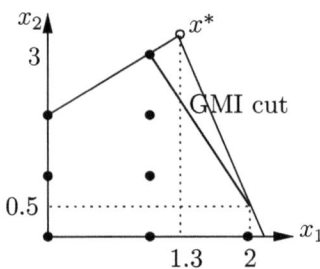

Figure 5.5: Formulation strengthened by a Gomory mixed integer cut

Adding this cut to the linear programming relaxation, we get the following formulation (see Fig. 5.5).

$$\begin{array}{rl} \max\ z = & 5.5x_1 + 2.1x_2 \\ & -x_1 + x_2 \leq 2 \\ & 8x_1 + 2x_2 \leq 17 \\ & 5x_1 + 2x_2 \leq 11 \\ & x_1, x_2 \geq 0. \end{array}$$

5.3. GOMORY'S MIXED INTEGER CUTS

Note that the Gomory mixed integer cut $5x_1 + 2x_2 \leq 11$ is stronger than the Gomory fractional cut $x_2 \leq 3$ generated from the same row (recall the solution of our example in Sect. 1.2.2). This is not surprising since, as noted in Remark 5.20, Gomory mixed integer cuts are always at least as strong as the fractional cuts generated from the same rows.

Solving the linear programming relaxation, we get the optimal tableau:

$$\begin{array}{rrrrcl} z & & & +1/12 x_4 & +29/30 x_5 & = & 12.05 \\ & & x_3 & +7/6 x_4 & -5/3 x_5 & = & 3.5 \\ & x_1 & & +1/3 x_4 & -1/3 x_5 & = & 2 \\ & & x_2 & -5/6 x_4 & +4/3 x_5 & = & 0.5 \\ & & & x_1, x_2, x_3, x_4, x_5 \geq 0. \end{array}$$

The equation

$$x_3 + 7/6 x_4 - 5/3 x_5 = 3.5$$

found in the above tableau produces the following Gomory mixed integer cut.

$$\frac{1/6}{0.5} x_4 + \frac{1/3}{0.5} x_5 \geq 1, \quad \text{i.e.} \quad x_4 + 2x_5 \geq 3.$$

Expressing this cut in the original variables x_1, x_2 we get the inequality

$$3x_1 + x_2 \leq 6.$$

Adding this inequality and resolving the linear relaxation, we find the basic solution $x_1 = 1$, $x_2 = 3$ and $z = 11.8$. Since x_1 and x_2 are integer, this is the optimal solution to the integer program. ∎

Implementing Gomory's Mixed Integer Cuts

Gomory presented his results on fractional cuts in 1958 and it had an enormous immediate impact: reducing integer linear programming to a sequence of linear programs was a great theoretical breakthrough. However, when Gomory programmed his fractional cutting plane algorithm later that year, he was disappointed by the computational results. Convergence was often very slow.

Gomory [176] extended his approach to mixed integer linear programs in 1960, inventing the Gomory mixed integer cuts. Three years later, in 1963, Gomory [177] states that these cuts were "almost completely computationally untested." Surprisingly this statement was still true three decades later! During that period, the general view was that the Gomory cuts are mathematically elegant but impractical, even though there was scant evidence in the literature to justify this negative attitude. Gomory's mixed integer cuts were revived in 1996 [30], based on an implementation that added several cuts from the optimal simplex tableau at a time (instead of just one cut, as tried by Gomory when testing fractional cuts), reoptimized the resulting strengthened linear program, performed a few *rounds* of such cut generation, and incorporated this procedure in a branch-and-cut framework (instead of applying a pure cutting plane approach as Gomory had done). Incorporated in this way, Gomory's mixed integer cuts became an attractive component of integer programming solvers. In addition, linear programming solvers had become more stable by the 1990s.

Commercial integer programming solvers, such as Cplex, started incorporating the Gomory mixed integer cuts in 1999. Bixby et al. [57] give a fascinating account of the evolution of the Cplex solver. They view 1999 as the transition year from the "old generation" of Cplex to the "new generation." Their paper lists some of the key features in a 2002 "new generation" solver and compares the speedup in computing time obtained by enabling one feature versus disabling it, while keeping everything else unchanged. The average speedups obtained for each feature on a set of 106 instances are summarized in the next table (we refer the reader to [57] for information on the choice of the test set).

Feature	Speedup factor
Cuts	54
Preprocessing	11
Branching variable selection	3
Heuristics	1.5

The clear winner in these tests was cutting planes. Eight types of cutting planes were implemented in Cplex in 2002. Performing a similar experiment, disabling only one of the cut generators at a time, they obtained the following speedups in computing time.

Cut type	Speedup factor
Gomory mixed integer	2.5
Mixed integer rounding	1.8
Knapsack cover	1.4
Flow cover	1.2
Implied bounds	1.2
Path	1.04
Clique	1.02
GUB cover	1.02

5.3. GOMORY'S MIXED INTEGER CUTS

Even when all the other cutting planes are used in Cplex (2002 version), the addition of Gomory mixed integer cuts by itself produces a solver that is two and a half times faster! As Bixby and his co-authors conclude "Gomory cuts are the clear winner by this measure." In the above table, the Gomory mixed integer cuts are those generated from rows of optimal simplex tableaux. Note also the excellent performance of the mixed integer rounding inequalities. These are obtained using formula (5.16) where the inequality (5.15) is obtained by aggregating the constraints in $Ax \leq b$ using various heuristics. Knapsack cover and flow cover inequalities will be presented in Chap. 7.

Note, however, that the textbook formulas for generating Gomory mixed integer and mixed integer rounding cuts are not used directly in open-source and commercial software that use finite numerical precision in the computations. These solvers perform additional steps to avoid the generation of invalid cuts, and of cuts that could substantially slow down the solution of the linear programs. These steps come in two flavors: some modify the cut coefficients slightly while others simply discard the cut. We will discuss briefly both types of steps, starting with the first type. Consider a bounded variable x_j with upper and lower bounds no greater than L in absolute value (for example $L = 10^4$). When the coefficient of x_j has a very small absolute value (say below 10^{-12}) in a Gomory mixed integer cut, such a coefficient is set to 0 and the right-hand side of the cut is adjusted accordingly (using the upper bound when the coefficient of x_j is positive, and the lower bound when it is negative). The resulting inequality is a slight weakening of the Gomory mixed integer cut, but it is numerically more stable. For the second issue, several parameters of a Gomory mixed integer cut are checked before adding it to the formulation. One such parameter is the value of f in formula (5.13): if f or $1-f$ is too small, the cut is discarded. A reasonable cut off point is 10^{-2}, i.e., only add Gomory mixed integer cuts for which $0.01 \leq f \leq 0.99$. One also usually discards cuts that have too large a ratio between the absolute values of the largest and smallest nonzero coefficients (this ratio is sometimes called the *dynamism* of the cut). A reasonable rule might be to discard Gomory mixed integer cuts with a dynamism in excess of 10^6. Furthermore, in order to avoid fill-in of the basis inverse when solving the linear programming relaxations, one also discards cuts that are too dense. The first two parameters help reduce the generation of invalid cuts while the third helps solving the linear programs. A paper of Cook et al. [88] addresses the issue of always rounding coefficients in the "right" direction to keep valid cuts. Despite the various steps to make the Gomory mixed integer cut generation safer, it should be clear that any integer programming solver based on finite precision arithmetic will fail on some instances.

Another issue that has attracted attention but still needs further investigation is the choice of the equations used to generate the Gomory mixed integer cuts: Gomory proposed to use the rows of the optimal simplex tableau but other equations can also be used. Balas and Saxena [34], and Dash et al. [107] showed that integer programming formulations can typically be strengthened very significantly by generating Gomory cuts from a well chosen set of equations. However, finding such a good family of equations efficiently remains a challenge.

5.4 Lift-and-Project

In this section, we consider mixed 0,1 linear programs.

Given $I \subseteq \{1, \ldots, n\}$, we consider a polyhedron $P := \{x \in \mathbb{R}^n : Ax \leq b\}$ which is contained in $\{x \in \mathbb{R}^n : 0 \leq x_j \leq 1, j \in I\}$ and we let $S := P \cap \{x \in \mathbb{R}^n : x_j \in \{0,1\}, j \in I\}$. Given $j \in I$, let

$$P_j := \text{conv}\left((P \cap \{x \in \mathbb{R}^n : x_j = 0\}) \cup (P \cap \{x \in \mathbb{R}^n : x_j = 1\})\right).$$

Note that since $0 \leq x_j \leq 1$ are valid inequalities for P, the set P_j is the convex hull of two faces of P (Fig. 5.6). A *lift-and-project* inequality is a split inequality for P relative to some disjunction $x_j = 0$ or $x_j = 1$, $j \in I$. That is, $\alpha x \leq \beta$ is a lift-and-project inequality if and only if, for some $j \in I$, it is valid for P_j.

The term *lift-and-project* refers to the description of P_j as the projection of a polyhedron in a "lifted" space, namely P_j is the projection of the extended formulation (5.2) where $\pi = e^j$ and $\pi_0 = 0$. The *lift-and-project closure* of P is the set of points satisfying all lift-and-project inequalities, that is

$$P^{\text{lift}} := \bigcap_{j \in I} P_j.$$

Since P_j is a polyhedron for all $j \in I$, it follows that P^{lift} is a polyhedron as well. Furthermore, it follows from the definition that $P^{\text{split}} \subseteq P^{\text{lift}}$.

5.4. LIFT-AND-PROJECT

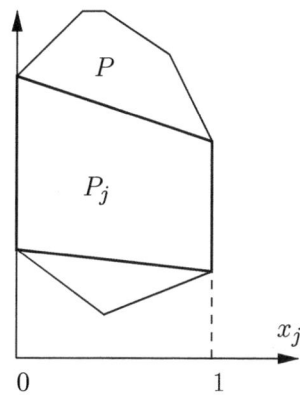

Figure 5.6: P_j

5.4.1 Lift-and-Project Rank for Mixed 0,1 Linear Programs

Unlike general mixed integer linear sets, for which the split rank might not be finite as seen in Example 5.11, mixed $0,1$ linear sets have the nice property that the convex hull can be obtained by iteratively adding lift-and-project inequalities. Indeed, a much stronger property holds, that we describe here.

Given $I \subseteq \{1,\ldots,n\}$, consider a polyhedron P contained in $\{x \in \mathbb{R}^n : 0 \leq x_j \leq 1,\ j \in I\}$ and let $S := P \cap \{x \in \mathbb{R}^n : x_j \in \{0,1\}, j \in I\}$. Possibly by permuting the indices, we assume $I = \{1,\ldots,p\}$ and, for $t = 1,\ldots,p$, define
$$(P)^t := ((P_1)_2\ldots)_t.$$
The next theorem shows that $(P)^p = \text{conv}(S)$. Since $P^{\text{split}} \subseteq P^{\text{lift}} \subseteq (P)^1$, this implies that the split rank and the lift-and-project rank of P are at most equal to the number p of $0, 1$ variables.

Theorem 5.22 (Sequential Convexification Theorem, Balas [24]). *Let $I = \{1,\ldots,p\}$, and let P be a polyhedron contained in $\{x \in \mathbb{R}^n : 0 \leq x_j \leq 1,\ j \in I\}$. Then, for $t = 1,\ldots,p$, $(P)^t = \text{conv}(\{x \in P : x_j \in \{0,1\}, j = 1,\ldots,t\})$. In particular, $(P)^p = \text{conv}(S)$.*

Proof. Let $S_t := \{x \in P : x_j \in \{0,1\}, j = 1,\ldots,t\}$. We need to show $(P)^t = \text{conv}(S_t)$ for $t = 1,\ldots,p$. We prove this result by induction. The result holds for $t = 1$ since $(P)^1 = P_1 = \text{conv}(S_1)$ where the second equality follows from the definition of P_1. Assume inductively that $(P)^{t-1} = \text{conv}(S_{t-1})$. Then

$$\begin{aligned}(P)^t &= ((P)^{t-1})_t \\ &= \operatorname{conv}\big(((P)^{t-1} \cap \{x \in \mathbb{R}^n : x_t = 0\}) \cup ((P)^{t-1} \cap \{x \in \mathbb{R}^n : x_t = 1\})\big) \\ &= \operatorname{conv}\big((\operatorname{conv}(S_{t-1}) \cap \{x \in \mathbb{R}^n : x_t = 0\}) \cup (\operatorname{conv}(S_{t-1}) \cap \{x \in \mathbb{R}^n : x_t = 1\})\big).\end{aligned}$$

We will need the following claim.

Claim. *Consider a set $A \subset \mathbb{R}^n$ and a hyperplane $H := \{x \in \mathbb{R}^n : \gamma x = \gamma_0\}$ such that $\gamma x \leq \gamma_0$ for every $x \in A$. Then $\operatorname{conv}(A) \cap H = \operatorname{conv}(A \cap H)$.*

Clearly $\operatorname{conv}(A \cap H) \subseteq \operatorname{conv}(A) \cap H$. To show $\operatorname{conv}(A) \cap H \subseteq \operatorname{conv}(A \cap H)$, consider $x \in \operatorname{conv}(A) \cap H$. Then $x = \sum_{i=1}^k \lambda_i x^i$ where $x^i \in A$, $\lambda_i > 0$ for $i = 1, \ldots k$, and $\sum_{i=1}^k \lambda_i = 1$. Since $x^i \in A$, we have $\gamma x^i \leq \gamma_0$. Since $x \in H$, we have $\gamma x = \gamma_0$. This implies $x^i \in H$. Therefore $x^i \in A \cap H$, proving the claim.

Applying the claim to the set S_{t-1} and the hyperplanes $\{x \in \mathbb{R}^n : x_t = 0\}$ and $\{x \in \mathbb{R}^n : x_t = 1\}$, we obtain

$$(P)^t = \operatorname{conv}\big((\operatorname{conv}(S_{t-1} \cap \{x \in \mathbb{R}^n : x_t = 0\})) \cup (\operatorname{conv}(S_{t-1} \cap \{x \in \mathbb{R}^n : x_t = 1\}))\big).$$

For any two sets A, B, it follows from the characterization of convex hulls that $\operatorname{conv}(\operatorname{conv}(A) \cup \operatorname{conv}(B)) = \operatorname{conv}(A \cup B)$. This implies

$$\begin{aligned}(P)^t &= \operatorname{conv}\big((S_{t-1} \cap \{x \in \mathbb{R}^n : x_t = 0\}) \cup (S_{t-1} \cap \{x \in \mathbb{R}^n : x_t = 1\})\big) \\ &= \operatorname{conv}(S_t).\end{aligned}$$

$\square$

Theorem 5.22 does not extend to the general mixed integer case (see Exercise 5.21).

Example 5.23. The purpose of this example is to show that the split rank of mixed 0,1 linear programs might indeed be as large as the number of binary variables. The example is due to Cornuéjols and Li [94]. Consider the following polytope, studied by Chvátal, Cook, and Hartmann [75]

$$P := \{x \in [0,1]^n : \sum_{j \in J} x_j + \sum_{j \notin J}(1 - x_j) \geq \frac{1}{2}, \text{ for all } J \subseteq \{1, 2, \cdots, n\}\}.$$

Note that $P \cap \{0,1\}^n = \emptyset$. We will show that the $(n-1)$th split closure P^{n-1} of P is nonempty, implying that the split rank of P is n.

For $j = 1, \ldots, n$, let F_j be the set of all vectors $x \in \mathbb{R}^n$ such that j components of x are $\frac{1}{2}$ and each of the remaining $n-j$ components are in $\{0,1\}$. Note that $F_1 \subseteq P$ (indeed, P is the convex hull of F_1). Let $P^0 := P$. We will show that P^k contains F_{k+1}, $k = 0, \ldots, n-1$. Thus, $P^{n-1} \neq \emptyset$.

We proceed by induction on k. The statement holds for $k = 0$, thus we assume that $k \geq 1$ and that $F_k \subseteq P^{k-1}$. We need to show that, for every

5.4. LIFT-AND-PROJECT

split $(\pi, \pi_0) \in \mathbb{Z}^n \times \mathbb{Z}$, F_{k+1} is contained in $(P^{k-1})^{(\pi,\pi_0)}$. Let $v \in F_{k+1}$. We show that $v \in (P^{k-1})^{(\pi,\pi_0)}$. We assume that $\pi_0 < \pi v < \pi_0 + 1$, otherwise $v \in (P^{k-1})^{(\pi,\pi_0)}$ by definition. Since all fractional components of v equal $\frac{1}{2}$, it follows that $\pi v = \pi_0 + \frac{1}{2}$. This implies that there exists $j \in \{1, \ldots, n\}$ such that $v_j = \frac{1}{2}$ and $|\pi_j| \geq 1$. Assume $\pi_j \geq 1$ and let $v^0, v^1 \in \mathbb{R}^n$ be equal to v except for the jth component, which is 0 and 1 respectively. By construction $v^0, v^1 \in F_k$, therefore $v^0, v^1 \in P^{k-1}$. Observe that $\pi v^0 = \pi v - \frac{1}{2}\pi_j \leq \pi_0$, while $\pi v^1 = \pi v + \frac{1}{2}\pi_j \geq \pi_0 + 1$. Thus $v^0, v^1 \in (P^{k-1})^{(\pi,\pi_0)}$, which implies that $v \in (P^{k-1})^{(\pi,\pi_0)}$ since $v = \frac{v^0 + v^1}{2}$. If $\pi_j \leq -1$ the proof is identical, with the roles of v^0, v^1 interchanged. ∎

In view of Example 5.11 showing that no bound may exist on the split rank when the integer variables are not restricted to be 0,1, and Theorem 5.22 showing that the rank is always bounded when they are 0,1 valued, one is tempted to convert general integer variables into 0,1 variables. For a bounded integer variable $0 \leq x \leq u$, there are several natural transformations:

(i) a binary expansion of x (see Owen and Mehrotra [295]);

(ii) $x = \sum_{i=1}^{u} i z_i$, $\sum z_i \leq 1$, $z_i \in \{0, 1\}$ (see Sherali and Adams [331] and Köppe et al. [241]);

(iii) $x = \sum_{i=1}^{u} z_i$, $z_i \leq z_{i-1}$, $z_i \in \{0, 1\}$ (see Roy [320] and Bonami and Margot [61]).

More studies are needed to determine whether any practical benefit can be gained from such transformations.

5.4.2 A Finite Cutting Plane Algorithm for Mixed 0, 1 Linear Programming

Theorem 5.22 implies that, for mixed 0, 1 linear programs, the convex hull of the feasible solutions can be described by a finite number of lift-and-project cuts. However, the result does not immediately provide a finite cutting plane algorithm for this type of problems. Next we describe such an algorithm, due to Balas et al. [29].

We assume that we are given mixed 0, 1 programming problems in the form

$$\begin{aligned} \max \quad & cx \\ & Ax \leq b \\ & x_j \in \{0, 1\} \quad j \in I \end{aligned} \quad (5.32)$$

where $Ax \leq b$ is a linear system of m constraints in n variables which includes the constraints $0 \leq x_j \leq 1$, $j \in I$, and where $I \subseteq \{1,\ldots,n\}$.

At each iteration we strengthen the formulation by introducing a lift-and-project cut, until the optimal solution of the linear relaxation satisfies the integrality conditions. We denote by $A^k x \leq b^k$ the system after introducing k cuts, where $A^0 = A$ and $b^0 = b$, and we let $P^k := \{x \in \mathbb{R}^n : A^k x \leq b^k\}$. At iteration k, we compute an optimal solution $\bar{x}$ for the linear programming relaxation $\max\{cx : x \in P^k\}$. If $\bar{x}$ does not satisfy the integrality conditions, we select an index $j \in I$ such that $0 < \bar{x}_j < 1$ and a suitable subsystem $\tilde{A}x \leq \tilde{b}$ of the system $A^k x \leq b^k$, and compute an optimal basic solution (u^1, u^2) of the cut-generating linear program (5.8)

$$\begin{aligned} \min \ & (u^1 - u^2)\tilde{b} + u^2 \frac{\tilde{b} - \tilde{A}\bar{x}}{\bar{x}_j} \\ & (u^1 - u^2)\tilde{A} = e^j \\ & u^1, u^2 \geq 0. \end{aligned} \qquad (5.33)$$

Let $u := u^1 - u^2$. The lift-and-project inequality

$$u^1 \frac{\tilde{b} - \tilde{A}x}{u\tilde{b}} + u^2 \frac{\tilde{b} - \tilde{A}x}{1 - u\tilde{b}} \geq 1 \qquad (5.34)$$

is added to the formulation $A^k x \leq b^k$, and the process repeated.

For every $j \in I$, the cuts generated by solving (5.33) with respect to index j are called j-cuts. In the remainder, we assume that $I = \{1,\ldots,p\}$.

For any iteration k and for $j = 1,\ldots,p$, let $A^{k,j} x \leq b^{k,j}$ denote the system of linear inequalities comprising $Ax \leq b$ and all the h-cuts of $A^k x \leq b^k$ for $h = 1,\ldots,j$. We define $A^{k,0} := A$, $b^{k,0} := b$. Let $P^{k,j} := \{x : A^{k,j} x \leq b^{k,j}\}$, $j = 0,\ldots,p$. Note that $P^{k,0} = P$ and $P^{k,p} = P^k$.

Specialized Lift-and-Project Algorithm

Start with $k := 0$.

1. Compute an optimal basic solution $\bar{x}$ of the linear program $\max\{cx : x \in P^k\}$.

2. If $\bar{x}_1, \ldots, \bar{x}_p \in \{0,1\}$, then $\bar{x}$ is optimal for (5.32). Otherwise

3. Let $j \in I$ be the largest index such that $0 < \bar{x}_j < 1$. Let $\tilde{A} := A^{k,j-1}$ and $\tilde{b} := b^{k,j-1}$.
   Compute an optimal basic solution (u^1, u^2) to the linear program (5.33), and add the j-cut (5.34) to $A^k x \leq b^k$.

4. Set $k := k + 1$ and go to 1.

Theorem 5.24. *The specialized lift-and-project algorithm terminates after a finite number of iterations for every mixed $0,1$ linear program.*

Proof. The proof is in two steps.

(i) We prove that, at each iteration k, the j-cut computed in Step 3 of the algorithm cuts off the solution $\bar{x}$ computed in Step 1. We first show that $\bar{x}$ is a vertex of $P^{k,j}$. Let $H := \{x : x_t = \bar{x}_t, t = j+1,\ldots,p\}$. By the choice of j, $\bar{x}_t \in \{0,1\}$ for $t = j+1,\ldots,p$, thus $P^k \cap H$ is a face of P^k. Let $(P^{k,j})_{j+1,\ldots,p} := ((P^{k,j}_{j+1})_{j+2}\ldots)_p$ and observe that $(P^{k,j})_{j+1,\ldots,p} \subseteq P^k \subseteq P^{k,j}$. By Theorem 5.22, $(P^{k,j})_{j+1,\ldots,p} = \text{conv}(P^{k,j} \cap \{x : x_t \in \{0,1\}, t = j+1,\ldots,p\})$, therefore $(P^{k,j})_{j+1,\ldots,p} \cap H = P^k \cap H = P^{k,j} \cap H$. Since $\bar{x}$ is a vertex of P^k, it follows that it is a vertex of $P^k \cap H = P^{k,j} \cap H$, and thus it is a vertex of $P^{k,j}$. Since $0 < \bar{x}_j < 1$, $\bar{x}$ is not a vertex of $(P^{k,j-1})_j$. Given that $\bar{x}$ is a vertex of $P^{k,j}$ and that $(P^{k,j-1})_j \subseteq P^{k,j}$, it follows that $\bar{x} \notin (P^{k,j-1})_j$. Thus the j-cut computed in Step 3 cuts off $\bar{x}$.

(ii) We show that, for $j = 1,\ldots,p$, the number of j-cuts generated is finite. Observe that no cut can be generated twice, since by (i) at every iteration we cut off at least one vertex of the current relaxation. Inductively, it suffices to show that, for $j = 1,\ldots,p$ and an iteration $\bar{k}$ such that no h-cut with $h \leq j-1$ is added after iteration $\bar{k}$, only a finite number of j-cuts are added after iteration $\bar{k}$. Indeed, for $k \geq \bar{k}$, $A^{k,j-1} = A^{\bar{k},j-1}$, thus every j-cut added after iteration $\bar{k}$ corresponds to a basic solution of the system $uA^{\bar{k},j-1} = e^j$. Since there are only a finite number of such vectors u, it follows that a finite number of j-cuts are added after iteration $\bar{k}$. □

5.5 Further Readings

Balas' work on disjunctive programming and the union of polyhedra [24] provided the initial framework for studying split inequalities. Following this point of view, the polyhedron $P^{(\pi,\pi_0)}$ is first formulated in an extended space and this extended formulation is then projected back onto the original space. The approach of Sect. 5.1.1 is new, providing a linear description of $P^{(\pi,\pi_0)}$ directly in the original space, see Conforti et al. [83]. The proof uses a necessary and sufficient condition for a point to be in $P^{(\pi,\pi_0)}$, due to Bonami [58].

This perspective on split inequalities is geometric. Earlier, Gomory [176] had introduced mixed integer inequalities in a paper whose flavor is mainly arithmetic. It turns out that the two notions are equivalent. This equivalence was observed by Nemhauser and Wolsey [286] who also introduced another equivalent notion, that of mixed integer rounding inequalities.

The term of split inequality was coined by Cook et al. [90]. They showed that the split closure is a polyhedron. Simpler proofs were later given by Andersen et al. [11], Vielma [346] and Dash et al. [107].

Event though, as seen in Example 5.11, mixed integer linear programming formulations may have infinite split rank, Owen and Merhotra [294] showed that, for polytopes, iteratively taking the split closure does yield conv(S) in the limit. Del Pia and Weismantel [113] extended the result to general rational polyhedra.

Chvátal's paper [73] was very influential. In particular, the notion of Chvátal rank has become a common approach to understanding formulations for pure integer linear programs. Eisenbrand and Schulz [132] give bounds on the Chvátal rank for 0,1 polytopes. Contrary to the lift-and-project and split ranks, which are at most n for mixed 0,1 linear sets with n binary variables, they show that the Chvátal rank can be greater than n. Chvátal et al. [75] study the rank of classes of inequalities for combinatorial optimization problems. Caprara and Fischetti [67] study $\{0, \frac{1}{2}\}$ Chvátal inequalities.

Cut-Generating Linear Programs and Normalizations

Given a polyhedron $P := \{x \in \mathbb{R}^n : Ax \leq b\}$ where A is an $m \times n$-matrix, and a split (π, π_0), let $P^{(\pi, \pi_0)} := \text{conv}(\Pi_1 \cup \Pi_2)$, where $\Pi_1 := P \cap \{x : \pi x \leq \pi_0\}$ and $\Pi_2 := P \cap \{x : -\pi x \leq -(\pi_0 + 1)\}$.

An inequality $\alpha x \leq \beta$ is valid for $P^{(\pi, \pi_0)}$ if and only if it is valid for both Π_1 and Π_2. Hence assuming that both Π_1 and Π_2 are nonempty, by Theorem 3.22, $\alpha x \leq \beta$ is valid for $P^{(\pi, \pi_0)}$ if and only if there exist $u, v \in \mathbb{R}^m$ and $u_0, v_0 \in \mathbb{R}$ such that

$$\alpha = uA + u_0\pi = vA - v_0\pi, \ \beta \leq ub + u_0\pi_0, \ \beta \leq vb - v_0(\pi_0 + 1), \ u, u_0, v, v_0 \geq 0. \tag{5.35}$$

Let C be the set of points $(\alpha, \beta, u, u_0, v, v_0)$ satisfying (5.35). A given point $\bar{x}$ is in $P^{(\pi, \pi_0)}$ if and only if the linear program

$$\min\{\beta - \alpha\bar{x} : (\alpha, \beta, u, u_0, v, v_0) \in C\} \tag{5.36}$$

admits a solution with negative value. Since C is a cone, the value of the linear program (5.36) is either 0 or $-\infty$. Hence a *normalization* is needed, i.e., the addition of an inequality that guarantees that the above linear

5.5. FURTHER READINGS

program always has a finite optimum. Note that, if the inequality $\alpha x \leq \beta$ is not valid for P, then $u_0, v_0 > 0$, therefore a possible normalization is

$$u_0 + v_0 = 1$$

With the addition of this constraint, the linear program (5.36) is equivalent to the linear program (5.8), see, e.g., [59].

One of the most widely used (and effective) truncation condition, called the "standard normalization condition" is

$$u + u_0 + v + v_0 = 1.$$

This latter condition was proposed in Balas [27]. The choice of the normalization condition turns out to be crucial for an effective selection of a "strong" disjunctive cut. This is discussed by Fischetti et al. [142].

Computations

On the computational front, the strength of the Chvátal closure has been investigated on instances from the MIPLIB library [56] (a publicly available library of integer programming instances originating from various applications). Based on pure instances from MIPLIB 3, Fischetti and Lodi [141] found that the Chvátal closure closes around 63% of the integrality gap on average (the *integrality gap* is the difference between the values of the objective function when optimized over conv(S) and over P respectively). A similar experiment was performed by Balas and Saxena [34] for the split closure. They found that the split closure closes 72% of the integrality gap on average on the MIPLIB instances. These experiments show that the Chvátal and split closures are surprisingly strong. Recall however that optimizing over these closures is NP-hard. It is therefore not surprising that both experiments were very computational intensive. Finding deep split inequalities efficiently remains a challenging practical issue.

Gomory's mixed integer cuts [177] from the optimal simplex tableau are easy to compute and turn out to be surprisingly good in practice (Bixby et al. [57]). On MIPLIB 3 instances, adding these cuts already reduces the integrality gap by 24% on average [60]. Marchand and Wolsey [266] implemented an aggregation heuristic to generate mixed integer rounding cuts with excellent computational results. For example, on the MIPLIB 3 instances, the Marchand–Wolsey aggregation heuristic (as available in the COIN-OR repository) reduces the integrality gap by 23% on average [60].

The term lift-and-project was coined by Balas et al. [29]. Balas and Jeroslow [31] show how to strengthen a lift-and-project inequality in the manner described in Sect. 5.1.4. Balas and Perregaard [32] showed how to generate lift-and-project inequalities directly in the space of the original variables: starting from a Gomory mixed integer cut from the optimal tableau, a deeper lift-and-project cut can then be obtained by pivoting. Balas and Bonami [28] made this cut available through an open-source implementation.

5.6 Exercises

Exercise 5.1. Consider $S_1, S_2 \subseteq \mathbb{R}^n_+$. For $i = 1, 2$, let $\sum_{j=1}^n \alpha_j^i x_j \leq \alpha_0^i$ be a valid inequality for S_i. Prove that $\sum_{j=1}^n \min(\alpha_j^1, \alpha_j^2) x_j \leq \max(\alpha_0^1, \alpha_0^2)$ is a valid inequality for $S_1 \cup S_2$.

Exercise 5.2. Let Π_1 and Π_2 be defined as in (5.1). Assume Π_1 and Π_2 have n_k^1 and n_k^2 faces of dimension k respectively, for $k = 1, \ldots, n+p-2$. Give an upper bound on the number of facets of $\operatorname{conv}(\Pi_1 \cup \Pi_2)$. Can you construct a family of polyhedra P with m constraints such that the number of facets of $\operatorname{conv}(\Pi_1 \cup \Pi_2)$ grows more than linearly with m?

Exercise 5.3. Let $P := \{(x_1, x_2, y) \in \mathbb{R}^3 : x_1 \geq y, x_2 \geq y, x_1+x_2+2y \leq 2, y \geq 0\}$ and $S := P \cap (\mathbb{Z}^2 \times \mathbb{R})$. Prove that $x_1 \geq 3y$ and $x_2 \geq 3y$ are split inequalities.

Exercise 5.4. Let $P := \{(x_1, x_2, y) \in \mathbb{R}^3 : x_1 \geq y, x_2 \geq y, x_1+x_2+2y \leq 2, y \geq 0\}$ and $S := P \cap (\mathbb{Z}^2 \times \mathbb{R})$. Prove that $P^{\text{split}} = \{(x_1, x_2, y) \in \mathbb{R}^3 : x_1 \geq 3y, x_2 \geq 3y, x_1 + x_2 + 2y \leq 2, y \geq 0\}$.

Exercise 5.5. Let $P = \{x \in \mathbb{R}^n_+ : Ax = b\}$ and let $\alpha x = \beta$ be a linear combination of the equations in $Ax = b$. Derive explicitly the Gomory mixed integer inequality (5.13) using the mixed integer rounding procedure explained in Sect. 5.1.5.

Exercise 5.6. Let $c \in \mathbb{R}^n$, $g \in \mathbb{R}^p$, $b \in \mathbb{R}$ and $S := \{(x, y) \in \mathbb{Z}^n \times \mathbb{R}^p : cx + gy \leq b + \alpha x_k, cx + gy \leq b + \beta(1 - x_k)\}$ where $1 \leq k \leq n$ and $\alpha, \beta > 0$. Prove that $cx + gy \leq b$ is a mixed integer rounding inequality for S.

Exercise 5.7. Let $P := \{(x, y) \in \mathbb{R}^2 : 2x \geq y, 2x + y \leq 2, y \geq 0\}$ and $S := P \cap (\mathbb{Z} \times \mathbb{R})$. Show that $\operatorname{conv}(S) \neq P$ and that the Chvátal closure of P is P itself.

Exercise 5.8. Let $(a_1, \ldots, a_n) \in \mathbb{Z}^n \setminus \{0\}$, $b \in \mathbb{R}$ and $S := \{x \in \mathbb{Z}^n : \sum_{j=1}^n a_j x_j \leq b\}$. Show that $\operatorname{conv}(S) = \{x \in \mathbb{R}^n : \sum_{j=1}^n \frac{a_j}{k} x_j \leq \lfloor \frac{b}{k} \rfloor\}$ where k is the greatest common divisor of $a_1, \ldots, a_n$.

5.6. EXERCISES

Exercise 5.9. Given $a \in \mathbb{Z}^n$, $g \in \mathbb{R}^p$, $b \in \mathbb{R}$, let $S = \{(x,y) \in \mathbb{Z}^n \times \mathbb{R}^p : ax + gy \leq b\}$. Give a perfect formulation for $\text{conv}(S)$.

Exercise 5.10. Let $S := \{(x,y) \in \mathbb{Z}^n \times \mathbb{R}_+^p : \sum_{j=1}^n a_j x_j + \sum_{j=1}^p g_j y_j \leq b\}$ where $a_1, \ldots, a_n \in \mathbb{Z}$ are not all equal to 0 and are relatively prime, $g_1, \ldots, g_p \in \mathbb{R}$ and $b \in \mathbb{R} \setminus \mathbb{Z}$. Let $f := b - \lfloor b \rfloor$ and $J^- := \{j \in \{1, \ldots, p\} : g_j < 0\}$.

1. Prove that the inequality $\sum_{j=1}^n \lfloor a_j \rfloor x_j + \frac{1}{1-f} \sum_{j \in J^-} g_j y_j \leq \lfloor b \rfloor$ is a valid for S.

2. Prove that the above inequality defines a facet of $\text{conv}(S)$.

Exercise 5.11. Let $B \subseteq \{1, \ldots, n+t\}$ be a feasible basis of (5.27). Prove that, if B is a lexicographically optimal basis, then the basic solution $(\bar{x}_0, \ldots, \bar{x}_{n+t})$ associated with B is the lexicographically largest feasible solution for (5.27). Prove that, if a lexicographically optimal basis exists, it can be computed with the simplex algorithm.

Exercise 5.12. Consider a pure integer program. Suppose that Gomory fractional cuts are added in an iterative fashion as outlined after Eq. (5.25). Furthermore assume that Rules 1–3 of the Gomory's lexicographic cutting plane method are applied periodically, at iterations $k, 2k, \ldots, pk, \ldots$ where k is a positive integer. Prove that such a cutting plane method terminates in a finite number of iterations.

Exercise 5.13. Consider a mixed integer linear program where the objective function value is integer. Modify Gomory's lexicographic cutting plane method in a straightforward fashion by using the Gomory mixed integer cuts instead of the Gomory fractional cuts. Prove that such a cutting plane method terminates in a finite number of iterations.

Exercise 5.14.

1. Let $a_1, \ldots, a_n \in \mathbb{Q}$, $b \in \mathbb{Q}_+ \setminus \mathbb{Z}$, and $S := \{x \in \mathbb{Z}_+^n : \sum_{j=1}^n a_j x_j = b\}$. Show that $\sum_{j \in J} x_j \geq 1$ is a valid inequality for S where $J := \{j \in \{1, \ldots, n\} : a_j \notin \mathbb{Z}\}$.

2. For a pure integer program, consider a lexicographic cutting plane algorithm based on the above cuts instead of the Gomory fractional cuts. Prove that such a cutting plane method terminates in a finite number of iterations.

Exercise 5.15. Consider the following mixed integer linear program

$$
\begin{array}{rlrlrlrl}
z = \max & 7x_1 & +5x_2 & +x_3 & +y_1 & & & \\
 & x_1 & +3x_2 & & +4y_1 & +y_2 & & = 11 \\
 & 5x_1 & +x_2 & +3x_3 & & & +y_3 & = 12 \\
 & & & 2x_3 & +2y_1 & & -y_4 & = 3 \\
\end{array}
$$

$$x_1, \quad x_2, \quad x_3 \in \mathbb{Z}_+$$
$$y_1, \quad y_2, \quad y_3, \quad y_4 \in \mathbb{R}_+.$$

The optimal tableau of the linear programming relaxation is:

$$
\begin{array}{rlrlrl}
z & +0.357x_3 & +1.286y_2 & +1.143y_3 & +2.071y_4 & = 21.643 \\
x_1 & +0.786x_3 & -0.071y_2 & +0.214y_3 & -0.143y_4 & = 2.214 \\
x_2 & -0.929x_3 & +0.357y_2 & -0.071y_3 & +0.714y_4 & = 0.929 \\
y_1 & +0.500x_3 & & & -0.500y_4 & = 1.500 \\
\end{array}
$$

1. The optimal linear programming solution is $x_1 = 2.214$, $x_2 = 0.929$, $y_1 = 1.5$ and $x_3 = y_2 = y_3 = y_4 = 0$. Use the equations where x_1 and x_2 are basic to derive two Gomory mixed integer inequalities that cut off this fractional solution.

2. The coefficients in the above optimal simplex tableau are rounded to three decimals. Discuss how this may affect the validity of the Gomory mixed integer inequalities you generated above.

Exercise 5.16. Assume that $P := \{x \in \mathbb{R}^n : Ax \leq b\}$ is a rational polyhedron and $S := P \cap (\mathbb{Z}^I \times \mathbb{R}^C)$ is a mixed integer set. Define a restricted Chvátal inequality as in (5.18) with the additional condition that $u < 1$. Prove or disprove that P^{Ch} is the intersection of P with all restricted Chvátal inequalities.

Exercise 5.17. Let $P = \{x \in \mathbb{R}^n : Ax \leq b\}$ where $Ax \leq b$ is a TDI system, and let $S = P \cap \mathbb{Z}^n$. Show that $P^{Ch} = \{x \in \mathbb{R}^n : Ax \leq \lfloor b \rfloor\}$.

Exercise 5.18. Let $C = \{x \in \mathbb{R}^n : Ax = 0, x \geq 0\}$ where A be an integral matrix. Let Δ be the largest among the absolute values of the determinants of the square submatrices of A. Show that the extreme rays $r^1, \ldots, r^q$ of C can be chosen to be integral vectors satisfying $-\Delta \mathbf{1} \leq r^t \leq \Delta \mathbf{1}$, $1 \leq t \leq q$.

Exercise 5.19. Consider $P := \{x \in \mathbb{R}^n_+ : x_i + x_j \leq 1 \text{ for all } 1 \leq i < j \leq n\}$, and let $S = P \cap \{0,1\}^n$.

1. Show that the inequality $\sum_{j=1}^n x_j \leq 1$ has Chvátal rank $k \geq \lfloor \log_2 n \rfloor$ for $n \geq 3$.

2. Find an upper bound on the Chvátal rank of the inequality $\sum_{j=1}^n x_j \leq 1$.

5.6. EXERCISES

Exercise 5.20. Given a positive integer t, consider $P := \{x \in \mathbb{R}^2 : tx_1 + x_2 \leq 1+t,\ -tx_1 + x_2 \leq 1\}$ and $S := P \cap \mathbb{Z}^2$. Show that P has Chvátal rank at least t.

Exercise 5.21. Show that the sequential convexification theorem does not extend to $0, 1, 2$ variables. Specifically, consider $P := \{x \in \mathbb{R}^{n+p}_+ : Ax \geq b\}$ and $S := \{x \in \{0,1,2\}^n \times \mathbb{R}^p_+ : Ax \geq b\}$. Let $P_j := \text{conv}((P \cap \{x_j = 0\}) \cup (P \cap \{x_j = 1\}) \cup (P \cap \{x_j = 2\}))$ for $j \leq n$. Give an example where $(P_1)_2 \neq \text{conv}(\{x \in \{0,1,2\}^2 \times \mathbb{R}^{n-2+p}_+ : Ax \geq b\})$.
Do we always have $(P_1)_2 = (P_2)_1$?

Exercise 5.22. Consider a pure integer set $S := P \cap \mathbb{Z}^n$ where $P := \{x \in \mathbb{R}^n : Ax \leq b\}$ is a rational polyhedron. Define the *two-side-split closure* S^1 of P as the intersection of all split inequalities that are not one-side split inequalities, i.e., they are obtained from split disjunctions $\pi x \leq \pi_0$ or $\pi x \geq \pi_0 + 1$ such that both $\Pi_1 := P \cap \{x : \pi x \leq \pi_0\}$ and $\Pi_2 := P \cap \{x : \pi x \geq \pi_0+1\}$ are nonempty. We can iterate the closure process to obtain the tth *two-side-split closure* S^t for $t \geq 2$ integer, by taking the *two-side-split closure* of S^{t-1}. Using the following example, show that there is in general no finite k such that $S^k = \text{conv}(S)$.
$S := P \cap \mathbb{Z}^2$ where $P := \{x \in \mathbb{R}^2 : x_1 \geq 0,\ x_2 \geq 0,\ x_2 \leq 1 + \frac{1}{4}x_1,\ x_1 \leq 1 + \frac{1}{4}x_2\}$.

Exercise 5.23. Show that the lift-and-project closure is strictly contained in P whenever $P \neq \text{conv}(S)$.

Exercise 5.24. For the following two choices of polytope $P \subseteq \mathbb{R}^2$ and corresponding $S = P \cap \{0,1\}^2$, compute the lift-and-project and the Chvátal closures. In each case, determine which closure gives the tighter relaxation.

1. $P := \{x \in [0,1]^2 : 2x_1 + 2x_2 \leq 3\}$

2. $P := \{x \in [0,1]^2 : 2x_1 + x_2 \leq 2,\ x_2 \leq 2x_1\}$.

Exercise 5.25. Let $P \subseteq \mathbb{R}^n$ be a polyhedron and let $S := P \cap (\{0,1\}^p \times \mathbb{R}^{n-p})$. Show that the kth lift-and-project closure of P is the set

$$\bigcap_{\substack{J \subseteq \{1,\ldots,p\} \\ |J|=k}} \text{conv}\{x \in P : x_j \in \{0,1\} \text{ for } j \in J\}.$$

Exercise 5.26. Consider the polytope $P := \{x \in \mathbb{R}^n_+ : x_i + x_j \leq 1 \text{ for all } 1 \leq i < j \leq n\}$ and $S := P \cap \{0,1\}^n$. Show that the kth lift-and-project closure of P is equal to

$$\{x \in \mathbb{R}^n_+ : \sum_{j \in J} x_j \leq 1 \text{ for all } J \text{ such that } |J| = k+2\}.$$

The authors taking a break

Chapter 6

Intersection Cuts and Corner Polyhedra

In this chapter, we present two classical points of view for approximating a mixed integer linear set: Gomory's corner polyhedron and Balas' intersection cuts. It turns out that they are equivalent: the nontrivial valid inequalities for the corner polyhedron are exactly the intersection cuts. Within this framework, we stress two ideas: the best possible intersection cuts are generated from maximal lattice-free convex sets, and formulas for these cuts can be interpreted using the so-called infinite relaxation.

6.1 Corner Polyhedron

We consider a mixed integer linear set defined by the following constraints

$$\begin{aligned} Ax &= b \\ x_j &\in \mathbb{Z} \quad \text{for } j = 1, \ldots, p \\ x_j &\geq 0 \quad \text{for } j = 1, \ldots, n \end{aligned} \quad (6.1)$$

where $p \leq n$, $A \in \mathbb{Q}^{m \times n}$ and $b \in \mathbb{Q}^m$ is a column vector. We assume that the matrix A has full row rank m. Given a feasible basis B, let $N := \{1, \ldots, n\} \setminus B$ index the nonbasic variables. We rewrite the system $Ax = b$ as

$$x_i = \bar{b}_i - \sum_{j \in N} \bar{a}_{ij} x_j \quad \text{for } i \in B \quad (6.2)$$

where $\bar{b}_i \geq 0$, $i \in B$. The corresponding basic solution is $\bar{x}_i = \bar{b}_i$, $i \in B$, $\bar{x}_j = 0, j \in N$. If $\bar{b}_i \in \mathbb{Z}$ for all $i \in B \cap \{1, \ldots, p\}$, then $\bar{x}$ is a feasible solution to (6.1).

If this is not the case, we address the problem of finding valid inequalities for the set (6.1) that are violated by the point $\bar{x}$. Typically, $\bar{x}$ is an optimal solution of the linear programming relaxation of a mixed integer linear program having (6.1) as feasible set.

The key idea is to work with the corner polyhedron introduced by Gomory [178], which is obtained from (6.1) by dropping the nonnegativity restriction on all the basic variables x_i, $i \in B$. Note that in this relaxation we can drop the constraints $x_i = \bar{b}_i - \sum_{j \in N} \bar{a}_{ij} x_j$ for all $i \in B \cap \{p+1, \ldots, n\}$ because these variables x_i are continuous and they only appear in one equation and no other constraint (recall that we dropped the nonnegativity constraint on these variables). Therefore from now on we assume that all basic variables in (6.2) are integer variables, i.e., $B \subseteq \{1, \ldots, p\}$.

Under this assumption, the relaxation of (6.1) introduced by Gomory is

$$\begin{array}{rll} x_i &= \bar{b}_i - \sum_{j \in N} \bar{a}_{ij} x_j & \text{for } i \in B \\ x_i &\in \mathbb{Z} & \text{for } i = 1, \ldots, p \\ x_j &\geq 0 & \text{for } j \in N. \end{array} \qquad (6.3)$$

The convex hull of the feasible solutions to (6.3) is called the *corner polyhedron* relative to the basis B and it is denoted by corner(B) (Fig. 6.1). Any valid inequality for the corner polyhedron is valid for the set (6.1).

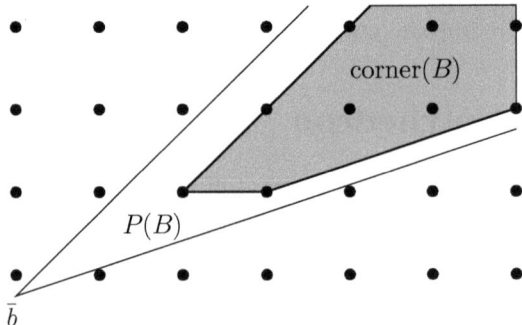

Figure 6.1: Corner polyhedron viewed in the space of the basic variables

Let $P(B)$ be the linear relaxation of (6.3). $P(B)$ is a polyhedron whose vertices and extreme rays are simple to describe, a property that will be useful in generating valid inequalities for corner(B). Indeed, the point $\bar{x}$ defined by $\bar{x}_i = \bar{b}_i$, for $i \in B$, $\bar{x}_j = 0$, for $j \in N$, is the unique vertex of $P(B)$. In particular $P(B)$ is a translate of its recession cone, that is $P(B) = \{\bar{x}\} + \text{rec}(P(B))$. The recession cone of $P(B)$ is defined by the following linear system.

6.1. CORNER POLYHEDRON

$$x_i = -\sum_{j \in N} \bar{a}_{ij} x_j \quad \text{for } i \in B$$
$$x_j \geq 0 \quad \text{for } j \in N.$$

Since the projection of this cone onto $\mathbb{R}^N$ is defined by the inequalities $x_j \geq 0$, $j \in N$, and the variables x_i, $i \in B$, are defined by the above equations, its extreme rays are the vectors satisfying at equality all but one of the nonnegativity constraints. Thus there are $|N|$ extreme rays, $\bar{r}^j$ for $j \in N$, defined by

$$\bar{r}_h^j = \begin{cases} -\bar{a}_{hj} & \text{if } h \in B, \\ 1 & \text{if } h = j, \\ 0 & \text{if } h \in N \setminus \{j\}. \end{cases} \tag{6.4}$$

Remark 6.1. *The vectors $\bar{r}^j$, $j \in N$, are linearly independent. Hence $P(B)$ is an $|N|$-dimensional polyhedron whose affine hull is defined by the equations $x_i = \bar{b}_i - \sum_{j \in N} \bar{a}_{ij} x_j$ for $i \in B$.*

The rationality assumption of the matrix A will be used in the proof of the next lemma.

Lemma 6.2. *If the affine hull of $P(B)$ contains a point in $\mathbb{Z}^p \times \mathbb{R}^{n-p}$, then corner$(B)$ is an $|N|$-dimensional polyhedron. Otherwise corner(B) is empty.*

Proof. Since corner(B) is contained in the affine hull of $P(B)$, corner(B) is empty when the affine hull of $P(B)$ contains no point in $\mathbb{Z}^p \times \mathbb{R}^{n-p}$.

Next we assume that the affine hull of $P(B)$ contains a point in $\mathbb{Z}^p \times \mathbb{R}^{n-p}$, and we show that corner(B) is an $|N|$-dimensional polyhedron. We first show that corner(B) is nonempty.

Let $x' \in \mathbb{Z}^p \times \mathbb{R}^{n-p}$ belong to the affine hull of $P(B)$. By Remark 6.1 $x'_i = \bar{b}_i - \sum_{j \in N} \bar{a}_{ij} x'_j$ for $i \in B$.

Let $N^- := \{j \in N : x'_j < 0\}$. If N^- is empty, then $x' \in \text{corner}(B)$. Let $D \in \mathbb{Z}_+$ be such that $D\bar{a}_{ij} \in \mathbb{Z}$ for all $i \in B$ and $j \in N^-$. Define the point x'' as follows

$$x''_j := x'_j, \, j \in N \setminus N^-; \quad x''_j := x'_j - D\lfloor \frac{x'_j}{D} \rfloor, \, j \in N^-; \quad x''_i := \bar{b}_i - \sum_{j \in N} \bar{a}_{ij} x''_j, \, i \in B.$$

By construction, $x''_j \geq 0$ for all $j \in N$ and x''_i is integer for $i = 1, \ldots, p$. Since x'' satisfies $x''_i = \bar{b}_i - \sum_{j \in N} \bar{a}_{ij} x''_j$, x'' belongs to corner(B). This shows that corner(B) is nonempty.

The recession cones of $P(B)$ and corner(B) coincide by Theorem 4.30, because $P(B)$ is a rational polyhedron. By Remark 6.1, this implies that the dimension of corner(B) is $|N|$. □

Example 6.3. Consider the pure integer program

$$\begin{aligned}
\max \tfrac{1}{2}x_2 + x_3 & \\
x_1 + x_2 + x_3 &\leq 2 \\
x_1 - \tfrac{1}{2}x_3 &\geq 0 \\
x_2 - \tfrac{1}{2}x_3 &\geq 0 \\
x_1 + \tfrac{1}{2}x_3 &\leq 1 \\
-x_1 + x_2 + x_3 &\leq 1 \\
x_1, x_2, x_3 &\in \mathbb{Z} \\
x_1, x_2, x_3 &\geq 0.
\end{aligned} \quad (6.5)$$

This problem has four feasible solutions $(0,0,0)$, $(1,0,0)$, $(0,1,0)$, and $(1,1,0)$, all satisfying $x_3 = 0$. These four points are shown in the (x_1, x_2)-space in Fig. 6.2.

We first write the problem in standard form (6.1) by introducing continuous slack or surplus variables $x_4, \ldots, x_8$. Solving the linear programming relaxation, we obtain

$$\begin{aligned}
x_1 &= \tfrac{1}{2} + \tfrac{1}{4}x_6 - \tfrac{3}{4}x_7 + \tfrac{1}{4}x_8 \\
x_2 &= \tfrac{1}{2} + \tfrac{3}{4}x_6 - \tfrac{1}{4}x_7 - \tfrac{1}{4}x_8 \\
x_3 &= 1 - \tfrac{1}{2}x_6 - \tfrac{1}{2}x_7 - \tfrac{1}{2}x_8 \\
x_4 &= 0 - \tfrac{1}{2}x_6 + \tfrac{3}{2}x_7 + \tfrac{1}{2}x_8 \\
x_5 &= 0 + \tfrac{1}{2}x_6 - \tfrac{1}{2}x_7 + \tfrac{1}{2}x_8
\end{aligned}$$

The optimal basic solution is $x_1 = x_2 = \tfrac{1}{2}$, $x_3 = 1$, $x_4 = \ldots = x_8 = 0$.

Relaxing the nonnegativity of the basic variables and dropping the two constraints relative to the continuous basic variables x_4 and x_5, we obtain the following realization of formulation (6.3) for this example:

$$\begin{aligned}
x_1 &= \tfrac{1}{2} + \tfrac{1}{4}x_6 - \tfrac{3}{4}x_7 + \tfrac{1}{4}x_8 \\
x_2 &= \tfrac{1}{2} + \tfrac{3}{4}x_6 - \tfrac{1}{4}x_7 - \tfrac{1}{4}x_8 \\
x_3 &= 1 - \tfrac{1}{2}x_6 - \tfrac{1}{2}x_7 - \tfrac{1}{2}x_8 \\
x_1, x_2, x_3 &\in \mathbb{Z} \\
x_6, x_7, x_8 &\geq 0.
\end{aligned} \quad (6.6)$$

Let $P(B)$ be the linear relaxation of (6.6). The projection of $P(B)$ in the space of original variables x_1, x_2, x_3 is a polyhedron with unique vertex $\bar{b} = (\tfrac{1}{2}, \tfrac{1}{2}, 1)$. The extreme rays of its recession cone are $v^1 = (\tfrac{1}{2}, \tfrac{3}{2}, -1)$, $v^2 = (-\tfrac{3}{2}, -\tfrac{1}{2}, -1)$ and $v^3 = (\tfrac{1}{2}, -\tfrac{1}{2}, -1)$. In Fig. 6.2, the shaded region (both light and dark) is the intersection of $P(B)$ with the plane $x_3 = 0$.

The last equation in (6.6) and the facts that $x_6 + x_7 + x_8 > 0$ and $x_3 \in \mathbb{Z}$ in every solution of (6.6) imply that $x_3 \leq 0$ is a valid inequality for

6.1. CORNER POLYHEDRON

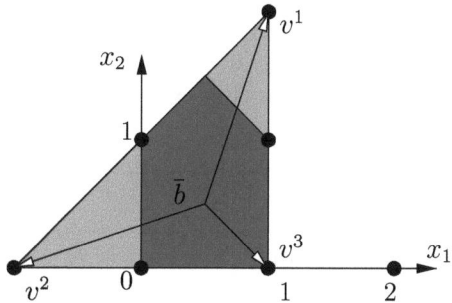

Figure 6.2: Intersection of the corner polyhedron with the plane $x_3 = 0$

corner(B). In fact, corner(B) is exactly the intersection of $P(B)$ with $x_3 \leq 0$ since this later polyhedron has integral vertices and the same recession cone as $P(B)$. Therefore corner(B) is entirely defined by the inequalities $x_3 \leq 0$ and $x_6, x_7, x_8 \geq 0$. Equivalently, in the original (x_1, x_2, x_3)-space, corner(B) is entirely defined by $x_3 \leq 0$, $x_2 - \frac{1}{2}x_3 \geq 0$, $x_1 + \frac{1}{2}x_3 \leq 1$, $-x_1 + x_2 + x_3 \leq 1$. In Fig. 6.2, the shaded region (both light and dark) is therefore also the intersection of corner(B) with the plane $x_3 = 0$.

Let P be the polyhedron defined by the inequalities of (6.5) that are satisfied at equality by the point $\bar{b} = (\frac{1}{2}, \frac{1}{2}, 1)$. The intersection of P with the plane $x_3 = 0$ is the dark shaded region in Fig. 6.2. Thus P is strictly contained in $P(B)$. ■

Using the fact that every basic variable is a linear combination of nonbasic ones, note that every valid linear inequality for corner(B) can be written in terms of the nonbasic variables x_j for $j \in N$ only, as $\sum_{j \in N} \gamma_j x_j \geq \delta$. We say that a valid inequality $\sum_{j \in N} \gamma_j x_j \geq \delta$ for corner(B) is *trivial* if it is implied by the nonnegativity constraints $x_j \geq 0$, $j \in N$. This is the case if and only if $\gamma_j \geq 0$ for all $j \in N$ and $\delta \leq 0$. A valid inequality is said to be *nontrivial* otherwise.

Lemma 6.4. *Assume* corner(B) *is nonempty. Every nontrivial valid inequality for* corner(B) *can be written in the form* $\sum_{j \in N} \gamma_j x_j \geq 1$ *where* $\gamma_j \geq 0$ *for all* $j \in N$.

Proof. We already observed that every valid linear inequality for corner(B) can be written as $\sum_{j \in N} \gamma_j x_j \geq \delta$. We argue next that $\gamma_j \geq 0$ for all $j \in N$. Indeed, if $\gamma_k < 0$ for some $k \in N$, then consider $\bar{r}^k$ defined in (6.4). We have $\sum_{j \in N} \gamma_j \bar{r}_j^k = \gamma_k < 0$, hence $\min\{\sum_{j \in N} \gamma_j x_j : x \in \text{corner}(B)\}$ is unbounded, because $\bar{r}^k$ is in the recession cone of corner(B), contradicting the fact that $\sum_{j \in N} \gamma_j x_j \geq \delta$ is valid for corner(B).

If $\delta \leq 0$, the inequality $\sum_{j \in N} \gamma_j x_j \geq \delta$ is trivial since it is implied by the nonnegativity constraints $x_j \geq 0$, $j \in N$. Hence $\delta > 0$ and, up to multiplying by δ^{-1}, we may assume that $\delta = 1$. □

Since the variables x_i, $i \in B$, are free integer variables, (6.3) can be reformulated as follows:

$$\begin{aligned} \sum_{j \in N} \bar{a}_{ij} x_j &\equiv \bar{b}_i \mod 1 &&\text{for } i \in B \\ x_j &\in \mathbb{Z} &&\text{for } j \in \{1, \ldots, p\} \cap N \\ x_j &\geq 0 &&\text{for } j \in N. \end{aligned} \qquad (6.7)$$

This point of view was introduced by Gomory and extensively studied by Gomory and Johnson. We will discuss it in Sect. 6.3.

6.2 Intersection Cuts

We describe a paradigm introduced by Balas [22] for constructing valid inequalities for the corner polyhedron cutting off the basic solution $\bar{x}$.

Consider a closed convex set $C \subseteq \mathbb{R}^n$ such that the interior of C contains the point $\bar{x}$. (Recall that $\bar{x}$ belongs to the interior of C if C contains an n-dimensional ball centered at $\bar{x}$. This implies that C is full-dimensional). Assume that the interior of C contains no point in $\mathbb{Z}^p \times \mathbb{R}^{n-p}$. In particular C does not contain any feasible point of (6.3) in its interior. For each of the $|N|$ extreme rays of corner(B), define

$$\alpha_j := \max\{\alpha \geq 0 : \bar{x} + \alpha \bar{r}^j \in C\}. \qquad (6.8)$$

Since $\bar{x}$ is in the interior of C, $\alpha_j > 0$. When the half-line $\{\bar{x} + \alpha \bar{r}^j : \alpha \geq 0\}$ intersects the boundary of C, then α_j is finite, the point $\bar{x} + \alpha_j \bar{r}^j$ belongs to the boundary of C and the semi-open segment $\{\bar{x} + \alpha \bar{r}^j, 0 \leq \alpha < \alpha_j\}$ is contained in the interior of C. When $\bar{r}_j$ belongs to the recession cone of C, we have $\alpha_j = +\infty$. Define $\frac{1}{+\infty} := 0$. The inequality

$$\sum_{j \in N} \frac{x_j}{\alpha_j} \geq 1 \qquad (6.9)$$

is the *intersection cut* defined by C for corner(B).

Theorem 6.5. *Let $C \subset \mathbb{R}^n$ be a closed convex set whose interior contains the point $\bar{x}$ but no point in $\mathbb{Z}^p \times \mathbb{R}^{n-p}$. The intersection cut (6.9) defined by C is a valid inequality for corner(B).*

6.2. INTERSECTION CUTS

Proof. The set of points of the linear relaxation $P(B)$ of corner(B) that are cut off by (6.9) is $S := \{x \in P(B) : \sum_{j \in N} \frac{x_j}{\alpha_j} < 1\}$. We will show that S is contained in the interior of C. Since the interior of C does not contain a point in $\mathbb{Z}^p \times \mathbb{R}^{n-p}$, the result will follow.

Consider polyhedron $\bar{S} := \{x \in P(B) : \sum_{j \in N} \frac{x_j}{\alpha_j} \leq 1\}$. By Remark 6.1, $\bar{S}$ is a $|N|$-dimensional polyhedron with vertices $\bar{x}$ and $\bar{x} + \alpha_j \bar{r}^j$ for α_j finite, and extreme rays $\bar{r}^j$ for $\alpha_j = +\infty$. Since the vertices of $\bar{S}$ that lie on the hyperplane $\{x \in \mathbb{R}^n : \sum_{j \in N} \frac{x_j}{\alpha_j} = 1\}$ are the points $\bar{x} + \alpha_j \bar{r}^j$ for α_j finite, every point in S can be expressed as a convex combination of points in the segments $\{\bar{x} + \alpha \bar{r}^j, 0 \leq \alpha < \alpha_j\}$ for α_j finite, plus a conic combination of extreme rays $\bar{r}^j$, for $\alpha_j = +\infty$. By the definition of α_j, the interior of C contains the segments $\{\bar{x} + \alpha \bar{r}^j, 0 \leq \alpha < 1\}$ when α_j is finite, and the rays $\bar{r}^j$ belong to the recession cone of C when $\alpha_j = +\infty$. Therefore, the set S is contained in the interior of C. □

The intersection cut has a simple geometric interpretation. Denoting the interior of C by int(C), it follows that $P(B) \setminus \text{int}(C)$ contains all points of $P(B) \cap (\mathbb{Z}^p \times \mathbb{R}^{n-p})$. If we define Q to be the closed convex hull of $P(B) \setminus \text{int}(C)$, then corner$(B) \subseteq Q$. One can show that Q is also a polyhedron, and indeed

$$Q = \{x \in P(B) : \sum_{j \in N} \frac{x_j}{\alpha_j} \geq 1\},$$

where α_j, $j \in N$, are defined as in (6.8) (see Exercise 6.6). In other words, the intersection cut is the only inequality one needs to add to the description of $P(B)$ in order to obtain Q.

Note that Corollary 5.7 and Remark 5.8 imply that split inequalities are a special case of intersection cuts, where the convex set C is of the form $C = \{x : \pi_0 \leq \pi x \leq \pi_0 + 1\}$ for some split (π, π_0).

Consider two valid inequalities $\sum_{j \in N} \gamma_j x_j \geq 1$ and $\sum_{j \in N} \gamma'_j x_j \geq 1$ for corner(B). We say that the first inequality *dominates* the second if every point $x \in \mathbb{R}^n_+$ satisfying the second inequality also satisfies the first. Note that $\sum_{j \in N} \gamma_j x_j \geq 1$ dominates $\sum_{j \in N} \gamma'_j x_j \geq 1$ if and only if $\gamma_j \leq \gamma'_j$ for all $j \in N$.

Remark 6.6. *Let C_1, C_2 be two closed convex sets whose interiors contain $\bar{x}$ but no point of $\mathbb{Z}^p \times \mathbb{R}^{n-p}$. If C_1 is contained in C_2, then the intersection cut defined by C_2 dominates the intersection cut defined by C_1.*

A closed convex set C whose interior contains $\bar{x}$ but no point of $\mathbb{Z}^p \times \mathbb{R}^{n-p}$ is *maximal* if C is not strictly contained in a closed convex set with the same properties. Any closed convex set whose interior contains $\bar{x}$ but no point of $\mathbb{Z}^p \times \mathbb{R}^{n-p}$ is contained in a maximal such set [41]. This property and Remark 6.6 imply that it is enough to consider intersection cuts defined by maximal closed convex sets whose interior contains $\bar{x}$ but no point of $\mathbb{Z}^p \times \mathbb{R}^{n-p}$.

A set $K \subset \mathbb{R}^p$ that contains no point of $\mathbb{Z}^p$ in its interior is called $\mathbb{Z}^p$-*free* or *lattice-free*.

Remark 6.7. *One way of constructing a closed convex set C whose interior contains $\bar{x}$ but no point of $\mathbb{Z}^p \times \mathbb{R}^{n-p}$ is as follows. In the space $\mathbb{R}^p$, construct a $\mathbb{Z}^p$-free closed convex set K whose interior contains the orthogonal projection of $\bar{x}$ onto $\mathbb{R}^p$. The cylinder $C = K \times \mathbb{R}^{n-p}$ is a closed convex set whose interior contains $\bar{x}$ but no point of $\mathbb{Z}^p \times \mathbb{R}^{n-p}$.*

Example 6.8. Consider the following 4-variable mixed integer linear set

$$\begin{aligned} x_1 &= b_1 + a_{11} y_1 + a_{12} y_2 \\ x_2 &= b_2 + a_{21} y_1 + a_{22} y_2 \\ x &\in \mathbb{Z}^2 \\ y &\geq 0 \end{aligned} \quad (6.10)$$

where the rays $r^1 = \begin{pmatrix} a_{11} \\ a_{21} \end{pmatrix}, r^2 = \begin{pmatrix} a_{12} \\ a_{22} \end{pmatrix} \in \mathbb{R}^2$ are not collinear and $b = \begin{pmatrix} b_1 \\ b_2 \end{pmatrix} \notin \mathbb{Z}^2$.

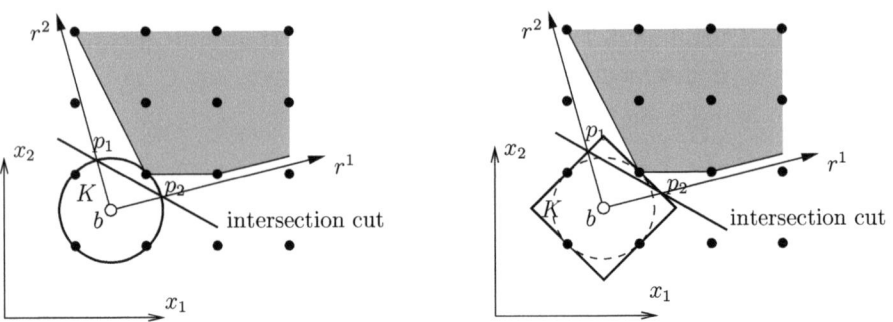

Figure 6.3: Intersection cuts determined by lattice-free convex sets

Figure 6.3 represents the projection of the feasible region of (6.10) in the space of the variables x_1, x_2. The set of feasible points $x \in \mathbb{R}^2$ for the linear relaxation of (6.10) is the cone with apex b and extreme rays r^1, r^2. The feasible points $x \in \mathbb{Z}^2$ for (6.10) are represented by the black dots in this

6.2. INTERSECTION CUTS

cone. The shaded region represents the projection of the corner polyhedron in the (x_1, x_2)-space. The figure depicts two examples of lattice-free convex sets $K \subset \mathbb{R}^2$ containing b in their interior, a disk in the left example and a square that contains this disk on the right.

Because there are two nonbasic variables in this example, the intersection cut can be represented by a line in the space of the basic variables, namely the line passing through the intersection points p^1, p^2 of the boundary of K with the half lines $\{b + \alpha r^1 : \alpha \geq 0\}$, $\{b + \alpha r^2 : \alpha \geq 0\}$.

The coefficients α_1, α_2 defining the intersection cut $\frac{y_1}{\alpha_1} + \frac{y_2}{\alpha_2} \geq 1$ are $\alpha_j = \frac{\|p^j - b\|}{\|r^j\|}$, $j = 1, 2$, using the definition of α_j in (6.8). Note that the intersection cut on the right dominates the one on the left, as observed in Remark 6.6, because the lattice-free set on the right contains the one on the left. ∎

Example 6.9. (Intersection Cut Defined by a Split)

Given $\pi \in \mathbb{Z}^p$ and $\pi_0 \in \mathbb{Z}$, let $K := \{x \in \mathbb{R}^p : \pi_0 \leq \pi x \leq \pi_0 + 1\}$. The set K is a $\mathbb{Z}^p$-free convex set since either $\pi \bar{x} \leq \pi_0$ or $\pi \bar{x} \geq \pi_0 + 1$, for any $\bar{x} \in \mathbb{Z}^p$. Furthermore it is easy to verify that if the entries of π are relatively prime, both hyperplanes $\{x \in \mathbb{R}^p : \pi x = \pi_0\}$ and $\{x \in \mathbb{R}^p : \pi x = \pi_0 + 1\}$ contain points in $\mathbb{Z}^p$ (see Exercise 1.20). Therefore K is a maximal $\mathbb{Z}^p$-free convex set in this case. Consider the cylinder $C := K \times \mathbb{R}^{n-p} = \{x \in \mathbb{R}^n : \pi_0 \leq \sum_{j=1}^{p} \pi_j x_j \leq \pi_0 + 1\}$. Such a set C is called a *split set*. By Remark 6.7, C is a convex set whose interior contains no point of $\mathbb{Z}^p \times \mathbb{R}^{n-p}$.

Given a corner polyhedron $\text{corner}(B)$, let $\bar{x}$ be the unique vertex of its linear relaxation $P(B)$. If $\bar{x}_j \notin \mathbb{Z}$ for some $j = 1, \ldots, p$, there exist $\pi \in \mathbb{Z}^p$, $\pi_0 \in \mathbb{Z}$ such that $\pi_0 < \sum_{j=1}^{p} \pi_j \bar{x}_j < \pi_0 + 1$. Let $\pi_j := 0$ for $j = p+1, \ldots, n$. Then the split set C defined above contains $\bar{x}$ in its interior. We apply formula (6.8) to C. Define $\epsilon := \pi \bar{x} - \pi_0$. Since $\pi_0 < \pi \bar{x} < \pi_0 + 1$, we have $0 < \epsilon < 1$. Also, for $j \in N$, define scalars:

$$\alpha_j := \begin{cases} -\frac{\epsilon}{\pi \bar{r}^j} & \text{if } \pi \bar{r}^j < 0, \\ \frac{1-\epsilon}{\pi \bar{r}^j} & \text{if } \pi \bar{r}^j > 0, \\ +\infty & \text{otherwise}, \end{cases} \quad (6.11)$$

where $\bar{r}^j$ is defined in (6.4).

As observed earlier, the interpretation of α_j is the following. Consider the half-line $\bar{x} + \alpha \bar{r}^j$, where $\alpha \geq 0$, starting from $\bar{x}$ in the direction $\bar{r}^j$. The value α_j is the largest $\alpha \geq 0$ such that $\bar{x} + \alpha \bar{r}^j$ belongs to C. In other words, when the above half-line intersects one of the hyperplanes $\pi x = \pi_0$ or $\pi x = \pi_0 + 1$, this intersection point $\bar{x} + \alpha_j \bar{r}^j$ defines α_j (see Fig. 6.4) and

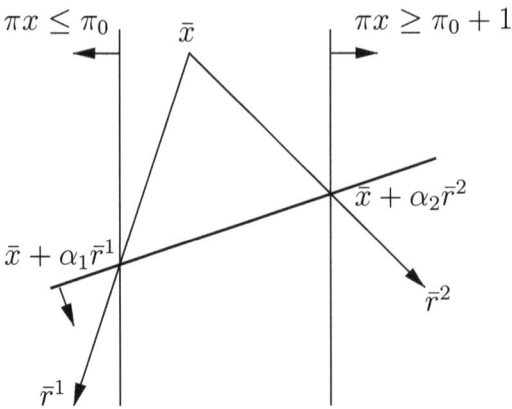

Figure 6.4: Intersection cut defined by a split set

when the direction $\bar{r}^j$ is parallel to the hyperplane $\pi x = \pi_0$, $\alpha_j = +\infty$. The intersection cut defined by the split set C is given by:

$$\sum_{j \in N} \frac{x_j}{\alpha_j} \geq 1. \tag{6.12}$$

∎

Example 6.10. (Gomory's Mixed Integer Cuts from the Tableau)

We already mentioned that split cuts are intersection cuts. We can interpret the formula of a Gomory mixed integer cut derived from a row of the simplex tableau (6.2) in the context of an intersection cut defined by a split set. The argument is as follows. Consider a simplex tableau (6.2), the corresponding basic solution $\bar{x}$, and the corner polyhedron corner(B) described by the system (6.3). Let $x_i = \bar{b}_i - \sum_{j \in N} \bar{a}_{ij} x_j$ be an equation where $\bar{b}_i$ is fractional. Let $f := \bar{b}_i - \lfloor \bar{b}_i \rfloor$ and $f_j := \bar{a}_{ij} - \lfloor \bar{a}_{ij} \rfloor$ for $j \in N \cap \{1, \ldots, p\}$. Define $\pi_0 := \lfloor \bar{b}_i \rfloor$, and for $j = 1, \ldots, p$, define

$$\pi_j := \begin{cases} \lfloor \bar{a}_{ij} \rfloor & \text{if } j \in N \text{ and } f_j \leq f, \\ \lceil \bar{a}_{ij} \rceil & \text{if } j \in N \text{ and } f_j > f, \\ 1 & \text{if } j = i, \\ 0 & \text{otherwise.} \end{cases} \tag{6.13}$$

For $j = p+1, \ldots, n$, define $\pi_j := 0$. Note that $\pi_0 < \pi \bar{x} < \pi_0 + 1$.

Next we derive the intersection cut from the split set $C := \{x \in \mathbb{R}^n : \pi_0 \leq \pi x \leq \pi_0 + 1\}$ following Example 6.9. We will compute α_j for $j \in N$ using formula (6.11). To do this, we need to compute ϵ and $\pi \bar{r}^j$.

6.2. INTERSECTION CUTS

$$\epsilon = \pi\bar{x} - \pi_0 = \sum_{i \in B} \pi_i \bar{x}_i - \pi_0 = \bar{x}_i - \lfloor \bar{x}_i \rfloor = f.$$

Let $j \in N$. Using (6.4) and (6.13), we get $\pi \bar{r}^j = \sum_{h \in N} \pi_h \bar{r}_h^j + \sum_{h \in B} \pi_h \bar{r}_h^j = \pi_j \bar{r}_j^j + \pi_i \bar{r}_i^j$ because $\bar{r}_h^j = 0$ for all $h \in N \setminus \{j\}$ and $\pi_h = 0$ for all $h \in B \setminus \{i\}$. This gives us

$$\pi \bar{r}^j = \begin{cases} \lfloor \bar{a}_{ij} \rfloor - \bar{a}_{ij} = -f_j & \text{if } 1 \leq j \leq p \text{ and } f_j \leq f, \\ \lceil \bar{a}_{ij} \rceil - \bar{a}_{ij} = 1 - f_j & \text{if } 1 \leq j \leq p \text{ and } f_j > f, \\ -\bar{a}_{ij} & \text{if } j \geq p+1. \end{cases} \quad (6.14)$$

Now α_j follows from formula (6.11). Therefore the intersection cut (6.12) defined by the split set C is

$$\sum_{\substack{j \in N,\, j \leq p \\ f_j \leq f}} \frac{f_j}{f} x_j + \sum_{\substack{j \in N,\, j \leq p \\ f_j > f}} \frac{1 - f_j}{1 - f} x_j + \sum_{\substack{p+1 \leq j \leq n \\ \bar{a}_{ij} > 0}} \frac{\bar{a}_{ij}}{f} x_j - \sum_{\substack{p+1 \leq j \leq n \\ \bar{a}_{ij} < 0}} \frac{\bar{a}_{ij}}{1 - f} x_j \geq 1. \quad (6.15)$$

This is exactly the Gomory mixed integer cut (5.31).

The Gomory formula looks complicated, and it may help to think of it as an inequality of the form

$$\sum_{j=1}^p \pi(-\bar{a}_{ij}) x_j + \sum_{j=p+1}^n \psi(-\bar{a}_{ij}) x_j \geq 1$$

where the functions π and ψ, associated with the integer and continuous variables, respectively, are defined by

$$\pi(r) := \min\left\{ \frac{r - \lfloor r \rfloor}{1 - f}, \frac{1 + \lfloor r \rfloor - r)}{f} \right\}, \quad \psi(r) := \max\left\{ \frac{r}{1 - f}, \frac{-r}{f} \right\}. \quad (6.16)$$

These two functions are illustration in Fig. 6.5. They produce the Gomory mixed integer cut. Section 6.3.3 studies properties that general functions π and ψ must satisfy in order to produce valid inequalities for corner(B). ∎

The next example shows that intersection cuts can be much stronger than split inequalities.

Example 6.11. (Intersection Cuts Can Have an Arbitrarily Large Split Rank)

We refer the reader to Sect. 5.1.3 for the definition of split rank of a valid inequality. Consider the polytope $P := \{(x_1, x_2, y) \in \mathbb{R}^3_+ \ :\ x_1 \geq$

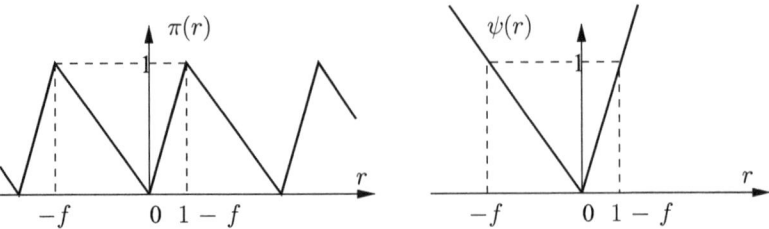

Figure 6.5: Gomory functions

y, $x_2 \geq y$, $x_1 + x_2 + 2y \leq 2\}$, and let $S := \{(x_1, x_2, y) \in P : x_1, x_2 \in \mathbb{Z}\}$. Example 5.11 shows that the inequality $y \leq 0$ does not have a finite split rank. We show next that $y \leq 0$ can be obtained as an intersection cut. By adding slack or surplus variables, the system defining P is equivalent to

$$\begin{aligned} -x_1 + y + s_1 &= 0 \\ -x_2 + y + s_2 &= 0 \\ x_1 + x_2 + 2y + s_3 &= 2 \\ x_1, x_2, y, s_1, s_2, s_3 &\geq 0. \end{aligned}$$

The tableau relative to the basis B defining the vertex $x_1 = \frac{1}{2}$, $x_2 = \frac{1}{2}$, $y = \frac{1}{2}$, $s_1 = s_2 = s_3 = 0$ is

$$\begin{aligned} x_1 &= \tfrac{1}{2} + \tfrac{3}{4}s_1 - \tfrac{1}{4}s_2 - \tfrac{1}{4}s_3 \\ x_2 &= \tfrac{1}{2} - \tfrac{1}{4}s_1 + \tfrac{3}{4}s_2 - \tfrac{1}{4}s_3 \\ y &= \tfrac{1}{2} - \tfrac{1}{4}s_1 - \tfrac{1}{4}s_2 - \tfrac{1}{4}s_3 \\ x_1, x_2, y, s_1, s_2, s_3 &\geq 0. \end{aligned}$$

Since y is a continuous basic variable, we drop the corresponding tableau row. The corner polyhedron corner(B) is the convex hull of the points satisfying

$$\begin{aligned} x_1 &= \tfrac{1}{2} + \tfrac{3}{4}s_1 - \tfrac{1}{4}s_2 - \tfrac{1}{4}s_3 \\ x_2 &= \tfrac{1}{2} - \tfrac{1}{4}s_1 + \tfrac{3}{4}s_2 - \tfrac{1}{4}s_3 \\ s_1, s_2, s_3 &\geq 0 \\ x_1, x_2 &\in \mathbb{Z}. \end{aligned}$$

The extreme rays of corner(B) are the vectors $(\frac{3}{4}, -\frac{1}{4}, 1, 0, 0)$, $(-\frac{1}{4}, \frac{3}{4}, 0, 1, 0)$ and $(-\frac{1}{4}, -\frac{1}{4}, 0, 0, 1)$. Let K be the triangle conv$\{(0,0), (2,0), (0,2)\}$, and $C := K \times \mathbb{R}^3$. One can readily observe that K is lattice-free. We may therefore consider the intersection cut defined by C. The largest α such that $(\frac{1}{2}, \frac{1}{2}, 0, 0, 0) + \alpha(\frac{3}{4}, -\frac{1}{4}, 1, 0, 0)$ belongs to C is $\alpha_1 = 2$, the largest α

6.2. INTERSECTION CUTS

such that $(\frac{1}{2}, \frac{1}{2}, 0, 0, 0) + \alpha(-\frac{1}{4}, \frac{3}{4}, 0, 1, 0)$ belongs to C is $\alpha_2 = 2$ and the largest α such that $(\frac{1}{2}, \frac{1}{2}, 0, 0, 0) + \alpha(-\frac{1}{4}, -\frac{1}{4}, 0, 0, 1)$ belongs to C is $\alpha_3 = 2$. The intersection cut defined by C is therefore $\frac{1}{2}s_1 + \frac{1}{2}s_2 + \frac{1}{2}s_3 \geq 1$. Since $y = \frac{1}{2} - \frac{1}{4}s_1 - \frac{1}{4}s_2 - \frac{1}{4}s_3$, the intersection cut is equivalent to $y \leq 0$. Adding this single inequality to the initial formulation, we obtain conv(S). But, as mentioned above, the intersection cut $y \leq 0$ does not have finite split rank.

Dey and Louveaux [115] study the split rank of intersection cuts for problems with two integer variables. Surprisingly, they show that all intersection cuts have finite split rank except for the ones defined by lattice-free triangles with integral vertices and an integral point in the middle of each side. These triangles are all unimodular transformations of the triangle K defined above. ∎

Theorem 6.5 shows that intersection cuts are valid for corner(B). The following theorem provides a converse statement, namely that corner(B) is completely defined by the trivial inequalities and intersection cuts. We assume here that corner(B) is nonempty.

Theorem 6.12. *Every nontrivial facet-defining inequality for* corner(B) *is an intersection cut.*

Proof. We prove the theorem in the pure integer case, that is, when $p = n$ (see [78] for the general case). Consider a nontrivial valid inequality for corner(B). By Lemma 6.4 it is of the form $\sum_{j \in N} \gamma_j x_j \geq 1$. We show that it is an intersection cut.

Consider the polyhedron $S := P(B) \cap \{x \in \mathbb{R}^n : \sum_{j \in N} \gamma_j x_j \leq 1\}$. Since $\sum_{j \in N} \gamma_j x_j \geq 1$ is a valid inequality for corner(B), all points of $\mathbb{Z}^n \cap S$ satisfy $\sum_{j \in N} \gamma_j x_j = 1$.

Since $P(B)$ is a rational polyhedron, $P(B) = \{x \in \mathbb{R}^n : Cx \leq d\}$ for some integral matrix C and vector d. Let

$$T := \{x \in \mathbb{R}^n : Cx \leq d+1, \sum_{j \in N} \gamma_j x_j \leq 1\}.$$

We first show that T is a $\mathbb{Z}^n$-free convex set. Assume that the interior of T contains an integral point $\tilde{x}$. That is, $\tilde{x}$ satisfies all inequalities defining T strictly. Since $Cx \leq d+1$ is an integral system, then $C\tilde{x} \leq d$ and $\sum_{j \in N} \gamma_j \tilde{x}_j < 1$. This contradicts the fact that all points of $\mathbb{Z}^n \cap S$ satisfy $\sum_{j \in N} \gamma_j x_j = 1$.

Since the basic solution $\bar{x}$ belongs to S and $\sum_{j \in N} \gamma_j \bar{x}_j = 0$, T is a $\mathbb{Z}^n$-free convex set containing $\bar{x}$ in its interior. Note that the intersection cut defined by T is $\sum_{j \in N} \gamma_j \bar{x}_j \geq 1$. □

6.2.1 The Gauge Function

Intersection cuts have a nice description in the language of convex analysis. Let $K \subseteq \mathbb{R}^n$ be a closed, convex set with the origin in its interior. A standard concept in convex analysis [201, 316] is that of *gauge* (also known as Minkowski function), which is the function γ_K defined by

$$\gamma_K(r) := \inf\{t > 0 : \frac{r}{t} \in K\}, \quad \text{for all } r \in \mathbb{R}^n.$$

Since the origin is in the interior of K, $\gamma_K(r) < +\infty$ for all $r \in \mathbb{R}^n$. Furthermore $\gamma_K(r) \leq 1$ if and only if $r \in K$ (Exercise 6.7).

The coefficients α_j of the intersection cut defined in (6.8) can be expressed in terms of the gauge of $K := C - \bar{x}$, namely $\frac{1}{\alpha_j} = \gamma_K(\bar{r}^j)$.

Remark 6.13. *The intersection cut defined by a $\mathbb{Z}^p \times \mathbb{R}^{n-p}$-free convex set C is precisely $\sum_{j \in N} \gamma_K(\bar{r}^j) x_j \geq 1$, where $K := C - \bar{x}$.*

Next we discuss some important properties of the gauge function. A function $g : \mathbb{R}^n \to \mathbb{R}$ is *subadditive* if $g(r^1) + g(r^2) \geq g(r^1 + r^2)$ for all $r^1, r^2 \in \mathbb{R}^n$. The function g is *positively homogeneous* if $g(\lambda r) = \lambda g(r)$ for every $r \in \mathbb{R}^n$ and every $\lambda > 0$. The function g is *sublinear* if it is both subadditive and positively homogeneous.

Note that if $g : \mathbb{R}^n \to \mathbb{R}$ is positively homogeneous, then $g(0) = 0$. Indeed, for any $\lambda > 0$, we have that $g(0) = g(\lambda 0) = \lambda g(0)$, which implies that $g(0) = 0$.

Lemma 6.14. *Given a closed convex set K with the origin in its interior, the gauge γ_K is a nonnegative sublinear function.*

Proof. It follows from the definition of gauge that γ_K is positively homogeneous and nonnegative. Since K is a closed convex set, γ_K is a convex function. We now show that γ_K is subadditive. We have that $\gamma_K(r^1) + \gamma_K(r^2) \geq 2\gamma_K(\frac{r^1+r^2}{2}) = \gamma_K(r^1 + r^2)$, where the inequality follows by convexity and the equality follows by positive homogeneity. □

Lemma 6.15. *Every sublinear function $g : \mathbb{R}^n \to \mathbb{R}$ is convex, and therefore continuous.*

Proof. Let g be a sublinear function. The convexity of g follows from $\frac{1}{2}(g(r^1) + g(r^2)) = g(\frac{r^1}{2}) + g(\frac{r^2}{2}) \geq g(\frac{r^1+r^2}{2})$ for every $r^1, r^2 \in \mathbb{R}^n$, where the equality follows by positive homogeneity and the inequality by subadditivity. Every convex function is continuous, see, e.g., Rockafellar [316]. □

6.2. INTERSECTION CUTS

Theorem 6.16. *Let $g : \mathbb{R}^n \to \mathbb{R}$ be a nonnegative sublinear function and let $K := \{r \in \mathbb{R}^n : g(r) \leq 1\}$. Then K is a closed convex set with the origin in its interior and g is the gauge of K.*

Proof. By Lemma 6.15, g is continuous and convex, therefore K is a closed convex set. Since the interior of K is $\{r \in \mathbb{R}^n : g(r) < 1\}$ and $g(0) = 0$, the origin is in the interior of K.

Let $r \in \mathbb{R}^n$. We need to show that $g(r) = \gamma_K(r)$. If the half-line $\{\alpha r : \alpha \geq 0\}$ intersects the boundary of K, let $\alpha^* > 0$ be such that $g(\alpha^* r) = 1$. Since g is positively homogeneous, $g(r) = \frac{1}{\alpha^*} = \inf\{t > 0 : \frac{r}{t} \in K\} = \gamma_K(r)$. If $\{\alpha r : \alpha \geq 0\}$ does not intersect the boundary of K, then, since g is positively homogeneous, $g(\alpha r) = \alpha g(r) \leq 1$ for all $\alpha > 0$, therefore $g(r) = 0$ because g is nonnegative. Hence $g(r) = 0 = \inf\{t > 0 : \frac{r}{t} \in K\} = \gamma_K(r)$. □

6.2.2 Maximal Lattice-Free Convex Sets

For a good reference on lattices and convexity, we recommend Barvinok [39]. Here we will only work with the integer lattice $\mathbb{Z}^p$. By Remark 6.6, the undominated intersection cuts are the ones defined by full-dimensional *maximal* $\mathbb{Z}^p \times \mathbb{R}^{n-p}$-*free convex sets* in $\mathbb{R}^n$, that is, full-dimensional subsets of $\mathbb{R}^n$ that are convex, have no point of $\mathbb{Z}^p \times \mathbb{R}^{n-p}$ in their interior, and are inclusion maximal with respect to these two properties.

Lemma 6.17. *Let C be a full-dimensional maximal $\mathbb{Z}^p \times \mathbb{R}^{n-p}$-free convex set and let K be its orthogonal projection onto $\mathbb{R}^p$. Then K is a maximal $\mathbb{Z}^p$-free convex set and $C = K \times \mathbb{R}^{n-p}$.*

Proof. A classical result in convex analysis implies that the interior of K is the orthogonal projection onto $\mathbb{R}^p$ of the interior of C (see Theorem 6.6 in Rockafellar [316]). Since C is a $\mathbb{Z}^p \times \mathbb{R}^{n-p}$-free convex set, it follows that K is a $\mathbb{Z}^p$-free convex set. Let K' be a maximal $\mathbb{Z}^p$-free convex set containing K. Then the set $K' \times \mathbb{R}^{n-p}$ is a $\mathbb{Z}^p \times \mathbb{R}^{n-p}$-free convex set and $C \subseteq K \times \mathbb{R}^{n-p} \subseteq K' \times \mathbb{R}^{n-p}$. Since C is maximal, these three sets coincide and the result follows. □

The above lemma shows that it suffices to study $\mathbb{Z}^p$-free convex sets. Next we state a characterization of lattice-free sets due to Lovász [261]. We recall that the *relative interior* of a set $S \subseteq \mathbb{R}^n$ is the set of all points $x \in S$ for which there exists a ball $B \subseteq \mathbb{R}^n$ centered at x such that $B \cap \text{aff}(S)$ is contained in S.

Theorem 6.18 (Lovász [261]). *Let $K \subset \mathbb{R}^p$ be a full-dimensional set. Then K is a maximal lattice-free convex set if and only if K is a polyhedron that does not contain any point of $\mathbb{Z}^p$ in its interior and there is at least one point of $\mathbb{Z}^p$ in the relative interior of each facet of K.*

Furthermore, if K is a maximal lattice-free convex set, then $\text{rec}(K) = \text{lin}(K)$.

We prove the theorem under the assumption that K is a bounded set. A complete proof of the above theorem appears in [41].

Proof of Theorem 6.18 in the bounded case. Let K be a maximal lattice-free convex set and assume that K is bounded. Then there exist vectors l, u in $\mathbb{Z}^p$ such that K is contained in the box $B := \{x \in \mathbb{R}^p : l \leq x \leq u\}$. Since K is a lattice-free convex set, for each $v \in B \cap \mathbb{Z}^p$ there exists a halfspace $\{x \in \mathbb{R}^p : \alpha^v x \leq \beta^v\}$ containing K such that $\alpha^v v = \beta^v$ (see the separation theorem for convex sets [39]). Since B is a bounded set, $B \cap \mathbb{Z}^p$ is a finite set. Therefore the set

$$P := \{x \in \mathbb{R}^p : l \leq x \leq u, \, \alpha^v x \leq \beta^v \text{ for all } v \in B \cap \mathbb{Z}^p\}$$

is a polytope. By construction, P is lattice-free and $K \subseteq P$, thus $K = P$ by maximality of K.

We now show that each facet of K contains a lattice point in its relative interior. Assume $K = \{x : a^i x \leq b_i, \, i \in M\}$, where $a^i x \leq b_i, \, i \in M$, are all distinct facet-defining inequalities for K. Assume by contradiction that the facet $F_t := \{x \in K : a^t x = b_t\}$ does not contain a point of $\mathbb{Z}^p$ in its relative interior. Given $\varepsilon > 0$, let $K' := \{x : a^i x \leq b_i, \, i \in M \setminus \{t\}, \, a^t x \leq b_t + \varepsilon\}$. Since the recession cones of K and K' coincide, K' is a polytope. Since K is a maximal lattice-free convex set and $K \subset K'$, K' contains points of $\mathbb{Z}^p$ in its interior. Since K' is a polytope, the number of points in $\text{int}(K') \cap \mathbb{Z}^p$ is finite, hence there exists one such point minimizing $a^t x$, say z. Note that $a^t z > b_t$. By construction, the polytope $K'' := \{x : a^i x \leq b_i, \, i \in M \setminus \{t\}, \, a^t x \leq a^t z\}$ does not contain any point of $\mathbb{Z}^p$ in its interior, and the inclusion $K'' \supset K$ is strict. This contradicts the maximality of K. □

Doignon [121], Bell [45] and Scarf [322] show the following.

Theorem 6.19. *Any full-dimensional maximal lattice-free convex set $K \subseteq \mathbb{R}^p$ has at most 2^p facets.*

Proof. By Theorem 6.18, each facet F contains an integral point x^F in its relative interior. If there are more than 2^p facets, then there exist two

6.2. INTERSECTION CUTS

distinct facets F, F' such that x^F and $x^{F'}$ are congruent modulo 2. Now their middle point $\frac{1}{2}(x^F + x^{F'})$ is integral and it is in the interior of K, contradicting the fact that K is lattice-free. □

In $\mathbb{R}^2$, Theorem 6.19 implies that full-dimensional maximal lattice-free convex sets have at most 4 facets (Fig. 6.6). Using Theorem 6.18, one can show that they are either:

1. Splits, namely sets of the form $\{x \in \mathbb{R}^2 : \pi_0 \leq \pi_1 x_1 + \pi_2 x_2 \leq \pi_0 + 1\}$, where $\pi_0, \pi_1, \pi_2 \in \mathbb{Z}$ and π_1, π_2 are relatively prime;

2. Triangles with an integral point in the relative interior of each facet and no integral point in the interior of the triangle;

3. Quadrilaterals with an integral point in the relative interior of each facet and no integral point in the interior of the quadrilateral.

A sharpening of the above classification is given in Exercises 6.11 and 6.12.

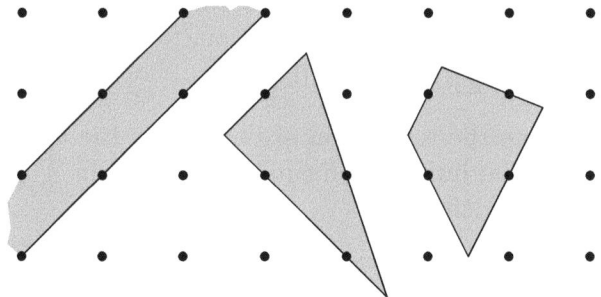

Figure 6.6: Maximal lattice-free convex sets with nonempty interior in $\mathbb{R}^2$

Consider the corner polyhedron corner(B) and the linear relaxation $P(B)$ of (6.3). As in Sect. 6.1, we denote by $\bar{x}$ the apex of $P(B)$ and by $\bar{r}^j$, $j \in N$ its extreme rays (recall (6.4)). By Remark 6.6, undominated intersection cuts for corner(B) are defined by maximal $\mathbb{Z}^p \times \mathbb{R}^{n-p}$-free convex sets containing $\bar{x}$ in their interior. By Lemma 6.17 and Theorem 6.18, these sets are polyhedra of the form $K \times \mathbb{R}^{n-p}$, where K is a maximal lattice-free polyhedron in $\mathbb{R}^p$. The next theorem shows how to compute the coefficients of the intersection cut from a facet description of K.

Theorem 6.20. *Let K be a $\mathbb{Z}^p$-free polyhedron containing $(\bar{x}_1, \ldots, \bar{x}_p)$ in its interior. Then K can be uniquely written in the form $K = \{x \in \mathbb{R}^p : \sum_{h=1}^{p} d_h^i (x_h - \bar{x}_h) \leq 1, i = 1, \ldots, t\}$, where t is the number of facets of K*

and $d^1, \ldots, d^t \in \mathbb{R}^p$. The coefficients in the intersection cut (6.9) defined by $C := K \times \mathbb{R}^{n-p}$ are

$$\frac{1}{\alpha_j} = \max_{i=1,\ldots,t} \sum_{h=1}^{p} d_h^i \bar{r}_h^j \qquad j \in N. \tag{6.17}$$

Proof. Every facet-defining inequality for K can be written in the form $\sum_{h=1}^{p} d_h(x_h - \bar{x}_h) \leq \delta$. Since $(\bar{x}_1, \ldots, \bar{x}_p)$ is in the interior of K, it follows that $\sum_{h=1}^{p} d_h(\bar{x}_h - \bar{x}_h) < \delta$, thus $\delta > 0$. Possibly by multiplying by δ^{-1}, every facet-defining inequality for K can be written in the form $\sum_{h=1}^{p} d_h(x_h - \bar{x}_h) \leq 1$.

We next show (6.17). Since $\alpha_j := \max\{\alpha \geq 0 : \bar{x} + \alpha \bar{r}^j \in C\}$ and $C = \{x \in \mathbb{R}^n : \sum_{h=1}^{p} d_h^i(x_h - \bar{x}_h) \leq 1, i = 1, \ldots, t\}$, it follows that $\frac{1}{\alpha_j} = \max\{0, \sum_{h=1}^{p} d_h^i \bar{r}_h^j \ i = 1, \ldots, t\}$. We only need to show that there exists $i \in \{1, \ldots, t\}$ such that $\sum_{h=1}^{p} d_h^i \bar{r}_h^j \geq 0$.

Since K is contained in a maximal $\mathbb{Z}^p$-free convex set, it follows from the last part of Theorem 6.18 that the recession cone of K has dimension less than p, hence it has empty interior. Thus, the system of strict inequalities $\sum_{h=1}^{p} d_h^i r_h < 0$, $i = 1, \ldots, t$ admits no solution. This shows that there exists $i \in \{1, \ldots, t\}$ such that $\sum_{h=1}^{p} d_h^i \bar{r}_h^j \geq 0$. □

Let $\rho^j \in \mathbb{R}^p$ denote the restriction of $\bar{r}^j \in \mathbb{R}^n$ to the first p components. Theorem 6.20 states that intersection cuts are of the form $\sum_{j \in N} \psi(\rho^j) x_j \geq 1$, where $\psi : \mathbb{R}^p \to \mathbb{R}_+$ is defined by

$$\psi(\rho) := \max_{i=1,\ldots,t} d^i \rho. \tag{6.18}$$

Note that ψ is the gauge of the set $K - (\bar{x}_1, \ldots, \bar{x}_p)$. The definition of ψ depends only on the number p of integer variables and the values $\bar{b}_i \in \mathbb{Q}$, $i \in B$ in (6.3). If these are fixed, then $\sum_{j \in N} \psi(\rho^j) x_j \geq 1$ is valid for corner(B) regardless of the number of continuous variables or of the values of the coefficients $\bar{a}_{ij}$, $i \in B, j \in N$. So ψ gives a formula for generating valid inequalities that is independent of the specific data of the problem.

In the next section we will establish a framework to study functions with such a property, even when the number of integer variables is not fixed.

Example 6.21. Consider the following instance of (6.3) with no integer nonbasic variable.

$$\begin{aligned} x_1 &= \tfrac{1}{2} + \tfrac{1}{4} x_3 - \tfrac{3}{4} x_4 - \tfrac{1}{4} x_5 + x_6 \\ x_2 &= \tfrac{1}{2} + \tfrac{3}{4} x_3 - \tfrac{1}{4} x_4 + \tfrac{3}{4} x_5 - \tfrac{3}{4} x_6 \end{aligned}$$

$$x_1, x_2 \in \mathbb{Z}$$
$$x_3, x_4, x_5, x_6 \geq 0.$$

6.3. INFINITE RELAXATIONS

Let B be the triangle with vertices $(0,0), (2,0), (0,2)$. B is a maximal $\mathbb{Z}^2$-free convex set, since it contains no integral point in its interior and all three sides have integral middle points, namely, $(1,0), (0,1), (1,1)$ (Fig. 6.7). Note that $\bar{b} = \begin{pmatrix} \frac{1}{2} \\ \frac{1}{2} \end{pmatrix}$ is in the interior of B, and B can be written in the form

$$B = \{x \in \mathbb{R}^2 : -2(x_1 - \tfrac{1}{2}) \leq 1,\ -2(x_2 - \tfrac{1}{2}) \leq 1,\ (x_1 - \tfrac{1}{2}) + (x_2 - \tfrac{1}{2}) \leq 1\}.$$

The gauge of the set $K := B - \bar{b}$ is the function defined by

$$\psi(\rho) = \max\{-2\rho_1, -2\rho_2, \rho_1 + \rho_2\}, \qquad \rho \in \mathbb{R}^2$$

Because $\rho^3 = \begin{pmatrix} \frac{1}{4} \\ \frac{3}{4} \end{pmatrix}$, $\rho^4 = \begin{pmatrix} -\frac{3}{4} \\ -\frac{1}{4} \end{pmatrix}$, $\rho^5 = \begin{pmatrix} -\frac{1}{4} \\ -\frac{3}{4} \end{pmatrix}$, $\rho^6 = \begin{pmatrix} 1 \\ -\frac{3}{4} \end{pmatrix}$, the intersection cut defined by B is therefore

$$x_3 + \frac{3}{2}x_4 + \frac{1}{2}x_5 + \frac{3}{2}x_6 \geq 1.$$

∎

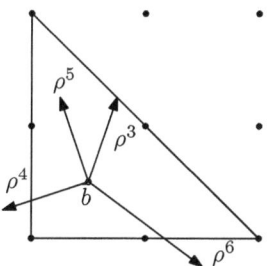

Figure 6.7: Maximal $\mathbb{Z}^2$-free triangle B and vectors $\rho^3 \ldots, \rho^6$

6.3 Infinite Relaxations

Theorem 6.20 gives a formula for computing the coefficients of an intersection cut, namely $\frac{1}{\alpha_j} = \psi(\rho^j)$, where ψ is the function defined in (6.18). As we pointed out, the definition of ψ does not depend on the number of continuous variables nor on the vectors ρ^js. Any function with such properties can therefore be used as a "black box" to generate cuts from the tableau of any integer program. Similarly, the functions π and ψ defined in (6.16) provide a formula to generate valid inequalities from one equation of the simplex tableau $x_i + \sum_{j \in N} \bar{a}_{ij} x_j = \bar{b}_i$, namely the inequality $\sum_{j \in N,\ j \leq p} \pi(-\bar{a}_{ij})x_j + \sum_{p+1 \leq j \leq n} \psi(-\bar{a}_{ij}) \geq 1$.

Gomory and Johnson [179] introduced a convenient setting to formalize and study these functions. In this framework, one works with a model with a fixed number of basic variables, but an infinite number of nonbasic ones, namely one for every possible choice of variables coefficients in (6.3).

Consider the constraints (6.7). We rename $f_i := \bar{b}_i$ and $r_i^j := -\bar{a}_{ij}$ for $i \in B$ and $j \in N$. In other words, defining $q := |B|$, the vector $r^j \in \mathbb{R}^q$ is the restriction of the vector $\bar{r}^j \in \mathbb{R}^n$ to the components $i \in B$. Renaming the variables so that the nonbasic integer variables are x_j, $j \in I$, and the nonbasic continuous variables are y_j, $j \in C$, (6.7) is written in the form

$$f_i + \sum_{j \in I} r_i^j x_j + \sum_{j \in C} r_i^j y_j \in \mathbb{Z} \quad i = 1, \ldots, q$$
$$x_j \in \mathbb{Z}_+ \quad \text{for all } j \in I \quad (6.19)$$
$$y_j \geq 0 \quad \text{for all } j \in C.$$

Gomory and Johnson [179] suggested relaxing the space of variables x_j, $j \in I$, y_j, $j \in C$, to an infinite-dimensional space, where an integer variable x_r and a continuous variable y_r are introduced for every $r \in \mathbb{R}^q$. We obtain the following *infinite relaxation*

$$f + \sum_{r \in \mathbb{R}^q} r x_r + \sum_{r \in \mathbb{R}^q} r y_r \in \mathbb{Z}^q$$
$$x_r \in \mathbb{Z}_+ \quad \text{for } r \in \mathbb{R}^q \quad (6.20)$$
$$y_r \geq 0 \quad \text{for } r \in \mathbb{R}^q$$
$$x, y \text{ have a finite support.}$$

The infinite dimensional vectors x, y having *finite support* means that the sets $\{r \in \mathbb{R}^q : x_r > 0\}$ and $\{r \in \mathbb{R}^q : y_r > 0\}$ are finite.

Every problem of the type (6.19) can be obtained from (6.20) by setting to 0 all but a finite number of variables. This is why x and y are restricted to have finite support in the above model. Furthermore, the study of model (6.20) yields information on (6.19) that is independent of the data in (6.19), but depends only on the vector $f \in \mathbb{R}^q$.

We denote by $M_f \subset \mathbb{Z}^{\mathbb{R}^q} \times \mathbb{R}^{\mathbb{R}^q}$ the set of feasible solutions to (6.20). Note that $M_f \neq \emptyset$ since defining $x_r = 1$ for $r = -f$, $x_r = 0$ otherwise, and setting $y = 0$, yields a feasible solution to (6.20).

A function $(\pi, \psi) : \mathbb{R}^q \times \mathbb{R}^q \to \mathbb{R}$ is *valid* for M_f if $\pi \geq 0$ and the linear inequality

$$\sum_{r \in \mathbb{R}^q} \pi(r) x_r + \sum_{r \in \mathbb{R}^q} \psi(r) y_r \geq 1 \quad (6.21)$$

is satisfied by all vectors in M_f.

6.3. INFINITE RELAXATIONS

The relevance of the above definition rests on the fact that any valid function (π, ψ) yields a valid inequality for the original set defined in (6.19), namely

$$\sum_{j \in I} \pi(r^j) x_j + \sum_{j \in C} \psi(r^j) y_j \geq 1.$$

Observe that, if we are given valid functions (π', ψ') and (π'', ψ'') for M_f, such that $\psi' \leq \psi''$ and $\pi' \leq \pi''$, then the inequality (6.21) defined by $(\pi, \psi) := (\pi', \psi')$ is stronger than that defined by $(\pi, \psi) := (\pi'', \psi'')$. This observation naturally leads to the following definition: a valid function (π, ψ) for M_f is *minimal* if there is no valid function (π', ψ'), distinct from (π, ψ), where $\pi' \leq \pi$ and $\psi' \leq \psi$.

We remark, omitting the proof, that for every valid function (π, ψ) there exists a minimal valid function (π', ψ') such that $\pi' \leq \pi$ and $\psi' \leq \psi$. It follows that we only need to focus our attention on minimal valid functions.

While the concept of valid function is natural, the assumption that $\pi \geq 0$ in the definition might, however, seem artificial. Indeed, if we omitted this assumption in the definition, then there would be valid functions for which π takes negative values. However, we next show that any valid function should be nonnegative over the rational vectors. Thus, since data in integer programming problems are usually rational and valid functions should be nonnegative over rational vectors, it makes sense to assume that $\pi \geq 0$.

To show that π should be nonnegative over the rational vectors, consider a function (π, ψ) such that (6.21) holds for every $(x, y) \in M_f$, and suppose $\pi(\tilde{r}) < 0$ for some $\tilde{r} \in \mathbb{Q}^q$. Let $D \in \mathbb{Z}_+$ be such that $D\tilde{r}$ is an integral vector, and let $(\bar{x}, \bar{y}) \in M_f$. Define $\tilde{x}$ by $\tilde{x}_{\tilde{r}} := \bar{x}_{\tilde{r}} + MD$ where M is a positive integer, and $\tilde{x}_r := \bar{x}_r$ for $r \neq \tilde{r}$. It follows that also $(\tilde{x}, \bar{y})$ is an element of M_f. We have $\sum \pi(r)\tilde{x}_r + \sum \psi(r)\bar{y}_r = \sum \pi(r)\bar{x}_r + \pi(\tilde{r})MD + \sum \psi(r)\bar{y}_r$. If we choose $M > (\sum \pi(r)\bar{x}_r + \sum \psi(r)\bar{y}_r - 1)/(D|\pi(\tilde{r})|)$, then $\sum \pi(r)\tilde{x}_r + \sum \psi(r)\bar{y}_r < 1$, contradicting the fact that (π, ψ) is valid.

In the next section we start by considering the pure integer case, namely the case where $y_r = 0$ for all $r \in \mathbb{R}^q$. We will then focus on the "continuous case," where $x_r = 0$ for all $r \in \mathbb{R}^q$, and finally we will give a characterization of minimal valid functions for the set M_f.

6.3.1 Pure Integer Infinite Relaxation

If in (6.20) we disregard the continuous variables, we obtain the following *pure integer infinite relaxation*.

$$f + \sum_{r \in \mathbb{R}^q} r x_r \in \mathbb{Z}^q$$
$$x_r \in \mathbb{Z}_+ \quad \text{for all } r \in \mathbb{R}^q \quad (6.22)$$
$$x \quad \text{has a finite support.}$$

Denote by G_f the set of feasible solutions to (6.22). Note that $G_f \neq \emptyset$ since the vector x defined by $x_r = 1$ for $r = -f$ and $x_r = 0$ otherwise is a feasible solution to (6.22).

A function $\pi : \mathbb{R}^q \to \mathbb{R}$ is *valid* for G_f if $\pi \geq 0$ and the linear inequality

$$\sum_{r \in \mathbb{R}^q} \pi(r) x_r \geq 1 \quad (6.23)$$

is satisfied by all feasible solutions of (6.22).

A valid function for G_f, $\pi : \mathbb{R}^q \to \mathbb{R}_+$, is *minimal* if there is no valid function $\pi' \neq \pi$ such that $\pi'(r) \leq \pi(r)$ for all $r \in \mathbb{R}^q$.

Note that any minimal valid function π must satisfy $\pi(r) \leq 1$ for all $r \in \mathbb{R}^q$ because every $x \in G_f$ has integral components, and therefore for all $r \in \mathbb{R}^q$ either $x_r = 0$ or $x_r \geq 1$. Furthermore, π must satisfy $\pi(-f) = 1$, since the vector defined by $x_{-f} = 1$, $x_r = 0$ for all $r \neq -f$ is in G_f.

Observe that, given $\bar{r} \in \mathbb{R}^q$, the vector x defined by $x_{\bar{r}} = x_{-f-\bar{r}} = 1$, $x_r = 0$ for all $r \neq \bar{r}, -f-\bar{r}$, is an element of G_f, therefore $\pi(\bar{r}) + \pi(-f-\bar{r}) \geq 1$. A function $\pi : \mathbb{R}^q \to \mathbb{R}$ is said to satisfy the *symmetry condition* if $\pi(r) + \pi(-f - r) = 1$ for all $r \in \mathbb{R}^q$.

A function $\pi : \mathbb{R}^q \to \mathbb{R}$ is *periodic* if $\pi(r) = \pi(r + w)$, for every $w \in \mathbb{Z}^q$. Therefore a periodic function is entirely defined by its values in $[0, 1[^q$. The next theorem shows that minimal valid functions are completely characterized by subadditivity, symmetry, and periodicity.

Theorem 6.22 (Gomory and Johnson [179]). *A function $\pi : \mathbb{R}^q \to \mathbb{R}_+$ is a minimal valid function for G_f if and only if $\pi(0) = 0$, π is subadditive, periodic and satisfies the symmetry condition.*

Proof. We first prove the "only if" part of the statement. Assume that π is a minimal valid function for G_f. We need to show the following four facts.

(a) $\pi(0) = 0$. If $\bar{x}$ is a feasible solution of G_f, then so is $\tilde{x}$ defined by $\tilde{x}_r := \bar{x}_r$ for $r \neq 0$, and $\tilde{x}_0 = 0$. Therefore the function π' defined by $\pi'(r) = \pi(r)$ for $r \neq 0$ and $\pi'(0) = 0$ is also valid. Since π is minimal and nonnegative, it follows that $\pi(0) = 0$.

(b) π is subadditive. Let $r^1, r^2 \in \mathbb{R}^q$. We need to show $\pi(r^1) + \pi(r^2) \geq \pi(r^1 + r^2)$. This inequality holds when $r^1 = 0$ or $r^2 = 0$ because $\pi(0) = 0$. Assume now that $r^1 \neq 0$ and $r^2 \neq 0$. Define the function π' as follows.

$$\pi'(r) := \begin{cases} \pi(r^1) + \pi(r^2) & \text{if } r = r^1 + r^2 \\ \pi(r) & \text{if } r \neq r^1 + r^2. \end{cases}$$

We show that π' is valid. Consider any $\bar{x} \in G_f$. We need to show that $\sum_r \pi'(r)\bar{x}_r \geq 1$. Define $\tilde{x}$ as follows

$$\tilde{x}_r := \begin{cases} \bar{x}_{r^1} + \bar{x}_{r^1+r^2} & \text{if } r = r^1 \\ \bar{x}_{r^2} + \bar{x}_{r^1+r^2} & \text{if } r = r^2 \\ 0 & \text{if } r = r^1 + r^2 \\ \bar{x}_r & \text{otherwise.} \end{cases}$$

Note that $\tilde{x} \geq 0$ and $f + \sum r\tilde{x}_r = f + \sum r\bar{x}_r \in \mathbb{Z}^q$, thus $\tilde{x} \in G_f$. Using the definitions of π' and $\tilde{x}$, it is easy to verify that $\sum_r \pi'(r)\bar{x}_r = \sum_r \pi(r)\tilde{x}_r \geq 1$, where the last inequality follows from the facts that π is valid and $\tilde{x} \in G_f$. This shows that π' is valid. Since π is minimal, we get $\pi(r^1 + r^2) \leq \pi'(r^1 + r^2) = \pi(r^1) + \pi(r^2)$.

(c) π is periodic. Suppose not. Then $\pi(\tilde{r}) > \pi(\tilde{r} + w)$ for some $\tilde{r} \in \mathbb{R}^q$ and $w \in \mathbb{Z}^q \setminus \{0\}$. Define the function π' by $\pi'(\tilde{r}) := \pi(\tilde{r} + w)$ and $\pi'(r) = \pi(r)$ for $r \neq \tilde{r}$. We show that π' is valid. Consider any $\bar{x} \in G_f$. Let $\tilde{x}$ be defined by

$$\tilde{x}_r := \begin{cases} \bar{x}_r & \text{if } r \neq \tilde{r}, \tilde{r} + w \\ 0 & \text{if } r = \tilde{r} \\ \bar{x}_{\tilde{r}} + \bar{x}_{\tilde{r}+w} & \text{if } r = \tilde{r} + w. \end{cases}$$

Since $\bar{x} \in G_f$ and $w\bar{x}_{\tilde{r}} \in \mathbb{Z}^q$, we have that $\tilde{x} \in G_f$. By the definition of π' and $\tilde{x}$, $\sum \pi'(r)\bar{x}_r = \sum \pi(r)\tilde{x}_r \geq 1$, where the last inequality follows from the facts that π is valid and $\tilde{x} \in G_f$. This contradicts the fact that π is minimal, since $\pi' \leq \pi$ and $\pi'(\tilde{r}) < \pi(\tilde{r})$.

(d) π satisfies the symmetry condition. Suppose there exists $\tilde{r} \in \mathbb{R}^q$ such that $\pi(\tilde{r}) + \pi(-f - \tilde{r}) \neq 1$. Since π is valid, $\pi(\tilde{r}) + \pi(-f - \tilde{r}) = 1 + \delta$ where $\delta > 0$. Note that, since $\pi(r) \leq 1$ for all $r \in \mathbb{R}^q$, it follows that $\pi(\tilde{r}) > 0$. Define the function π' by

$$\pi'(r) := \begin{cases} \frac{1}{1+\delta}\pi(\tilde{r}) & \text{if } r = \tilde{r}, \\ \pi(r) & \text{if } r \neq \tilde{r}, \end{cases} \quad r \in \mathbb{R}^q.$$

We show that π' is valid. Consider any $\bar{x} \in G_f$. Note that

$$\sum_{r \in \mathbb{R}^q} \pi'(r)\bar{x}_r = \sum_{\substack{r \in \mathbb{R}^q \\ r \neq \tilde{r}}} \pi(r)\bar{x}_r + \frac{1}{1+\delta}\pi(\tilde{r})\bar{x}_{\tilde{r}}$$

If $\bar{x}_{\tilde{r}} = 0$ then $\sum_{r \in \mathbb{R}^q} \pi'(r)\bar{x}_r = \sum_{r \in \mathbb{R}^q} \pi(r)\bar{x}_r \geq 1$ because π is valid. If $\bar{x}_{\tilde{r}} \geq (1+\delta)/\pi(\tilde{r})$ then $\sum_{r \in \mathbb{R}^q} \pi'(r)\bar{x}_r \geq 1$. Thus we can assume that $1 \leq \bar{x}_{\tilde{r}} < (1+\delta)/\pi(\tilde{r})$.

Observe that $\sum_{r \in \mathbb{R}^q, r \neq \tilde{r}} \pi(r)\bar{x}_r + \pi(\tilde{r})(\bar{x}_{\tilde{r}} - 1) \geq \sum_{r \in \mathbb{R}^q, r \neq \tilde{r}} \pi(r\bar{x}_r) + \pi(\tilde{r}(\bar{x}_{\tilde{r}} - 1)) \geq \pi(\sum_{r \in \mathbb{R}^q, r \neq \tilde{r}} r\bar{x}_r + \tilde{r}(\bar{x}_{\tilde{r}} - 1)) = \pi(-f - \tilde{r})$, where the inequalities follow by the subadditivity of π and the equality follows by the periodicity of π. Therefore

$$\begin{aligned} \sum_{r \in \mathbb{R}^q} \pi'(r)\bar{x}_r &= \sum_{\substack{r \in \mathbb{R}^q \\ r \neq \tilde{r}}} \pi(r)\bar{x}_r + \pi(\tilde{r})(\bar{x}_{\tilde{r}} - 1) + \pi(\tilde{r}) - \frac{\delta}{1+\delta}\pi(\tilde{r})\bar{x}_{\tilde{r}} \\ &\geq \pi(-f - \tilde{r}) + \pi(\tilde{r}) - \delta \\ &= 1 + \delta - \delta = 1, \end{aligned}$$

This shows that π' is valid, contradicting the minimality of π.

We now prove the "if" part of the statement. Assume that $\pi(0) = 0$, π is subadditive, periodic, and satisfies the symmetry condition.

We first show that π is valid. The symmetry condition implies $\pi(0) + \pi(-f) = 1$. Since $\pi(0) = 0$, we have $\pi(-f) = 1$. Any $\bar{x} \in G_f$ satisfies $\sum r\bar{x}_r = -f + w$ for some $w \in \mathbb{Z}^q$. We have that $\sum \pi(r)\bar{x}_r \geq \pi(\sum r\bar{x}_r) = \pi(-f + w) = \pi(-f) = 1$, where the inequality comes from subadditivity and the second to last equality comes from periodicity. Thus π is valid.

To show that π is minimal, suppose by contradiction that there exists a valid function $\pi' \leq \pi$ such that $\pi'(\tilde{r}) < \pi(\tilde{r})$ for some $\tilde{r} \in \mathbb{R}^q$. Then $\pi(\tilde{r}) + \pi(-f - \tilde{r}) = 1$ implies $\pi'(\tilde{r}) + \pi'(-f - \tilde{r}) < 1$, contradicting the validity of π'. □

6.3. INFINITE RELAXATIONS

Below, we give examples of minimal valid functions for the case $q = 1$ in (6.22). By periodicity it suffices to describe them in $[0, 1]$. These examples share the following property.

A function $\pi : [0, 1] \to \mathbb{R}$ is *piecewise-linear* if there are finitely many values $0 = r_0 < r_1 < \ldots < r_k = 1$ such that the function is of the form $\pi(r) = a_j r + b_j$ in interval $]r_{j-1}, r_j[$, for $j = 1, \ldots, k$. The r_js for $j = 0, \ldots, k$ are the *breakpoints*. The *slopes* of a piecewise-linear function are the different values of a_j for $j = 1, \ldots, k$. Note that a piecewise-linear function $\pi : [0, 1] \to \mathbb{R}$ is continuous if and only if $\pi(r_0) = a_0 r_0 + b_0$, $\pi(r_k) = a_k r_k + b_k$ and, for $j = 1, \ldots, k-1$, $\pi(r_j) = a_j r_j + b_j = a_{j+1} r_j + b_{j+1}$.

Example 6.23. Let $q = 1$ and $0 < t < f < 1$. Consider the Gomory function π defined in Example 6.10, and the functions π_1, π_2 defined in $[0, 1]$ as follows

$$\pi_1(r) := \begin{cases} \frac{r}{1-f} & \text{if } 0 \leq r \leq 1-f \\ \frac{1-r+t-f}{t} & \text{if } 1-f \leq r \leq 1-f+\frac{t/2}{1-f+t} \\ \frac{r-1/2}{1-f} & \text{if } 1-f+\frac{t/2}{1-f+t} \leq r \leq 1-\frac{t/2}{1-f+t} \\ \frac{1-r}{t} & \text{if } 1-\frac{t/2}{1-f+t} \leq r \leq 1 \end{cases}$$

$$\pi_2(r) := \begin{cases} \frac{r}{1-f} & \text{if } 0 \leq r \leq 1-f \\ \frac{1-r+t-f}{t} & \text{if } 1-f \leq r \leq 1-f+\frac{t}{2-f+t} \\ \frac{r}{2-f} & \text{if } 1-f+\frac{t}{2-f+t} \leq r \leq 1-\frac{t}{2-f+t} \\ \frac{1-r}{t} & \text{if } 1-\frac{t}{2-f+t} \leq r \leq 1 \end{cases}$$

and elsewhere by periodicity. The three functions are illustrated in Fig. 6.8. Note the symmetry relative to the points $(\frac{1-f}{2}, \frac{1}{2})$ and $(1 - \frac{f}{2}, \frac{1}{2})$.

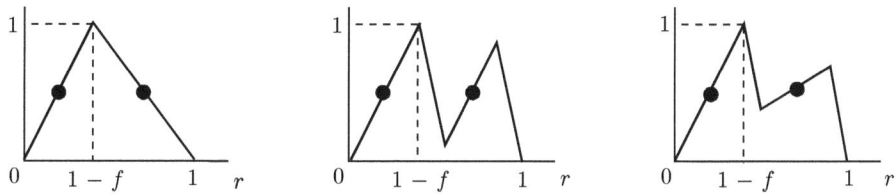

Figure 6.8: Minimal valid functions π, π_1, π_2

Consider a continuous nonnegative periodic function $\pi : \mathbb{R} \to \mathbb{R}_+$ that is piecewise-linear in the interval $[0, 1]$ and satisfies $\pi(0) = 0$. By Theorem 6.22, such a function π is minimal if it is subadditive and satisfies the symmetry condition. Checking whether the symmetry condition $\pi(r) + \pi(-f - r) = 1$ holds for all $r \in \mathbb{R}$ is easy: It suffices to check it at the breakpoints of the function in the interval $[0,1]$. Checking subadditivity of a function is a nontrivial task in general. Gomory and Johnson [180] showed that, for a nonnegative continuous periodic piecewise-linear function π that is symmetric, it is enough to check that $\pi(a) + \pi(b) \geq \pi(a+b)$ for all pairs of breakpoints a, b (possibly $a = b$) in the interval $[0,1]$ where the function is locally convex. Using this, the reader can verify that all three functions given above are minimal. ∎

Extreme Valid Functions and the Two-Slope Theorem

We have seen in Chap. 3 that, in order to describe a full-dimensional polyhedron in $\mathbb{R}^n$, the facet-defining inequalities suffice and they cannot be written as nonnegative combinations of inequalities defining distinct faces. This concept can be immediately generalized to the infinite-dimensional set G_f.

A valid function π for G_f is *extreme* if it cannot be expressed as a convex combination of two distinct valid functions. That is, if π is extreme and $\pi = \frac{1}{2}\pi_1 + \frac{1}{2}\pi_2$ where π_1 and π_2 are valid functions, then $\pi = \pi_1 = \pi_2$.

It follows from the definition that one is interested only in extreme valid functions for G_f, since the inequality (6.23) defined by a valid function π that is not extreme is implied by the two inequalities defined by π_1 and π_2, where $\pi = \frac{1}{2}\pi_1 + \frac{1}{2}\pi_2$, $\pi_1, \pi_2 \neq \pi$. It also follows easily from the definition that extreme valid inequalities are minimal.

Example 6.24. The first two functions of Fig. 6.8 are extreme; this will follow from the two-slope theorem (see below). The third function is extreme when $f \geq t + 1/2$; the proof is left as an exercise (Exercise 6.18). We remark that extreme valid functions are not always continuous. Indeed, Dey, Richard, Li and Miller [117] show that, for $0 < 1 - f < .5$, the following discontinuous valid function is extreme (see Fig. 6.9).

$$\pi(r) := \begin{cases} \frac{r}{1-f} & \text{for } 0 \leq r \leq 1-f \\ \frac{r}{2-f} & \text{for } 1-f < r < 1 \end{cases}$$

∎

Given a minimal valid function π, we recall that π must be subadditive (by Theorem 6.22). We denote by $E(\pi)$ the (possibly infinite) set of all possible inequalities $\pi(r^1) + \pi(r^2) \geq \pi(r^1 + r^2)$ that are satisfied as an equality.

6.3. INFINITE RELAXATIONS

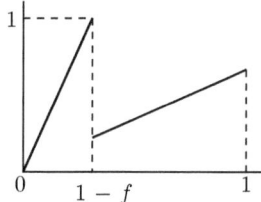

Figure 6.9: A discontinuous extreme valid function

Lemma 6.25. *Let π be a minimal valid function. Assume $\pi = \frac{1}{2}\pi_1 + \frac{1}{2}\pi_2$, where π_1 and π_2 are valid functions. Then π_1 and π_2 are minimal functions and $E(\pi) \subseteq E(\pi_1) \cap E(\pi_2)$.*

Proof. Suppose π_1 is not minimal. Let $\pi_1' \neq \pi$ be a valid function, such that $\pi_1' \leq \pi_1$. Then $\pi' = \frac{1}{2}\pi_1' + \frac{1}{2}\pi_2$ is a valid function, distinct from π, and $\pi' \leq \pi$. This contradicts the minimality of π.

Suppose $E(\pi) \not\subseteq E(\pi_1) \cap E(\pi_2)$. We may assume $E(\pi) \not\subseteq E(\pi_1)$. That is, there exist r^1, r^2 such that $\pi(r^1) + \pi(r^2) = \pi(r^1 + r^2)$ and $\pi_1(r^1) + \pi_1(r^2) > \pi_1(r^1 + r^2)$. Since π_2 is minimal, it is subadditive by Theorem 6.22 and therefore $\pi_2(r^1) + \pi_2(r^2) \geq \pi_2(r^1 + r^2)$. This contradicts the assumption that $\pi = \frac{1}{2}\pi_1 + \frac{1}{2}\pi_2$. □

No general characterization of the extreme valid functions is known. In fact, checking that a valid function is extreme, or proving that a certain class of valid functions are extreme, can be challenging. Gomory and Johnson [180] give an interesting class of extreme valid functions for the case of a single row problem ($q = 1$ in model (6.22)). We present their result in Theorem 6.27, the "two-slope theorem." A useful tool for showing that a given valid function is extreme is the so-called interval lemma, which we prove next.

Lemma 6.26 (Interval Lemma). *Let $f : \mathbb{R} \to \mathbb{R}$ be a function that is bounded on every bounded interval. Let $a_1 < a_2$ and $b_1 < b_2$. Consider the intervals $A := [a_1, a_2]$, $B := [b_1, b_2]$ and $A + B := [a_1 + b_1, a_2 + b_2]$. If $f(a) + f(b) = f(a+b)$ for all $a \in A$ and $b \in B$, then f is an affine function in each of the sets A, B and $A + B$, and it has the same slope in each of these sets.*

Proof. We first show the following.

Claim 1. Let $a \in A$, and let $\varepsilon > 0$ such that $b_1 + \varepsilon \in B$. For every nonnegative integer p such that $a + p\varepsilon \in A$, we have $f(a+p\varepsilon) - f(a) = p(f(b_1+\varepsilon) - f(b_1))$.

For $h = 1, \ldots, p$, by hypothesis $f(a + h\varepsilon) + f(b_1) = f(a + h\varepsilon + b_1) = f(a+(h-1)\varepsilon)+f(b_1+\varepsilon)$. Thus $f(a+h\varepsilon)-f(a+(h-1)\varepsilon) = f(b_1+\varepsilon)-f(b_1)$, for $h = 1, \ldots, p$. By summing these p equations, we obtain $f(a+p\varepsilon)-f(a) = p(f(b_1 + \varepsilon) - f(b_1))$. This concludes the proof of Claim 1.

Let $\bar{a}, \bar{a}' \in A$ such that $\bar{a} - \bar{a}' \in \mathbb{Q}$ and $\bar{a} > \bar{a}'$. Define $c := \frac{f(\bar{a})-f(\bar{a}')}{\bar{a}-\bar{a}'}$.

Claim 2. For every $a, a' \in A$ such that $a - a' \in \mathbb{Q}$, we have $f(a) - f(a') = c(a - a')$.

We may assume $a > a'$. Choose a rational $\varepsilon > 0$ such that $b_1 + \varepsilon \in B$ and the numbers $\bar{p} := \frac{\bar{a}-\bar{a}'}{\varepsilon}$ and $p = \frac{a-a'}{\varepsilon}$ are both integer. By Claim 1,

$$f(\bar{a})-f(\bar{a}') = \bar{p}(f(b_1+\varepsilon)-f(b_1)) \quad \text{and} \quad f(a)-f(a') = p(f(b_1+\varepsilon)-f(b_1)).$$

Dividing the last equality by $a-a' = p\varepsilon$ and the second to last by $\bar{a}-\bar{a}' = \bar{p}\varepsilon$, we obtain

$$\frac{f(b_1 + \varepsilon) - f(b_1)}{\varepsilon} = \frac{f(\bar{a}) - f(\bar{a}')}{\bar{a} - \bar{a}'} = \frac{f(a) - f(a')}{a - a'} = c.$$

Thus $f(a) - f(a') = c(a - a')$. This concludes the proof of Claim 2.

Claim 3. For every $a \in A$, $f(a) = f(a_1) + c(a - a_1)$.

Let $\delta(x) := f(x) - cx$. Since f is bounded on every bounded interval, δ is bounded over A, B and $A+B$. Let M be a number such that $|\delta(x)| \leq M$ for all $x \in A \cup B \cup (A + B)$.

We will show that $\delta(a) = \delta(a_1)$ for all $a \in A$, which proves the claim. Suppose by contradiction that, for some $a^* \in A$, $\delta(a^*) \neq \delta(a_1)$. Let N be a positive integer such that $N|\delta(a^*) - \delta(a_1)| > 2M$.

By Claim 2, $\delta(a^*) = \delta(a)$ for every $a \in A$ such that $a^* - a$ is rational. Thus there exists $\bar{a}$ such that $\delta(\bar{a}) = \delta(a^*)$, $a_1 + N(\bar{a} - a_1) \in A$ and $b_1 + \bar{a} - a_1 \in B$. Let $\varepsilon := \bar{a} - a_1$. By Claim 1,

$$\delta(a_1 + N\varepsilon) - \delta(a_1) = N(\delta(b_1 + \varepsilon) - \delta(b_1)) = N(\delta(a_1 + \varepsilon) - \delta(a_1)) = N(\delta(\bar{a}) - \delta(a_1))$$

Thus $|\delta(a_1+N\varepsilon)-\delta(a_1)| = N|\delta(\bar{a})-\delta(a_1)| = N|\delta(a^*)-\delta(a_1)| > 2M$, which implies $|\delta(a_1 + N\varepsilon)| + |\delta(a_1)| > 2M$, a contradiction. This concludes the proof of Claim 3.

By symmetry between A and B, Claim 3 implies that there exists some constant c' such that, for every $b \in B$, $f(b) = f(b_1) + c'(b - b_1)$. We show $c' = c$. Indeed, given $\varepsilon > 0$ such that $a_1 + \varepsilon \in A$ and $b_1 + \varepsilon \in B$, $c\varepsilon = f(a_1 + \varepsilon) - f(a_1) = f(b_1 + \varepsilon) - f(b_1) = c'\varepsilon$, where the second equality follows from Claim 1.

6.3. INFINITE RELAXATIONS

Therefore, for every $b \in B$, $f(b) = f(b_1) + cf(b - b_1)$. Finally, since $f(a) + f(b) = f(a+b)$ for every $a \in A$ and $b \in B$, it follows that for every $w \in A+B$, $f(w) = f(a_1 + b_1) + c(w - a_1 - b_1)$. □

Theorem 6.27 (Two-Slope Theorem). *Let $\pi : \mathbb{R} \to \mathbb{R}$ be a minimal valid function. If the restriction of π to the interval $[0,1]$ is a continuous piecewise-linear function with only two slopes, then π is extreme.*

Proof. Consider valid functions π_1, π_2 such that $\pi = \frac{1}{2}\pi_1 + \frac{1}{2}\pi_2$. By Lemma 6.25, π_1 and π_2 are minimal valid functions. Since π, π_1, π_2 are minimal, by Theorem 6.22 they are nonnegative and $\pi(0) = \pi_1(0) = \pi_2(0) = 0$, $\pi(1) = \pi_1(1) = \pi_2(1) = 0$, $\pi(1-f) = \pi_1(1-f) = \pi_2(1-f) = 1$. We will prove $\pi = \pi_1 = \pi_2$. We recall that minimal valid functions can only take values between 0 and 1, thus π, π_1, π_2 are bounded everywhere.

Consider $0 = r_0 < r_1 < \cdots < r_{k-1} < r_k = 1$, where $r_1, \ldots, r_{k-1}$ are the points in $[0,1]$ where the slope of π changes. Since π is continuous and $\pi(0) = \pi(1) = 0$, one of the slopes must be positive and the other negative. Let s^+ and s^- be the positive and negative slopes of π. Therefore $\pi(r) = s^+ r$ for $0 \le r \le r_1$ and $\pi(r) = \pi(r_{k-1}) + s^-(r - r_{k-1})$ for $r_{k-1} \le r \le r_k = 1$. Furthermore π has slope s^+ in interval $[r_i, r_{i+1}]$ if i is even and slope s^- if i is odd, $i = 0, \ldots, k-1$.

We next show the following. π_1, π_2 *are continuous piecewise-linear functions with two slopes. In intervals* $[r_i, r_{i+1}]$, i *even*, π_1, π_2 *have positive slopes* s_1^+, s_2^+. *In intervals* $[r_i, r_{i+1}]$, i *odd*, π_1, π_2 *have negative slopes* s_1^-, s_2^-.

Let $i \in \{0, \ldots, k\}$. Assume first i even. Let ϵ be a sufficiently small positive number and define $A = [0, \epsilon]$, $B = [r_i, r_{i+1} - \epsilon]$. Then $A + B = [r_i, r_{i+1}]$ and π has slope s^+ in all three intervals. Since $\pi(0) = 0$, then $\pi(a) + \pi(b) = \pi(a+b)$ for every $a \in A$ and $b \in B$. By Lemma 6.25, $\pi_1(a) + \pi_1(b) = \pi_1(a+b)$ and $\pi_2(a) + \pi_2(b) = \pi_2(a+b)$ for every $a \in A$ and $b \in B$. Thus, by the Interval lemma (Lemma 6.26), π_1 and π_2 are affine functions in each of the closed intervals A, B and $A + B$, where π_1 has positive slope s_1^+ and π_2 has positive slope s_2^+ in each of these sets. The proof for the case i odd is identical, only one needs to choose intervals $A = [r_i + \epsilon, r_{i+1}]$, $B = [1 - \epsilon, 1]$ and use the fact that $\pi(1) = 0$. This shows that, for i even, $\pi_1(r) = \pi_1(r_j) + s_1^+(r - r_j)$ and $\pi_2(r) = \pi_2(r_j) + s_2^+(r - r_j)$, while, for i odd, $\pi_1(r) = \pi_1(r_j) + s_1^-(r - r_j)$ and $\pi_2(r) = \pi_2(r_j) + s_2^-(r - r_j)$. In particular π_1 and π_2 are continuous piecewise-linear functions.

Define L_ℓ^+ and L_r^+ as the sum of the lengths of the intervals of positive slope included in $[0, 1-f]$ and $[1-f, 1]$, respectively. Define L_ℓ^- and L_r^- as the sum of the lengths of the intervals of negative slope included in $[0, 1-f]$ and $[1-f, 1]$, respectively. Note that $L_\ell^+ > 0$ and $L_r^- > 0$.

By the above claim, since $\pi(0) = \pi_1(0) = \pi_2(0) = 0$, $\pi(1) = \pi_1(1) = \pi_2(1) = 0$ and $\pi(1-f) = \pi_1(1-f) = \pi_2(1-f) = 1$, it follows that the vectors (s^+, s^-), (s_1^+, s_1^-), (s_2^+, s_2^-) all satisfy the system

$$\begin{aligned} L_\ell^+ \sigma^+ + L_\ell^- \sigma^- &= 1 \\ L_r^+ \sigma^+ + L_r^- \sigma^- &= -1. \end{aligned}$$

Suppose the constraint matrix of the above system is singular. Then the vector (L_r^+, L_r^-) is a multiple of (L_ℓ^+, L_ℓ^-), so it must be a nonnegative multiple, but this is impossible since the right-hand side of the two equations are one positive and one negative. Thus the constraint matrix is nonsingular, so the system has a unique solution. This implies that $\sigma^+ = s^+ = s_1^+ = s_2^+$ and $\sigma^- = s^- = s_1^- = s_2^-$, and therefore $\pi = \pi_1 = \pi_2$. $\square$

6.3.2 Continuous Infinite Relaxation

If in (6.20) we disregard the integer variables, we obtain the following *continuous infinite relaxation*

$$\begin{aligned} f + \sum_{r \in \mathbb{R}^q} r y_r &\in \mathbb{Z}^q \\ y_r &\geq 0 \quad \text{for all } r \in \mathbb{R}^q \\ y &\text{ has a finite support.} \end{aligned} \tag{6.24}$$

Denote by R_f the set of feasible solutions to (6.24). A function $\psi : \mathbb{R}^q \to \mathbb{R}$ is *valid* for R_f if the linear inequality

$$\sum_{r \in \mathbb{R}^q} \psi(r) y_r \geq 1 \tag{6.25}$$

is satisfied by all vectors in R_f.

A valid function $\psi : \mathbb{R}^q \to \mathbb{R}$ for R_f is *minimal* if there is no valid function $\psi' \neq \psi$ such that $\psi'(r) \leq \psi(r)$ for all $r \in \mathbb{R}^q$.

Note that the notions of valid function and minimal valid function defined above are closely related to the notions introduced at the end of Sect. 6.2. In particular, we will show a one-to-one correspondence between minimal valid functions for R_f and maximal $\mathbb{Z}^q$-free convex sets containing f in their interior.

The next lemma establishes how $\mathbb{Z}^q$-free convex sets with f in their interior naturally yield valid functions for R_f.

6.3. INFINITE RELAXATIONS

Lemma 6.28. *Let B be a $\mathbb{Z}^q$-free closed convex set with f in its interior. Let ψ be the gauge of $B - f$. Then ψ is a valid function.*

Proof. By Lemma 6.14, ψ is sublinear. Consider $\bar{y} \in R_f$. Then $\sum r\bar{y}_r = \bar{x} - f$, for some $\bar{x} \in \mathbb{Z}^q$.

$$\sum \psi(r)\bar{y}_r = \sum \psi(r\bar{y}_r) \geq \psi(\sum r\bar{y}_r) = \psi(\bar{x} - f) \geq 1$$

where the first equality follows by positive homogeneity, the first inequality by subadditivity, and the last from the fact that B is a $\mathbb{Z}^q$-free convex set and that ψ is the gauge of $B - f$. □

On the other hand, we will prove that all minimal valid functions are gauges of maximal $\mathbb{Z}^q$-free convex sets containing f in their interior. First, we need to prove the following.

Lemma 6.29. *If $\psi : \mathbb{R}^q \to \mathbb{R}$ is a minimal valid function for R_f, then ψ is nonnegative and sublinear.*

Proof. We first note that $\psi(0) \geq 0$. Indeed, consider any $\bar{y} \in R_f$. Let $\tilde{y} = \bar{y}$ except for the component $\tilde{y}_0$ which is set to an arbitrarily large value M. We have $\tilde{y} \in R_f$. Therefore $\sum \psi(r)\tilde{y}_r \geq 1$. For this inequality to hold for all $M > 0$, we must have $\psi(0) \geq 0$.

(a) ψ *is sublinear.* We first prove that ψ is subadditive. When $r^1 = 0$ or $r^2 = 0$, $\psi(r^1) + \psi(r^2) \geq \psi(r^1 + r^2)$ follows from $\psi(0) \geq 0$. The proof for the case that $r^1, r^2 \neq 0$ is identical to part (b) of the proof of Theorem 6.22.

We next show that ψ is positively homogeneous. Suppose there exists $\tilde{r} \in \mathbb{R}^q$ and $\lambda > 0$ such that $\psi(\lambda\tilde{r}) \neq \lambda\psi(\tilde{r})$. Without loss of generality we may assume that $\psi(\lambda\tilde{r}) < \lambda\psi(\tilde{r})$, else we can consider $\lambda\tilde{r}$ instead of $\tilde{r}$ and λ^{-1} instead of λ. Define a function ψ' by $\psi'(\tilde{r}) := \lambda^{-1}\psi(\lambda\tilde{r})$, $\psi'(r) := \psi(r)$ for all $r \neq \tilde{r}$. We will show that ψ' is valid. Consider any $\bar{y} \in R_f$. We need to show that $\sum_r \psi'(r)\bar{y}_r \geq 1$. Define $\tilde{y}$ as follows

$$\tilde{y}_r := \begin{cases} 0 & \text{if } r = \tilde{r} \\ \bar{y}_{\lambda\tilde{r}} + \lambda^{-1}\bar{y}_{\tilde{r}} & \text{if } r = \lambda\tilde{r} \\ \bar{y}_r & \text{otherwise.} \end{cases}$$

Note that $f + \sum r\tilde{y}_r = f + \sum r\bar{y}_r \in \mathbb{Z}^q$ and $\tilde{y} \geq 0$, thus $\tilde{y} \in R_f$. Using the definitions of ψ' and $\tilde{y}$, it follows that $\sum_r \psi'(r)\bar{y}_r = \sum_r \psi(r)\tilde{y}_r \geq 1$, where the latter inequality follows from the facts that ψ is valid and $\tilde{y} \in R_f$. This shows that ψ' is valid, contradicting the fact that ψ is minimal. Therefore ψ is positively homogeneous.

(b) *ψ is nonnegative*. Suppose $\psi(\tilde r) < 0$ for some $\tilde r \in \mathbb{Q}^q$. Let $D \in \mathbb{Z}_+$ such that $D\tilde r$ is an integral vector, and let $\bar y$ be a feasible solution of R_f (for example $\bar y_r = 1$ for $r = -f$, $\bar y_r = 0$ otherwise). Let $\tilde y$ be defined by $\tilde y_{\tilde r} := \bar y_{\tilde r} + MD$ where M is a positive integer, and $\tilde y_r := \bar y_r$ for $r \neq \tilde r$. It follows that $\tilde y$ is a feasible solution of R_f.

We have $\sum \psi(r)\tilde y_r = \sum \psi(r)\bar y_r + \psi(\tilde r)MD$. Choose the integer M large enough, namely $M > \frac{\sum \psi(r)\bar y_r - 1}{D|\psi(\tilde r)|}$. Then $\sum \psi(r)\tilde y_r < 1$, contradicting the fact that $\tilde y$ is feasible.

Since ψ is sublinear, by Lemma 6.15 it is continuous. Thus, since ψ is nonnegative over $\mathbb{Q}^q$ and $\mathbb{Q}^q$ is dense in $\mathbb{R}^q$, ψ is nonnegative over $\mathbb{R}^q$. □

Theorem 6.30. *A function ψ is a minimal valid function for R_f if and only if there exists some maximal $\mathbb{Z}^q$-free convex set B such that ψ is the gauge of $B - f$.*

Proof. For the "only if" part, let ψ be a minimal valid function. Define $B := \{x \in \mathbb{R}^q : \psi(x - f) \leq 1\}$. Since ψ is a minimal valid function, by Lemma 6.29, ψ is a nonnegative sublinear function. Thus, by Theorem 6.16, B is a closed convex set with f in its interior and ψ is the gauge of $B - f$. Furthermore, B is a $\mathbb{Z}^q$-free convex set because, given that ψ is valid, $\psi(\bar x - f) \geq 1$ for every $\bar x \in \mathbb{Z}^q$. We only need to prove that B is a maximal $\mathbb{Z}^q$-free convex set. Suppose not, and let B' be a $\mathbb{Z}^q$-free convex set properly containing B. Let ψ' be the gauge of $B' - f$. Then by definition of gauge $\psi' \leq \psi$, and $\psi' \neq \psi$ since $B' \neq B$. By Lemma 6.28, ψ' is a valid function, a contradiction to the minimality of ψ.

To prove the "if" part, assume that ψ is the gauge of $B - f$ for some maximal $\mathbb{Z}^q$-free convex set B. By Lemma 6.28 ψ is valid for R_f. Suppose there exists a valid function ψ' such that $\psi' \leq \psi$ and $\psi' \neq \psi$. Then $B' := \{x : \psi'(x - f) \leq 1\}$ is a $\mathbb{Z}^q$-free convex set and $B' \supset B$, contradicting the maximality of B. □

A function $\psi : \mathbb{R}^q \to \mathbb{R}$ is *piecewise-linear* if $\mathbb{R}^q$ can be covered by a finite number of polyhedra $P_1, \ldots, P_t$ whose interiors are pairwise disjoint, so that the restriction of ψ to the interior of P_i, $i = 1, \ldots, t$ is an affine function. The restrictions of ψ to P_i, $i = 1, \ldots, t$ are the *pieces* of ψ.

Given a maximal $\mathbb{Z}^q$-free convex set B containing f in its interior, if follows from Theorem 6.18 that B is a polyhedron, thus by Theorem 6.20 B can be written in the form

$$B = \{x \in \mathbb{R}^q : d^i(x - f) \leq 1, \ i = 1, \ldots, t\}$$

6.3. INFINITE RELAXATIONS

for some $d^1, \ldots, d^t \in \mathbb{R}^q$, and that the gauge of $B - f$ is the function defined by
$$\psi(r) := \max_{i=1,\ldots,t} d^i r. \tag{6.26}$$

Note that the function ψ defined by (6.26) is piecewise-linear, and it has as many pieces as the number of facets of B. This discussion and Theorems 6.19, 6.30 imply the following.

Corollary 6.31. *Every minimal valid function for R_f is a nonnegative sublinear piecewise-linear function with at most 2^q pieces.*

Example 6.32. Consider the maximal $\mathbb{Z}^2$-free set B defined in Example 6.21, and let $f = \begin{pmatrix} \frac{1}{2} \\ \frac{1}{2} \end{pmatrix}$. The corresponding function ψ has three pieces, corresponding to the three polyhedral cones P_1, P_2, P_3 shown in Fig. 6.10, and we have

$$\psi(r) = \begin{cases} -2r_1 & \text{for } r \in P_1 \\ -2r_2 & \text{for } r \in P_2 \\ r_1 + r_2 & \text{for } r \in P_3 \end{cases} \quad r \in \mathbb{R}^2.$$

■

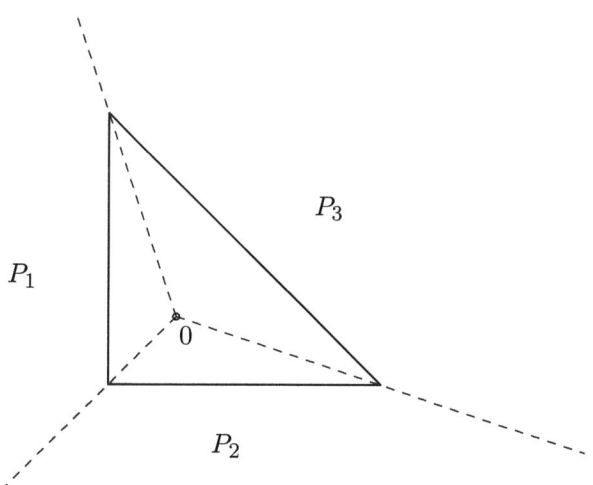

Figure 6.10: Set $B - f$, and the three pieces P_1, P_2, P_3 of the associated gauge ψ

6.3.3 The Mixed Integer Infinite Relaxation

We finally return to the infinite relaxation (6.20) defined at the beginning of Sect. 6.3:

$$f + \sum_{r \in \mathbb{R}^q} r x_r + \sum_{r \in \mathbb{R}^q} r y_r \in \mathbb{Z}^q$$
$$x_r \in \mathbb{Z}_+ \qquad \text{for all } r \in \mathbb{R}^q$$
$$y_r \geq 0 \qquad \text{for all } r \in \mathbb{R}^q$$
$$x, y \qquad \text{have a finite support.}$$

Recall that M_f denotes the set of feasible solutions to (6.20). The purpose of this section is to provide a characterization of the minimal valid inequalities for M_f.

Lemma 6.33. *Let (π, ψ) be a minimal valid function for M_f. Then $\pi \leq \psi$ and ψ is a nonnegative sublinear function.*

Proof. The same proof as that in Lemma 6.29 shows that ψ is nonnegative and sublinear. We next show that $\pi \leq \psi$. Suppose not, and let $\tilde{r} \in \mathbb{R}^q$ such that $\pi(\tilde{r}) > \psi(\tilde{r})$. Let π' be the function defined by $\pi'(\tilde{r}) := \psi(\tilde{r})$, $\pi'(r) := \pi(r)$ for $r \neq \tilde{r}$. We will show that (π', ψ) is valid for M_f. Given $(\bar{x}, \bar{y}) \in M_f$, define

$$\tilde{x}_r := \begin{cases} 0 & \text{for } r = \tilde{r} \\ \bar{x}_r & \text{for } r \neq \tilde{r} \end{cases} \qquad \tilde{y}_r := \begin{cases} \bar{x}_r + \bar{y}_r & \text{for } r = \tilde{r} \\ \bar{y}_r & \text{for } r \neq \tilde{r}. \end{cases}$$

It is immediate to check that $(\tilde{x}, \tilde{y}) \in M_f$. It follows that $\sum_{r \in \mathbb{R}^q} \pi'(r) \bar{x}_r + \sum_{r \in \mathbb{R}^q} \psi(r) \bar{y}_r = \sum_{r \in \mathbb{R}^q} \pi(r) \tilde{x}_r + \sum_{r \in \mathbb{R}^q} \psi(r) \tilde{y}_r \geq 1$. This shows that (π', ψ) is a valid function, contradicting the minimality of (π, ψ). □

The next theorem, due to Johnson [215], provides a characterization of minimal valid functions, and it shows that in a minimal valid function (π, ψ) for M_f, the function ψ is uniquely determined by π.

Theorem 6.34. *Let (π, ψ) be a valid function for M_f. The function (π, ψ) is minimal for M_f if and only if π is a minimal valid function for G_f and ψ is defined by*

$$\psi(r) = \limsup_{\epsilon \to 0^+} \frac{\pi(\epsilon r)}{\epsilon} \qquad \text{for every } r \in \mathbb{R}^q. \qquad (6.27)$$

Proof. Using the same arguments as in points (a)–(d) of the proof of Theorem 6.22, one can show that, if (π, ψ) is minimal, then the function $\pi : \mathbb{R}^q \to \mathbb{R}$

6.3. INFINITE RELAXATIONS

is subadditive, periodic and satisfies the symmetry condition, and $\pi(0) = 0$. Thus, by Theorem 6.22, if (π, ψ) is a minimal valid function for M_f then π is a minimal valid function for G_f.

Therefore, we only need to show that, given a function (π, ψ) such that π is minimal for G_f, (π, ψ) is a minimal valid function for M_f if and only if ψ is defined by (6.27).

Let us define the function ψ' by

$$\psi'(r) := \limsup_{\epsilon \to 0^+} \frac{\pi(\epsilon r)}{\epsilon} \quad \text{for every } r \in \mathbb{R}^q.$$

We will show that ψ' is well defined, (π, ψ') is valid for M_f, and that $\psi' \leq \psi$. This will imply that (π, ψ) is minimal if and only if $\psi = \psi'$, and the statement will follow.

We now show that ψ' is well defined. This amounts to showing that the lim sup in (6.27) is always finite. We recall that

$$\limsup_{\epsilon \to 0^+} \frac{\pi(\epsilon r)}{\epsilon} := \lim_{\alpha \to 0^+} \sup\{\frac{\pi(\epsilon r)}{\epsilon} : 0 < \epsilon \leq \alpha\} = \inf_{\alpha > 0} \sup\{\frac{\pi(\epsilon r)}{\epsilon} : 0 < \epsilon \leq \alpha\}.$$

Let ψ'' be a function such that $\psi'' \leq \psi$ and (π, ψ'') is a minimal valid function for M_f (as mentioned at the beginning of Sect. 6.3, such a function exists). By Lemma 6.33, $\pi \leq \psi''$ and ψ'' is a sublinear function. Thus, for every $\epsilon > 0$ and every $r \in \mathbb{R}^q$, it follows that

$$\frac{\pi(\epsilon r)}{\epsilon} \leq \frac{\psi''(\epsilon r)}{\epsilon} = \psi''(r)$$

thus

$$\limsup_{\epsilon \to 0^+} \frac{\pi(\epsilon r)}{\epsilon} \leq \psi''(r).$$

This shows that ψ' is well defined and $\psi' \leq \psi'' \leq \psi$. Furthermore, it follows from the definition of ψ' and the definition of lim sup that ψ' is sublinear.

We conclude the proof by showing that (π, ψ') is valid for M_f. Let $(\bar{x}, \bar{y}) \in M_f$. Suppose by contradiction that

$$\sum_{r \in \mathbb{R}^q} \pi(r) \bar{x}_r + \sum_{r \in \mathbb{R}^q} \psi'(r) \bar{y}_r = 1 - \delta$$

where $\delta > 0$. Define $\bar{r} := \sum_{r \in \mathbb{R}^q} r \bar{y}_r$. By definition of ψ', it follows that, for some $\bar{\alpha} > 0$ sufficiently small,

$$\frac{\pi(\epsilon \bar{r})}{\epsilon} < \psi'(\bar{r}) + \delta \quad \text{for all } 0 < \epsilon \leq \bar{\alpha}. \tag{6.28}$$

Choose $D \in \mathbb{Z}$ such that $1/D \leq \bar{\alpha}$, and define, for all $r \in \mathbb{R}^q$,

$$\tilde{x}_r = \begin{cases} \bar{x}_r & r \neq \frac{\bar{r}}{D} \\ \bar{x}_r + D & r = \frac{\bar{r}}{D} \end{cases}$$

Note that all entries of $\tilde{x}$ are nonnegative integers and that $\sum_{r \in \mathbb{R}^q} r\tilde{x}_r = \sum_{r \in \mathbb{R}^q} r\bar{x}_r + \sum_{r \in \mathbb{R}^q} r\bar{y}_r$, thus $\tilde{x}$ is in G_f. Now

$$\begin{aligned}
\sum_{r \in \mathbb{R}^q} \pi(r)\tilde{x}_r &= \sum_{r \in \mathbb{R}^q} \pi(r)\bar{x}_r + \frac{\pi(\bar{r}/D)}{1/D} \\
&< \sum_{r \in \mathbb{R}^q} \pi(r)\bar{x}_r + \psi'(\bar{r}) + \delta \quad \text{(by (6.28) because } 1/D \leq \bar{\alpha}\text{)} \\
&\leq \sum_{r \in \mathbb{R}^q} \pi(r)\bar{x}_r + \sum_{r \in \mathbb{R}^q} \psi'(r)\bar{y}_r + \delta = 1, \quad \text{(by sublinearity of } \psi'\text{)}
\end{aligned}$$

contradicting the fact that π is valid for G_f. □

If (π, ψ) is a minimal valid function, then, by Theorem 6.34, π is a minimal valid function for G_f. However, the next example illustrates that in general it is not the case that ψ is a minimal valid function for R_f.

Example 6.35. Consider the three functions π, π_1, π_2 of Fig. 6.8, where $t > 0$ and $t + 1/2 < f < 1$. As discussed in Example 6.24, these functions are extreme for G_f. For ease of notation, let $\pi_0 := \pi$. For $i = 0, 1, 2$, let s_i^+ be the positive slope of π_i at 0 and s_i^- be the negative slope at 1 (or at 0, since the function is periodic). By Theorem 6.34, for each π_i, the function ψ_i for which (π_i, ψ_i) is minimal for M_f is the function defined by

$$\psi_i(r) := \begin{cases} s_i^+ r & \text{if } r \geq 0 \\ s_i^- r & \text{if } r < 0 \end{cases} \quad r \in \mathbb{R}.$$

The positive slopes are identical ($s_i^+ = (1-f)^{-1}$ for $i = 0, 1, 2$), while the most negative slope is $s_0^- = f^{-1}$, thus ψ_0 is pointwise smaller than the other two functions. In particular ψ_i is not minimal for R_f for $i = 1, 2$. Note that ψ_0 is minimal for R_f, since it is the gauge of the set $B - f$, where $B := [0, 1]$ is a maximal $\mathbb{Z}$-free set. ∎

6.3.4 Trivial and Unique Liftings

While Theorem 6.34 implies that minimal valid functions (π, ψ) are entirely determined by the function π, and that they are in one-to-one correspondence with the minimal valid functions for G_f, verifying that a function π is valid for G_f, let alone minimal, is a difficult task in general.

On the other hand, the function ψ has a nice geometric characterization. Indeed, for any minimal valid function (π, ψ) for M_f, Lemma 6.33 and Theorem 6.16 imply that the set $B := \{x \in \mathbb{R}^q : \psi(x - f) \leq 1\}$ is a

6.3. INFINITE RELAXATIONS

$\mathbb{Z}^q$-free convex set and ψ is the gauge of $B - f$. Conversely, Lemma 6.28 show that $\mathbb{Z}^q$-free convex sets define valid functions for R_f. Therefore, it may be desirable to start from a valid sublinear function ψ for R_f, and construct a function π such that (π, ψ) is valid for M_f. We say that any such function π is a *lifting* for ψ, and that π is a *minimal lifting* for ψ if there is no lifting π' for ψ such that $\pi' \leq \pi$, $\pi' \neq \pi$.

Note that, if we start from a minimal valid function ψ for R_f, then, for every minimal lifting π of ψ, the function (π, ψ) is a minimal valid function for M_f. Also, it follows from the definition that, given a valid function ψ for R_f and a minimal lifting π for ψ, there exists some function $\psi' \leq \psi$ such that (π, ψ') is a minimal valid function for M_f. In particular, by Theorem 6.34, every minimal lifting π for ψ is a minimal valid function for G_f. It follows from Lemma 6.33 that $\pi \leq \psi$ for every minimal lifting π of ψ. Moreover, since by Theorem 6.22 π is periodic over the unit hypercube, it must be the case that $\pi(r) \leq \psi(r + w)$ for all $r \in \mathbb{R}^q$ and every $w \in \mathbb{Z}^q$.

Remark 6.36. Let ψ be a valid function for R_f. Define the function $\bar{\pi}$ by

$$\bar{\pi}(r) := \inf_{w \in \mathbb{Z}^q} \psi(r + w) \qquad r \in \mathbb{R}^q. \tag{6.29}$$

Then $\pi \leq \bar{\pi}$ for every minimal lifting π of ψ. In particular, $\bar{\pi}$ is a lifting for ψ.

The function $\bar{\pi}$ defined in (6.29) is called the *trivial lifting* of ψ [31, 179].

Example 6.37. Let us consider the case $q = 1$. Assume that $0 < f < 1$, and Let $B = [0, 1]$. Let ψ be the gauge of $B - f$. As one can easily check,

$$\psi(r) = \max\left\{\frac{r}{1-f}, -\frac{r}{f}\right\}.$$

One can verify that the trivial lifting $\bar{\pi}$ for ψ is the following

$$\bar{\pi}(r) = \begin{cases} \frac{r - \lfloor r \rfloor}{1 - f} & \text{if } r - \lfloor r \rfloor \leq 1 - f \\ \frac{\lceil r \rceil - r}{f} & \text{if } r - \lfloor r \rfloor > 1 - f \end{cases}$$

Observe that ψ and $\bar{\pi}$ are the functions, given in (6.16), that define the Gomory mixed-integer inequalities.

It follows from the discussion in Example 6.23 that $\bar{\pi}$, in this case, is a minimal valid function for G_f, therefore $\bar{\pi}$ is in this case a minimal lifting. In particular, it follows from Remark 6.36 that, in this example, $\bar{\pi}$ is the unique minimal lifting of ψ. ∎

In general, the trivial lifting is not minimal. However, we can argue that, if we start from a minimal valid function ψ, there always exists an infinite region of $\mathbb{R}^q$ within which all minimal liftings of ψ coincide with the trivial lifting $\bar{\pi}$.

Lemma 6.38. *Let (π, ψ) be a minimal valid function for M_f. Given $r^* \in \mathbb{R}^q$, if*

$$\psi(r^*) + \psi(z - f - r^*) = \psi(z - f) = 1 \quad \text{for some } z \in \mathbb{Z}^q, \quad (6.30)$$

then $\pi(r^*) = \psi(r^*) = \inf_{w \in \mathbb{Z}^q} \psi(r^* + w)$.

Proof. Given $z \in \mathbb{Z}^q$, define

$$x_r := \begin{cases} 1 & \text{for } r = r^* \\ 0 & \text{for } r \neq r^* \end{cases} \qquad y_r := \begin{cases} 1 & \text{for } r = z - f - r^* \\ 0 & \text{for } r \neq z - f - r^* \end{cases}$$

It is straightforward to check that $(x, y) \in M_f$. Therefore we have

$$1 \leq \pi(r^*) + \psi(z - f - r^*) \leq \psi(r^*) + \psi(z - f - r^*) = \psi(z - f) = 1 \quad (6.31)$$

where the first inequality follows from the fact that $(x, y) \in M_f$ and that (π, ψ) is a valid function for M_f, the second inequality follows because $\pi(r^*) \leq \psi(r^*)$ by Lemma 6.33. Now (6.31) implies $\pi(r^*) = \psi(r^*)$. Finally, by Remark 6.36, $\pi(r^*) \leq \inf_{w \in \mathbb{Z}^q} \psi(r^* + w) \leq \psi(r^*)$, thus we have equality throughout. □

Given a minimal valid function ψ for R_f, if we let $R(\psi) := \{r \in \mathbb{R}^q : \psi(r^*) + \psi(\bar{z} - f - r) = \psi(\bar{z} - f) = 1 \text{ for some } \bar{z} \in \mathbb{Z}^q\}$, it follows from the above lemma that all minimal valid liftings coincide with the trivial lifting over the region $R(\psi) + \mathbb{Z}^q = \{r + w : r \in R^f, w \in \mathbb{Z}^q\}$. In particular, whenever $R(\psi) + \mathbb{Z}^q = \mathbb{R}^q$, the trivial lifting is the unique minimal lifting for ψ.

In [40] a converse of the above statement is proven. Namely, if $R(\psi) + \mathbb{Z}^q \subset \mathbb{R}^q$, then there exist more than one minimal lifting, so in particular the trivial lifting is not minimal.

Example 6.39. (Dey and Wolsey [118]) Let $q = 2$. Consider the maximal lattice-free triangle $B = \text{conv}(\binom{0}{0}, \binom{2}{0}, \binom{0}{2})$, and let f be a point in the interior of B (see Fig. 6.11). Let ψ be the gauge of $B - f$. By Theorem 6.30, ψ is a minimal valid function for R_f.

For each of the three points $z_1 = \binom{1}{0}$, $z_2 = \binom{0}{1}$, $z_3 = \binom{1}{1}$ on the boundary of B, we have that $\psi(z_i - f) = 1$. For $i = 1, 2, 3$, define $R(z_i) := \{r \in \mathbb{R}^2 :$

$\psi(r) + \psi(z_i - f - r) = \psi(z_i - f) = 1\}$, and let $R := R(z_1) \cup R(z_2) \cup R(z_3)$. Observe that $R \subseteq R(\psi)$. We will show that $R + \mathbb{Z}^q = \mathbb{R}^q$, so in this case the trivial lifting $\bar{\pi}$ is the unique minimal lifting of ψ.

Since B is a maximal lattice-free convex set, the function ψ is given by (6.26). Therefore the regions $R(z_1)$, $R(z_2)$, $R(z_3)$ are three quadrilaterals, namely they are obtained by translating the grey quadrilaterals depicted in Fig. 6.11 by $-f$.

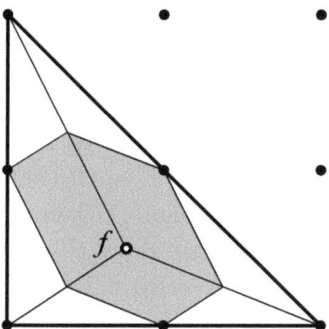

Figure 6.11: Lattice free triangle giving an inequality with a unique minimal lifting. The shaded region depicts $f + R$

For $r \in \mathbb{R}^2$, $r - \lfloor r \rfloor$ is in the unit box $[0,1] \times [0,1]$. Thus it suffices to show that every point in $[0,1] \times [0,1]$ can be translated by an integral vector into $f + R$. Note that $[0,1] \times [0,1] \setminus (f + R)$ is the union of the two triangles $\text{conv}(\binom{0}{0}, \binom{1}{0}, \frac{f}{2})$ and $\text{conv}(\binom{0}{0}, \binom{0}{1}, \frac{f}{2})$. The first one can be translated into $f + R$ by adding the vector $\binom{0}{1}$ and the second can be translated into $f + R$ by adding the vector $\binom{1}{0}$. The above argument shows that integral translations of R cover $\mathbb{R}^2$. Since the area of R is equal to 1, integral translations of R actually define a tiling of $\mathbb{R}^2$. This discussion implies that the trivial lifting can be computed efficiently. Indeed, for any $r \in \mathbb{R}^2$, it gives a construction for an integral vector $\bar{w}$ such that $r + \bar{w} \in R$ and, by Lemma 6.38, $\inf_{w \in \mathbb{Z}^2} \psi(r + w) = \psi(r + \bar{w})$. ∎

6.4 Further Readings

The material of this section follows mostly [80]. We refer the reader to the surveys of Del Pia and Weismantel [112] and Richard and Dey [315] for alternative accounts.

For pure integer programming problems, Gomory [178] gives an algorithm to optimize linear functions over the corner polyhedron. Gomory's algorithm consists of computing a shortest path in a suitably constructed network. If the constraint matrix A is integral and we let $D := \det(A_B)$, where A_B is the matrix formed by the columns of A indexed by the basis B, the running time of the algorithm is $O(|N|D^2)$.

The notion of intersection cut was developed by Balas in the early 1970s [22, 23].

Gomory and Johnson [180] also define a notion of facet for the infinite group problem (6.22). Intuitively, a facet is a valid inequality whose contact with G_f is maximal. For any valid function π, let $P(\pi)$ denote the set of points in G_f that satisfy $\sum_{r \in \mathbb{Q}^q} \pi(r)x_r = 1$. Function π defines a facet of G_f if there is no other valid function π^* such that $P(\pi) \subseteq P(\pi^*)$.

In earlier work, Gomory and Johnson [179] emphasized extreme functions rather than facets. It is not hard to show that, if a valid function defines a facet, then it is extreme (see Exercise 6.21). Gomory and Johnson [180] prove the following theorem, where $E(\pi)$ denotes the set introduced before Lemma 6.25.

Theorem 6.40 (Facet Theorem). *Let π be a minimal valid function. If the set of equalities $E(\pi)$ has no other solution than π itself, then π defines a facet of G_f.*

Basu, Hildebrand, and Köppe [44] give a multi-dimensional version of the interval lemma and prove properties on extreme functions in this setting. Basu, Hildebrand, Köppe, and Molinaro [43] extend the 2-slope theorem (Theorem 6.27) to multi-dimensional functions $\mathbb{R}^q \to \mathbb{R}$.

The continuous infinite relaxation (6.24) was introduced by Andersen, Louveaux, Weismantel and Wolsey [12] for the case of two equations (i.e., $q = 2$). Theorem 6.30 is due to Borozan and Cornuéjols [63].

Results on the corner set (6.3) have been extended to the case where the vector determined by the basic components is required to belong to a set $S := \mathbb{Z} \cap P$, where P is a rational polyhedron. This allows one to retain the nonnegativity constraints on the basic variables, for example. Johnson [216] discusses the case where P is a polytope. Similarly, Basu et al. [42] consider the extension of (6.24) where $f + \sum_r ry_r \in S$ in the case that P is a rational polyhedron, showing a one-to-one correspondence between minimal valid functions and maximal S-free convex sets, and that every maximal S-free convex set is a polyhedron. Dey and Wolsey [119] prove a similar result, while Conforti, Cornuéjols, Daniilidis, Lemaréchal, and Malick [76] use tools of convex analysis to present a formal theory of cut-generating functions and

6.5. EXERCISES

S-free convex sets. Dey and Morán [116] prove that, if $S = C \cap \mathbb{Z}^d$ where $C \subseteq \mathbb{R}^d$ is a convex set, then every maximal S-free convex set is a polyhedron with at most 2^d facets. This has applications in the theory of cutting planes for nonlinear integer programming. Averkov [18] gives conditions on S that assure that maximal S-free sets are polyhedra.

Dey and Wolsey [118] study liftings of minimal valid functions for the infinite continuous relaxation in the two dimensional case, while [77] and [40] consider the multi-dimensional case. Averkov and Basu [19] show that the existence of a unique lifting is a property of the underlying maximal lattice-free convex set B but is independent of the position of f in the interior of B.

Dash, Dey, and Günlük [106] study the relation between lattice-free convex sets and asymmetric disjunctions.

6.5 Exercises

Exercise 6.1. Show that the Gomory mixed integer inequality generated from a tableau row (6.2) is valid for corner(B).

Exercise 6.2. Give a minimal system of linear inequalities describing the corner polyhedron for the integer program of Example 6.3 when the choice of basic variables is x_1, x_2, x_3, x_5, x_7.

Exercise 6.3. Reformulate (6.6) in the modular form (6.7). Show that the solution set of this modular problem is contained in the union of the simplices $S_k := \{(x_6, x_7, x_8) \in \mathbb{R}^3_+ : x_6 + x_7 + x_8 = 2k\}$ where k is a positive integer. Show that the modular problem admits solutions (x_6, x_7, x_8) of the form $(2k, 0, 0)$, $(0, 2k, 0)$ and $(0, 0, 2k)$ for every positive integer k. Describe the corner polyhedron corner(B) in the space of the variables x_6, x_7, x_8. Deduce a description of the corner polyhedron corner(B) in the space of the variables x_1, x_2, x_3.

Exercise 6.4. Consider the problem

$$\begin{array}{ll} \min \sum_{j \in N} c_j x_j & \\ \sum_{j \in N} a_{ij} x_j \equiv b_i \mod 1 & \text{for } i \in B \\ x_j \in \mathbb{Z}_+ & \text{for } j \in N. \end{array}$$

where $c_j \geq 0$, a_{ij} and b_i are rational data, for $i \in B$ and $j \in N$.

(a) Show that there can only be a finite number of different sums $\sum_{j \in N} a_{ij} x_j$ in the above integer program when the sums are computed modulo 1.

(b) Use (a) to formulate the above integer program as a shortest path problem from a source node v_0 to a sink node v_b in an appropriate directed graph. Describe clearly the nodes and arcs in your graph.

Exercise 6.5. Consider a pure integer linear program

$$\begin{aligned} \max \quad & cx \\ & Ax = b \\ & x_j \in \mathbb{Z}_+ \quad \text{for } j = 1, \ldots, n \end{aligned}$$

where we assume that A has full row rank m. Let B be a dual feasible basis, that is $\bar{c}_N := c_B A_B^{-1} A_N - c_N \geq 0$ when $c = (c_B, c_N)$ and $A = (A_B, A_N)$ are partitioned into their basic and nonbasic parts. This exercise proposes a sufficient condition under which any optimal solution of the Gomory relaxation

$$\begin{aligned} \min \quad & \sum_{j \in N} \bar{c}_j x_j \\ x_i &= \bar{b}_i - \sum_{j \in N} \bar{a}_{ij} x_j \quad \text{for } i \in B \\ x_i &\in \mathbb{Z} \quad \text{for } i \in B \\ x_j &\in \mathbb{Z}_+ \quad \text{for } j \in N \end{aligned}$$

is a nonnegative vector and therefore solves the original integer program. Let D denote the absolute value of the determinant of A_B. Let $m_i := \max_{j \in N} \bar{a}_{ij}$. Prove that, if $\bar{b}_i \geq (D-1)m_i$ for all $i \in B$, then every optimal solution of the Gomory relaxation has the property that $x_i \geq 0$ for $i \in B$.

Exercise 6.6. Consider a closed convex set $C \subseteq \mathbb{R}^n$ whose interior contains $\bar{x}$ but no point of $\mathbb{Z}^p \times \mathbb{R}^{n-p}$. Let Q be the closed convex hull of $P(B) \setminus \text{int}(C)$, where $\text{int}(C)$ denotes the interior of C. Show that Q is the set of points in $P(B)$ satisfying the intersection cut defined by C.

Exercise 6.7. Let $K \subseteq \mathbb{R}^n$ be a closed convex set with the origin in its interior. Prove that the gauge function satisfies $\gamma_K(r) \leq 1$ if and only if $r \in K$.

Exercise 6.8. Prove that the following functions $g : \mathbb{R} \to \mathbb{R}$ are subadditive.

(a) $g(x) := \lceil x \rceil$

(b) $g(x) := x \mod t$, where t is a given positive integer.

Exercise 6.9.

(a) Show that, if $g : \mathbb{R}^n \to \mathbb{R}$ is a subadditive function, $g(0) \geq 0$.

(b) Show that, if $f, g : \mathbb{R}^n \to \mathbb{R}$ are two subadditive functions, $\max(f, g)$ is also a subadditive function.

6.5. EXERCISES

Exercise 6.10. Let $g : \mathbb{R}^m \to \mathbb{R}$ be a nondecreasing subadditive function such that $g(0) = 0$. (Function g is *nondecreasing* if for any $a, a' \in \mathbb{R}^m$ such that $a \leq a'$, we have $g(a) \leq g(a')$.)

Let $S := P \cap \mathbb{Z}^n$ where $P := \{x \in \mathbb{R}^n_+ : Ax \geq b\}$. We denote by a^j the jth column of the $m \times n$ matrix A. Prove that $\sum_{j=1}^n g(a^j) x_j \geq g(b)$ is a valid inequality for $\text{conv}(S)$.

Exercise 6.11. Let T be a maximal lattice-free convex set in $\mathbb{R}^2$ which is a triangle. Show that T satisfies one of the following.

(a) All vertices of T are integral points and T contains exactly one integral point in the relative interior of each facet.

(b) At least one vertex of T is not an integral point and the opposite facet contains at least two integral points in its relative interior.

(c) T contains exactly three integral points, one in the relative interior of each facet.

Exercise 6.12. Let Q be a maximal lattice-free convex set in $\mathbb{R}^2$ which is a quadrilateral.

(a) Show that Q contains *exactly* four integral points on its boundary.

(b) Show that these four integral points are the vertices of a parallelogram of area one.

Exercise 6.13. Let T be a maximal lattice-free triangle in $\mathbb{R}^2$ with the property that all vertices of T are integral points and T contains exactly one integral point in the relative interior of each facet. Prove that there exists a transformation $\phi : \mathbb{R}^2 \to \mathbb{R}^2$ of the form $\phi(x) = c + Mx$ where $c \in \mathbb{Z}^2$ and $M \in \mathbb{Z}^{2\times 2}$ is a unimodular matrix such that $\phi(T)$ is the triangle with vertices $(0,0)$, $(2,0)$ and $(0,2)$.

Exercise 6.14. A convex set in $\mathbb{R}^2$ is $\mathbb{Z}^2_+$-free if it contains no point of $\mathbb{Z}^2_+$ in its interior. Characterize the maximal $\mathbb{Z}^2_+$-free convex sets that are not $\mathbb{Z}^2$-free convex sets.

Exercise 6.15. Let $A \in \mathbb{Q}^{m \times n}$ and $b \in \mathbb{Q}^m$, such that the system $Ax \leq b$ has no integral solution but any system obtained from it by removing one of the m inequalities has an integral solution. Show that $m \leq 2^n$.

Exercise 6.16. In $\mathbb{R}^p$, let $f_i = \frac{1}{2}$ for $i = 1, \ldots, p$. Define the octahedron Ω_f centered at f with vertices $f \pm \frac{p}{2} e^i$, where e^i denotes the ith unit vector. Ω_f has 2^p facets, each of which contains a 0,1 point in its center.

(a) Show that the intersection cut $\sum \pi(r) y_r \geq 1$ obtained from the octahedron Ω_f is obtained from the function $\pi(r) := \frac{2}{p}(|r_1| + \ldots + |r_p|)$.

(b) Show that the above intersection cut from the octahedron is implied by p split inequalities.

Exercise 6.17. Consider a continuous nonnegative periodic function $\pi : \mathbb{R} \to \mathbb{R}_+$ that is piecewise-linear in the interval $[0, 1]$ and satisfies $\pi(0) = 0$.

(a) Show that, in order to check whether the symmetry condition $\pi(r) + \pi(-f - r) = 1$ holds for all $r \in \mathbb{R}$, it suffices to check it at the breakpoints of the function in the interval $[0, 1]$.

(b) Assume that, in addition to the above properties, π satisfies the symmetry condition. Show that, in order to check whether subadditivity $\pi(a) + \pi(b) \geq \pi(a + b)$ holds for all $a, b \in \mathbb{R}$, it is enough to check the inequality $\pi(a) + \pi(b) \geq \pi(a + b)$ at all the breakpoints a, b in the interval $[0, 1]$ where the function is locally convex.

Exercise 6.18. Assume $t > 0$ and $1/2 + t \leq f < 1$. Show that the function π_2 in Example 6.23 is extreme.

Exercise 6.19. Consider the model

$$f + \sum_{r \in \mathbb{R}^q} r y_r \in \mathbb{Z}_+^q$$

$$y_r \geq 0 \quad \text{for all } r \in \mathbb{R}^q$$

$$y \quad \text{has a finite support}$$

A convex set in $\mathbb{R}^q$ is $\mathbb{Z}_+^q$-free if it contains no point of $\mathbb{Z}_+^q$ in its interior. A function $\psi : \mathbb{R}^q \to \mathbb{R}$ is *valid* for the above model if the inequality $\sum_{r \in \mathbb{R}^q} \psi(r) y_r \geq 1$ is satisfied by every feasible vector y.

(a) Show that if $\psi : \mathbb{R}^q \to \mathbb{R}$ is sublinear and the set $B_\psi := \{x \in \mathbb{R}^q : \psi(x - f) \leq 1\}$ is $\mathbb{Z}_+^q$-free, then ψ is valid.

(b) Let $q = 2$ and $f = \binom{1/4}{1/2}$. Show that the function $\psi : \mathbb{R}^2 \to \mathbb{R}$ defined by $\psi(r) = \max\{4r_1 + 4r_2, 4r_1 - 4r_2\}$ ($r \in \mathbb{R}^2$) is valid, by showing that ψ is sublinear and B_ψ $\mathbb{Z}_+^q$-free.

Exercise 6.20. Let $q = 2$. Consider the triangle K with vertices $(-\frac{1}{2}, 0)$, $(\frac{3}{2}, 0)$, $(\frac{1}{2}, 2)$ and the point $f = (\frac{1}{2}, \frac{1}{2})$.

6.5. EXERCISES

(a) Show that K is a maximal lattice-free convex set;

(b) Compute the function ψ_K given by (6.26);

(c) Let π_K be any minimal lifting of ψ_K. Determine the region $R := \{r \in \mathbb{R}^2 : \pi_K(r) = \psi_K(r)\}$;

(d) Show that ψ_K has a unique minimal lifting π_K (Hint: Show that $R+\mathbb{Z}^2$ covers the plane).

Exercise 6.21. Show that if a valid function defines a facet of G_f (according to the definition in Sect. 6.4), then it is extreme.

Exercise 6.22. Let $\psi : \mathbb{R}^q \to \mathbb{R}$ be a minimal valid function for (6.24). Given $d \in \mathbb{R}^q$, consider the model

$$f + \sum_{r \in \mathbb{R}^q} r y_r + dz \in \mathbb{Z}^q$$

$$\begin{array}{ll} y_r \geq 0 & \text{for all } r \in \mathbb{R}^q \\ y & \text{has a finite support} \\ z \in \mathbb{Z}_+. \end{array} \qquad (6.32)$$

Let $\pi_\ell(d)$ be the minimum scalar λ such that the inequality $\sum_{r \in \mathbb{R}^q} \psi(r) y_r + \lambda z \geq 1$ is valid for (6.32). Prove that when (π_ℓ, ψ) is valid for (6.20), then π_ℓ is the unique minimal lifting of ψ.

Michele Conforti riding his bicycle

Chapter 7
Valid Inequalities for Structured Integer Programs

In Chaps. 5 and 6 we have introduced several classes of valid inequalities that can be used to strengthen integer programming formulations in a cutting plane scheme. All these valid inequalities are "general purpose," in the sense that their derivation does not take into consideration the structure of the specific problem at hand. Many integer programs have an underlying combinatorial structure, which can be exploited to derive "strong" valid inequalities, where the term "strong" typically refers to the fact that the inequality is facet-defining for the convex hull of feasible solutions.

In this chapter we will present several examples. We will introduce the *cover* and *flow cover* inequalities, which are valid whenever the constraints exhibit certain combinatorial structures that often arise in integer programming. We will introduce lifting, which is a procedure for generating facet-defining inequalities starting from lower-dimensional faces, and a particularly attractive variant known as sequence-independent lifting. When applied to the above inequalities, we obtain lifted cover inequalities and lifted flow cover inequalities, which are standard features of current branch-and-cut solvers. We also discuss the traveling salesman problem, for which the polyhedral approach has produced spectacular results. Finally we present the equivalence between separation and optimization.

7.1 Cover Inequalities for the 0,1 Knapsack Problem

Consider the 0,1 knapsack set

$$K := \left\{ x \in \{0,1\}^n : \sum_{j=1}^n a_j x_j \leq b \right\}$$

where $b > 0$ and $a_j > 0$ for $j \in N := \{1, \ldots, n\}$.

Recall from Example 3.19 that the dimension of $\mathrm{conv}(K)$ is $n - |J|$ where $J = \{j \in N : a_j > b\}$. In the remainder, we assume that $a_j \leq b$ for all $j \in N$, so that $\mathrm{conv}(K)$ has dimension n.

In Sect. 2.2 we introduced the concept of minimal covers. Recall that a *cover* is a subset $C \subseteq N$ such that $\sum_{j \in C} a_j > b$ and it is *minimal* if $\sum_{j \in C \setminus \{k\}} a_j \leq b$ for all $k \in C$. For any cover C, the *cover inequality* associated with C is

$$\sum_{j \in C} x_j \leq |C| - 1,$$

and it is valid for $\mathrm{conv}(K)$.

Proposition 7.1. *Let C be a cover for K. The cover inequality associated with C is facet-defining for $P_C := \mathrm{conv}(K) \cap \{x \in \mathbb{R}^n : x_j = 0, \ j \in N \setminus C\}$ if and only if C is a minimal cover.*

Proof. Note that $\dim(P_C) = |C|$. Assume C is a minimal cover. For all $j \in C$, let x^j be the point defined by $x_i^j = 1$ for all $i \in C \setminus \{j\}$ and $x_i^j = 0$ for all $i \in (N \setminus C) \cup \{j\}$. These are $|C|$ affinely independent points in P_C that satisfy the cover inequality associated with C at equality. This shows that the cover inequality associated with C is a facet of P_C.

Conversely, suppose that C is not a minimal cover, and let $C' \subset C$ be a cover contained in C. The cover inequality associated with C is the sum of the cover inequality associated with C' and the inequalities $x_j \leq 1$, $j \in C \setminus C'$. Since these inequalities are valid for P_C, the cover inequality associated with C is not facet-defining for P_C. □

Proposition 7.1 shows that minimal cover inequalities define facets of $\mathrm{conv}(K) \cap \{x \in \mathbb{R}^n : x_j = 0, \ j \in N \setminus C\}$. In the next section we will discuss the following problem: given a minimal cover C, how can one compute coefficients α_j, $j \in N \setminus C$, so that the inequality $\sum_{j \in C} x_j + \sum_{j \in N \setminus C} \alpha_j x_j \leq |C| - 1$ is facet-defining for $\mathrm{conv}(K)$?

Separation

To use cover inequalities in a cutting plane scheme, one is faced with the *separation problem*, that is, given a vector $\bar{x} \in [0,1]^n$, find a cover inequality for K that is violated by $\bar{x}$, or show that none exists. Note that a cover inequality relative to C is violated by $\bar{x}$ if and only if $\sum_{j \in C}(1 - \bar{x}_j) < 1$. Thus, deciding whether a violated cover inequality exists reduces to solving the problem

$$\zeta = \min\{\sum_{j \in C}(1 - \bar{x}_j) \,:\, C \text{ is a cover for } K\}. \tag{7.1}$$

If $\zeta \geq 1$, then $\bar{x}$ satisfies all the cover inequalities for K. If $\zeta < 1$, then an optimal cover for (7.1) yields a violated cover inequality. Note that (7.1) always has an optimal solution that is a minimal cover.

Assuming that $a_1, \ldots, a_n$ and b are integer, problem (7.1) can be formulated as the following integer program

$$\begin{array}{rl} \zeta = & \min \sum_{j=1}^n (1 - \bar{x}_j) z_j \\ & \sum_{j=1}^n a_j z_j \geq b + 1 \\ & z \in \{0,1\}^n. \end{array} \tag{7.2}$$

It is worth noting that the separation problem (7.1) is NP-hard in general [239]. In practice one is interested in fast heuristics to detect violated cover inequalities. A simple example is the following: find a basic optimal solution z^* of the linear programming relaxation of (7.2) (see Exercise 3.3); if the optimal objective value of the linear programming relaxation is ≥ 1, then also $\zeta \geq 1$ and there is no violated cover inequality. Otherwise (observing that z^* has at most one fractional coordinate) output the cover $C := \{j \in N \,:\, z_j^* > 0\}$. Note that this heuristic does not guarantee that the inequality associated with C cuts off the fractional point $\bar{x}$, even if there exists a cover inequality cutting off $\bar{x}$.

7.2 Lifting

Consider a mixed integer set $S := \{x \in \mathbb{Z}_+^n \times \mathbb{R}_+^p \,:\, Ax \leq b\}$. Given a subset C of $N := \{1, \ldots, n+p\}$, and a valid inequality $\sum_{j \in C} \alpha_j x_j \leq \beta$ for $\text{conv}(S) \cap \{x \in \mathbb{R}^{n+p} \,:\, x_j = 0, j \in N \setminus C\}$, an inequality $\sum_{j=1}^{n+p} \alpha_j x_j \leq \beta$ is called a *lifting* of $\sum_{j \in C} \alpha_j x_j \leq \beta$ if it is valid for $\text{conv}(S)$. In the remainder of this section we will focus on the case where $S \subseteq \{0,1\}^n$.

Proposition 7.2. *Consider a set $S \subseteq \{0,1\}^n$ such that $S \cap \{x : x_n = 1\} \neq \emptyset$, and let $\sum_{i=1}^{n-1} \alpha_i x_i \leq \beta$ be a valid inequality for $S \cap \{x : x_n = 0\}$. Then*

$$\alpha_n := \beta - \max \left\{ \sum_{i=1}^{n-1} \alpha_i x_i : x \in S, \, x_n = 1 \right\} \tag{7.3}$$

is the largest coefficient such that $\sum_{i=1}^{n-1} \alpha_i x_i + \alpha_n x_n \leq \beta$ is valid for S.

Furthermore, if $\sum_{i=1}^{n-1} \alpha_i x_i \leq \beta$ defines a d-dimensional face of $\mathrm{conv}(S) \cap \{x_n = 0\}$, then $\sum_{i=1}^{n} \alpha_i x_i \leq \beta$ defines a face of $\mathrm{conv}(S)$ of dimension at least $d+1$.

Proof. The inequality $\sum_{i=1}^{n} \alpha_i x_i \leq \beta$ is valid for $S \cap \{x : x_n = 0\}$ by assumption, and it is valid for $S \cap \{x : x_n = 1\}$ by definition of α_n. Thus $\sum_{i=1}^{n} \alpha_i x_i \leq \beta$ is valid for S, and α_n is the largest coefficient with such property.

Consider $d+1$ affinely independent points of $\mathrm{conv}(S) \cap \{x_n = 0\}$ satisfying $\sum_{i=1}^{n-1} \alpha_i x_i \leq \beta$ at equality. These points also satisfy $\sum_{i=1}^{n} \alpha_i x_i \leq \beta$ at equality. Any point $\bar{x} \in S$ with $\bar{x}_n = 1$ achieving the maximum in (7.3) gives one more point satisfying $\sum_{i=1}^{n} \alpha_i x_i \leq \beta$ at equality, and it is affinely independent of the previous ones since it satisfies $x_n = 1$. Thus $\sum_{i=1}^{n} \alpha_i x_i \leq \beta$ defines a face of $\mathrm{conv}(S)$ of dimension at least $d+1$. □

Sequential Lifting. Consider a set $S := \{x \in \{0,1\}^n : Ax \leq b\}$ of dimension n, where A is a nonnegative matrix. Proposition 7.2 suggests the following way of lifting a facet-defining inequality $\sum_{j \in C} \alpha_j x_j \leq \beta$ of $\mathrm{conv}(S) \cap \{x : x_j = 0, \, j \in N \setminus C\}$ into a facet-defining inequality $\sum_{j=1}^{n} \alpha_j x_j \leq \beta$ of $\mathrm{conv}(S)$.

> Choose an ordering $j_1, \ldots, j_\ell$ of the indices in $N \setminus C$. Let $C_0 = C$ and $C_h = C_{h-1} \cup \{j_h\}$ for $h = 1, \ldots, \ell$.
>
> For $h = 1$ up to $h = \ell$, compute
>
> $$\alpha_{j_h} := \beta - \max \left\{ \sum_{j \in C_{h-1}} \alpha_j x_j : x \in S, \, x_j = 0, \, j \in N \setminus C_h, \, x_{j_h} = 1 \right\}. \tag{7.4}$$

By Proposition 7.2 the inequality $\sum_{j=1}^{n} \alpha_j x_j \leq \beta$ obtained this way is facet-defining for $\mathrm{conv}(S)$.

The recursive procedure outlined above is called *sequential lifting*. Note that the assumption that $A \geq 0$ implies that, for every $\bar{x} \in S$, $\{x \in \{0,1\}^n :$

7.2. LIFTING

$x \leq \bar{x}\} \subseteq S$. This and the fact that $\dim(S) = n$ guarantee that (7.4) is feasible. We remark that different orderings of $N \setminus C$ may produce different lifted inequalities. Furthermore, not all possible liftings can be derived from the above procedure, as the next example illustrates.

Example 7.3. Consider the 0,1 knapsack set

$$8x_1 + 7x_2 + 6x_3 + 4x_4 + 6x_5 + 6x_6 + 6x_7 \leq 22$$
$$x_j \in \{0,1\} \quad \text{for } j = 1, \ldots, 7.$$

The index set $C := \{1, 2, 3, 4\}$ is a minimal cover. The corresponding minimal cover inequality is $x_1 + x_2 + x_3 + x_4 \leq 3$.

We perform sequential lifting according to the order $5, 6, 7$. According to Proposition 7.2, the largest lifting coefficient for x_5 is

$$\alpha_5 = 3 - \max\{x_1 + x_2 + x_3 + x_4 : 8x_1 + 7x_2 + 6x_3 + 4x_4 \leq 22 - 6, \, x_1, x_2, x_3, x_4 \in \{0,1\}\}.$$

It is easily verified that $\alpha_5 = 1$. The lifting coefficient of x_6 is

$$\alpha_6 = 3 - \max\{x_1 + x_2 + x_3 + x_4 + x_5 : 8x_1 + 7x_2 + 6x_3 + 4x_4 + 6x_5 \leq 16, \, x_1, \ldots, x_5 \in \{0,1\}\}.$$

It follows that $\alpha_6 = 0$. Similarly $\alpha_7 = 0$. This sequence yields the inequality $x_1 + x_2 + x_3 + x_4 + x_5 \leq 3$. By symmetry, the sequences $6, 5, 7$ and $7, 5, 6$ yield the inequalities $x_1 + x_2 + x_3 + x_4 + x_6 \leq 3$ and $x_1 + x_2 + x_3 + x_4 + x_7 \leq 3$, respectively. By Propositions 7.1 and 7.2, all these inequalities are facet-defining.

Not all possible facet-defining lifted inequalities can be obtained sequentially. As an example, consider the following lifted inequality:

$$x_1 + x_2 + x_3 + x_4 + 0.5x_5 + 0.5x_6 + 0.5x_7 \leq 3.$$

We leave it to the reader to show that the inequality is valid and facet-defining for the knapsack set. However, it cannot be obtained by sequential lifting since its lifting coefficients are fractional. ∎

7.2.1 Lifting Minimal Cover Inequalities

The following theorem was proved by Balas [25].

Theorem 7.4. *Let $K := \{x \in \{0,1\}^n : \sum_{j=1}^n a_j x_j \leq b\}$, where $b \geq a_j > 0$ for all $j \in N$. Let C be a minimal cover for K, and let*

$$\sum_{j \in C} x_j + \sum_{j \in N \setminus C} \alpha_j x_j \leq |C| - 1 \tag{7.5}$$

be a lifting of the cover inequality associated with C. Up to permuting the indices, assume that $C = \{1, \ldots, t\}$ and $a_1 \geq a_2 \geq \ldots \geq a_t$. Let $\mu_0 := 0$ and $\mu_h := \sum_{\ell=1}^h a_\ell$ for $h = 1, \ldots, t$. Let $\lambda := \mu_t - b$ (note that $\lambda > 0$).

If (7.5) defines a facet of $\mathrm{conv}(K)$, then the following hold for every $j \in N \setminus C$.

(i) If, for some h, $\mu_h \leq a_j \leq \mu_{h+1} - \lambda$, then $\alpha_j = h$.

(ii) If, for some h, $\mu_{h+1} - \lambda < a_j < \mu_{h+1}$, then $h \leq \alpha_j \leq h+1$.

Furthermore, for every $j \in N \setminus C$, if $\mu_{h+1} - \lambda < a_j < \mu_{h+1}$, then there exists a facet-defining inequality of the form (7.5) such that $\alpha_j = h+1$.

Proof. Assume that (7.5) is facet-defining for $\mathrm{conv}(K)$ and let $j \in N \setminus C$. Since $0 < a_j \leq b < \mu_t$, there exists an index h, $0 \leq h \leq t-1$, such that $\mu_h \leq a_j < \mu_{h+1}$.

By Proposition 7.2, $\alpha_j \leq |C| - 1 - \theta$, where

$$\theta := \max\left\{\sum_{i=1}^t x_i : \sum_{i=1}^t a_i x_i \leq b - a_j, \ x \in \{0,1\}^t\right\}.$$

Observe that, since $a_1 \geq a_2 \geq \ldots \geq a_t$, $\theta = |C| - k + 1$, where k is the smallest index such that $\sum_{\ell=k}^t a_\ell \leq b - a_j$. Therefore $\alpha_j \leq k - 2$.

Since $\sum_{\ell=k}^t a_\ell = \mu_t - \mu_{k-1} = b + \lambda - \mu_{k-1}$, it follows that k is the smallest index such that $a_j \leq \mu_{k-1} - \lambda$. Therefore k is the index such that $\mu_{k-2} - \lambda < a_j \leq \mu_{k-1} - \lambda$.

It follows that

$$\alpha_j \leq k - 2 = \begin{cases} h & \text{when } \mu_h \leq a_j \leq \mu_{h+1} - \lambda \\ h+1 & \text{when } \mu_{h+1} - \lambda < a_j < \mu_{h+1}. \end{cases} \tag{7.6}$$

Next we show that $\alpha_j \geq h$. We apply Proposition 7.2 to the inequality

$$\sum_{i \in C} x_i + \sum_{i \in N \setminus (C \cup \{j\})} \alpha_i x_i \leq |C| - 1. \tag{7.7}$$

Since (7.5) is facet-defining, it follows that

$$\alpha_j = |C| - 1 - \max\left\{\sum_{i \in C} x_i + \sum_{i \in N \setminus (C \cup \{j\})} \alpha_i x_i : \sum_{i \in N \setminus \{j\}} a_i x_i \leq b - a_j, \ x \in \{0,1\}^{N \setminus \{j\}}\right\}. \tag{7.8}$$

7.2. LIFTING

Observe that, since $a_1 \geq a_2 \geq \ldots \geq a_t$, (7.8) admits an optimal solution $x^* \in \{0,1\}^{N \setminus \{j\}}$ such that $x_1^* \leq x_2^* \leq \ldots \leq x_t^*$. Since $\sum_{\ell=h+1}^{t} a_\ell = \mu_t - \mu_h \geq b + \lambda - a_j > b - a_j$, we have that $x_\ell^* = 0$ for some $\ell \in \{h+1, \ldots, t\}$. It follows that $x_1^* = \ldots = x_h^* = 0$. Let $\bar{x}$ be the vector in $\{0,1\}^n$ defined by $\bar{x}_j = 0$, $\bar{x}_i = 1$ for $i = 1, \ldots, h$, and $\bar{x}_i = x_i^*$ otherwise. We have that $\bar{x} \in K$ because $\sum_{i \in N} a_i \bar{x}_i = \sum_{i \in N \setminus \{j\}} a_i x_i^* + \mu_h \leq b - a_j + \mu_h \leq b$. Since (7.5) is valid for K, it follows that $\sum_{i \in C} \bar{x}_i + \sum_{i \in N \setminus C} \alpha_i \bar{x}_i \leq |C| - 1$. Therefore

$$\alpha_j = |C| - 1 - \left(\sum_{i \in C} x_i^* + \sum_{i \in N \setminus (C \cup \{j\})} \alpha_i x_i^* \right) \geq \sum_{i \in C} (\bar{x}_i - x_i^*) + \sum_{i \in N \setminus (C \cup \{j\})} \alpha_i (\bar{x}_i - x_i^*) = h.$$

This proves (i) and (ii). We prove the last statement of the theorem. Assume $a_j > \mu_{h+1} - \lambda$. If we do sequential lifting in which we lift x_j first, it follows from the proof of (7.6) that the coefficient of x_j in the resulting inequality is $h + 1$. By Proposition 7.2 this inequality is facet-defining for $\mathrm{conv}(K)$. □

Remark 7.5. Let K and C be as in Theorem 7.4. For every $j \in N \setminus C$, let $h(j)$ be the index such that $\mu_{h(j)} \leq a_j < \mu_{h(j)+1}$. The inequality $\sum_{j \in C} x_j + \sum_{j \in N \setminus C} h(j) x_j \leq |C| - 1$ is a lifting of the minimal cover inequality associated with C. Furthermore, if $a_j \leq \mu_{h(j)+1} - \lambda$ for all $j \in N \setminus C$, then the above is the unique facet-defining lifting.

Example 7.6. We illustrate the above theorem on the knapsack set

$$K := \{x \in \{0,1\}^5 \ : \ 5x_1 + 4x_2 + 3x_3 + 2x_4 + x_5 \leq 5\}.$$

The set $C := \{3,4,5\}$ is a minimal cover. We would like to lift the inequality $x_3 + x_4 + x_5 \leq 2$ into a facet of $\mathrm{conv}(K)$. We have $\mu_0 = 0$, $\mu_1 = 3$, $\mu_2 = 5$, $\mu_3 = 6$ and $\lambda = 1$. Therefore $\alpha_1 = 2$ since $\mu_2 \leq a_1 \leq \mu_3 - \lambda$. Similarly $\alpha_2 = 1$ since $\mu_1 \leq a_2 \leq \mu_2 - \lambda$. By Theorem 7.4, the inequality $2x_1 + x_2 + x_3 + x_4 + x_5 \leq 2$ defines a facet of $\mathrm{conv}(K)$. Furthermore, by Remark 7.5, this is the unique facet-defining lifting. ■

7.2.2 Lifting Functions, Superadditivity, and Sequence Independent Lifting

Let $S := \{x \in \{0,1\}^n \ : \ Ax \leq b\}$, where we assume that $A \geq 0$ and $\dim(S) = n$. Therefore $b \geq 0$. Let $C \subset N := \{1, \ldots, n\}$, and let $\sum_{j \in C} \alpha_j x_j \leq \beta$ be a valid inequality for $S \cap \{x \ : \ x_j = 0 \text{ for } j \in N \setminus C\}$.

Consider any lifting of the above inequality,

$$\sum_{j=1}^{n} \alpha_j x_j \leq \beta. \tag{7.9}$$

Let a^j denote the jth column of A. By Proposition 7.2, for all $j \in N\setminus C$, inequality (7.9) must satisfy $\alpha_j \leq f(a^j)$ (because $A \geq 0$ and $\{x \in S : x_j = 0\} \neq \emptyset$), where $f : [0,b] \to \mathbb{R}$ is the function defined by

$$f(z) := \beta - \max \sum_{i \in C} \alpha_i x_i$$

$$\sum_{i \in C} a^i x_i \leq b - z \tag{7.10}$$

$$x_i \in \{0,1\} \text{ for } i \in C.$$

The function $f : [0,b] \to \mathbb{R}$ is the *lifting function* of the inequality $\sum_{j \in C} \alpha_j x_j \leq \beta$.

A function $g : U \to \mathbb{R}$ is *superadditive* if $g(u+v) \geq g(u) + g(v)$ for all $u, v \in U$ such that $u + v \in U$.

Theorem 7.7. *Let $g : [0,b] \to \mathbb{R}$ be a superadditive function such that $g \leq f$. Then $\sum_{j \in C} \alpha_j x_j + \sum_{j \in N\setminus C} g(a^j) x_j \leq \beta$ is a valid inequality for S. In particular, if f is superadditive, then the inequality $\sum_{j \in C} \alpha_j x_j + \sum_{j \in N\setminus C} f(a^j) x_j \leq \beta$ is the unique maximal lifting of $\sum_{j \in C} \alpha_j x_j \leq \beta$.*

Proof. For the first part of the statement, let $\alpha_j := g(a^j)$ for $j \in N\setminus C$. Let $t := n - |C|$. Given an ordering $j_1, \ldots, j_t$ of the indices in $N\setminus C$, let $C_0 := C$ and $C_i := C_{i-1} \cup \{j_i\}$, $i = 1, \ldots, t$, and define the function $f_i : [0,b] \to \mathbb{R}$ by

$$f_i(z) := \beta - \max \sum_{j \in C_{i-1}} \alpha_j x_j$$

$$\sum_{j \in C_{i-1}} a^j x_j \leq b - z \tag{7.11}$$

$$x_j \in \{0,1\} \text{ for } j \in C_{i-1}.$$

Note that $f_1 = f$ and, by definition, $f_1 \geq f_2 \geq \ldots \geq f_t$. By Proposition 7.2, the inequality $\sum_{j=1}^{n} \alpha_j x_j \leq \beta$ is valid for S if $\alpha_{j_i} \leq f_i(a^{j_i})$ for $i = 1, \ldots, t$. We will show that $g \leq f_i$ for $i = 1 \ldots, t$, implying that $\alpha_{j_i} = g(a^{j_i}) \leq f_i(a^{j_i})$.

7.2. LIFTING

The proof is by induction on i. By assumption $g \leq f_1$. Consider $2 \leq i \leq t$, and assume by induction that $g \leq f_{i-1}$. Given $z \in [0, b]$, we need to prove that $g(z) \leq f_i(z)$. Let x^* be an optimal solution of (7.11), and define $u^* := a^{j_{i-1}} x^*_{j_{i-1}}$. It follows that

$$\begin{aligned}
f_i(z) &= \beta - \sum_{j \in C_{i-2}} \alpha_j x^*_j - \alpha_{j_{i-1}} x^*_{j_{i-1}} \\
&= \beta - \max\left\{ \sum_{j \in C_{i-2}} \alpha_j x_j : \begin{array}{l} \sum_{j \in C_{i-2}} a^j x_j \leq b - z - u^* \\ x_j \in \{0,1\} \text{ for } j \in C_{i-2} \end{array} \right\} - \alpha_{j_{i-1}} x^*_{j_{i-1}} \\
&= f_{i-1}(z + u^*) - g(a^{j_{i-1}}) x^*_{j_{i-1}} \\
&\geq g(z + u^*) - g(a^{j_{i-1}}) x^*_{j_{i-1}} \quad \text{(because } g \leq f_{i-1}) \\
&\geq g(z + u^*) - g(a^{j_{i-1}} x^*_{j_{i-1}}) \quad \text{(because } g \text{ is superadditive and } x^*_{j_{i-1}} \in \mathbb{Z}_+) \\
&= g(z + u^*) - g(u^*) \\
&\geq g(z) \quad \text{(because } g \text{ is superadditive).}
\end{aligned}$$

For the last part of the statement, assume that $f = f_1$ is superadditive. By the first part of the statement, $\sum_{j \in C} \alpha_j x_j + \sum_{j \in N \setminus C} f(a^j) x_j \leq \beta$ is valid for S. It follows from the first part of the proof that $f_1 \leq f_i$ for $i = 1, \ldots, t$. Since $f_1 \geq f_2 \geq \ldots \geq f_t$, we have $f_1 = f_2 = \ldots = f_t$. This shows that $\alpha_j \leq f(a^j)$, $j \in N \setminus C$, for every lifting $\sum_{j=1}^n \alpha_j x_j \leq \beta$ of $\sum_{j \in C} \alpha_j x_j \leq \beta$. □

Note that the inequality $\sum_{j \in C} \alpha_j x_j + \sum_{j \in N \setminus C} g(a^j) x_j \leq \beta$ defined in the first part of the statement of Theorem 7.7 is valid even when $S \cap \{x : x_j = 1\} = \emptyset$ for some $j \in N \setminus C$ (If (7.10) is infeasible, we set $f(z) = +\infty$).

7.2.3 Sequence Independent Lifting for Minimal Cover Inequalities

Consider the 0,1 knapsack set $K := \{x \in \{0,1\}^n : \sum_{j=1}^n a_j x_j \leq b\}$ where $0 < a_j \leq b$ for all $j = 1, \ldots, n$. Let C be a minimal cover. We present a sequence independent lifting of the cover inequality $\sum_{j \in C} x_j \leq |C| - 1$.

The lifting function f defined in (7.10) becomes

$$f(z) = |C| - 1 - \max \sum_{j \in C} x_j$$

$$\sum_{j \in C} a_j x_j \leq b - z$$

$$x_j \in \{0, 1\} \text{ for } j \in C.$$

We assume without loss of generality that $C = \{1, \ldots, t\}$ with $a_1 \geq \ldots \geq a_t$. Let $\mu_0 := 0$ and, for $h = 1, \ldots, t$, let $\mu_h := \sum_{\ell=1}^h a_\ell$. Let $\lambda := \mu_t - b > 0$.

The first part of the proof of Theorem 7.4 shows that

$$f(z) = \begin{cases} 0 & \text{if } 0 \leq z \leq \mu_1 - \lambda \\ h & \text{if } \mu_h - \lambda < z \leq \mu_{h+1} - \lambda, \text{ for } h = 1, \ldots, t-1. \end{cases}$$

The function f is not superadditive in general. Consider the function g defined by

$$g(z) := \begin{cases} 0 & \text{if } z = 0 \\ h & \text{if } \mu_h - \lambda + \rho_h < z \leq \mu_{h+1} - \lambda, \text{ for } h = 0, \ldots, t-1 \\ h - \frac{\mu_h - \lambda + \rho_h - z}{\rho_1} & \text{if } \mu_h - \lambda < z \leq \mu_h - \lambda + \rho_h, \text{ for } h = 1, \ldots, t-1 \end{cases} \quad (7.12)$$

where $\rho_h = \max\{0, a_{h+1} - (a_1 - \lambda)\}$ for $h = 0, \ldots, r-1$. Note that $g \leq f$. It can be shown that the function g is superadditive (see [192]). Hence by Theorem 7.7 the inequality

$$\sum_{j \in C} x_j + \sum_{j \in N \setminus C} g(a^j) x_j \leq |C| - 1.$$

is a lifting of the minimal cover inequality.

Example 7.8. Consider the 0,1 knapsack set from Example 7.3

$$8x_1 + 7x_2 + 6x_3 + 4x_4 + 6x_5 + 6x_6 + 6x_7 \leq 22$$
$$x_j \in \{0, 1\} \quad \text{for } j = 1, \ldots, 7.$$

We consider the minimal cover $C := \{1, 2, 3, 4\}$ of Example 7.3 and the corresponding minimal cover inequality is $x_1 + x_2 + x_3 + x_4 \leq 3$. We lift it with the superadditive function g defined in (7.12). Figure 7.1 plots the function. The lifted minimal cover inequality is $x_1 + x_2 + x_3 + x_4 + 0.5x_5 + 0.5x_6 + 0.5x_7 \leq 3$. ∎

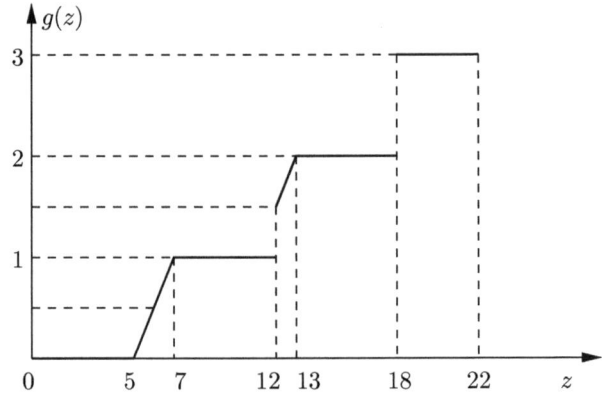

Figure 7.1: A sequence independent lifting function

7.3 Flow Cover Inequalities

The *single-node flow set* is the mixed integer linear set defined as follows

$$T := \left\{ (x,y) \in \{0,1\}^n \times \mathbb{R}^n_+ : \begin{array}{l} \sum_{j=1}^n y_j \leq b \\ y_j \leq a_j x_j \quad \text{for } j = 1, \ldots, n \end{array} \right\} \quad (7.13)$$

where $0 < a_j \leq b$ for all $j = 1, \ldots, n$. This structure appears in many integer programming formulations that model fixed charges (Sect. 2.10). The elements of the set T can be interpreted in terms of a network consisting of n arcs with capacities $a_1, \ldots, a_n$ entering the same node, and one arc of capacity b going out. The variable x_j indicates whether arc j is open, while y_j is the flow through arc j, $j = 1, \ldots, n$. Note that $\dim(T) = 2n$.

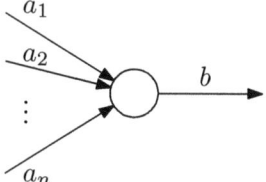

Let $N := \{1, \ldots, n\}$. A set $C \subseteq N$ is a *flow cover* of T if $\sum_{j \in C} a_j > b$. Let $\lambda := \sum_{j \in C} a_j - b$. The inequality

$$\sum_{j \in C} y_j + \sum_{j \in C} (a_j - \lambda)^+ (1 - x_j) \leq b \quad (7.14)$$

is the *flow cover inequality* defined by C.

Theorem 7.9 (Padberg et al. [303]). *Let C be a flow cover for the single-node flow set T, and let $\lambda := \sum_{j \in C} a_j - b$. The flow cover inequality defined by C is valid for T. Furthermore, it defines a facet of $\mathrm{conv}(T)$ if $\lambda < \max_{j \in C} a_j$.*

Proof. The flow cover inequality defined by C is valid for T since

$$\sum_{j \in C} y_j \leq \min\{b, \sum_{j \in C} a_j x_j\} = b - (b - \sum_{j \in C} a_j x_j)^+ = b - (\sum_{j \in C} a_j (1 - x_j) - \lambda)^+$$
$$\leq b - \sum_{j \in C} (a_j - \lambda)^+ (1 - x_j),$$

where the last inequality holds because x is a 0,1 vector.

Let F be the face of $\mathrm{conv}(T)$ defined by (7.14). Note that F is a proper face, since the point defined by $x_j = 1$, $y_j = 0$ for $j = 1, \ldots, n$ is in $T \setminus F$.

Assume that $\lambda < \max_{j \in C} a_j$. We will show that F is a facet. It suffices to provide a set $X \subseteq F$ of points such that $\dim(X) = 2n - 1$.

Without loss of generality, assume that $C = \{1, \ldots, k\}$, and $a_1 \geq \ldots \geq a_t \geq \lambda$, $a_{t+1}, \ldots, a_k < \lambda$ where $1 \leq t \leq k$.

Define the point $\tilde{x} \in \{0,1\}^n$ by $\tilde{x}_j = 1$ for $j \in C$, $\tilde{x}_j = 0$ for $j \in N \setminus C$. For $i \in C$, let $x^i := \tilde{x} - e^i$, where e^i denotes the ith unit vector. For $i \in N \setminus C$, let $x^i := \tilde{x} - e^1 + e^i$.

For $i = 1, \ldots, t$, define the points $y^i, \tilde{y}^i \in \mathbb{R}^n$ by

$$y^i_j := \begin{cases} a_j & j \in C \setminus \{i\} \\ 0 & j \in (N \setminus C) \cup \{i\} \end{cases} \quad , \quad \tilde{y}^i_j := \begin{cases} y^i_j & j \in N \setminus \{i\} \\ a_i - \lambda & j = i. \end{cases}$$

For $i = t+1, \ldots, k$, define the point $y^i \in \mathbb{R}^n$ by

$$y^i_j := \begin{cases} a_1 + a_i - \lambda & j = 1 \\ a_j & j \in C \setminus \{1, i\} \\ 0 & j \in (N \setminus C) \cup \{i\}. \end{cases}$$

Finally, for $i \in N \setminus C$, let $y^i \in \mathbb{R}^n$ be defined by

$$y^i_j := \begin{cases} y^1_j & j \in N \setminus \{i\} \\ \min\{a_i, a_1 - \lambda\} & j = i. \end{cases}$$

Let X be the following set of $2n$ points in $\{0,1\}^n \times \mathbb{R}^n$: (x^i, y^i) for $i \in N$; (x^i, y^1) for $i \in N \setminus C$; $(\tilde{x}, \tilde{y}^i)$ for $i = 1, \ldots, t$; $(\tilde{x}, y^i)$ for $i = t+1, \ldots, k$.

One can verify that $X \subseteq T \cap F$. We will conclude by showing that $\dim(X) = 2n - 1$. It suffices to show that the system

$$\alpha x + \beta y = \gamma \quad \text{for all } (x, y) \in X, \tag{7.15}$$

in the variables $(\alpha, \beta, \gamma) \in \mathbb{R}^n \times \mathbb{R}^n \times \mathbb{R}$, has a unique nonzero solution up to scalar multiplication. Consider such a nonzero solution (α, β, γ).

Let $i \in N \setminus C$. Then $\alpha_i = (\alpha x^i + \beta y^1) - (\alpha x^1 + \beta y^1) = \gamma - \gamma = 0$. Similarly, $\min\{a_i, a_1 - \lambda\}\beta_i = (\alpha x^i + \beta y^i) - (\alpha x^i + \beta y^1) = 0$, implying $\beta_i = 0$. This shows $\alpha_i = \beta_i = 0$ for all $i \in N \setminus C$.

For $i = 2, \ldots, t$, $\lambda(\beta_1 - \beta_i) = (\alpha \tilde{x} + \beta \tilde{y}^i) - (\alpha \tilde{x} + \beta \tilde{y}^1) = 0$. For $i = t+1, \ldots, k$, $a_i(\beta_1 - \beta_i) = (\alpha \tilde{x} + \beta y^i) - (\alpha \tilde{x} + \beta \tilde{y}^1) = 0$. This shows that $\beta_i = \beta_1$ for all $i \in C$.

7.3. FLOW COVER INEQUALITIES

For $i = t+1, \ldots, k$, $\alpha_i = (\alpha \tilde{x} + \beta y^i) - (\alpha x^i + \beta y^i) = 0$. For $i = 1, \ldots, t$, $\alpha_i + \beta_i(a_i - \lambda) = (\alpha \tilde{x} + \beta \tilde{y}^i) - (\alpha x^i + \beta y^i) = 0$, thus $\alpha_i = -\beta_1(a_i - \lambda)$. Since (α, β) is not the zero vector, it follows that $\beta_1 \neq 0$, and up to rescaling we may assume that $\beta_1 = 1$.

Finally, substituting $(\tilde{x}^1, \tilde{y}^1)$ into (7.15) gives $\gamma = b - \sum_{j \in C}(a_j - \lambda)^+$. Therefore the points in F defined above generate the affine space

$$\sum_{j \in C} y_j + \sum_{j \in C}(a_j - \lambda)^+(1 - x_j) = b.$$

This proves that (7.14) defines a facet of $\text{conv}(T)$. □

Note that when the inclusion $C \subset N$ is strict, the condition $\lambda < \max_{j \in C} a_j$ is also necessary for the flow cover inequality (7.14) to define a facet of $\text{conv}(T)$ (Exercise 7.14).

Example 7.10. (Minimal Knapsack Covers Are a Special Case of Flow Cover Inequalities) Consider the knapsack set $K := \{x \in \{0,1\}^n : \sum_{j=1}^n a_j x_j \leq b\}$. Note that $\text{conv}(K)$ is isomorphic to the face of the single-node flow set $\text{conv}(T)$ defined in (7.13), namely the face $\text{conv}(T) \cap \{(x, y) : y_j = a_j x_j, j = 1, \ldots, n\}$.

Let C be a minimal cover for K. Then C is a flow cover for T. Substituting $a_j x_j$ for y_j, for all $j = 1, \ldots, n$, in the expression (7.14) of the flow cover inequality relative to C, we obtain the following valid inequality for K

$$\sum_{j \in C} a_j x_j + \sum_{j \in C}(a_j - \lambda)^+(1 - x_j) \leq b.$$

Note that, since C is a minimal cover, $a_j > \lambda$ for $j = 1, \ldots, n$, thus $(a_j - \lambda)^+ = a_j - \lambda$. Rearranging the terms in the expression above, we obtain

$$\lambda \sum_{j \in C} x_j \leq b - \sum_{j \in C} a_j + |C|\lambda = (|C| - 1)\lambda.$$

The above is the knapsack cover inequality relative to C multiplied by λ. ∎

Example 7.11. (Application to Facility Location) Consider the facility location problem described in Sect. 2.10.1. The problem can be written in the form

$$\min \sum_{i=1}^{m}\sum_{j=1}^{n} c_{ij} y_{ij} + \sum_{j=1}^{n} f_j x_j$$

$$\sum_{j=1}^{n} y_{ij} = d_i \quad i=1,\ldots,m$$

$$\sum_{i=1}^{m} y_{ij} \leq u_j x_j \quad j=1,\ldots,n$$

$$y \geq 0$$

$$x \in \{0,1\}^n.$$

Note that the above formulation differs slightly from the one in Sect. 2.10.1, in that here y_{ij} represents the amount of goods transported from facility i to client j, whereas in Sect. 2.10.1 y_{ij} represented the fraction of demand of customer i satisfied by facility j. Nonetheless the two formulations are obviously equivalent. Let us introduce the variables z_j, $j=1,\ldots,n$, where

$$z_j = \sum_{i=1}^{m} y_{ij}.$$

If we define $b := \sum_{i=1}^{m} d_i$, then the points $(x,z) \in \{0,1\} \times \mathbb{R}^n$ corresponding to feasible solutions must satisfy the constraints

$$\sum_{j=1}^{n} z_j \leq b$$

$$z_j \leq u_j x_j \quad j=1,\ldots,n$$

$$z_j \geq 0 \quad j=1,\ldots,n$$

$$x_j \in \{0,1\} \quad j=1,\ldots,n.$$

This defines a single-node flow set. Any known family of valid inequalities for the single-node flow set, such as the flow cover inequalities, can therefore be adopted to strengthen the formulation of the facility location problem. ∎

Theorem 7.9 shows that, whenever $\lambda < \max_{j \in C} a_j$, the inequality $\sum_{j \in C} y_j + \sum_{j \in C} (a_j - \lambda)^+ (1 - x_j) \leq b$ can be lifted into a facet of $\mathrm{conv}(T)$ by simply setting to 0 the coefficients of the variables x_j, y_j, for $j \in N \setminus C$. The next section provides other ways of lifting the coefficients of x_j, y_j, for $j \in N \setminus C$.

Lifted Flow Cover Inequalities

Let C be a flow cover for the single-node flow set T defined in (7.13), where $0 < a_j \leq b$ for all $j = 1, \ldots, n$. Let $\lambda := \sum_{j \in C} a_j - b$. Throughout this section, we assume that $\lambda < \max_{j \in C} a_j$.

By Theorem 7.9, the flow cover inequality defined by C is facet-defining for $\mathrm{conv}(T)$. We intend to characterize the pairs of coefficients (α_j, β_j), $j \in N \setminus C$, such that the inequality

$$\sum_{j \in C} y_j + \sum_{j \in C}(a_j - \lambda)^+(1 - x_j) + \sum_{j \in N \setminus C}(\alpha_j y_j + \beta_j x_j) \leq b \qquad (7.16)$$

is facet-defining for $\mathrm{conv}(T)$.

Let $C := \{j_1, \ldots, j_t\}$ and assume $a_{j_1} \geq a_{j_2} \geq \ldots \geq a_{j_t}$. Let $\mu_0 := 0$ and $\mu_h := \sum_{\ell=1}^{h} a_{j_\ell}$, $h = 1, \ldots, t$. Assume also that $N \setminus C = \{1, \ldots, n-t\}$ and let $T^i := T \cap \{(x, y) \in \mathbb{R}^{2n} : x_j = y_j = 0, \ j = i+1, \ldots, n-t\}$, $i = 0, \ldots, n-t$.

Suppose we want to sequentially lift the pairs of variables (x_i, y_i) starting from $i = 1$ up to $i = n - t$. That is, once we have determined pairs of coefficients $(\alpha_1, \beta_1), \ldots, (\alpha_{i-1}, \beta_{i-1})$ so that

$$\sum_{j \in C} y_j + \sum_{j \in C}(a_j - \lambda)^+(1 - x_j) + \sum_{j=1}^{i-1}(\alpha_j y_j + \beta_j x_j) \leq b \qquad (7.17)$$

is facet-defining for $\mathrm{conv}(T^{i-1})$, we want to find coefficients (α_i, β_i) such that

$$\sum_{j \in C} y_j + \sum_{j \in C}(a_j - \lambda)^+(1 - x_j) + \sum_{j=1}^{i}(\alpha_j y_j + \beta_j x_j) \leq b \qquad (7.18)$$

is facet-defining for $\mathrm{conv}(T^i)$.

Let $f_i : [0, b] \to \mathbb{R}$ be the function defined by

$$f_i(z) := b - \max \sum_{j \in C} y_j + \sum_{j \in C}(a_j - \lambda)^+(1 - x_j) + \sum_{j=1}^{i-1}(\alpha_j y_j + \beta_j x_j)$$

$$\sum_{j \in C} y_j + \sum_{j=1}^{i-1} y_j \leq b - z \qquad (7.19)$$

$$0 \leq y_j \leq a_j x_j, \ x_j \in \{0,1\} \quad j \in C \cup \{1, \ldots, i-1\}.$$

Note that, since (7.17) is valid for $\mathrm{conv}(T^{i-1})$, $f_i(z) \geq 0$ for $z \in [0, b]$. It follows from the definition that $f_1 \geq f_2 \geq \ldots, f_{n-t}$. The function $f := f_1$ is called the *lifting function* for C.

Lemma 7.12. *Assume that* (7.17) *is valid for* $\mathrm{conv}(T^{i-1})$. *Then* (7.18) *is valid for* $\mathrm{conv}(T^i)$ *if and only if* (α_i, β_i) *satisfies*

$$\alpha_i y_i + \beta_i \leq f_i(y_i) \text{ for all } y_i \in [0, a_i].$$

Furthermore, if (7.17) *defines a facet of* $\mathrm{conv}(T^{i-1})$, *then* (7.18) *is a facet of* $\mathrm{conv}(T^i)$ *if and only if it is valid for* $\mathrm{conv}(T^i)$ *and there exist* $y'_i, y''_i \in [0, a_i]$, $y'_i \neq y''_i$, *such that* $\alpha_i y'_i + \beta_i = f_i(y'_i)$ *and* $\alpha_i y''_i + \beta_i = f_i(y''_i)$.

Proof. By definition of the function f_i, (7.18) is valid for $\mathrm{conv}(T^i)$ if and only if (α_i, β_i) satisfies $\alpha_i y_i + \beta_i x_i \leq f_i(y_i)$ for all $(x_i, y_i) \in \{0, 1\} \times \mathbb{R}_+$ such that $y_i \leq a_i x_i$. Since $f_i \geq 0$, such condition is verified if and only if $\alpha_i y_i + \beta_i \leq f_i(y_i)$ for all $y_i \in [0, a_i]$.

For the second part of the lemma, assume that (7.17) defines a facet of $\mathrm{conv}(T^{i-1})$. So, in particular, there exists a set $X \subseteq T^{i-1}$ of points satisfying (7.17) to equality such that $\dim(X) = \dim(T^{i-1}) - 1$. Suppose there exist $y'_i, y''_i \in [0, a_i]$, $y'_i \neq y''_i$, such that $\alpha_i y'_i + \beta_i = f_i(y'_i)$ and $\alpha_i y''_i + \beta_i = f_i(y''_i)$. Then there exist points $(\bar{x}', \bar{y}')$ and $(\bar{x}'', \bar{y}'')$ in T^i that are optimal solutions to (7.19) for $z = y'_i$ and $z = y''_i$, respectively, and where $(\bar{x}'_i, \bar{y}'_i) = (1, y'_i)$ and $(\bar{x}''_i, \bar{y}''_i) = (1, y''_i)$. Then the points in $X \cup \{(\bar{x}', \bar{y}'), (\bar{x}'', \bar{y}'')\} \subseteq T^i$ satisfy (7.18) at equality and $\dim(X \cup \{(\bar{x}', \bar{y}'), (\bar{x}'', \bar{y}'')\}) = \dim(X) + 2 = \dim(T^i) - 1$.

Conversely, assume that (7.18) defines a facet of $\mathrm{conv}(T^i)$. Then there exist two linearly independent points $(x', y'), (x'', y'')$ in T_i satisfying (7.18) at equality such that $(x'_i, y'_i) \neq (0, 0)$ and $(x''_i, y''_i) \neq (0, 0)$. If follows that $x'_i = x''_i = 1$, $y'_i \neq y''_i$, $\alpha_i y'_i + \beta_i = f_i(y'_i)$ and $\alpha_i y''_i + \beta_i = f_i(y''_i)$. □

Lemma 7.13. *Let* $r := \max\{i \in C : a_{j_i} > \lambda\}$. *For* $z \in [0, b]$, *the lifting function for* C *evaluated at* z *is*

$$f(z) = \begin{cases} h\lambda, & \text{if } \mu_h \leq z < \mu_{h+1} - \lambda, \ h = 0, \ldots, r-1 \\ z - \mu_h + h\lambda, & \text{if } \mu_h - \lambda \leq z < \mu_h, \ h = 1, \ldots, r-1, \\ z - \mu_r + r\lambda, & \text{if } \mu_r - \lambda \leq z \leq b. \end{cases}$$

Proof. Recall that

$$f(z) := b - \max\{\sum_{j \in C}(y_j + (a_j - \lambda)^+(1 - x_j)) : \sum_{j \in C} y_j \leq b - z, \ y_j \leq a_j x_j, \ x_j \in \{0, 1\}, j \in C\}.$$

Consider a point (x, y) achieving the maximum in the above equation. For $i = r+1, \ldots, t$, we can assume that $x_{j_i} = 1$, since $(a_{j_i} - \lambda)^+ = 0$.

Assume that $\sum_{i=r+1}^{t} a_{j_i} \geq b - z$, which is the case if and only if $\mu_r - \lambda \leq z$. Then the maximum is achieved by setting $x_{j_i} = 0$ for $i = 1, \ldots, r$ and setting the value of y_{j_i}, $i = r+1, \ldots, t$, so that $\sum_{i=r+1}^{t} y_{j_i} = b - z$. The value of the objective function is then $b - (b - z + \sum_{i=1}^{r}(a_{j_i} - \lambda)) = z - \mu_r + r\lambda$.

Assume next that $z < \mu_r - \lambda$. Then $\mu_h - \lambda \leq z < \mu_{h+1} - \lambda$ for some h, $0 \leq h \leq r - 1$. Observe that we can assume that $x_{j_1} \leq x_{j_2} \leq \cdots \leq x_{j_r}$. Indeed, given $i < \ell \leq r$, if $x_{j_i} = 1$ and $x_{j_\ell} = 0$, then the solution (x', y') obtained from (x, y) by setting $x'_{j_i} = 0$, $x'_{j_\ell} = 1$, $y'_{j_i} = 0$, $y'_{j_\ell} = (y_{j_\ell} - a_{j_i} + a_{j_\ell})^+$ has value greater than or equal to that of (x, y), and it is feasible because $a_{j_i} \geq a_{j_\ell}$.

Note that it is optimal to set $x_{j_\ell} = 1$, $y_{j_\ell} = a_{j_\ell}$ for $\ell = h+2, \ldots, t$ because $\sum_{\ell=h+2}^{t} a_{j_\ell} = b + \lambda - \mu_{h+1} < b - z$; and it is optimal to set $x_{j_\ell} = 0$, $y_{j_\ell} = 0$ for $\ell = 1, \ldots, h$ because $\sum_{\ell=h+1}^{t} a_{j_\ell} = b + \lambda - \mu_h \geq b - z$. It remains to determine optimal values for $x_{j_{h+1}}$ and $y_{j_{h+1}}$.

If $z \geq \mu_h$, then $b - z - \sum_{\ell=h+2}^{t} a_{j_\ell} \leq a_{j_{h+1}} - \lambda$, so an optimal solution is

$$x_{j_i} = \begin{cases} 0 & i = 1, \ldots, h+1, \\ 1 & i = h+2, \ldots, t, \end{cases} \qquad y_{j_i} = \begin{cases} 0 & i = 1, \ldots, h+1 \\ a_{j_i} & i = h+2, \ldots, t. \end{cases}$$

Thus $f(z) = b - \sum_{i=h+2}^{t} a_{j_i} - \sum_{i=1}^{h+1}(a_{j_i} - \lambda) = h\lambda$.

If $z < \mu_h$, then $b - z - \sum_{\ell=h+2}^{t} a_{j_\ell} > a_{j_{h+1}} - \lambda$, so an optimal solution is

$$x_{j_i} = \begin{cases} 0 & i = 1, \ldots, h, \\ 1 & i = h+1, \ldots, t, \end{cases} \qquad y_{j_i} = \begin{cases} 0 & i = 1, \ldots, h \\ b - z - \sum_{\ell=h+2}^{t} a_{j_\ell} & i = h+1 \\ a_{j_i} & i = h+2, \ldots, t. \end{cases}$$

Thus $f(z) = b - (b - z - \sum_{\ell=h+2}^{t} a_{j_\ell}) - \sum_{i=h+2}^{t} a_{j_i} - \sum_{i=1}^{h}(a_{j_i} - \lambda) = z - \mu_h + h\lambda$. $\square$

Lemma 7.14. *The function f is superadditive in the interval $[0, b]$.*

The proof of the above lemma can be found in [192]. Lemma 7.14 implies that the lifting of flow cover inequalities is always sequence independent, as explained in the next lemma, which closely resembles Theorem 7.7.

Lemma 7.15. *Let C be a flow cover of T. For $i = 1, \ldots, n-t$, the function f_i defined in (7.19) coincides with the lifting function f.*

Proof. Let $i \geq 2$ and assume by induction that $f = f_1 = \cdots = f_{i-1}$. Let $z \in [0, b]$ and let (x^*, y^*) be an optimal solution for (7.19). It follows from the definition of $f_i(z)$ that

$$0 \leq f_i(0) \leq f_{i-1}(y^*_{i-1}) - (\alpha_{i-1} y^*_{i-1} + \beta_{i-1} x^*_{i-1}),$$

thus $\alpha_{i-1}y_{i-1}^* + \beta_{i-1}x_{i-1}^* \leq f(y_{i-1}^*)$. By the choice of (x^*, y^*), it follows that
$$f_i(z) = f_{i-1}(z + y_{i-1}^*) - (\alpha_{i-1}y_{i-1}^* + \beta_{i-1}x_{i-1}^*) \geq f(z + y_{i-1}^*) - f(y_{i-1}^*) \geq f(z),$$
where the last inequality follows by the superadditivity of the function f. Since the definition of f_i implies $f \geq f_i$, it follows that $f_i = f$. □

Lemma 7.15 shows that each pair (α_i, β_i), $i \in N \setminus C$, can be lifted independently of the others.

Theorem 7.16 (Gu et al. [191]). *Let C be a flow cover for T such that $\lambda < \max_{i \in C} a_j$. Let $r := \max\{i \in C : a_{j_i} > \lambda\}$. The inequality (7.16) is facet-defining for T if and only if, for each $i \in N \setminus C$, one of the following holds*

(i) $\alpha_i = 0$, $\beta_i = 0$;

(ii) $\alpha_i = \frac{\lambda}{a_{j_h}}$, $\beta_i = \lambda(h - 1 - \frac{\mu_h - \lambda}{a_{j_h}})$ *for some* $h \in \{2, \ldots, r\}$ *such that* $\mu_h - \lambda \leq a_i$;

(iii) $\alpha_i = 1$, $\beta_i = \ell\lambda - \mu_\ell$ *where* $a_i > \mu_\ell - \lambda$ *and either* $\ell = r$ *or* $\ell < r$ *and* $a_i \leq \mu_\ell$;

(iv) $\alpha_i = \frac{\lambda}{a_i + \lambda - \mu_\ell}$, $\beta_i = \ell\lambda - \frac{\lambda a_i}{a_i + \lambda - \mu_\ell}$ *where ℓ is such that* $\mu_\ell < a_i \leq \mu_{\ell+1} - \lambda$ *and* $\ell < r$.

Proof. By Lemmas 7.12 and 7.15, the inequality (7.16) is facet-defining for conv(T) if and only if, for every $i \in N \setminus C$, the line of equation $v = \alpha_i u + \beta_i$ lies below the graph of the function f in the interval $[0, a_i]$ (i.e., $\{(u, v) \in [0, a_i] \times \mathbb{R} : v = f(u)\}$), and it intersects such graph in at least two points in $[0, a_i]$. Since $a_{j_1} \geq a_{j_2} \geq \ldots \geq a_{j_t}$, then all possible such lines are a) the line passing through $(0, 0)$ and $(\mu_1 - \lambda, 0)$, b) the line passing through $(\mu_{h-1} - \lambda, f(\mu_{h-1} - \lambda))$ and $(\mu_h - \lambda, f(\mu_h - \lambda))$, if $\mu_h - \lambda \leq a_i$ and $h \leq r$, c) the line passing through $(\mu_\ell - \lambda, f(\mu_\ell - \lambda))$ and $(a_i, f(a_i))$ where ℓ is the largest index such that $0 \leq \ell \leq r$ and $a_i > \mu_\ell - \lambda$. The line of equation $v = \alpha_i u + \beta_i$ satisfies a) or b) if (α_i, β_i) satisfy (i) or (ii), respectively. If $v = \alpha_i u + \beta_i$ satisfies c), then (α_i, β_i) satisfy (iii) if $\ell = r$ or $\ell < r$ and $\mu_\ell - \lambda < a_i \leq \mu_\ell$ and (iv) if $\ell < r$ and $\mu_\ell < a_i \leq \mu_{\ell+1} - \lambda$. □

Example 7.17. Consider the single-node flow set
$$T := \left\{ \begin{array}{l} (x, y) \in \{0, 1\}^6 \times \mathbb{R}_+^6 \ : \ y_1 + y_2 + y_3 + y_4 + y_5 + y_6 \leq 20 \\ \phantom{(x, y) \in \{0, 1\}^6 \times \mathbb{R}_+^6 \ :\ } y_1 \leq 17x_1,\ y_2 \leq 9x_2,\ y_3 \leq 8x_3 \\ \phantom{(x, y) \in \{0, 1\}^6 \times \mathbb{R}_+^6 \ :\ } y_4 \leq 6x_4,\ y_5 \leq 5x_5,\ y_6 \leq 4x_6 \end{array} \right\}$$

7.4. FACES OF THE SYMMETRIC TRAVELING SALESMAN... 299

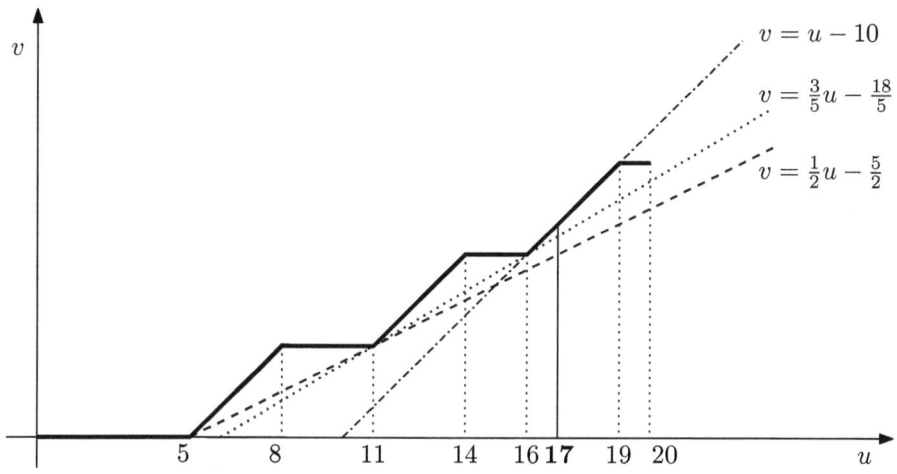

Figure 7.2: Lifting function f and possible lifting coefficients for (y_1, x_1)

Consider the flow cover $C := \{3, 4, 5, 6\}$. Note that $\mu_1 = 8$, $\mu_2 = 14$, $\mu_3 = 19$, $\mu_4 = 23$, $\lambda = 3$ and $r = 4$. For $a_1 = 17$, Case (ii) of the theorem holds for $h = 2$ and $h = 3$, and Case (iii) holds for $\ell = 3$. For $a_2 = 9$, Case (iv) holds for $\ell = 1$. Therefore it follows from Theorem 7.16 that the lifted flow cover inequality

$$\alpha_1 y_1 + \beta_1 x_1 + \alpha_2 y_2 + \beta_2 x_2 + y_3 + y_4 + y_5 + y_6 - 5x_3 - 3x_4 - 2x_5 - x_6 \leq 9$$

is facet-defining for conv(T) if and only if $(\alpha_1, \beta_1) \in \{(0,0), (\frac{1}{2}, -\frac{5}{2}), (\frac{3}{5}, -\frac{18}{5}), (1, -10)\}$ and $(\alpha_2, \beta_2) \in \{(0,0), (\frac{3}{4}, -\frac{15}{4})\}$ (see Fig. 7.2).
∎

7.4 Faces of the Symmetric Traveling Salesman Polytope

In this section we consider the symmetric traveling salesman problem, introduced in Sect. 2.7. Among the formulations we presented, the most successful in practice has been the Dantzig–Fulkerson–Johnson formulation (2.15), which we restate here. Let $G = (V, E)$ be the complete graph on n nodes, where $V := \{1, \ldots, n\}$.

$$\min \sum_{e \in E} c_e x_e$$
$$\sum_{e \in \delta(i)} x_e = 2 \quad \text{for } i \in V \qquad (7.20)$$
$$\sum_{e \in \delta(S)} x_e \geq 2 \quad \text{for } S \subset V \text{ s.t. } 2 \leq |S| \leq n-2$$
$$x_e \in \{0,1\} \quad \text{for } e \in E.$$

The convex hull of feasible solutions to (7.20) is the *traveling salesman polytope*, which will be denoted by P_{tsp}. The constraints $\sum_{e \in \delta(i)} x_e = 2$ are the *degree constraints*, while the constraints $\sum_{e \in \delta(S)} x_e \geq 2$ are the *subtour elimination constraints*.

Theorem 7.18. *The affine hull of the traveling salesman polytope on $n \geq 3$ nodes is $\{x \in \mathbb{R}^{\binom{n}{2}} : \sum_{e \in \delta(i)} x_e = 2\}$. Furthermore, $\dim(P_{\text{tsp}}) = \binom{n}{2} - n$.*

Proof. Note that every point in P_{tsp} must satisfy the n degree constraints $\sum_{e \in \delta(i)} x_e = 2$ for $i \in V$. We first note that such constraints are linearly independent. Indeed, let $Ax = \mathbf{2}$ be the system formed by the n degree constraints. Let A' be the $n \times n$ submatrix of A obtained by the columns corresponding to edges $1j$, $j = 2, \ldots, n$ and edge 23. It is routine to show that $\det(A') = \pm 2$. Therefore $\dim(P_{\text{tsp}}) \leq \binom{n}{2} - n$. To show equality, consider the Hamiltonian-path polytope P of the complete graph on nodes $\{1, \ldots, n-1\}$. We showed in Example 3.21 that $\dim(P) = \binom{n-1}{2} - 1 = \binom{n}{2} - n$, thus there exists a family $\mathcal{Q}$ of $\binom{n}{2} - n + 1$ Hamiltonian paths on $n-1$ nodes whose incidence vectors are affinely independent. Let $\mathcal{T}$ be the family of $\binom{n}{2} - n + 1$ Hamiltonian tours on nodes $\{1, \ldots, n\}$ obtained by completing each Hamiltonian path $Q \in \mathcal{Q}$ to a Hamiltonian tour by adding the two edges between node n and the two endnodes of Q. Since the incidence vectors of elements in $\mathcal{Q}$ are affinely independent, the incidence vectors of the elements of $\mathcal{T}$ are $\binom{n}{2} - n + 1$ affinely independent points in P_{tsp}. □

Theorem 7.19. *For $S \subset V$ with $2 \leq |S| \leq n-2$, the subtour elimination constraint $\sum_{e \in \delta(S)} x_e \geq 2$ defines a facet of the traveling salesman polytope on $n \geq 4$ nodes.*

Proof. Given $S \subset V$, $2 \leq |S| \leq n-2$, let F be the face defined by $\sum_{e \in \delta(S)} x_e \geq 2$. Then there exists some valid inequality $\alpha x \leq \beta$ for P_{tsp} which defines a facet $\bar{F}$ such that $F \subseteq \bar{F}$. We want to show that $F = \bar{F}$.

We first show that, up to linear combinations with the degree constraints, we may assume that $\alpha_e = 0$ for all $e \in \delta(S)$. Indeed, assume w.l.o.g. $1 \in S$.

7.4. FACES OF THE SYMMETRIC TRAVELING SALESMAN... 301

By subtracting from $\alpha x \leq \beta$ the constraint $\sum_{i \in \bar{S}} \alpha_{1i} \sum_{e \in \delta(i)} x_e = 2 \sum_{i \in \bar{S}} \alpha_{1i}$, we may assume that $\alpha_{1i} = 0$ for all $i \in \bar{S}$. Let $k \in S \setminus \{1\}$, and let $i, j \in \bar{S}$, $i \neq j$. Let H be a tour containing edges $1i$ and kj, and no other edge in $\delta(S)$. Note that $H' := H \cup \{1j, ki\} \setminus \{1i, kj\}$ is also a Hamiltonian tour. If $\bar{x}$ and $\bar{x}'$ are the incidence vectors of H and H', then $\bar{x}, \bar{x}' \in F \subseteq \bar{F}$, thus $\alpha \bar{x} = \alpha \bar{x}' = \beta$. It follows that $\alpha_{1i} + \alpha_{kj} = \alpha_{1j} + \alpha_{ki}$, thus $\alpha_{kj} = \alpha_{ki}$ for all $i, j \in \bar{S}$. This shows that, for all $k \in S$, there exists λ_k such that $\alpha_{ki} = \lambda_k$ for all $i \in \bar{S}$. Subtracting from $\alpha x \leq \beta$ the constraint $\sum_{k \in S} \lambda_k \sum_{e \in \delta(k)} x_e = 2 \sum_{k \in S} \lambda_k$ we may assume that $\alpha_e = 0$ for all $e \in \delta(S)$.

Next, we show that there exist constants λ and $\bar{\lambda}$ such that $\alpha_e = \lambda$ for all $e \in E[S]$ and $\alpha_e = \bar{\lambda}$ for all $e \in E[\bar{S}]$. Indeed, given distinct edges $e, e' \in E[S]$, there exist Hamiltonian tours H and H' such that $|H \cap \delta(S)| = |H' \cap \delta(S)| = 2$, and $(H \triangle H') \setminus \delta(S) = \{e, e'\}$.

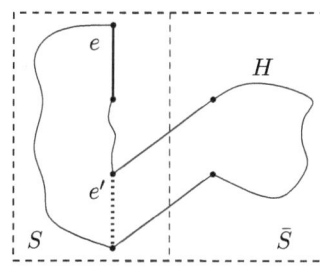
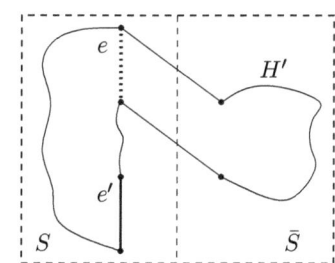

Let $\bar{x}$ and $\bar{x}'$ be the incidence vectors of H and H' respectively. Since $\bar{x}, \bar{x}' \in F \subseteq \bar{F}$, it follows that $\alpha \bar{x} = \alpha \bar{x}' = \beta$. Thus $\alpha_e = \alpha_{e'}$, because $\alpha_e = 0$ for all $e \in \delta(S)$ and $(H \triangle H') \setminus \delta(S) = \{e, e'\}$.

Since every tour H such that $|H \cap \delta(S)| = 2$ satisfies $|H \cap E[S]| = |S| - 1$ and $|H \cap E[\bar{S}]| = |\bar{S}| - 1$, and since $F \subseteq \bar{F}$, it follows that the equation $\alpha x = \beta$ is equivalent to $\lambda \sum_{e \in E[S]} x_e + \bar{\lambda} \sum_{e \in E[\bar{S}]} x_e = \lambda(|S| - 1) + \bar{\lambda}(|\bar{S}| - 1)$. Since the inequalities $\sum_{e \in E[S]} x_e \leq |S| - 1$ and $\sum_{e \in E[\bar{S}]} x_e \leq |\bar{S}| - 1$ both define the face F, it follows that $F = \bar{F}$. □

Because there are exponentially many subtour elimination constraints, solving the linear programming relaxation of (7.20) is itself a challenge.

$$\begin{aligned} \min \ & \sum_{e \in E} c_e x_e \\ & \sum_{e \in \delta(i)} x_e = 2 && \text{for } i \in V \\ & \sum_{e \in \delta(S)} x_e \geq 2 && \text{for } S \subset V \text{ s.t. } 2 \leq |S| \leq n - 2 \\ & 0 \leq x_e \leq 1 && \text{for } e \in E. \end{aligned} \quad (7.21)$$

The feasible set of (7.21) is called the *subtour elimination polytope*. It is impossible to input all the subtour elimination constraints in a solver for medium or large instances (say $n \geq 30$); they must be generated as needed. One starts by solving the following linear programming relaxation.

$$\begin{aligned}
\min \quad & \sum_{e \in E} c_e x_e \\
& \sum_{e \in \delta(i)} x_e = 2 \quad \text{for } i \in V \\
& 0 \leq x_e \leq 1 \quad \text{for } e \in E.
\end{aligned} \quad (7.22)$$

One then adds inequalities that are valid for the subtour elimination polytope but violated by the current linear programming solution $\bar{x}$. The linear program is strengthened iteratively until an optimal solution of (7.21) is found (we will explain how to do this shortly). But solving (7.21) is usually not enough. The formulation is strengthened further by generating additional inequalities that are valid for the traveling salesman polytope but violated by the current linear programming solution $\bar{x}$. This idea was pioneered by Dantzig et al. [103], who solved a 49-city instance in 1954. It was improved in the 1980s by Grötschel [184] and Padberg and Rinaldi [301] who solved instances with hundreds of cities, and refined by Applegate et al. [13] in the 2000s, who managed to solve instances with thousands and even tens of thousands of cities. The formulation strengthening approach mentioned above is typically combined with some amount of enumeration performed within the context of a branch-and-cut algorithm. However the generation of cutting planes is absolutely crucial. This involves solving the separation problem: given a points $\bar{x} \in \mathbb{R}^E$, find a valid inequality for the traveling salesman polytope that is violated by $\bar{x}$, or show that no such inequality exists.

7.4.1 Separation of Subtour Elimination Constraints

Assume that we have a solution $\bar{x}$ of the linear program (7.22) or of some strengthened linear program. The separation problem for subtour elimination inequalities is the following: Prove that $\bar{x}$ is in the subtour elimination polytope, or find one or more subtour elimination constraints that are violated by $\bar{x}$. Note that $\sum_{e \in \delta(S)} \bar{x}_e$ is the weight of the cut $\delta(S)$ in the graph $G = (V, E)$ with edge weights $\bar{x}_e$, $e \in E$. There are efficient polynomial-time algorithms for finding a minimum weight cut in a graph (see Sect. 4.11). If the algorithm finds that the minimum weight of a cut is 2 or more, then all subtour elimination constraints are satisfied, i.e., $\bar{x}$ is in the subtour elimination polytope. On the other hand, if the algorithm finds a cut $\delta(S^*)$ of weight strictly less than 2, the corresponding subtour elimination constraint

7.4. FACES OF THE SYMMETRIC TRAVELING SALESMAN... 303

$\sum_{e \in \delta(S^*)} x_e \geq 2$ is violated by $\bar{x}$. One then adds $\sum_{e \in \delta(S^*)} x_e \geq 2$ to the linear programming formulation, finds an improved solution $\bar{x}$, and repeats the process.

In order to make the separation of subtour elimination inequalities more efficient, fast procedures are typically applied first before resorting to the more expensive minimum weight cut algorithm. For example, let $\bar{E} := \{e \in E : \bar{x}_e > 0\}$. If the graph $(V, \bar{E})$ has at least two connected components, any node set S^* that induces a connected component provides a violated subtour elimination constraint $\sum_{e \in \delta(S^*)} x_e \geq 2$. Identifying the connected components of a graph can be done extremely fast [335].

7.4.2 Comb Inequalities

A solution $\bar{x}$ in the subtour elimination polytope is not necessarily in the traveling salesman polytope as shown by the following example with $n = 6$ nodes. The cost between each pair of nodes is defined as follows. For the edges represented in Fig. 7.3 the costs are shown on the graph (left figure), and the cost of any edge ij not represented in the figure is the cost of a shortest path between i and j in the graph. It is easy to verify that every tour has cost at least 4, but the fractional solution $\bar{x}$ shown on the right figure has cost 3 (the value $\bar{x}_e$ on any edge not represented in Fig. 7.3 is 0). One can check directly that $\bar{x}$ satisfies all the subtour elimination constraints. We will describe a valid inequality for the traveling salesman polytope that separates $\bar{x}$.

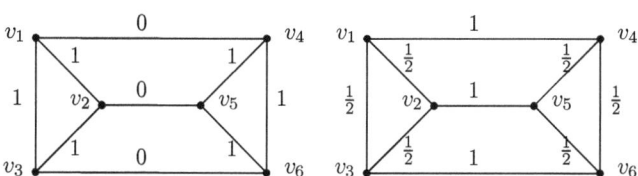

Figure 7.3: Traveling salesman problem on 6 nodes, and a fractional vertex of the subtour elimination polytope

For $k \geq 3$ odd, let $S_0, S_1, \ldots, S_k \subseteq V$ be such that $S_1, \ldots, S_k$ are pairwise disjoint, and for each $i = 1, \ldots, k$, $S_i \cap S_0 \neq \emptyset$ and $S_i \setminus S_0 \neq \emptyset$. The inequality

$$\sum_{i=0}^{k} \sum_{e \in E[S_i]} x_e \leq \sum_{i=0}^{k} |S_i| - \frac{3k+1}{2} \qquad (7.23)$$

is called a *comb inequality*.

Proposition 7.20. *The comb inequality (7.23) is valid for the traveling salesman polytope.*

Proof. We show that (7.23) is a Chvátal inequality for the subtour elimination polytope. Consider the following inequalities, valid for the subtour elimination polytope.

$$\begin{aligned}
\sum_{e \in \delta(v)} x_e &= 2 & v \in S_0; \\
-x_e &\leq 0 & e \in \delta(S_0) \setminus \cup_{i=1}^{k} E[S_i]; \\
\sum_{e \in E[S_i]} x_e &\leq |S_i| - 1 & i = 1, \ldots, k; \\
\sum_{e \in E[S_i \setminus S_0]} x_e &\leq |S_i \setminus S_0| - 1 & i = 1, \ldots, k; \\
\sum_{e \in E[S_i \cap S_0]} x_e &\leq |S_i \cap S_0| - 1 & i = 1, \ldots, k.
\end{aligned}$$

Summing the above inequalities multiplied by $\frac{1}{2}$, one obtains the inequality

$$\sum_{i=0}^{k} \sum_{e \in E[S_i]} x_e \leq \sum_{i=0}^{k} |S_i| - \frac{3k}{2}.$$

Observe that, since k is odd, $\lfloor -\frac{3k}{2} \rfloor = -\frac{3k+1}{2}$, therefore rounding down the right-hand side of the previous inequality one obtains (7.23). □

Grötschel and Padberg [189] showed that the comb inequalities define facets of the traveling salesman polytope for $n \geq 6$.

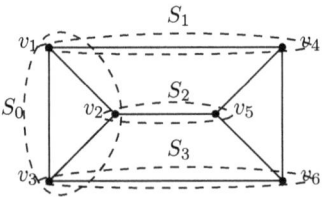

Figure 7.4: A comb

In the example of Fig. 7.3, let $S_0 = \{v_1, v_2, v_3\}$, and $S_1 = \{v_1, v_4\}$, $S_2 = \{v_2, v_5\}$, $S_3 = \{v_3, v_6\}$ (see Fig. 7.4). The corresponding comb inequality is $x_{12} + x_{13} + x_{23} + x_{14} + x_{25} + x_{36} \leq 4$. However $\bar{x}_{12} + \bar{x}_{13} + \bar{x}_{23} + \bar{x}_{14} + \bar{x}_{25} + \bar{x}_{36} = 4.5$, showing that the above comb inequality cuts off $\bar{x}$. Unlike for the subtour elimination inequalities, no polynomial algorithm is known for separating comb inequalities in general. In the special case where $|S_i| = 2$ for $i = 1, \ldots, k$, comb inequalities are known as *blossom inequalities*, and there is a polynomial separation algorithm for this class (Padberg and Rao [300]). In addition to the separation of blossom inequalities,

7.4. FACES OF THE SYMMETRIC TRAVELING SALESMAN...

state-of-the-art software for the traveling salesman problem have sophisticated heuristics to separate more general comb inequalities. Note that, even if all comb inequalities could be separated, we would still not be done in general since the traveling salesman polytope has many other types of facets. In fact, Billera and Sarangarajan [53] showed that any 0,1 polytope is affinely equivalent to a face of an asymmetric traveling salesman polytope of sufficiently large dimension. We are discussing the symmetric traveling salesman polytope in this section, but the Billera–Sarangarajan result is a good indication of how complicated the traveling salesman polytope is. The following idea tries to bypass understanding its structure.

7.4.3 Local Cuts

In their solver for the symmetric traveling salesman problem, Applegate et al. [13] separate subtour elimination constraints and comb inequalities. But then, instead of going on separating other classes of inequalities with known structure, they introduce an interesting approach, the separation of *local cuts*. To get a sense of the contribution of each of these three steps, they considered an Euclidean traveling salesman problem in the plane with 100,000 cities (the cities were generated randomly in a square, the costs were the Euclidean distance between cities up to a small rounding to avoid irrationals), and they constructed a good feasible solution using a heuristic. The lower bound obtained using subtour elimination constraints was already less than 1 % from the heuristic solution. After adding comb inequalities, the gap was reduced to less than 0.2 %, and after adding local cuts, the gap was reduced to below 0.1 %. We now discuss the generation of local cuts.

Let $\mathcal{S} \subset \{0,1\}^E$ denote the set of incidence vectors of tours, and let $\bar{x} \in \mathbb{R}^E$ be a fractional solution that we would like to separate from $\mathcal{S}$. The idea is to map the space $\mathbb{R}^E$ to a space of much lower dimension by a linear mapping Φ and then, using general-purpose methods, to look for linear inequalities $ay \leq b$ that are satisfied by all points $y \in \Phi(\mathcal{S})$ and violated by $\bar{y} := \Phi(\bar{x})$. Every such inequality yields a cut $a\Phi(x) \leq b$ separating $\bar{x}$ from $\mathcal{S}$. For the traveling salesman problem, Applegate, Bixby, Chvátal, and Cook chose Φ as follows. Partition V into pairwise disjoint nonempty sets $V_1, \ldots, V_k$, let $H = (U, F)$ be the graph obtained from G by shrinking each set V_i into a single node u_i, and let $y = \Phi(x) \in \{0,1\}^{|F|}$ be defined by $y_{ij} = \sum_{v \in V_i} \sum_{w \in V_j} x_{vw}$ for all $ij \in F$. This mapping transforms a tour x into a vector y with the following properties.

- $y_e \in \mathbb{Z}_+$ for all $e \in F$,
- $\sum_{e \in \delta(i)} y_e$ is even for all $i \in U$,
- the subgraph of H induced by the edge set $\{e \in F : y_e > 0\}$ is connected.

The convex hull of such vectors is known as the *graphical traveling salesman polyhedron*. Let us denote it by $GTSP^k$ for a graph on k nodes. The goal is to find an inequality that separates $\bar{y}$ from the graphical traveling salesman polyhedron $GTSP^k$, or prove that $\bar{y} \in GTSP^k$. Because k is chosen to be relatively small, this separation can be done by brute force. To simplify the exposition, let us intersect $GTSP^k$ with $\sum_{e \in F} y_e \leq n$ (every $y := \Phi(x)$ satisfies this inequality since $\sum_{e \in E} x_e = n$ for $x \in S$). Let $GTSP^{k,n}$ denote this polytope. We want to solve the following separation problem: Find an inequality that separates $\bar{y}$ from the polytope $GTSP^{k,n}$, or prove that $\bar{y} \in GTSP^{k,n}$. More generally, we want to solve the following separation problem.

Let $\mathcal{Y}$ be a finite set of points in $\mathbb{R}^t$. Given a point $\bar{y} \in \mathbb{R}^t$, either find an inequality that separates $\bar{y}$ from the polytope $\mathrm{conv}(\mathcal{Y})$, or prove that $\bar{y} \in \mathrm{conv}(\mathcal{Y})$.

This can be done by *delayed column generation*.

At a general iteration i, we have a set S_i of points in $\mathcal{Y}$.

At the first iteration, we initialize $S_1 := \{y^1\}$ where y^1 is an arbitrary point in $\mathcal{Y}$.

At iteration i, we check whether $\bar{y} \in \mathrm{conv}(S_i)$ (this amounts to checking the existence of a vector $u \geq 0$ satisfying $\sum_{h=1}^i u_h = 1$ and $\bar{y} = \sum_{h=1}^i y^h u_h$, which can be done by linear programming). If this is the case we have proved that $\bar{y} \in \mathrm{conv}(\mathcal{Y})$. Otherwise we find a linear inequality $ay \leq b$ separating $\bar{y}$ from $\mathrm{conv}(S_i)$ (see Proposition 7.21 below). We then solve $\max\{ay : y \in \mathcal{Y}\}$ (this is where brute force may be needed). If the solution y^{i+1} found satisfies $ay^{i+1} \leq b$, then the inequality $ay \leq b$ separates $\bar{y}$ from $\mathrm{conv}(\mathcal{Y})$. Otherwise we set $S^{i+1} := S^i \cup \{y^{i+1}\}$ and we perform the next iteration.

Proposition 7.21. *If $\bar{y} \notin \mathrm{conv}(S_i)$, an inequality $ay \leq b$ separating $\bar{y}$ from $\mathrm{conv}(S_i)$ can be found by solving a linear program.*

Proof. If $\bar{y} \notin \mathrm{conv}(S_i)$, the linear program

$$\begin{aligned}
\min \quad & 0 \\
\sum_{h=1}^i y^h u_h &= \bar{y} \\
\sum_{h=1}^i u_h &= 1 \\
u &\geq 0
\end{aligned}$$

has no solution. Therefore its dual

$$\begin{aligned}\max \quad & a\bar{y} - b \\ & ay^h - b \leq 0 \quad h = 1, \ldots, i\end{aligned}$$

has an unbounded solution (a, b). □

Applegate, Bixby, Chvátal and Cook call *local cuts* the inequalities generated by this procedure. In their implementation, they refined the procedure so that it only generates facets of the graphical traveling salesman polyhedron. Different choices of the shrunk node sets $V_1, \ldots, V_k$ are used to try to generate several inequalities cutting off the current fractional solution $\bar{x}$. The interested reader is referred to [13] for details.

7.5 Equivalence Between Optimization and Separation

By Meyer's theorem (Theorem 4.30), solving an integer program is equivalent to solving a linear program with a potentially very large number of constraints. In fact, several integer programming formulations, such as the subtour elimination formulation of the traveling salesman polytope or the single-node flow set formulation given by all flow cover inequalities, already have a number of constraints that is exponential in the data size of the problem, so solving the corresponding linear programming relaxations is not straightforward. We would like to solve these linear programs without generating explicitly all the constraints. A fundamental result of Grötschel et al. [186] establishes the equivalence of *optimization* and *separation*: solving a linear programming problem is as hard as finding a constraint cutting off a given point, or deciding that none exists.

Optimization Problem. *Given a polyhedron $P \subset \mathbb{R}^n$ and an objective $c \in \mathbb{R}^n$, find $x^* \in P$ such that $cx^* = \max\{cx : x \in P\}$, or show $P = \emptyset$, or find a direction z in P for which cz is unbounded.*

Separation Problem. *Given a polyhedron $P \subset \mathbb{R}^n$ and a point $\bar{x} \in \mathbb{R}^n$, either show that $\bar{x} \in P$ or give a valid inequality $\alpha x \leq \alpha_0$ for P such that $\alpha \bar{x} > \alpha_0$.*

We are particularly interested in solving the above separation problem when the inequalities defining P are not given explicitly. This is typically the case in integer programming, where P is given as the convex hull of a mixed integer set $\{(x, y) \in \mathbb{Z}_+^p \times \mathbb{R}_+^q : Ax + Gy \leq b\}$ with data A, G, b.

An important theorem of Grötschel et al. [186] states that the optimization problem can be solved in polynomial time if and only if the separation problem can be solved in polynomial time. Similar results were obtained by Padberg and Rao [299] and Karp and Papadimitriou [233]. Of course, P needs to be described in a reasonable fashion for the polynomiality statement to make sense. We will return to this issue later. First, we introduce the main tool needed for proving the equivalence, namely the *ellipsoid algorithm*. We only give a brief outline here. The reader is referred to [188] for a detailed treatment.

Ellipsoid Algorithm

Input. A matrix $A \in \mathbb{Q}^{m \times n}$ and a vector $b \in \mathbb{Q}^m$.
Output. A point of $P := \{x \in \mathbb{R}^n : Ax \leq b\}$ or a proof that P is not full dimensional.

Initialize with a large enough integer t^* and an ellipsoid E_0 that is guaranteed to contain P. Set $t = 0$.

Iteration t. If the center x^t of E^t is in P, stop. Otherwise find a constraint $a^i x \leq b_i$ from $Ax \leq b$ such that $a^i x^t > b_i$. Find the smallest ellipsoid E_{t+1} containing $E_t \cap \{a^i x \leq b_i\}$. Increment t by 1. If $t < t^*$, perform the next iteration. If $t = t^*$, stop: P is not full-dimensional.

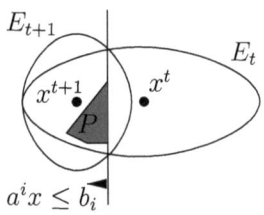

Figure 7.5: Illustration of the ellipsoid algorithm

Figure 7.5 illustrates an iteration of the ellipsoid algorithm. Khachiyan [235] showed that the ellipsoid algorithm can be made to run in polynomial time.

Theorem 7.22. *The ellipsoid algorithm terminates with a correct output if E_0 and t^* are chosen large enough. Furthermore this choice can be made so that the number of iterations is polynomial.*

7.5. EQUIVALENCE BETWEEN OPTIMIZATION...

The following observations about volumes are key to proving that only a polynomial number of iterations are required. We state them without proof.

- The smallest ellipsoid E_{t+1} containing $E_t \cap \{a^i x \leq b_i\}$ can be computed in closed form.

- $\text{Vol}(E_{t+1}) \leq \rho \text{Vol}(E_t)$, where $\rho < 1$ is a constant that depends only on n.

- There exists $\epsilon > 0$, whose encoding size is polynomial in n and in the size of the coefficients of (A, b), such that either P has no interior, or $\text{Vol}(P) \geq \epsilon$.

- $\text{Vol}(E_0) \leq \Delta$, where the encoding size of Δ is polynomial in n and in the size of the coefficients of (A, b).

Since $\text{Vol}(E_t) \leq \rho^t \text{Vol}(E_0)$, the ellipsoid algorithm requires at most $t^* = \log \frac{\Delta}{\epsilon}$ iterations before one can conclude that P has an empty interior. Thus the number of iterations is polynomial. To turn the ellipsoid algorithm into a polynomial algorithm, one needs to keep a polynomial description of the ellipsoids used in the algorithm. This can be achieved by working with slightly larger ellipsoids, instead of the family E_t defined above. We skip the details.

The ellipsoid algorithm returns a point in P whenever P is full dimensional. Dealing with non-full dimensional polyhedra is tricky. Grötschel et al. [187] describe a polynomial-time algorithm that, upon termination of the ellipsoid algorithm with the outcome that P is not full-dimensional, determines an equation $\alpha x = \beta$ satisfied by all $x \in P$. Once such equation is known, one can reduce the dimension of the problem by one, and iterate. A detailed description can be found in [188].

Another issue is the optimization of a linear function cx over P, instead of just finding a feasible point, as described in the above algorithm. This can be done in polynomial time by using binary search on the objective value, or a "sliding objective." Again, we refer to [188] for a description of these techniques.

Finally, we note a beautiful aspect of the ellipsoid algorithm: It does not require an explicit description of P as $\{x \in \mathbb{R}^n : Ax \leq b\}$, but instead it can rely on a separation algorithm that, given the point x^t, either shows that this point is in P, or produces a valid inequality $a^i x \leq b_i$ for P such that $a^i x^t > b_i$.

As a consequence, if we have a separation algorithm at our disposal, the ellipsoid algorithm with a sliding objective solves the optimization problem.

Example 7.23. Consider the traveling salesman problem in an undirected graph $G = (V, E)$. As observed in Sect. 7.4.1, the separation problem for the subtour elimination polytope can be solved in polynomial time (as it amounts to finding a minimum cut in G). Therefore, by applying the ellipsoid algorithm, one can optimize over the subtour elimination polytope in polynomial time. ∎

The complexity of the separation algorithm depends on how P is given to us. We will need P to be "well-described" in the following sense.

Definition 7.24. *A polyhedron $P \subset \mathbb{R}^n$ belongs to a* well-described *family if the length L of the input needed to describe P satisfies $n \leq L$, and there exists a rational matrix (A, b) such that $P = \{x \in \mathbb{R}^n : Ax \leq b\}$ and the encoding size of each of the entries in the matrix (A, b) is polynomially bounded by L.*

Examples of well-described polyhedra are

- $P := \{x \in \mathbb{R}^n : Ax \leq b\}$, where A, b have rational entries and are given as input.

- $P := \mathrm{conv}\{x \in \mathbb{Z}^n : Ax \leq b\}$, where A, b have rational entries and are given as input.

- P is the subtour elimination polytope of a graph G, where G is given as input.

On the other hand, the subtour elimination polytope of a complete graph on n nodes in not well-described if the given input is just the positive integer n in binary encoding, because in this case the length of the input is $\lceil \log_2(n+1) \rceil$, which is smaller than n for $n \geq 3$.

Remark 7.25. *It follows from Theorems 3.38 and 3.39 that P belongs to a well-described family of polyhedra if and only if there exist rational vectors $x^1, \ldots, x^k, r^1, \ldots, r^t$ each of which has an encoding size that is polynomially bounded by the length L of the input used to describe P, and such that $P = \mathrm{conv}\{x^1, \ldots, x^k\} + \mathrm{cone}\{r^1, \ldots, r^t\}$.*

Theorem 7.26. *For well-described polyhedra, the separation problem is solvable in polynomial time if and only if the optimization problem is.*

Proof. We give the proof for the case when P is full-dimensional and bounded. The proof is more complicated when P is not full-dimensional, and we refer the reader to [188] in that case.

"Polynomial Separation $\Rightarrow$ Polynomial Optimization." This follows from the ellipsoid algorithm.

"Polynomial Optimization $\Rightarrow$ Polynomial Separation."

Claim 1: If Optimization can be solved in polynomial time on P, then an interior point of P can be found in polynomial time.

Indeed, find a first point x^0 by maximizing any objective function over P. Assume affinely independent points $x^0, \ldots, x^i$ have been found. Choose c orthogonal to the affine hull of $x^0, \ldots, x^i$. Solve $\max cx$ and $\max -cx$ over P, respectively. Al least one of these programs gives an optimal solution x^{i+1} that is affinely independent of $x^0, \ldots, x^i$. Repeat until $i = n$. Now $\bar{x} = \frac{1}{n+1} \sum_{i=0}^{n} x^i$ is an interior point of P. This proves Claim 1.

Translate P so that the origin is in the interior of P. By Claim 1, this can be done in polynomial time; indeed, if $\bar{x}$ is an interior point of P, $P - \bar{x}$ contains the origin in the interior.

Claim 2: If Optimization can be solved in polynomial time on P, then Separation can be solved in polynomial time on its polar P^*.

Given $\pi^* \in \mathbb{R}^n$, let x^* be an optimal solution to $\max\{\pi^* x : x \in P\}$. If $\pi^* x^* \leq 1$, then $\pi^* \in P^*$. If $\pi^* x^* > 1$, then $\pi x^* \leq 1$ is a valid inequality for P^* which cuts off π^*. Its description is polynomial in the input size of the separation problem on P^* (the input is the description of P (by Remark 7.25 P^* is well-described by the same input) and π^*). This proves the claim.

By Claim 2 and by the first part of the proof (Polynomial Separation $\Rightarrow$ Polynomial Optimization), it follows that Optimization can be solved in polynomial time on P^*. Applying Claim 2 to P^*, we get that Separation can be solved in polynomial time on P^{**}. Since P contain the origin in its interior, it follows from Corollary 3.50 that $P^{**} = P$. □

7.6 Further Readings

The solution of an instance of the traveling salesman problem on 49 cities, detailed by Dantzig et al. [103] in 1954, laid out the foundations of the cutting plane method, and has served as a template for tackling hard combinatorial problems using integer programming (see for example Grötschel

[183, 184], Crowder and Padberg [98], Grötschel et al. [185]). We refer the reader to Cook [85] for an account of the history of combinatorial integer programming, to the monograph by Applegate et al. [13] for a history of traveling salesman computations, and to Cook's book [86] for an expository introduction to the traveling salesman problem. On the theory side, the key insight of Dantzig, Fulkerson and Johnson that one can solve integer programs by introducing inequalities as needed, culminated in the proof of equivalence of separation and optimization by Grötschel et al. [186].

Several early works on valid inequalities for structured problems focused on packing problems, see for example Padberg [297], Nemhauser and Trotter [283, 284], and Wolsey [350]. Padberg [297] introduced the notion of sequential lifting in the context of odd-holes inequalities and generalized it in [298], where he described the sequential lifting procedure discussed in Sect. 7.2. The effectiveness of polyhedral methods in solving general 0, 1 problems was illustrated in the 1983 paper of Crowder et al. [99], where they reported successfully solving 10 pure 0,1 linear programs with up to 2750 variables, employing a variety of tools including lifted cover inequalities. Van Roy and Wolsey [340] reported computational experience in solving a variety of mixed 0,1 programming problems using strong valid inequalities. The paper formalized the *automatic reformulation* approach, that has since become a staple in integer programming: identify a suitable "structured relaxation" R of the feasible region (such as, for example, a single-node flow set), find a family of "strong" valid inequalities for R, and devise an efficient separation algorithm for the inequalities in the family.

The results of Sect. 7.2.2 on superadditive liftings were proved by Wolsey [351], and generalized to mixed 0,1 linear problems by Gu et al. [192]. The sequence independent liftings of cover and flow cover inequalities (Sects. 7.2.3 and 7.3) are given in [192]. Gu et al. [191] report on a successful application of lifted flow cover inequalities to solving mixed 0,1 linear problems. See Louveaux and Wolsey [258] for a survey on sequence-independent liftings.

Wolsey [352] showed that the subtour formulation of the symmetric traveling salesman problem has an integrality gap of 3/2 whenever the distances define a metric. Goemans [169] computed the worst-case improvement resulting from the addition of many of the known classes of inequalities for the graphical traveling salesman polyhedron, showing for example that the comb inequalities cannot improve the subtour bound by a factor greater than 10/9.

7.6. FURTHER READINGS

Equivalence of Separation and Optimization for Convex Sets

The ellipsoid method was first introduced by Yudin and Nemirovski [356] and Shor [332] for convex nonlinear programming, and was used by Khachiyan [235] in a seminal paper in 1979 to give the first polynomial-time algorithm for linear programming. Several researchers realized, soon after Khachiyan's breakthrough, that the method could be modified to run in polynomial time even if the polyhedron is implicitly described by a separation oracle. The strongest version of this result is given by Grötschel et al. [186] (see also [188]), and it can be extended to general convex sets, but similar results have also been discovered by Karp and Papadimitriou [233] and Padberg and Rao [299].

As mentioned above, the equivalence of linear optimization and separation holds also for general convex sets. However, given a convex set $K \subset \mathbb{R}^n$ and $c \in \mathbb{Q}^n$, it may very well be that the optimal solutions of $\max\{cx : x \in K\}$ have irrational components. Analogously, given $y \notin K$, there is no guarantee that a rational hyperplane separating y from K exists, in general. Therefore optimization and separation over K can only be solved in an approximate sense. Formally, given a convex set $K \subseteq \mathbb{R}^n$ and a number $\varepsilon > 0$, let $S(K, \varepsilon) := \{x \in \mathbb{R}^n : \|x - y\| \leq \varepsilon \text{ for some } y \in K\}$ and $S(K, -\varepsilon) := \{x \in K : S(\{x\}, \varepsilon) \subseteq K\}$.

The *weak optimization problem* is the following: given a vector $c \in \mathbb{Q}^n$, and a rational number $\varepsilon > 0$, either determine that $S(K, -\varepsilon)$ is empty, or find $y \in S(K, \varepsilon) \cap \mathbb{Q}^n$ such that $cx \leq cy + \varepsilon$ for all $x \in S(K, -\varepsilon)$.

The *weak separation problem* is the following: given a point $y \in \mathbb{Q}^n$, and a rational number $\delta > 0$, either determine that $y \in S(K, \delta)$, or find $c \in \mathbb{Q}^n$, $\|c\|_\infty = 1$, such that $cx \leq cy + \delta$ for all $x \in K$.

In order to state the equivalence of the two problems, one needs to specify how K is described. Furthermore, the equivalence holds under some restrictive assumptions. Namely, we say that a convex set K is *circumscribed* if the following information is given as part of the input: a positive integer n such that $K \subset \mathbb{R}^n$, and a rational positive number R such that K is contained in the ball of radius R centered at 0. A circumscribed convex set K is denoted by $(K; n, R)$.

We say that a circumscribed convex set $(K; n, R)$ is given by a *weak separation oracle* if we have access to an oracle that provides a solution c to the weak separation problem for every choice of y and δ, where the encoding size of c is polynomially bounded by n and the encoding sizes of R, y, and δ.

We say that a circumscribed convex set $(K; n, R)$ is given by a *weak optimization oracle* if we have access to an oracle providing a solution y

to the weak optimization problem for every choice of c and ε, where the encoding size of y is polynomially bounded by n and the encoding sizes of R, c and ε.

If $(K; n, R)$ is expressed by a weak separation or a weak optimization oracle, an algorithm involving K is said to be *oracle-polynomial time* if the total number of operations, including calls to the oracle, is bounded by a polynomial in n and the encoding sizes of R and of other input data (such as objective function c and tolerance ε).

Theorem 7.27 (Grötschel et al. [186]). *There exists an oracle-polynomial time algorithm that solves the weak optimization problem for every circumscribed convex set $(K; n, R)$ given by a weak separation oracle and every choice of $c \in \mathbb{Q}^n$ and $\varepsilon > 0$.*
There exists an oracle-polynomial time algorithm that solves the weak separation problem for every circumscribed convex set $(K; n, R)$ given by a weak optimization oracle and every choice of $y \in \mathbb{Q}^n$ and $\delta > 0$.

The equivalence hinges on an approximate version of the ellipsoid method. Below we give a high-level description of the method.

Input. A rational number $\varepsilon > 0$ and a circumscribed closed convex set $(K; n, R)$ given by a separation oracle.
Output. Either a rational point $y \in S(K, \varepsilon)$, or an ellipsoid E such that $K \subseteq E$ and $\text{vol}(E) < \varepsilon$.

Initialize with a large enough integer t^* and a small enough $\delta < \varepsilon$. Set $t = 0$, and let E^0 be the ball of radius R centered at 0.
Iteration t. Let x^t be the center of the current ellipsoid E^t containing K. Make a call to the separation oracle with $y = x^t$. If the oracle concludes that x^t is in $S(K, \delta)$, then $x^t \in S(K, \varepsilon)$, stop. If the oracle returns $c \in \mathbb{Q}^n$ such that $cx \leq cx^t + \delta$ for all $x \in K$, then find an ellipsoid E^{t+1} that is an appropriate approximation of the smallest ellipsoid containing $E^t \cap \{cx \leq cx^t + \delta\}$. Increment t by 1. If $t < t^*$, perform the next iteration. If $t = t^*$, stop: $\text{vol}(E^t) < \varepsilon$.

The algorithm described above is oracle-polynomial time because it can be shown that t^* and δ can be chosen so that their encoding size is polynomial in n and in the encoding sizes of R and ε. Furthermore, the ellipsoid E^{t+1} can be computed by a closed form formula.

7.7 Exercises

Exercise 7.1. Consider the 0,1 knapsack set $K := \{x \in \{0,1\}^n : \sum_{j=1}^n a_j x_j \leq b\}$ where $0 < a_j \leq b$ for all $j = 1, \ldots, n$.

(i) Show that $x_j \geq 0$ defines a facet of $\operatorname{conv}(K)$.

(ii) Give conditions for the inequality $x_j \leq 1$ to define a facet of $\operatorname{conv}(K)$.

Exercise 7.2. Consider the graph C_5 with five vertices v_i for $i = 1, \ldots, 5$ and five edges $v_1 v_2, \ldots, v_4 v_5, v_5 v_1$. Let $STAB(C_5)$ denote the stable set polytope of C_5, namely the convex hull of its stable sets.

(i) Show that $x_j \geq 0$ is a facet of $STAB(C_5)$.

(ii) Show that $x_j + x_k \leq 1$ is a facet of $STAB(C_5)$ whenever $v_j v_k$ is an edge of C_5.

(iii) Show that $\sum_{j=1}^5 x_j \leq 2$ is a facet of $STAB(C_5)$.

(iv) Let W_5 be the graph obtained from C_5 by adding a new vertex w adjacent to every v_j, $j = 1, \ldots, 5$. Show how each facet in (i), (ii) and (iii) is lifted to a facet of $STAB(W_5)$.

Exercise 7.3. A *wheel* W_n is the graph with $n+1$ vertices $v_0, v_1, \ldots, v_n$, and $2n$ edges $v_1 v_2, v_2 v_3, \ldots, v_{n-1} v_n, v_n v_1$ and $v_0 v_i$ for all $i = 1, \ldots n$. A Hamiltonian cycle is one that goes though each vertex exactly once. We represent each Hamiltonian cycle by a 0,1 vector in the edge space of the graph, namely $\mathbb{R}^{2n}$. Define $Hamilton(W_n)$ to be the convex hull of the 0,1 vectors representing Hamiltonian cycles of W_n.

(i) What is the dimension of $Hamilton(W_n)$? How many vertices does $Hamilton(W_n)$ have? How many facets?

(ii) Show that the inequalities $x_e \leq 1$ define facets of $Hamilton(W_n)$ for $e = v_1 v_2, \ldots, v_{n-1} v_n, v_n v_1$.

(iii) Give a minimal description of $Hamilton(W_n)$.

Exercise 7.4. Let $G = (V, E)$ be a graph.

1. Show that the blossom inequalities (4.17) for the matching polytope are Chvátal inequalities for the system $\sum_{e \in \delta(v)} x_e \leq 1$, $v \in V$, $x \geq 0$.

2. Show that, if G is not bipartite, then there is at least one blossom inequality that is facet-defining for the matching polytope of G.

In particular, the matching polytope has Chvátal rank zero or one, and the rank is one if and only if G is not bipartite.

Exercise 7.5. Consider $S \subseteq \{0,1\}^n$. Suppose $S \cap \{x_n = 0\} \neq \emptyset$ and $S \cap \{x_n = 1\} \neq \emptyset$. Let $\sum_{i=1}^{n-1} \alpha_i x_i \leq \beta$ be a valid inequality for $S \cap \{x_n = 1\}$. State and prove a result similar to Proposition 7.2 that lifts this inequality into a valid inequality for $\operatorname{conv}(S)$.

Exercise 7.6. Consider $S \subseteq \{0,1\}^n$. Suppose that $\operatorname{conv}(S) \cap \{x : x_k = 0$ for all $k = p+1, \ldots, n\}$ has dimension p, and that $\sum_{j=1}^{p} \alpha_j x_j \leq \beta$ defines one of its faces of dimension $p-2$ or smaller. Construct an example showing that a lifting may still produce a facet of $\operatorname{conv}(S)$.

Exercise 7.7. Consider the sequential lifting procedure. Prove that the largest possible value of the lifting coefficient α_j is obtained when x_j is lifted first in the sequence. Prove that the smallest value is obtained when x_j is lifted last.

Exercise 7.8. Consider the 0,1 knapsack set $K := \mathbb{Z}^n \cap P$ where $P := \{x \in \mathbb{R}^n : \sum_{j=1}^{n} a_j x_j \leq b, \ 0 \leq x \leq 1\}$. Let C be a minimal cover, and let $h \in C$ such that $a_h = \max_{j \in C} a_j$. Show that the inequality

$$\sum_{j \in C} x_j + \sum_{j \in N \setminus C} \lfloor \frac{a_j}{a_h} \rfloor x_j \leq |C| - 1$$

is a Chvátal inequality for P.

Exercise 7.9. Let K be a knapsack set where $b \geq a_1, \ldots, \geq a_n > 0$, Let $C = \{j_1, \ldots, j_t\}$ be a minimal cover of K. The *extension* of C is the set $E(C) := C \cup \{k \in N \setminus C : a_k \geq a_j \text{ for all } j \in C\}$. Let ℓ be the smallest index in $\{1, \ldots, n\} \setminus E(C)$ (if the latter is nonempty).

(i) Prove that, if $\sum_{j \in C \cup \{1\} \setminus \{j_1, j_2\}} a_j \leq b$ and $\sum_{j \in C \cup \{\ell\} \setminus \{j_1\}} a_j \leq b$, then the *extended cover inequality* $\sum_{j \in E(C)} x_j \leq |C| - 1$ defines a facet of $\operatorname{conv}(K)$.

(ii) Prove that extended cover inequalities are Chvátal inequalities.

Exercise 7.10. Consider the knapsack set $\{x \in \{0,1\}^4 : 8x_1 + 5x_2 + 3x_3 + 12x_4 \leq 14\}$. Given the minimal cover $C = \{1, 2, 3\}$, compute the best possible lifting coefficient of variable x_4 using Theorem 7.4. Is the corresponding lifted cover inequality a Chvátal inequality?

7.7. EXERCISES

Exercise 7.11. Let set S and the inequality $\sum_{j \in C} \alpha_j x_j \leq \beta$ be defined as in Sect. 7.2.2. Suppose that the lifting function defined in (7.10) is superadditive. Prove that $\sum_{j \in C} \alpha_j x_j + \sum_{j \in N \setminus C} f(a^j) x_j \leq \beta$ is valid for S and, for every valid inequality $\sum_{j \in N} \alpha_j x_j \leq \beta$, $\alpha_j \leq f(a^j)$ for all $j \in N \setminus C$.

Exercise 7.12. Show that the function g defined in (7.12) is superadditive.

Exercise 7.13. Show that the lifted minimal cover inequality of Example 7.8 induces a facet.

Exercise 7.14. Prove that, when the inclusion $C \subset N$ is strict, the condition $\lambda < \max_{j \in C} a_j$ is necessary for the flow cover inequality (7.14) to define a facet of $\mathrm{conv}(T)$.

Exercise 7.15. Consider the following mixed integer linear set.

$$T := \{x \in \{0,1\}^n, y \in \mathbb{R}_+^n : \quad \sum_{j=1}^k y_j - \sum_{j=k+1}^n y_j \leq b \\ y_j \leq a_j x_j \text{ for all } j = 1, \ldots, n\}$$

where $b > 0$ and $a_j > 0$ for all $j = 1, \ldots, n$. Consider $C \subseteq \{1, \ldots, k\}$ such that $\sum_{j \in C} a_j > b$. Let $\lambda := \sum_{j \in C} a_j - b$. Consider $L \subseteq \{k+1, \ldots, n\}$ and let $\bar{L} := \{k+1, \ldots, n\} \setminus L$. Prove that if $\max_{j \in C} a_j > \lambda$ and $a_j > \lambda$ for all $j \in L$, then

$$\sum_{j \in C} y_j - \sum_{j \in \bar{L}} y_j + \sum_{j \in C}(a_j - \lambda)^+(1 - x_j) - \sum_{j \in L} \lambda x_j \leq b$$

defines a facet of $\mathrm{conv}(T)$.

Exercise 7.16. Prove that the function f defined in Lemma 7.13 is superadditive in the interval $[0, b]$.

Exercise 7.17. Prove that flow cover inequalities (7.14) are Gomory mixed integer inequalities.

Exercise 7.18. Show that the comb inequality (7.23) can be written in the following equivalent form $\sum_{i=0}^{k} \sum_{e \in \delta(S_i)} x_e \geq 3k + 1$.

Exercise 7.19. Let $G = (V, E)$ be an undirected graph. Recall from Sect. 2.4.2 that $\mathrm{stab}(G)$ is the set of the incidence vectors of all the stable sets of G. The *stable set polytope of G* is $\mathrm{STAB}(G) = \mathrm{conv}(\mathrm{stab}(G))$. Let

$Q(G) := \{x \in \mathbb{R}^V : x_i + x_j \leq 1, ij \in E\}$ and $K(G) := \{x \in \mathbb{R}^V : \sum_{i \in K} \leq 1, K \text{ clique of } G\}$. Recall that $\text{stab}(G) = Q(G) \cap \mathbb{Z}^V = K(G) \cap \mathbb{Z}^V$.

(i) Prove that, given a clique K of G, the clique inequality $\sum_{v \in K} x_v \leq 1$ is facet-defining for $\text{STAB}(G)$ if and only if K is a maximal clique.

(ii) Given an odd cycle C of G, the *odd cycle inequality* is $\sum_{v \in V(C)} x_v \leq (|C|-1)/2$. The cycle C is *chordless* if and only if $E \setminus C$ has no edge with both endnodes in $V(C)$.

   – Show that the odd cycle inequality is a Chvátal inequality for $Q(G)$.

   – Show that the odd cycle inequality is facet-defining for $\text{STAB}(G) \cap \{x : x_i = 0, i \in V \setminus V(C)\}$ if and only if C is chordless.

(iii) A graph $H = (V(H), E(H))$ is an *antihole* if the nodes of H can be labeled $v_1, \ldots, v_h$ so that v_i is adjacent to v_j, $j \neq i$, if and only if both $i - j \pmod{h} \geq 2$ and $j - i \pmod{h} \geq 2$. The inequality $\sum_{i \in V(H)} x_i \leq 2$ is the *antihole inequality* relative to H. Let H be an antihole contained in G such that $|V(H)|$ is odd.

   – Show that the antihole inequality relative to H is a Chvátal inequality for $K(G)$.

   – Show that, if $E \setminus E(H)$ has no edge with both endnodes in $V(H)$, then the antihole inequality relative to H is facet-defining for $\text{STAB}(G) \cap \{x : x_i = 0, i \in V \setminus V(H)\}$.

(iv) Given positive integers n, k, $n \geq 2k+1$, a graph $W_n^k = (V(W_n^k), E(W_n^k))$ is a *web* if the nodes of W_n^k can be labeled $v_1, \ldots, v_n$ so that v_i is adjacent to v_j, $j \neq i$, if and only if $i - j \pmod{h} \leq k$ or $j - i \pmod{h} \leq k$. Show that, if W_n^k is a web contained in G and n is not divisible by $k+1$, then the *web inequality* $\sum_{i \in V(W_n^k)} x_i \leq \lfloor n/(k+1) \rfloor$ is a Chvátal inequality for $K(G)$.

Exercise 7.20. Given an undirected graph $G = (V, E)$, the *stability number* $\alpha(G)$ of G is the size of the largest stable set in G. An edge $e \in E$ is *α-critical* if $\alpha(G \setminus e) = \alpha(G) + 1$. Let $\tilde{E} \subseteq E$ be the set of α-critical edges in G. Show that, if the graph $\tilde{G} = (V, \tilde{E})$ is connected, then the inequality $\sum_{i \in V} x_i \leq \alpha(G)$ is facet-defining for $\text{STAB}(G)$.

7.7. EXERCISES

Exercise 7.21. Show that the mixing inequalities (4.29) and (4.30) are facet-defining for P^{mix} (defined in Sect. 4.8.1).

Exercise 7.22. Show that the separation problem for the mixing inequalities (4.29) can be reduced to a shortest path problem in a graph with $O(n)$ nodes.
Show that the separation problem for the mixing inequalities (4.30) can be reduced to finding a negative cost cycle in a graph with $O(n)$ nodes.

Chapter 8

Reformulations and Relaxations

To take advantage of the special structure occurring in a formulation of an integer program, it may be desirable to use a decomposition approach. For example, when the constraints can be partitioned into a set of nice constraints and the remaining set of complicating constraints, a Lagrangian approach may be appropriate. The Lagrangian dual provides a bound that can be stronger than that obtained by solving the usual linear programming relaxation; such a bound may be attractive in a branch-and-bound algorithm. An alternative to the Lagrangian approach is a Dantzig–Wolfe reformulation; when used within the context of an enumeration algorithm, this approach is known as branch-and-price. When it is the variables that can be partitioned into nice and complicating variables, a Benders reformulation may be appropriate.

8.1 Lagrangian Relaxation

Consider a mixed integer linear program

$$\begin{aligned} z_I := \max \quad & cx \\ & Ax \leq b \\ & x_j \in \mathbb{Z} \quad \text{for } j = 1, \ldots, p \\ & x_j \geq 0 \quad \text{for } j = 1, \ldots, n, \end{aligned} \tag{8.1}$$

where $p \leq n$ and all data are assumed to be rational. Following the convention introduced in Remark 3.10, define $z_I := +\infty$ when (8.1) is unbounded and $z_I := -\infty$ when (8.1) is infeasible.

Partition the constraints $Ax \leq b$ into two subsystems $A_1 x \leq b^1$, $A_2 x \leq b^2$ and let m_i denote the number of rows of A_i for $i = 1, 2$. Let

$$Q := \{x \in \mathbb{R}^n_+ : A_2 x \leq b^2, \ x_j \in \mathbb{Z} \text{ for } j = 1, \ldots, p\}.$$

For any $\lambda \in \mathbb{R}^{m_1}_+$, consider the problem LR(λ), called *Lagrangian relaxation*:

$$z_{LR}(\lambda) := \max_{x \in Q} cx + \lambda(b^1 - A_1 x) \tag{8.2}$$

Problem LR(λ) is a relaxation of (8.1) in the following sense.

Proposition 8.1. $z_{LR}(\lambda) \geq z_I$ for every $\lambda \in \mathbb{R}^{m_1}_+$.

Proof. The result holds when $z_I = -\infty$, so we may assume that a feasible solution of (8.1) exists. Let $\bar{x}$ be any feasible solution of (8.1). Since $\bar{x} \in Q$, $\bar{x}$ is feasible to LR(λ) and since $A_1 \bar{x} \leq b^1$ and $\lambda \geq 0$, we have $z_{LR}(\lambda) \geq c\bar{x} + \lambda(b^1 - A_1 \bar{x}) \geq c\bar{x}$. This implies $z_{LR}(\lambda) \geq z_I$. □

The choice of the partition of $Ax \leq b$ into $A_1 x \leq b^1$, $A_2 x \leq b^2$ depends on the application and is based on the structure of the problem. Typically, the constraints $A_2 x \leq b^2$ should be "nice" in the sense that one should be able to optimize a linear function over the set Q, and thus evaluate $z_{LR}(\lambda)$, efficiently, while $A_1 x \leq b^1$ are typically viewed as "complicating" constraints.

The smallest upper bound that we can obtain from Proposition 8.1 is

$$z_{LD} := \min_{\lambda \geq 0} z_{LR}(\lambda). \tag{8.3}$$

This problem is called the *Lagrangian dual* of the integer program (8.1).

Theorem 8.2. *Assume* $\{x : A_1 x \leq b^1, \ x \in \text{conv}(Q)\} \neq \emptyset$. *Then* $z_{LD} = \max\{cx : A_1 x \leq b^1, \ x \in \text{conv}(Q)\}$.

Proof. Since $A_2 x \leq b^2$ is a rational system, by Meyer's theorem (Theorem 4.30), conv(Q) is a rational polyhedron. Let $Cx \leq d$ be a system of linear inequalities such that conv(Q) = $\{x \in \mathbb{R}^n : Cx \leq d\}$. Since $Q \neq \emptyset$, by linear programming duality $z_{LR}(\lambda) = \max\{cx + \lambda(b^1 - A_1 x) : Cx \leq d\} = \min\{\lambda b^1 + \mu d : \mu C = c - \lambda A_1, \ \mu \geq 0\}$ for all $\lambda \geq 0$. It follows that

8.1. LAGRANGIAN RELAXATION

$z_{LD} = \min_{\lambda \geq 0} z_{LR}(\lambda) = \min\{\lambda b^1 + \mu d : \lambda A_1 + \mu C = c, \lambda \geq 0, \mu \geq 0\}$. The latter is precisely the dual of the linear program $\max\{cx : A_1 x \leq b^1, Cx \leq d\}$. Their optimal values coincide because the primal is feasible by assumption. □

Note that when the assumption stated in the theorem does not hold, it is possible that $z_{LD} = +\infty$ and $\max\{cx : A_1 x \leq b^1, x \in \text{conv}(Q)\} = -\infty$, as the reader can verify using the integer program $\max\{x_1+x_2 : x_1-x_2 \leq -1, -x_1 + x_2 \leq -1, x_1, x_2 \in \mathbb{Z}_+\}$. We leave the details as Exercise 8.4.

In the remainder of this chapter, we assume that $\{x : A_1 x \leq b^1, x \in \text{conv}(Q)\} \neq \emptyset$. This implies that $Q \neq \emptyset$.

Let $\{v^k\}_{k \in K}$ denote the extreme points of $\text{conv}(Q)$ and let $\{r^h\}_{h \in H}$ denote its extreme rays. The Lagrangian bound $z_{LR}(\lambda)$ is finite if and only if $(c - \lambda A_1) r \leq 0$ for every $r \in \text{rec}(\text{conv}(Q))$, that is, if and only if λ belongs to the polyhedron $P := \{\lambda \in \mathbb{R}_+^{m_1} : \lambda A_1 r^h \geq c r^h \text{ for all } h \in H\}$. In this case,

$$z_{LR}(\lambda) = \lambda b^1 + \max_{k \in K}(c - \lambda A_1) v^k \qquad (8.4)$$

Corollary 8.3. *The function z_{LR} defined in (8.2) is a piecewise linear convex function of λ over its domain.*

Proof. By (8.4) the function z_{LR} is the maximum of a finite number of affine functions, therefore it is convex and piecewise linear. □

Let z_{LP} denote the optimal value of the linear programming relaxation of (8.1). Theorem 8.2 implies the following.

Corollary 8.4. $z_I \leq z_{LD} \leq z_{LP}$.

Proof. $\text{conv}(S) \subseteq \text{conv}(Q) \cap \{x \in \mathbb{R}_+^n : A_1 x \leq b^1\} \subseteq \{x \in \mathbb{R}_+^n : Ax \leq b\}$. Maximizing the linear function cx over these three sets gives the desired inequalities. □

Thus the Lagrangian dual bound is always at least as tight as the linear programming bound obtained from the usual linear programming relaxation. However, Theorem 8.2 implies the following.

Corollary 8.5. $z_{LD} = z_{LP}$ *for all $c \in \mathbb{R}^n$ if $\text{conv}(Q) = \{x \in \mathbb{R}_+^n : A_2 x \leq b^2\}$.*

In particular, for pure integer programs, when A_2 is totally unimodular and b^2 is an integral vector, we have $z_{LD} = z_{LP}$, i.e., the Lagrangian dual bound is no better that the usual linear programming relaxation bound.

Remark 8.6. *The above theory easily extends to the case where some of the constraints of the integer program (8.1) are equality constraints. For example, if the constraints of the integer program partition into $A_1 x = b^1$ and $A_2 x \leq b^2$, then the theory goes through by replacing $\lambda \in \mathbb{R}_+^{m_1}$ by $\lambda \in \mathbb{R}^{m_1}$, i.e., the multiplier λ_i is unrestricted in sign when it is associated with an equality constraint.*

8.1.1 Examples

Uncapacitated Facility Location

We first illustrate the Lagrangian relaxation approach on the uncapacitated facility location problem:

$$\begin{array}{ll} \max & \sum_{i=1}^m \sum_{j=1}^n c_{ij} y_{ij} - \sum_{j=1}^n f_j x_j \\ & \sum_{j=1}^n y_{ij} = 1 & \text{for all } i \\ & y_{ij} \leq x_j & \text{for all } i, j \\ & x \in \{0,1\}^n,\ y \geq 0. \end{array} \quad (8.5)$$

In order to solve (8.5) by branch and bound, using linear programming bounds, one has to solve large linear programs at each node of the enumeration tree. On the other hand, it is easy to obtain bounds from the following Lagrangian relaxation [93].

$$\begin{array}{ll} z_{LR}(\lambda) = \max & \sum_{i=1}^m \sum_{j=1}^n (c_{ij} - \lambda_i) y_{ij} - \sum_{j=1}^n f_j x_j + \sum_{i=1}^m \lambda_i \\ & y_{ij} \leq x_j & \text{for all } i, j \\ & x \in \{0,1\}^n,\ y \geq 0. \end{array}$$

$$(8.6)$$

The Lagrangian dual is $\min_{\lambda \in \mathbb{R}^m} z_{LR}(\lambda)$. Note that λ is unrestricted in sign in this case (recall Remark 8.6). The following propositions show that $z_{LR}(\lambda)$ can be computed with a simple formula. The proof of the next proposition is easy and we leave it as an exercise.

Proposition 8.7. *An optimal solution of the Lagrangian relaxation (8.6) is*

$$y_{ij}(\lambda) = \begin{cases} 1 & \text{if } c_{ij} - \lambda_i > 0 \text{ and } \sum_\ell (c_{\ell j} - \lambda_\ell)^+ - f_j > 0, \\ 0 & \text{otherwise.} \end{cases}$$

$$x_j(\lambda) = \begin{cases} 1 & \text{if } \sum_\ell (c_{\ell j} - \lambda_\ell)^+ - f_j > 0, \\ 0 & \text{otherwise.} \end{cases}$$

8.1. LAGRANGIAN RELAXATION

Proposition 8.8. *For any given $\lambda \in \mathbb{R}^m$, the following hold.*

(i) $z_{LR}(\lambda) = \sum_j (\sum_i (c_{ij} - \lambda_i)^+ - f_j)^+ + \sum_i \lambda_i.$

(ii) $z_{LD} = z_{LP}.$

Proof. (i) follows from Proposition 8.7. (ii) follows from Corollary 8.5 and the fact that the constraints of (8.6) are totally unimodular. □

Traveling Salesman Problem

Consider the symmetric traveling salesman problem on an undirected graph $G = (V, E)$ with costs c_e, $e \in E$. Held and Karp [197, 198] proposed to construct a Lagrangian relaxation from the Dantzig–Fulkerson–Johnson formulation by relaxing all the degree constraints except for the one relative to a given node, say node 1, and including the redundant constraint $\sum_{e \in E} x_e = |V|$. The resulting relaxation is

$$\begin{aligned}
z_{LR}(\lambda) = \min \quad & \sum_{e \in E} c_e x_e + \sum_{i \in V \setminus \{1\}} \lambda_i (2 - \sum_{e \in \delta(i)} x_e) \\
& \sum_{e \in \delta(1)} x_e = 2 \\
& \sum_{e \in E[S]} x_e \leq |S| - 1 \quad \text{for } \emptyset \subset S \subset V \setminus \{1\} \quad (8.7)\\
& \sum_{e \in E} x_e = |V| \\
& x_e \in \{0, 1\} \quad \text{for } e \in E.
\end{aligned}$$

A *1-tree* in G is a subset T of edges such that $T \cap E[V \setminus \{1\}]$ is a spanning tree in $G \setminus \{1\}$ and T has exactly 2 edges incident with node 1. One can readily verify that the feasible solutions of (8.7) are precisely the incidence vectors of 1-trees. Therefore, given $\lambda \in \mathbb{R}^{V \setminus \{1\}}$, computing an optimal solution for (8.7) consists in computing a 1-tree of minimum cost. Note that a minimum-cost 1-tree can be computed efficiently as follows:

Compute a minimum-cost spanning tree T' in $G \setminus \{1\}$, and let e', e'' be two edges incident with node 1 of minimum cost; output $T = T' \cup \{e', e''\}$.

Recall that a minimum-cost spanning tree can be computed using Kruskal's algorithm (Sect. 4.5).

The *1-tree polytope of G* is the convex hull of incidence vectors of 1-trees.

Proposition 8.9. *The 1-tree polytope of G is described by the linear relaxation of the constraints in (8.7).*

Proof. Observe that, by subtracting the degree constraint $\sum_{e \in \delta(1)} x_e = 2$ from $\sum_{e \in E} x_e = |V|$, one obtains the equivalent constraint $\sum_{e \in E \setminus \delta(1)} x_e =$

$|V|-2$, which is expressed only in terms of the variables relative to edges of $G \setminus \{1\}$. It follows that the polyhedron P described by the constraints of (8.7) is the direct product $P = P_1 \times P_2$ of the polyhedra

$$P_1 := \left\{ x \in \mathbb{R}^{\delta(1)} : \sum_{e \in \delta(1)} x_e = 2,\ 0 \le x_e \le 1 \text{ for all } e \in \delta(1) \right\},$$

$$P_2 := \left\{ x \in \mathbb{R}_+^{E \setminus \delta(1)} : \sum_{e \in E[S]} x_e \le |S|-1 \text{ for } \emptyset \ne S \subset V \setminus \{1\},\ \sum_{e \in E \setminus \delta(1)} x_e = |V|-2 \right\}.$$

It is easy to see that the polyhedron P_1 is the convex hull of incidence vectors of subsets of $\delta(1)$ of cardinality 2 whereas, by Theorem 4.25, P_2 is the spanning tree polytope of $G \setminus \{1\}$. □

By Corollary 8.5 and Proposition 8.9, the lower bound given by the Lagrangian dual $z_{LD} = \max_{\lambda \ge 0} z_{LR}(\lambda)$ is the same as that given by the Dantzig–Fulkerson–Johnson relaxation.

We illustrate the above relaxation on the example given in Sect. 7.4.2, and represented in Fig. 7.3. We recall that the figure on the left represents the costs on the edges of G, while the figure on the right represents the components of an optimal solution of the subtour relaxation, which has value 3. We apply the Lagrangian relaxation described above, where we relax all degree constraints except the one relative to node v_1. Let us consider the problem $z(\lambda)$ when $\lambda_i = 1$, $i = 2, \ldots, 6$. The objective function of the Lagrangian relaxation is $10 - x_{14} - x_{23} - 2x_{25} - 2x_{36} - x_{45} - x_{46} - x_{56}$. The coefficients of the variables are represented in the figure below, together with an optimal 1-tree (depicted by boldface edges).

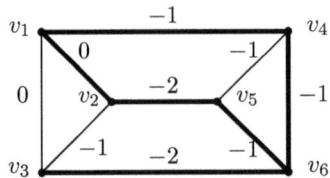

Note that the cost of the 1-tree is $10 - 7 = 3$, which coincides with that of the optimal subtour solution. We conclude that the above choice of λ is optimal.

8.1.2 Subgradient Algorithm

To use the Lagrangian dual bound in a branch-and-bound algorithm, we need to compute z_{LD} or at least approximate it from above. Let $\{v^k\}_{k \in K}$

8.1. LAGRANGIAN RELAXATION

denote the extreme points of $\text{conv}(Q)$ and let $\{r^h\}_{h \in H}$ denote its extreme rays. Let $P := \{\lambda \in \mathbb{R}_+^{m_1} : (c - \lambda A_1)r^h \leq 0 \text{ for all } h \in H\}$. By (8.4),

$$z_{LD} = \min_{\lambda \in P} \max_{k \in K} cv^k + \lambda(b^1 - A_1 v^k). \tag{8.8}$$

Therefore we need to solve a problem of the form

$$\min_{\lambda \in P} g(\lambda) \tag{8.9}$$

where P is a polyhedron and g is a convex nondifferentiable function (Corollary 8.3).

For any convex function $g : \mathbb{R}^n \to \mathbb{R}$, the vector $s \in \mathbb{R}^n$ is a *subgradient* of g at point $\lambda^* \in \mathbb{R}^n$ if

$$(\lambda - \lambda^*)s \leq g(\lambda) - g(\lambda^*) \quad \text{for all } \lambda \in \mathbb{R}^n.$$

Every convex function g has a subgradient at every point of its domain (the proof is left as an exercise, see Exercise 8.9). Note that, if g is differentiable at λ^*, then the subgradient of g at λ^* is uniquely defined and it is the gradient of g at λ^*. In general, the set of subgradients of g at a point λ^* is a convex set. Note that, if the zero vector is a subgradient of g at λ^*, then $g(\lambda^*) \leq g(\lambda)$ for all $\lambda \in \mathbb{R}^n$, and therefore λ^* is a minimizer of g. Conversely, if λ^* is a minimizer of g, then $g(\lambda) - g(\lambda^*) \geq 0$ for all $\lambda \in \mathbb{R}^n$, and therefore the zero vector is a subgradient at λ^*, but λ^* may have other nonzero subgradients.

Subgradients are easily available in our case, as shown by the following proposition.

Proposition 8.10. *If x^* is an optimal solution of $LR(\lambda^*)$ as defined in (8.2), then $s^* := b^1 - A_1 x^*$ is a subgradient of the function z_{LR} at point λ^*.*

Proof. For all $\lambda \geq 0$,

$$\begin{aligned} z_{LR}(\lambda) &\geq cx^* + \lambda(b^1 - A_1 x^*) = cx^* + \lambda^*(b^1 - A_1 x^*) + (\lambda - \lambda^*)(b^1 - A_1 x^*) \\ &= z_{LR}(\lambda^*) + (\lambda - \lambda^*)s^*. \end{aligned}$$

$\square$

For example, for the Lagrangian relaxation of the uncapacitated facility location problem given in (8.6), a subgradient of the function z_{LR} at λ is given by $s_i = 1 - \sum_j y_{ij}(\lambda)$, $i = 1, \ldots, m$, where $y(\lambda)$ is defined in Proposition 8.7. For the Lagrangian relaxation of the traveling salesman problem defined in (8.7), a subgradient of the function z_{LR} at λ is given by $s_i = 2 - |\delta(i) \cap T|$ for all $i \in V \setminus \{1\}$, where T is a minimum-cost 1-tree with respect to λ.

A simple idea to minimize $g(\lambda)$ is to take steps in a direction opposite to a current subgradient direction. Since we want $\lambda \in P$, we project onto P at each iteration. We denote by $\text{proj}_P(\lambda)$ the projection of λ onto P, that is, the point $\lambda' \in P$ at minimum Euclidean distance from λ. In practice, P is often a simple set such as the nonnegative orthant, in which case the projection step is straightforward. We will show that, for an appropriate choice of the step size, such a scheme converges to an optimum.

Subgradient algorithm for $\min_{\lambda \in P} g(\lambda)$:

Step 1 (initialization) Choose a starting point $\lambda^1 \in P$, set $g_{best}^1 := g(\lambda^1)$, and set the iteration counter $t := 1$.

Step 2 (finding a subgradient) Choose a subgradient s^t of g at point λ^t. If $s^t = 0$, stop: λ^t is optimal. Otherwise, go to Step 3.

Step 3 (step size) Let $\lambda^{t+1} = \text{proj}_P(\lambda^t - \alpha_t s^t)$ for some $\alpha_t > 0$. Set $g_{best}^{t+1} = \min\{g_{best}^t, g(\lambda^{t+1})\}$. Increment t by 1. If t is greater than some prespecified iteration limit, stop. Otherwise return to Step 2.

The choice of the step size α_t remains to be specified. We will consider several options, some guaranteeing convergence to the optimal value, others guaranteeing fast convergence, albeit not necessarily to an optimum.

We first consider a divergent series $\sum_{t=1}^\infty \alpha_t = +\infty$ where $\alpha_t > 0$ converges to 0 as t increases. For example $\alpha_t = \frac{1}{t}$ is such a series. Poljak [310] proved that the subgradient algorithm converges to the optimal value with such a choice of α_t.

Theorem 8.11 (Poljak [310]). *Assume that problem (8.9) has finite value g^*, and that the length of all subgradients of g is bounded by a constant $S \in \mathbb{R}_+$. If the sequence $(\alpha_t)_{t=1}^\infty$ converges to 0 and $\sum_{t=1}^{+\infty} \alpha_t = +\infty$, then the sequence $(g_{best}^t)_{t=1}^\infty$ generated by the subgradient algorithm converges to g^*.*

Proof. The key to the proof is to study the Euclidean distance of the current point λ to an optimal solution λ^* of (8.9).

$$\begin{aligned} \|\lambda^{t+1} - \lambda^*\|^2 &= \|\text{proj}_P(\lambda^t - \alpha_t s^t) - \lambda^*\|^2 \\ &\leq \|\lambda^t - \alpha_t s^t - \lambda^*\|^2 \quad &(8.10) \\ &= \|\lambda^t - \lambda^*\|^2 - 2\alpha_t s^t(\lambda^t - \lambda^*) + \alpha_t^2 \|s^t\|^2 \\ &\leq \|\lambda^t - \lambda^*\|^2 - 2\alpha_t(g(\lambda^t) - g(\lambda^*)) + \alpha_t^2 \|s^t\|^2 \quad &(8.11) \end{aligned}$$

8.1. LAGRANGIAN RELAXATION

where (8.10) follows from the fact that $\lambda^* \in P$ and that the projection onto a convex set does not increase the distance to points in the set, while (8.11) follows from the definition of subgradient.

Note that $g_{best}^k = \min_{t=1,\ldots,k} g(\lambda^t)$ and $g^* = g(\lambda^*)$. Combining the above inequalities for $t = 1, \ldots, k$, we get

$$\|\lambda^{k+1} - \lambda^*\|_2^2 \leq \|\lambda^1 - \lambda^*\|_2^2 - 2(\sum_{t=1}^k \alpha_t)(g_{best}^k - g^*) + (\sum_{t=1}^k \alpha_t^2)S^2.$$

This implies

$$g_{best}^k - g^* \leq \frac{\|\lambda^1 - \lambda^*\|_2^2 + S^2(\sum_{t=1}^k \alpha_t^2)}{2\sum_{t=1}^k \alpha_t}$$

Since $\sum_{t=1}^{+\infty} \alpha_t = +\infty$, one can show that $\lim_{k \to +\infty} \frac{\sum_{t=1}^k \alpha_t^2}{\sum_{t=1}^k \alpha_t} = 0$, thus $\lim_{k \to +\infty} g_{best}^k = g^*$. □

However Poljak [310] also proved that convergence is slow. To reach an objective value within a given $\epsilon > 0$ of the optimum value g^*, $O(\frac{1}{\epsilon^2})$ iterations are needed.

Nesterov [287] shows how to go from $O(\frac{1}{\epsilon^2})$ iterations to $O(\frac{1}{\epsilon})$ by exploiting the special form of the function g, which is the maximum of a finite number of affine functions. Although this direction is promising, we do not discuss it any further in this textbook.

In the context of branch-and-bound algorithms, it may be preferable to get bounds even more quickly than by Nesterov's approach. This motivates the next two schemes for choosing α_t.

(i) A geometric series $\alpha_t := \alpha_0 \rho^t$ where $0 < \rho < 1$. Poljak [310] proved that the subgradient algorithm converges fast in this case, but not necessarily to an optimum solution. The Lagrangian relaxation bound obtained this way is a valid bound for use in a branch-and-bound algorithm (Proposition 8.1). The following variation of (i) works well in practice:

(ii) $\alpha_t := \frac{(g(\lambda^t) - g^*)\rho_t}{\|s^t\|^2}$ where g^* is a target value, usually a lower bound (obtained, for example, from a feasible solution to the integer program), and $\rho_0 = 2$,

$$\rho_t := \begin{cases} \frac{\rho_{t-1}}{2} & \text{if the objective value } g(\lambda_t) \text{ did not improve for } K \text{ consecutive iterations,} \\ \rho_{t-1} & \text{otherwise.} \end{cases}$$

Choosing K around 7 seems to work well in practice but the best choice may depend on the application. There is clearly an empirical component in choosing a "good" step size α_t, the goal being to obtain good upper bounds fast.

8.2 Dantzig–Wolfe Reformulation

In this section we present an alternative approach to the Lagrangian relaxation. As in Sect. 8.1, we consider the integer program (8.1), a partition of the constraints $Ax \leq b$ into two subsystems $A_1 x \leq b^1$, $A_2 x \leq b^2$, and

$$Q := \{x \in \mathbb{R}^n_+ : A_2 x \leq b^2,\ x_j \in \mathbb{Z} \text{ for } j = 1, \ldots, p\}.$$

Theorem 8.2 shows that the Lagrangian dual bound z_{LD} is equal to

$$\max\{cx : A_1 x \leq b^1,\ x \in \mathrm{conv}(Q)\}. \tag{8.12}$$

The same bound can be achieved as follows. Let $\{v^k\}_{k \in K}$ be a finite set of points in $\mathrm{conv}(Q)$ containing all the extreme points of $\mathrm{conv}(Q)$ and $\{r^h\}_{h \in H}$ a finite set of rays of $\mathrm{conv}(Q)$ containing all its extreme rays. Since every point in $\mathrm{conv}(Q)$ is a convex combination of the extreme points plus a conic combination of the extreme rays, problem (8.12) can be reformulated as

$$\begin{aligned}
\max \sum_{k \in K}(cv^k)\lambda_k + \sum_{h \in H}(cr^h)\mu_h & \\
\sum_{k \in K}(A_1 v^k)\lambda_k + \sum_{h \in H}(A_1 r^h)\mu_h &\leq b^1 \\
\sum_{k \in K}\lambda_k &= 1 \\
\lambda \in \mathbb{R}^K_+,\ \mu \in \mathbb{R}^H_+. &
\end{aligned} \tag{8.13}$$

Formulation (8.13) is the *Dantzig–Wolfe relaxation* of (8.1). A reformulation of the integer program (8.1) is obtained from (8.13) by enforcing the integrality conditions

$$\sum_{k \in K} v_j^k \lambda_k + \sum_{h \in H} r_j^h \mu_h \in \mathbb{Z}, \quad \text{for } j = 1, \ldots, p. \tag{8.14}$$

The formulation given by (8.13), (8.14) is the *Dantzig–Wolfe reformulation* of the integer program (8.1). Note that, in general, the Dantzig–Wolfe relaxation and reformulation have a large number of variables, namely at least as many as the number of vertices and extreme rays of $\mathrm{conv}(Q)$. We address this issue in Sect. 8.2.2.

8.2. DANTZIG–WOLFE REFORMULATION

Example 8.12. Consider the Lagrangian relaxation of the traveling salesman problem proposed in (8.7). As discussed in Sect. 8.1.1, the bound provided by (8.7) is equal to

$$\begin{aligned} \min \quad & \sum_{e \in E} c_e x_e \\ & \sum_{e \in \delta(i)} x_e = 2 \quad i \in V \setminus \{1\} \\ & x \in \mathrm{conv}(Q) \end{aligned}$$

where Q is the set of incidence vectors of 1-trees. Let $\mathcal{T}$ be the family of 1-trees of G and for any 1-tree $T \in \mathcal{T}$ let $c(T) := \sum_{e \in T} c_e$ denote its cost. Held and Karp [197] give the following Dantzig–Wolfe relaxation of the traveling salesman problem.

$$\begin{aligned} \min \quad & \sum_{T \in \mathcal{T}} c(T) \lambda_T \\ & \sum_{T \in \mathcal{T}} |\delta(i) \cap T| \lambda_T = 2 \quad i \in V \setminus \{1\} \\ & \sum_{T \in \mathcal{T}} \lambda_T = 1 \\ & \lambda_T \geq 0 \quad T \in \mathcal{T}. \end{aligned} \quad (8.15)$$

∎

Remark 8.13. If Q is a pure integer set and $\mathrm{conv}(Q)$ is bounded, and we choose $\{v^k\}_{k \in K}$ to be the set of <u>all</u> points in Q, then enforcing the integrality conditions (8.14) is equivalent to enforcing $\lambda_k \in \{0, 1\}$ for all $k \in K$.

In Example 8.12 it suffices to replace $\lambda_T \geq 0$ by $\lambda_T \in \{0, 1\}$ for $T \in \mathcal{T}$ in (8.15) to obtain a Dantzig–Wolfe reformulation of the traveling salesman problem.

Relation with the Lagrangian Dual

The Lagrangian dual can be viewed as the linear programming dual of the Dantzig–Wolfe relaxation. Indeed, the dual of (8.13) is

$$\begin{aligned} \min \ & \pi b^1 + z \\ & z + \pi(A_1 v^k) \geq c v^k \quad k \in K \\ & \pi(A_1 r^h) \geq c r^h \quad h \in H \\ & \pi \geq 0 \end{aligned}$$

The dual constraints indexed by K can be written equivalently as $z \geq (c - \pi A) v^k$, therefore the dual problem is equivalent to

$$\min_{\pi \in P} \max_{k \in K} \pi b^1 + (c - \pi A) v^k,$$

where $P := \{\pi : \pi \geq 0,\ \pi(A_1 r^h) \geq c r^h \text{ for } h \in H\}$. This is exactly the formula (8.8) of the Lagrangian dual.

8.2.1 Problems with Block Diagonal Structure

Some problems present the following structure, in which several blocks of constraints on disjoint sets of variables are linked by some complicating constraints.

$$\begin{aligned}
\max \quad & c^1 x^1 + c^2 x^2 + \cdots + c^p x^p \\
& B_1 x^1 && \leq b^1 \\
& \quad\quad B_2 x^2 && \leq b^2 \\
& \quad\quad\quad\quad \ddots \\
& \quad\quad\quad\quad\quad\quad B_p x^p && \leq b^p \\
& D_1 x^1 + D_2 x^2 + \cdots + D_p x^p && \leq d \\
& x^j \in \{0,1\}^{n_j}, \quad j = 1, \ldots, p.
\end{aligned} \quad (8.16)$$

Here we consider pure binary problems for ease of notation, but the discussion can be extended to general mixed integer linear problems. Let n_j be the number of variables in block j, for $j = 1, \ldots, p$, and let $Q_j := \{z \in \{0,1\}^{n_j} : B_j z \leq b^j\}$. The Dantzig–Wolfe relaxation of problem (8.16) can be stated as follows

$$\begin{aligned}
\max \quad & \sum_{j=1}^p \sum_{v \in Q_j} (c^j v) \lambda_v^j \\
& \sum_{j=1}^p \sum_{v \in Q_j} (D_j v) \lambda_v^j \leq d \\
& \sum_{v \in Q_j} \lambda_v^j = 1 \quad j = 1, \ldots, p \\
& \lambda_v^j \geq 0 \quad j = 1, \ldots, p, \; v \in Q_j.
\end{aligned} \quad (8.17)$$

Since we are considering a pure binary problem, the Dantzig–Wolfe reformulation of (8.16) is obtained by enforcing $\lambda_v^j \in \{0,1\}$ for $j = 1, \ldots, p, v \in Q_j$ (Remark 8.13).

Example 8.14. Consider the generalized assignment problem

$$\begin{aligned}
\max \quad & \sum_i \sum_j c_{ij} x_{ij} \\
& \sum_j x_{ij} \leq 1 \quad i = 1, \ldots, m \\
& \sum_i t_{ij} x_{ij} \leq T_j \quad j = 1, \ldots, n \\
& x \in \{0,1\}^{m \times n}.
\end{aligned}$$

The problem exhibits a block diagonal structure with n blocks, one for each $j = 1, \ldots, n$, where the jth block depends on the m variables $x_{1j}, \ldots, x_{mj}$. Let $Q_j := \{z \in \{0,1\}^m : \sum_{i=1}^m t_{ij} z_i \leq T_j\}$, $j = 1, \ldots, n$. Note that Q_j is a 0,1 knapsack set. The corresponding Dantzig–Wolfe relaxation is given by

8.2. DANTZIG–WOLFE REFORMULATION

$$\max \ \sum_j \sum_{v \in Q_j} (\sum_i c_{ij} v_i) \lambda_v^j$$
$$\sum_j \sum_{v \in Q_j} v_i \lambda_v^j \leq 1 \quad i = 1, \ldots, m$$
$$\sum_{v \in Q_j} \lambda_v^j = 1 \quad j = 1, \ldots, n$$
$$\lambda_v^j \geq 0 \quad j = 1, \ldots, n, \ v \in Q_j.$$

Since in general the inclusion $\text{conv}(Q_j) \subset \{z \in \mathbb{R}^m : \sum_{i=1}^m t_{ij} z_i \leq T_j, 0 \leq z \leq 1\}$ is strict, the Dantzig–Wolfe relaxation is stronger than the original linear programming relaxation. ∎

An important special case is when all blocks are identical, that is, $c^1 = c^2 = \ldots = c^p := c$, $B_1 = B_2 = \ldots = B_p$, $b^1 = b^2 = \ldots = b^p$, $D_1 = D_2 = \ldots, D_p =: D$. This is the case, for example, in the operating room problem (2.17) when all the operating rooms are identical. In this case, we have $Q_1 = Q_2 = \ldots = Q_p =: Q$. If we define, for every $v \in Q$, $\lambda_v := \sum_{j=1}^p \lambda_v^j$, the objective function $\sum_{j=1}^p \sum_{v \in Q} (cv) \lambda_v^j$ simplifies to $\sum_{v \in Q} (cv) \lambda_v$, while the constraints $\sum_{j=1}^p \sum_{v \in Q} (Dv) \lambda_v^j \leq d$ simplify to $\sum_{v \in Q} (Dv) \lambda_v \leq d$. Therefore, in this case, the relaxation (8.17) can be simplified as follows

$$\max \ \sum_{v \in Q} (cv) \lambda_v$$
$$\sum_{v \in Q} (Dv) \lambda_v \leq d$$
$$\sum_{v \in Q} \lambda_v = p \quad (8.18)$$
$$\lambda_v \geq 0 \quad v \in Q.$$

The Dantzig–Wolfe reformulation of (8.16) is obtained by enforcing $\lambda_v \in \mathbb{Z}$, $v \in Q$. One advantage of this reformulation is that, when all blocks are identical, the symmetry in (8.16) disappears in the reformulation.

Example 8.15. Consider the cutting stock problem discussed in Sect. 2.3, where we need to cut at least b_i rolls of width w_i out of rolls of width W, $i = 1, \ldots, m$, while minimizing the number of rolls of width W used. Consider the formulation (2.1), where p is an upper bound on the number of rolls of width W used, binary variable y_j indicates if roll j is used, $j = 1, \ldots, p$, and variable z_{ij} represents the number of rolls of width w_i cut out of roll j.

$$\min \ \sum_{j=1}^p y_j$$
$$\sum_{i=1}^m w_i z_{ij} \leq W y_j \quad j = 1, \ldots, p$$
$$\sum_{j=1}^p z_{ij} \geq b_i \quad i = 1, \ldots, m \quad (8.19)$$
$$y_j \in \{0, 1\} \quad j = 1, \ldots, p$$
$$z_{ij} \in \mathbb{Z}_+ \quad i = 1, \ldots, m, \ j = 1, \ldots, p.$$

Observe that we have a block diagonal structure with p identical blocks, relative to the $m+1$ variables $y_j, z_{1j}, \ldots, z_{mj}$, $j = 1, \ldots, p$. Let $Q := \{(\eta, \zeta) \in \{0,1\} \times \mathbb{Z}^m : \sum_{i=1}^m w_i \zeta_i \leq W\eta\}$. As in Sect. 2.3, consider all possible cutting patterns $\mathcal{S} := \{s \in \mathbb{Z}_+^m : \sum_{i=1}^m w_i s_i \leq W\}$. Observe that the points of Q are the zero vector and the points $(1, s)$, $s \in \mathcal{S}$. Thus, the Dantzig–Wolfe reformulation (8.19) can be written as

$$\min \sum_{s \in \mathcal{S}} x_s$$
$$\sum_{s \in \mathcal{S}} s_i x_s \geq b_i \quad \text{for } i = 1, \ldots, m$$
$$x \geq 0 \quad \text{integral}.$$

Note that the constraint $\sum_{s \in \mathcal{S}} x_s \leq p$ is redundant, since p was chosen to be an upper-bound on the number of rolls needed. This is precisely the Gilmore–Gomory formulation (2.2) presented in Sect. 2.3. ∎

8.2.2 Column Generation

Column generation is a general technique to solve linear programming problems with a very large number of variables, which is typically the case for the Dantzig–Wolfe relaxation of an integer program. At each iteration, the method keeps a manageable subset of the variables, solves the linear programming problem restricted to these variables, and either concludes that the optimal solution of the restricted problem corresponds to an optimal solution of the whole problem, or finds one or more "candidate variables" to improve the current solution.

We describe the method in the context of the Dantzig–Wolfe relaxation (8.13). The extension to relaxations (8.17) and (8.18) for problems with a block diagonal structure is straightforward. Suppose we have determined subsets $K' \subseteq K$, $H' \subseteq H$ such that the problem

$$\begin{array}{ll} \max & \sum_{k \in K'} (cv^k) \lambda_k + \sum_{h \in H'} (cr^h) \mu_h \\ & \sum_{k \in K'} (A_1 v^k) \lambda_k + \sum_{h \in H'} (A_1 r^h) \mu_h \leq b^1 \\ & \sum_{k \in K'} \lambda_k = 1 \\ & \lambda \in \mathbb{R}_+^{K'}, \quad \mu \in \mathbb{R}_+^{H'} \end{array} \quad (8.20)$$

is feasible. Problem (8.20) is referred to as the *master problem*. If (8.20) is unbounded, then the Dantzig–Wolfe relaxation (8.13) is also unbounded. Otherwise, let $(\bar{\lambda}, \bar{\mu})$ be an optimal solution of (8.20), and let $(\bar{\pi}, \bar{\sigma}) \in \mathbb{R}^m \times \mathbb{R}$ be an optimal solution to its dual, which is

8.2. DANTZIG–WOLFE REFORMULATION

$$\begin{aligned}
\min \quad & \pi b^1 + \sigma \\
& \pi(A_1 v^k) + \sigma \geq cv^k \quad k \in K' \\
& \pi(A_1 r^h) \geq cr^h \quad h \in H' \\
& \pi \geq 0.
\end{aligned} \qquad (8.21)$$

- Observe that the reduced cost of variable λ_k, $k \in K$, is equal to $\bar{c}_k = cv^k - \bar{\pi}(A_1 v^k) - \bar{\sigma}$, while the reduced cost of variable μ_h, $h \in H$, is $\bar{c}_h = cr^h - \bar{\pi}(A_1 r^h)$.

- If $\bar{c}_k \leq 0$ for all $k \in K$ and $\bar{c}_h \leq 0$ for all $h \in H$, then $(\bar{\lambda}, \bar{\mu})$ is an optimal solution to the Dantzig–Wolfe relaxation (8.13), where the values of the variables λ_k, $k \in K \setminus K'$, and μ_h, $h \in H \setminus H'$, are set to 0. Otherwise, we include one or more variables with positive reduced cost in the restricted problem (8.20).

- A variable with positive reduced cost, if any exists, can be computed as follows. Solve the integer program, known as the *pricing problem*,

$$\zeta := -\bar{\sigma} + \max_{x \in Q}(c - \bar{\pi} A_1)x \qquad (8.22)$$

  - Problem (8.22) is unbounded if and only if there exists an extreme ray r^h of $\mathrm{conv}(Q)$ such that $(c - \bar{\pi} A_1)r^h > 0$, that is, if variable μ_h has a positive reduced cost. Note that, by Meyer's theorem (Theorem 4.30), the recession cone of $\mathrm{conv}(Q)$ is $\{y \in \mathbb{R}^n_+ : A_2 y \leq 0\}$, thus such an extreme ray can be computed by linear programming.
  - If (8.22) is bounded and $\zeta > 0$, then there exists a vertex v^k of $\mathrm{conv}(Q)$ such that $cv^k - \bar{\pi}(A_1 v^k) - \bar{\sigma} > 0$, that is, variable λ_k has a positive reduced cost. Here we assume that the constraints $A_2 x \leq b^2$ were chosen so that (8.22) can be solved efficiently.
  - Finally, if (8.22) is bounded and $\zeta \leq 0$, then there is no variable of positive reduced cost.

Note that the pricing problem (8.22) has the same form as the Lagrangian subproblem (8.2).

Example 8.16. In Example 8.15, at each iteration the master problem is defined by a subset $\mathcal{S}' \subseteq \mathcal{S}$ of the patterns.

$$\begin{aligned}
\min \quad & \sum_{a \in \mathcal{S}'} x_a \\
& \sum_{a \in \mathcal{S}'} a_i x_a \geq b_i \quad \text{for } i = 1, \ldots, m \\
& x \geq 0.
\end{aligned}$$

Given an optimal solution $\bar{\pi} \in \mathbb{R}^m$ to the dual of the master problem, namely,

$$\max \quad \sum_{i=1}^{m} b_i \pi_i$$
$$\sum_{i=1}^{m} a_i \pi_i \leq 1 \quad \text{for } a \in \mathcal{S}'$$
$$\pi \geq 0,$$

the pricing problem is

$$\zeta = \max \quad \sum_{i=1}^{m} \bar{\pi}_i z_i$$
$$\sum_{i=1}^{m} w_i z_i \leq W$$
$$z_i \in \mathbb{Z}_+ \quad i = 1, \ldots, m,$$

which is a knapsack problem. If $\zeta \leq 1$ then any optimal solution to the master problem is also optimal for the linear programming relaxation of the cutting stock problem. Otherwise, the optimal solution z^* to the pricing problem is a pattern that corresponds to a variable of negative reduced cost. ∎

Example 8.17. We consider the Dantzig–Wolfe relaxation of the generalized assignment problem discussed in Example 8.14. When solving by column generation, at each iteration the master problem is defined by subsets $S_j \subseteq Q_j$, $j = 1, \ldots, n$.

$$\max \quad \sum_j \sum_{v \in S_j} \left(\sum_i c_{ij} v_i \right) \lambda_v^j$$
$$\sum_j \sum_{v \in S_j} v_i \lambda_v^j \leq 1 \quad i = 1, \ldots, m$$
$$\sum_{v \in S_j} \lambda_v^j = 1 \quad j = 1, \ldots, n$$
$$\lambda_v^j \geq 0 \quad j = 1, \ldots, n, \; v \in S_j.$$

Consider an optimal solution $(\bar{\pi}, \bar{\sigma}) \in \mathbb{R}^m \times \mathbb{R}^n$ to the dual of the master problem, namely,

$$\min \quad \sum_{i=1}^{m} \pi_i + \sum_{j=1}^{n} \sigma_j$$
$$\pi v + \sigma_j \geq \sum_i c_{ij} v_i \quad j = 1, \ldots, n, \; v \in S_j$$
$$\pi \geq 0.$$

In this case we have n pricing problems, namely, for $j = 1, \ldots, n$

$$\zeta_j = -\bar{\sigma}_j + \max \sum_{i=1}^{m} (c_{ij} - \bar{\pi}_i) z_i$$
$$\sum_{i=1}^{m} t_{ij} z_i \leq T_j \qquad (8.23)$$
$$z \in \{0, 1\}^m$$

which are 0,1 knapsack problems. If $\zeta_j \leq 0$ for $j = 1, \ldots, n$ then any optimal solution to the master problem is also optimal for the Dantzig–Wolfe relaxation. Otherwise, if $\zeta_j > 0$ for some j and z^* is an optimal solution, then the variable $\lambda_{z^*}^j$ has positive reduced cost. ∎

8.2.3 Branch-and-Price

To solve the Dantzig–Wolfe reformulation, one can apply the branch-and-bound procedure, solving the Dantzig–Wolfe relaxation by column generation to compute a bound at each node. This approach is known as *branch-and-price*. We describe it in the context of the Dantzig–Wolfe reformulation (8.13), (8.14) of the integer program (8.1). It can be specialized to problems with a block-diagonal structure using reformulations based on (8.17) or (8.18).

- Given an optimal solution (λ^*, μ^*) to the Dantzig–Wolfe relaxation (8.13) at a given node of the branch-and-bound tree, let x^* be the corresponding solution to the linear programming relaxation of (8.1), that is, $x^* = \sum_{k \in K} v^k \lambda_k^* + \sum_{h \in H} r^h \mu_h^*$.

- If x^* satisfies the integrality constraints $x_j^* \in \mathbb{Z}$, $j = 1, \ldots, p$, then prune the current node.

- Otherwise, select a fractional entry x_j^*, for some $j \in \{1, \ldots, p\}$, and branch on x_j^* by creating two new nodes relative to the disjunction

$$\sum_{k \in K} v_j^k \lambda_k + \sum_{h \in H} r_j^h \mu_h \leq \lfloor x_j^* \rfloor \quad \text{or} \quad \sum_{k \in K} v_j^k \lambda_k + \sum_{h \in H} r_j^h \mu_h \geq \lceil x_j^* \rceil. \tag{8.24}$$

Each node is obtained by including one of the two inequalities (8.24) in the Dantzig–Wolfe relaxation (8.13).

Several variations on the above scheme have been considered. For the pure integer case when $\text{conv}(Q)$ is a bounded set, if we choose $\{v^k\}_{k \in K}$ to be the whole set Q, then by Remark 8.13 enforcing the integrality conditions on the original variables x_j, $j = 1, \ldots, n$, is equivalent to enforcing the conditions $\lambda_k \in \{0, 1\}$, $k \in K$. One might be tempted to branch directly on a variable λ_k, rather than on the original variables. This is usually not advisable. The reason is that enforcing the condition $\lambda_k = 0$ is typically difficult. To illustrate this, consider the generalized assignment problem, and the solution of its Dantzig–Wolfe reformulation by column generation discussed in Example 8.14. Suppose we intend to fix variable λ_v^j to 0. When solving the pricing problem (8.23) it could be the case that the optimal solution will be point v, meaning that the variable λ_v^j should be included in the master problem. In this case, then, one should look for the second best solution of (8.23), which increases the computational burden of the pricing problem.

Another option that one can consider when generating the subproblems is, instead of including the constraint $\sum_{k \in K} v_j^k \lambda_k + \sum_{h \in H} r_j^h \mu_h \leq \lfloor x_j^* \rfloor$ or $\sum_{k \in K} v_j^k \lambda_k + \sum_{h \in H} r_j^h \mu_h \geq \lceil x_j^* \rceil$ in the reformulation, to enforce the constraint $x_j \leq \lfloor x_j^* \rfloor$ or $x_j \geq \lceil x_j^* \rceil$ directly on the set Q,

$$Q_0 := Q \cap \{x : x_j \leq \lfloor x_j^* \rfloor\}, \qquad Q_1 := Q \cap \{x : x_j \geq \lceil x_j^* \rceil\}.$$

The two subproblems will then be, for $t = 0, 1$,

$$\begin{aligned}
\max \quad & \sum_{v \in V^t}(cv)\lambda_v + \sum_{r \in R^t}(cr)\mu_r \\
& \sum_{v \in V^t}(A_1 v)\lambda_v + \sum_{r \in R^t}(A_1 r)\mu_r \leq b^1 \\
& \sum_{v \in V^t} \lambda_v = 1 \\
& \lambda \in \mathbb{R}_+^{V^t}, \mu \in \mathbb{R}_+^{R^t},
\end{aligned} \qquad (8.25)$$

where V^t is a finite set including the vertices of $\mathrm{conv}(Q_t)$ and R^t is a finite set including its extreme rays. Note that the linear programming bound given by (8.25), for $t = 0, 1$, is tighter than the relaxation obtained by enforcing the constraint $\sum_{k \in K} v_j^k \lambda_k + \sum_{h \in H} r_j^h \mu_h \leq \lfloor x_j^* \rfloor$ and $\sum_{k \in K} v_j^k \lambda_k + \sum_{h \in H} r_j^h \mu_h \geq \lceil x_j^* \rceil$, respectively, on (8.13). However, while it is assumed that linear optimization over Q is easy, enforcing further constraints on Q might make the pricing problem for (8.25) harder or even intractable, thus solving (8.25) by column generation might be impractical.

8.3 Benders Decomposition

Benders decomposition is a classical approach in integer programming. Its main idea is to solve the original problem by iteratively solving two simpler problems: the master problem is a relaxation of the original problem, while the Benders subproblem provides inequalities that strengthen the master problem. Consider the integer program

$$\begin{aligned}
\max \quad & cx + hy \\
& Ax + Gy \leq b \\
& x \in \mathbb{Z}_+^n \\
& y \in \mathbb{R}_+^p,
\end{aligned}$$

8.3. BENDERS DECOMPOSITION

or more generally a problem of the form

$$
\begin{aligned}
z_I := \max \quad & cx + hy \\
& Ax + Gy \leq b \\
& x \in X \\
& y \in \mathbb{R}^p_+,
\end{aligned}
\tag{8.26}
$$

where $X \subset \mathbb{R}^n$.

Let $\{u^k\}_{k \in K}$ denote the set of extreme points of the polyhedron $Q := \{u \in \mathbb{R}^m_+ : uG \geq h\}$ and $\{r^j\}_{j \in J}$ the set of extreme rays of the cone $C := \{u \in \mathbb{R}^m_+ : uG \geq 0\}$. The cone C is the recession cone of Q when $Q \neq \emptyset$.

Theorem 8.18 (Benders). *Problem (8.26) can be reformulated as*

$$
\begin{aligned}
z_I = \max \quad & \eta + cx \\
& \eta \leq u^k(b - Ax) \quad \text{for all } k \in K \\
& r^j(b - Ax) \geq 0 \quad \text{for all } j \in J \\
& x \in X, \quad \eta \in \mathbb{R}.
\end{aligned}
\tag{8.27}
$$

Proof. Let $P := \{(x,y) \in \mathbb{R}^n \times \mathbb{R}^p : Ax + Gy \leq b, \ y \geq 0\}$. Then (8.26) can be rewritten as

$$
z_I = \max \ \{cx + z_{LP}(x) : x \in \operatorname{proj}_x(P) \cap X\}
\tag{8.28}
$$

where

$$
\begin{aligned}
z_{LP}(x) = \max \quad & hy \\
& Gy \leq b - Ax \\
& y \in \mathbb{R}^p_+.
\end{aligned}
\tag{8.29}
$$

By Theorem 3.46, we have that

$$
\operatorname{proj}_x(P) = \{x \in \mathbb{R}^n : r^j(b - Ax) \geq 0 \text{ for all } j \in J\}.
$$

For $\bar{x} \in \operatorname{proj}_x(P)$, (8.29) is a feasible linear program whose dual is $\min\{u(b - A\bar{x}) : u \in Q\}$. Therefore $z_{LP}(\bar{x})$ is either finite or $+\infty$. If $z_{LP}(\bar{x})$ is finite, by linear programming duality, Q is nonempty. Because Q is a pointed polyhedron, it follows that $K \neq \emptyset$ and $z_{LP}(\bar{x}) = \min_{k \in K} u^k(b - A\bar{x})$. If (8.29) is unbounded, $z_{LP}(\bar{x}) = +\infty$ and the dual is infeasible. That is, $K = \emptyset$. Therefore in both cases,

$$
z_{LP}(\bar{x}) = \max\{\eta : \eta \leq u^k(b - A\bar{x}), \text{ for all } k \in K\}.
$$

Since (8.28) is a reformulation of (8.26), the theorem follows. $\square$

Formulation (8.27) is called the *Benders reformulation* of problem (8.26). It typically has an enormous number of constraints. Instead of using the full-fledged Benders reformulation (8.27), it may be beneficial to use a small subset of these constraints as cuts in a cutting plane algorithm. We discuss some of the issues that arise in implementing such an algorithm, known under the name of *Benders decomposition*.

The algorithm alternates between solving a relaxation of the Benders reformulation (8.27), obtained by considering only a subset of the constraints, and solving the linear program (8.29) to generate additional inequalities from (8.27). The first of these two problems is called the *master problem*, and the linear program (8.29) is known as the *Benders subproblem*. The dual of the Benders subproblem is $\min\{u(b - A\bar{x}) : u \in Q\}$. If Q is empty, then for any $\bar{x}$ the Benders subproblem is either infeasible or unbounded, and thus the original problem (8.26) is either infeasible or unbounded. We will assume in the remainder that Q is nonempty.

At iteration i, the master problem is of the form

$$\begin{aligned}
\max \quad & \eta + cx \\
& \eta \leq u^k(b - Ax) \quad \text{for all } k \in K_i \\
& r^j(b - Ax) \geq 0 \quad \text{for all } j \in J_i \\
& x \in X, \quad \eta \in \mathbb{R},
\end{aligned} \qquad (8.30)$$

where $K_i \subseteq K$ and $J_i \subseteq J$. The choice of the initial sets K_1, J_1 can be important and will be discussed later. Solve the master problem (8.30), and let (x^*, η^*) be an optimal solution. (To guarantee that the master problem is bounded, we can impose a constraint $\eta + cx \leq M$, where M is an upper bound on the optimal value of the original integer program (8.26).) Solve the Benders subproblem $z_{LP}(x^*)$, or equivalently its dual $\min\{u(b - Ax^*) : u \in Q\}$. If the Benders subproblem is infeasible, then its dual is unbounded since $Q \neq \emptyset$, thus there exists an extreme ray r^j of C, $j \in J$, such that $r^j(b - Ax^*) < 0$. In this case, a new *feasibility cut* $r^j(b - Ax) \geq 0$ has been found. Note that $j \in J \setminus J_i$ because the inequality cuts off the current optimum x^*. If the Benders subproblem has an optimum solution, then there are two cases. If $z_{LP}(x^*) < \eta^*$, then consider a basic dual optimal solution u^k, where $k \in K$. In this case, a new *optimality cut* $\eta \leq u^k(b - Ax)$ has been found, where $k \in K \setminus K_i$. Finally, if $z_{LP}(x^*) = \eta^*$, then (x^*, y^*) is optimal to (8.26), where y^* is an optimal solution of the Benders subproblem.

At a given iteration i, there are typically many ways of generating violated Benders cuts and, in fact, it is important in practice to choose Benders cuts judiciously. See Sect. 8.4 for a few pointers.

8.4. FURTHER READINGS

We now discuss briefly the choice of the initial sets K_1, J_1 in the case of mixed integer linear programming, i.e., $X = \{x \in \mathbb{Z}_+^n : Dx \le d\}$. In principle, one could start with $K_1 = J_1 = \emptyset$. However, this may lead to a high number of iterations, each of which involves solving an integer program. In order to decrease the number of integer programs to solve, in a first phase one can solve their linear programming relaxation instead, until enough inequalities are generated to define an optimal solution (x^0, η^0) of the linear relaxation of the Benders reformulation (8.27). This first phase is identical to the Benders decomposition algorithm described above, but now each iteration involves solving the *linear programming relaxation* of the current master problem. We use the feasibility cuts and optimality cuts obtained in this first phase to define the initial sets K_1, J_1.

Benders decomposition has been used successfully in several application areas, such as in the energy sector to plan transmission of electricity from power plants to consumers (see for example [54]). The underlying model is a network design problem such as the one introduced in Sect. 2.10.2. In this model, the design variables x are used to construct the master problem. Fixing $\bar{x}$ corresponds to fixing the underlying network. For fixed $\bar{x}$, the Benders subproblem (8.29) is a multicommodity flow problem in the corresponding network.

In many applications the constraint matrix G in (8.26) has a block diagonal structure, namely

$$\begin{array}{rl} z_I = \max & cx + \sum_{i=1}^m h^i y^i \\ & A_i x + G_i y^i \le b^i \quad i = 1, \ldots, m \\ & x \in X \\ & y^i \in \mathbb{R}_+^{p_i} \quad i = 1, \ldots, m. \end{array}$$

In this case the Benders subproblem decomposes into m separate smaller linear programs, and the Benders reformulation may have considerably less constraints (see Exercise 8.19). Examples include stochastic programming where the variables x represent first-stage decisions, and the variables y^i represent the second-stage decisions under scenario i. See Birge and Louveaux for a good introduction to stochastic programming [55].

8.4 Further Readings

Lagrangian Relaxation

Geoffrion [162] wrote one of the early papers on Lagrangian relaxation in integer programming. Fisher [138] surveys some of the early results in this area. Guignard and Kim [194] introduced a Lagrangian decomposition approach.

Heuristics for computing a Lagrangian bound were proposed by Erlenkotter [133] and Wong [354].

As pointed out in Sect. 8.2, the Lagrangian dual is the dual of the Dantzig–Wolfe relaxation. However, the subgradient method applied to the Lagrangian does not provide a primal solution (i.e., a feasible solution to the Dantzig–Wolfe relaxation). An interesting idea to modify the subgradient algorithm in order to also provide primal solutions is the volume algorithm proposed by Barahona and Anbil [36].

Briant, Lemaréchal, Meurdesoif, Michel, Perrot, and Vanderbeck [64] compare bundle methods and classical column generation.

Cutting Stock

Consider the cutting stock problem discussed in Sect. 2.3 and Example 8.15. Since the Gilmore–Gomory formulation has a variable for each cutting pattern, the size of the formulation is not polynomial in the encoding size of the problem. A relevant question is whether there exists an optimal solution that uses a number of patterns that is polynomial in the encoding length. Eisenbrand and Shmonin [131] show that this is the case.

Theorem 8.19. *There exists an optimal solution for the cutting stock problem that uses at most $\sum_{i=1}^{m} \log(b_i + 1)$ patterns.*

Proof. Let $\mathcal{S} := \{s \in \mathbb{Z}_+^m : \sum_{i=1}^{m} w_i s_i \leq W\}$. Note that there always exists an optimal solution $x^* \in \mathbb{R}^{\mathcal{S}}$ to the cutting stock problem such that $\sum_{s \in \mathcal{S}} s x_s^* = b$. Among all such optimal solutions, let x^* be the one using the minimum number of patterns, that is, the set $T := \{s \in \mathcal{S} : x_s^* > 0\}$ has the smallest possible cardinality.

Assume first that, for every distinct $T_1, T_2 \subset T$, $\sum_{s \in T_1} s \neq \sum_{s \in T_2} s$. Since vectors $s \in T$ are nonnegative, we have that, for any $\tilde{T} \subseteq T$, $b = \sum_{s \in T} s x_s^* \geq \sum_{s \in T} s \geq \sum_{s \in \tilde{T}} s$. This shows that the number of distinct vectors that can be expressed as the sum of vectors in subsets of T is bounded by $\prod_{i=1}^{m}(b_i+1)$. Since the sum of the patterns in every subset of T is distinct, it follows that $2^{|T|} \leq \prod_{i=1}^{m}(b_i + 1)$, and therefore $|T| \leq \sum_{i=1}^{m} \log(b_i + 1)$.

We may therefore assume that there exist $T_1, T_2 \subset T$ such that $T_1 \neq T_2$ and $\sum_{s \in T_1} s = \sum_{s \in T_2} s$. We may further assume that T_1 and T_2 are disjoint, otherwise we can consider $T_1 \setminus T_2$ and $T_2 \setminus T_1$ instead. W.l.o.g., $|T_1| \geq |T_2|$. Let $\xi = \min_{s \in T_1} x_s^*$. Consider $\bar{x}$ defined by $\bar{x}_s = x_s^*$ for $s \in T \setminus (T_1 \cup T_2)$, $x_s^* - \xi$ for $s \in T_1$, $x_s^* + \xi$ for $s \in T_2$. Since $b = \sum_{s \in T} s x_s^*$ and $\sum_{s \in T_1} s = \sum_{s \in T_2} s$, we have that $b = \sum_{s \in T} s \bar{x}_s$. By the choice of ξ, all coefficients in the above combination are nonnegative and integral. Since $|T_1| \geq |T_2|$, we have that

8.4. FURTHER READINGS

$\sum_{s \in T} x_s^* \geq \sum_{s \in T} \bar{x}_s$. Therefore $\bar{x}$ is an optimal solution to the cutting stock problem. Furthermore by the choice of ξ, we have that $x_s^* - \xi = 0$ for some $s \in T_1$ and this contradicts the assumption on the minimality of $|T|$. □

Eisenbrand and Shmonin [131] also prove that here exists an optimal solution for the cutting stock problem that uses at most 2^m patterns.

An important question that has been open until recently is wether *for fixed dimension m*, the cutting stock problem can be solved in polynomial time. Goemans and Rothvoß [171] prove that this is indeed the case.

Theorem 8.20. *There exists an algorithm that computes an optimal solution to the cutting stock problem whose running time is $\log a \cdot 2^{O(m)}$, where a is the largest entry in the cutting stock formulation.*

Dantzig–Wolfe Reformulation and Column Generation

The seminal paper on Dantzig–Wolfe reformulation is [104]. On column generation, the paper of Gilmore and Gomory [168] on the cutting-stock problem was very influential. Many other application areas have benefited from the column generation approach. We just mention Desrosiers, Soumis, and Desrochers, [114] who used it to solve routing problems with time windows.

Benders Reformulation

Benders [47] proposed his decomposition scheme in 1962. Successful early contributions were made by Geoffrion and Graves [163] and Geoffrion [161].

An important question in practice is how to choose Benders cuts judiciously. This issue was investigated by Magnanti and Wong [265], Fischetti, Salvagnin, and Zanette [143], among others.

Given a point (x^*, η^*), there is a Benders cut that cuts it off if and only if the linear program (8.29) has a solution y strictly less than η^*, in other words the linear system

$$\begin{aligned} hy &\geq \eta^* \\ Gy &\leq b - Ax^* \\ y &\geq 0 \end{aligned}$$

is infeasible. Equivalently the following dual linear program is unbounded.

$$\begin{aligned} \min \quad & v(b - Ax^*) - v_0 \eta^* \\ & vG \geq v_0 h \\ & v, v_0 \geq 0. \end{aligned}$$

Using a normalization constraint to bound it, we obtain

$$\begin{aligned} \min \quad & v(b - Ax^*) - v_0 \eta^* \\ & vG \geq v_0 h \\ & \sum_{i=1}^m v_i + v_0 = 1 \\ & v, v_0 \geq 0. \end{aligned} \quad (8.31)$$

This is a *cut-generating linear program* for Benders cuts. An optimal solution v^k, v_0^k of (8.31) satisfies $v^k(b - Ax^*) - v_0^k \eta^* < 0$. Therefore

$$v^k(b - Ax) - v_0^k \eta \geq 0$$

is a Benders cut that cuts off (x^*, η^*). When $v_0^k > 0$ we get an optimality cut $\eta \leq u^k(b - Ax)$, and when $v_0^k = 0$ we get a feasibility cut $r^j(b - Ax) \geq 0$.

Fischetti, Salvagnin, and Zanette [143] conducted numerical experiments on instances of the MIPLIB library and concluded that it pays to optimize the generation of Benders cuts using (8.31) rather than taking the first Benders cut that can be produced.

Costa [97] surveys Benders decomposition applied to fixed-charge network design problems.

8.5 Exercises

Exercise 8.1. Modify the branch-and-bound algorithm of Sect. 1.2.1 to use Lagrangian bounds instead of linear programming bounds. Discuss the advantages and drawbacks of such an approach.

Exercise 8.2. Write a Lagrangian relaxation for

$$\begin{aligned} z_I := \max \quad & cx \\ & A_1 x = b^1 \quad \text{complicating constraints} \\ & A_2 x \leq b^2 \quad \text{nice constraints} \\ & x_j \in \mathbb{Z} \quad \text{for } j = 1, \ldots, p \\ & x_j \geq 0 \quad \text{for } j = 1, \ldots, n. \end{aligned}$$

and prove a proposition similar to Proposition 8.1.

8.5. EXERCISES

Exercise 8.3. Consider a binary problem with block diagonal structure as in (8.16), and let Q_j, $j = 1, \ldots, p$ be defined as in Sect. 8.2.1. Consider the constraints $B_j x^j \leq b^j$ ($j = 1, \ldots, p$) as the "nice" constraints, and the remaining as the "complicating" constraints. Prove that the Lagrangian dual value is

$$z_{LD} = \min_{\lambda \geq 0} \left\{ \lambda d + \sum_{j=1}^{p} \max_{v \in Q_j} (c^j - \pi D_j) v \right\}.$$

Exercise 8.4. Consider the integer program $\max\{x_1 + x_2 : x_1 - x_2 \leq -1, -x_1 + x_2 \leq -1, x_1, x_2 \in \mathbb{Z}_+\}$. Compute $z_{LR}(\lambda)$ for any $\lambda \geq 0$ using $Q := \{(x_1, x_2) \in \mathbb{Z}_+^2 : -x_1 + x_2 \leq -1\}$. Compute the Lagrangian dual bound z_{LD}. Compare to $\max\{cx : A_1 x \leq b^1, x \in \text{conv}(Q)\}$.

Exercise 8.5.

(i) Assuming that the x part of a feasible solution to (8.6) is given, show that a feasible solution y that maximizes the objective is

$$y_{ij} = \begin{cases} x_j & \text{if } c_{ij} - \lambda_i > 0, \\ 0 & \text{otherwise.} \end{cases}$$

(ii) Using (i), prove Proposition 8.7.

Exercise 8.6. Consider two different Lagrangian duals for the generalized assignment problem

$$z_I := \max \quad \sum_{i=1}^{m} \sum_{j=1}^{n} c_{ij} x_{ij}$$
$$\sum_{j=1}^{n} x_{ij} \leq 1 \quad \text{for } i = 1, \ldots, m$$
$$\sum_{i=1}^{m} a_i x_{ij} \leq b_j \quad \text{for } j = 1, \ldots, n$$
$$x_{ij} \in \{0, 1\} \quad \text{for } i = 1, \ldots, m, \ j = 1, \ldots, n.$$

Discuss the relative merits of these two duals based on (i) the strength of the bound, (ii) ease of solution of the subproblems.

Exercise 8.7. In the setting of Sect. 8.1, consider the following alternative relaxation to the integer program (8.1). For every $\lambda \in \mathbb{R}_+^{m_1}$ let $z_{SD}(\lambda) := \max\{cx : \lambda A_1 x \leq \lambda b^1, x \in Q\}$. Define $z_{SD} := \inf_{\lambda \geq 0} z_{SD}(\lambda)$. Prove that $z_I \leq z_{SD} \leq z_{LD}$.

Exercise 8.8. In the setting of Sect. 8.1, the integer program (8.1) can be written equivalently as

$$z_I = \max \quad cx$$
$$x - y = 0$$
$$A_1 x \leq b^1$$
$$A_2 y \leq b^2$$
$$x_j, y_j \in \mathbb{Z} \quad \text{for } j = 1, \ldots, p$$
$$x, y \geq 0$$

Let $\bar{z}$ be the optimal solution of the Lagrangian dual obtained by dualizing the constraints $x - y = 0$. Prove that

$$\bar{z} = \max\{cx \,:\, x \in \operatorname{conv}(Q_1) \cap \operatorname{conv}(Q_2)\},$$

where $Q_i := \{x \in \mathbb{Z}_+^p \times \mathbb{R}_+^{n-p} \,:\, A_i x \leq b^i\}$, $i = 1, 2$, assuming that $\operatorname{conv}(Q_1) \cap \operatorname{conv}(Q_2)$ is nonempty.

Exercise 8.9. Show that, for every convex function $g : \mathbb{R}^n \to \mathbb{R}$ and every $\lambda^* \in \mathbb{R}^n$, there exists a subgradient of g at λ^*.

Exercise 8.10. Construct an example of a convex function $g : \mathbb{R}^n \to \mathbb{R}$ such that some subgradients at a point $\lambda^* \in \mathbb{R}^n$ are directions of ascent, whereas other subgradients are directions of descent. (A *direction of ascent* (resp. *descent*) at λ^* is a vector $s \in \mathbb{R}^n$ for which there exists $\epsilon > 0$ such that $g(\lambda^* + ts) > g(\lambda^*)$ (resp. $g(\lambda^* + ts) < g(\lambda^*)$) for all $0 < t < \epsilon$.)

Exercise 8.11. Show that, if (α_t) is a nonnegative sequence such that $\sum_{t=1}^{+\infty} \alpha_t$ is finite, then the subgradient algorithm converges to some point.

Construct an example of a convex function, a sequence (α_t) as above, and a starting point for which the subgradient algorithm converges to a point that is not optimal.

Exercise 8.12. Suppose we apply the subgradient method to solve the Lagrangian dual $\min_{\lambda \in \mathbb{R}^m} z_{LR}(\lambda)$, where $z_{LR}(\lambda)$ is the Lagrangian relaxation (8.6) for the uncapacitated facility location problem.

1. Specialize each of the steps 1–3 of the subgradient algorithm to this case.

2. In each iteration t, let $(x(\lambda^t), y(\lambda^t))$ be the optimal solution for (8.6) given in Proposition 8.7. Describe the best solution of (8.5) when each x_j is fixed to $x_j(\lambda^t)$, $j = 1, \ldots, n$.

8.5. EXERCISES

3. Point (2) gives a lower bound for (8.5). Can you use it to introduce an additional stopping criterion in the subgradient algorithm.

Exercise 8.13. In the context of the uncapacitated facility location problem, consider the function z defined as follows for any $x \in [0,1]^n$ such that $\sum_{j=1}^{n} x_j \geq 1$.

$$
\begin{array}{ll}
z(x) := \max \sum_{i=1}^{m}\sum_{j=1}^{n} c_{ij}y_{ij} - \sum_{j=1}^{n} f_j x_j & \\
\sum_{j=1}^{n} y_{ij} = 1 & \text{for all } i \\
y_{ij} \leq x_j & \text{for all } i,j \\
y \geq 0. &
\end{array}
$$

1. Prove that the function z is concave in the domain over which it is defined.

2. Determine a subgradient of z for any point in the set $S := \{x \in [0,1]^n : \sum_{j=1}^{n} x_j \geq 1\}$.

3. Specialize the subgradient algorithm to solve $\max_{x \in S} z$.

4. Show that $\max_{x \in S} z$ is equal to z_{LP} obtained by solving the linear programming relaxation of (8.5).

Exercise 8.14. Give a Dantzig–Wolfe reformulation of the uncapacitated facility location problem (8.5) based on the set $Q := \{(x,y) \in \{0,1\}^n \times \mathbb{R}^{m \times n} : y_{ij} \leq x_j \text{ for all } i,j\}$. [Hint: For each nonempty set $S \subseteq \{1,\ldots,m\}$ and $j \in \{1,\ldots,n\}$, let $\lambda_S^j = 1$ if a facility located at site j satisfies the demand of all clients in the set S, and 0 otherwise.]

Exercise 8.15. Consider the formulation for the network design problem given in Sect. 2.10.2.

1. Use the block diagonal structure, where each block corresponds to an arc $a \in A$, to derive a Dantzig–Wolfe reformulation, as described in Sect. 8.2.1.

2. Use the block diagonal structure, where each block corresponds to a commodity $k = 1,\ldots,K$, to derive a different Dantzig–Wolfe reformulation. (The reformulation will have a variable for every possible s_k, t_k-path, $k = 1,\ldots,K$.)

3. For each of these Dantzig–Wolfe reformulations, describe the pricing problem to solve the corresponding relaxation using column generation.

Exercise 8.16. In (8.26), replace the inequality constraints $Ax + Gy \leq b$ by equality constraints $Ax + Gy = b$. Explain how Theorem 8.18 and its proof must be modified.

Exercise 8.17. Let $X \subset \mathbb{R}^n$. Given $f : X \to \mathbb{R}$ and $F : X \to \mathbb{R}^m$, prove that the optimization problem

$$z_I := \max \quad f(x) + hy$$
$$F(x) + Gy \leq b$$
$$x \in X$$
$$y \in \mathbb{R}_+^p$$

can be reformulated in a form similar to (8.26) with cx and Ax replaced by $f(x)$ and $F(x)$ respectively.

Exercise 8.18. Consider a problem of the form

$$z_I := \max \quad cx$$
$$Ax + Gy \leq b$$
$$x \in X$$
$$y \in \mathbb{R}_+^p.$$

where $X \subset \mathbb{R}^n$. Show that its Benders reformulation is of the form

$$z_I = \max \quad cx$$
$$r^j(b - Ax) \geq 0 \quad \text{for all } j \in J$$
$$x \in X.$$

where $\{r^j\}_{j \in J}$ is the set of extreme rays of the cone $C := \{u \in \mathbb{R}_+^m : uG \geq 0\}$.

Exercise 8.19. Consider a problem of the form

$$z_I := \max \quad cx + \sum_{i=1}^m h^i y^i$$
$$A_i x + G_i y^i \leq b^i \quad i = 1, \ldots, m \quad (8.32)$$
$$x \in X$$
$$y^i \in \mathbb{R}_+^{p_i} \quad i = 1, \ldots, m$$

where $X \subset \mathbb{R}^n$.

For $i = 1, \ldots, m$, let $\{u^{ik}\}_{k \in K_i}$ denote the set of extreme points of the polyhedron $Q_i := \{u^i \geq 0 : u^i G_i \geq h^i\}$, and let $\{r^{ij}\}_{j \in J_i}$ be the set of extreme rays of the cone $C_i := \{u^i \geq 0 : u^i G_i \geq 0\}$.

8.5. EXERCISES

(i) Prove that problem (8.32) can be reformulated as

$$z_I = \max \sum_i \eta_i + cx$$
$$\eta_i \leq u^{ik}(b^i - A_i x) \quad \text{for all } k \in K_i, \, i = 1, \ldots, m$$
$$r^{ij}(b^i - A_i x) \geq 0 \quad \text{for all } j \in J_i, \, i = 1, \ldots, m$$
$$x \in X, \quad \eta \in \mathbb{R}^m.$$

(ii) Prove that in the standard Benders reformulation (8.27) of (8.32), $|K| = |K_1| \times |K_2| \times \cdots \times |K_m|$ and $|J| = |J_1| + |J_2| + \cdots + |J_m|$.

Exercise 8.20. The goal of this exercise is to find a Benders reformulation of the uncapacitated facility location problem.

$$\min \sum\sum c_{ij} y_{ij} + \sum f_j x_j$$
$$\sum_j y_{ij} = 1 \quad i = 1, \ldots, m$$
$$y_{ij} \leq x_j \quad i = 1, \ldots, m, \, j = 1, \ldots, n$$
$$y \geq 0, \, x \in \{0, 1\}^n.$$

(i) Show that, for every $x \in \{0,1\}^n$, the Benders subproblem can be written as $z_{LP}(x) = \sum_{i=1}^m z_{LP}^i(x)$, where

$$z_{LP}^i(x) := \min \sum_j c_{ij} y_{ij}$$
$$\sum_j y_{ij} = 1$$
$$y_{ij} \leq x_j \quad j = 1, \ldots, n$$
$$y \geq 0.$$

(ii) Characterize the extreme points and extreme rays of the polyhedron $Q_i := \{(u_i, w_i) \in \mathbb{R} \times \mathbb{R}_+^n : u_i - w_{ij} \leq c_{ij}\}$, $i = 1, \ldots, m$.

(iii) Deduce from (i) and (ii) that the uncapacitated facility location problem can be reformulated as

$$\min \sum_i \eta_i + \sum_j f_j x_j$$
$$\eta_i \geq c_{ik} - \sum_j (c_{ik} - c_{ij})^+ x_j \quad i = 1, \ldots, m, \, k = 1, \ldots, n$$
$$\sum_j x_j \geq 1$$
$$x \in \{0, 1\}^n.$$

The authors on a hike

Chapter 9

Enumeration

The goal of this chapter is threefold. First we present a polynomial algorithm for integer programming in fixed dimension. This algorithm is based on elegant ideas such as basis reduction and the flatness theorem. Second we revisit branch-and-cut, the most successful approach in practice for a wide range of applications. In particular we address a number of implementation issues related to the enumerative aspects of branch-and-cut. Finally we present an approach for dealing with integer programs that have a high degree of symmetry.

9.1 Integer Programming in Fixed Dimension

The *integer feasibility problem* "Does there exist $x \in \mathbb{Z}^n$ such that $Ax \leq b$?" is NP-complete in general. However, Lenstra [256] showed that this problem can be solved in polynomial time when n is a fixed constant. The key idea is that one can find in polynomial time either an integer point in $P := \{x : Ax \leq b\}$, or a direction d in which the polyhedron is *flat*, meaning that there is at most a fixed number $k(n)$ of parallel hyperplanes $dx = \delta_1, \ldots, dx = \delta_{k(n)}$ that contain all the integral points in P (if any). Applying this idea recursively to each of the $(n-1)$-dimensional polyhedra $P \cap \{x : dx = \delta_j\}$, the algorithm enumerates a fixed number of polyhedra overall, since n is fixed. It is not obvious that a flat direction always exists for a polyhedron that does not contain integral points and that it can be found in polynomial time. The proof is based on two ingredients,

Lovász's basis reduction algorithm and a result stating that for every full-dimensional polytope P on can compute in polynomial time an ellipsoid E such that $E \subset P \subset (n+1)E$.

9.1.1 Basis Reduction

A subset Λ of $\mathbb{R}^n$ is a *lattice* if there exists a basis $a^1, \ldots, a^n$ of $\mathbb{R}^n$ such that $\Lambda = \{\sum_{i=1}^n \lambda_i a^i \,:\, \lambda \in \mathbb{Z}^n\}$. The set of vectors $a^1, \ldots, a^n$ is called a *basis* of Λ, and Λ is said to be *generated by* the basis $a^1, \ldots, a^n$. For example, the lattice generated by the n unit vectors of $\mathbb{R}^n$ is $\mathbb{Z}^n$.

For ease of notation, throughout the section we identify a basis $a^1, \ldots, a^n$ with the square matrix $A = (a^1, \ldots, a^n)$. Given a square nonsingular matrix A, when we say "the lattice generated by A," we refer to the lattice generated by the columns of A.

Theorem 9.1. *Two nonsingular square matrices A and B generate the same lattice if and only if there exists a unimodular matrix U such that $B = AU$.*

Proof. Assume $B = AU$ for some unimodular matrix U. Since U is integral, by definition every column of B is in the lattice generated by A. On the other hand, $A = BU^{-1}$ and, by Lemma 1.16, U^{-1} is unimodular. So every column of A is in the lattice generated by B. It follows that A and B generate the same lattice.

Conversely, assume that A and B generate the same lattice. Then $B = AX$ for some integral matrix X and $A = BY$ for some integral matrix Y. Setting $U = X$, it follows that $U^{-1} = Y$. Since U and U^{-1} are integral matrices, U is unimodular by Lemma 1.16. □

Given a lattice $\Lambda \subseteq \mathbb{R}^n$, Theorem 9.1 implies that all the bases of Λ have the same determinant. This determinant is therefore called the *determinant of the lattice* Λ, and is denoted by $\det(\Lambda)$.

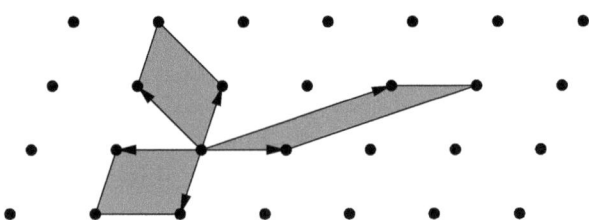

Figure 9.1: Three different bases in the plane; the three corresponding areas are equal

9.1. INTEGER PROGRAMMING IN FIXED DIMENSION

Any basis $B = (b^1, \ldots, b^n)$ of the lattice Λ satisfies *Hadamard's inequality*:

$$\det(\Lambda) \leq \|b^1\| \ldots \|b^n\|, \tag{9.1}$$

where $\|.\|$ denotes the Euclidean norm. Indeed, a classical result in linear algebra states that $|\det(B)|$ is the volume of the parallelepiped in $\mathbb{R}^n$ generated by the n vectors $b^1, \ldots, b^n$ (see [39] for example). Figure 9.1 illustrates this for three different bases of a lattice in the plane. Note that $\det(\Lambda) = \|b^1\| \ldots \|b^n\|$ if and only if the vectors b^i are pairwise orthogonal. Not every lattice has an orthogonal basis. However Hermite [200] showed that there always exists a basis that is fairly orthogonal in the sense that

$$\|b^1\| \ldots \|b^n\| \leq c(n)\det(\Lambda) \tag{9.2}$$

where the *orthogonality defect* $c(n)$ is a constant that only depends on the dimension n but not on the lattice Λ. Hermite's result implies that (9.2) holds if we choose $c(n) \geq (\frac{4}{3})^{n(n-1)/4}$.

Lovász showed that, for rational bases B, if one chooses $c(n) = 2^{n(n-1)/4}$, one can actually compute such an approximation to an orthogonal basis in polynomial time. The algorithm, attributed to Lovász by Lenstra in [256], was presented in a paper of Lenstra, Lenstra and Lovász [255], and is sometimes called the *LLL algorithm*. In this book we will call it *Lovász' basis reduction algorithm* or simply the *basis reduction algorithm*. The main objective of this section is to present this algorithm.

Consider a lattice Λ generated by n linearly independent vectors in $\mathbb{R}^n$. Lovász introduced the notion of a reduced basis, using a Gram–Schmidt orthogonal basis as a reference. The *Gram–Schmidt procedure* is as follows. Starting from a basis $B = (b^1, \ldots, b^n)$ of vectors in $\mathbb{R}^n$, define $g^1 := b^1$ and, recursively, for $j = 2, \ldots, n$, define g^j as the projection of b^j onto the orthogonal complement of the vector space spanned by $b^1, \ldots, b^{j-1}$. In other words,

$$\begin{aligned} g^1 &:= b^1 \\ g^j &:= b^j - \sum_{k=1}^{j-1} \mu_{jk} g^k \quad \text{for } 2 \leq j \leq n \end{aligned} \tag{9.3}$$

where

$$\mu_{jk} := \frac{(b^j)^T g^k}{\|g^k\|^2} \quad \text{for } 1 \leq k < j \leq n.$$

By construction, the *Gram–Schmidt basis* $G := (g^1, \ldots, g^n)$ is an orthogonal basis of $\mathbb{R}^n$ with the property that, for $j = 1, \ldots, n$, the vector spaces spanned by $b^1, \ldots, b^j$ and by $g^1, \ldots, g^j$ coincide. The coefficient μ_{jk}

is the kth coordinate of vector b^j relative to the orthogonal basis G. Note that $B = GR$ where $R = (r_{ij})_{i,j=1,\ldots,n}$ is the upper-triangular matrix whose diagonal elements are all 1 and where $r_{ij} = \mu_{ji}$ for $1 \leq i < j \leq n$. In particular, since all diagonal entries of R are equal to 1, it follows that $\det(R) = 1$ and $\det(B) = \det(G)$. Note that, since the vectors $g^1, \ldots, g^n$ are orthogonal, $G^T G$ is the diagonal matrix with entries $\|g^1\|^2, \ldots, \|g^n\|^2$. Since $\det(G^T G) = \det(G)^2$, we have $\|g^1\| \cdots \|g^n\| = |\det(G)| = |\det(B)| = \det(\Lambda)$. By construction, $\|b^j\| \geq \|g^j\|$ for $j = 1, \ldots, n$, thus Hadamard's inequality (9.1) holds.

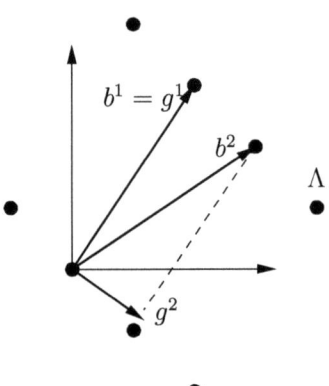

Figure 9.2: A basis (b^1, b^2) in $\mathbb{R}^2$ and its Gram–Schmidt orthogonal basis (g^1, g^2)

A basis $b^1, \ldots, b^n$ of the lattice Λ is said to be *reduced* if it satisfies the following two conditions

$$\begin{array}{lll}(i) & |\mu_{jk}| \leq \frac{1}{2} & \text{for } 1 \leq k < j \leq n, \\ (ii) & \|g^j + \mu_{j,j-1}g^{j-1}\|^2 \geq \frac{3}{4}\|g^{j-1}\|^2 & \text{for } 2 \leq j \leq n.\end{array} \qquad (9.4)$$

where $g^1, \ldots, g^n$ is the associated Gram–Schmidt orthogonal basis. Note that the basis (b^1, b^2) of Fig. 9.2 is not reduced because condition (i) is violated.

The next theorem shows that, if a basis of Λ is reduced, then it is "nearly orthogonal," in the sense that it satisfies (9.2) for $c(n) = 2^{n(n-1)/4}$. We will then explain Lovász' basis reduction algorithm, which given any basis B of Λ, produces a reduced basis in polynomial time.

Theorem 9.2. *Let $B = (b^1, \ldots, b^n)$ be a square nonsingular matrix. If B is reduced, then $\|b^1\| \cdots \|b^n\| \leq 2^{n(n-1)/4} \det(B)$.*

9.1. INTEGER PROGRAMMING IN FIXED DIMENSION

Proof. Let $g^1, \ldots, g^n$ be the Gram–Schmidt orthogonalization of $b^1, \ldots, b^n$. Since $g^1, \ldots, g^n$ are pairwise orthogonal, it follows that $\|g^j + \mu_{j,j-1} g^{j-1}\|^2 = \|g^j\|^2 + |\mu_{j,j-1}|^2 \|g^{j-1}\|^2$ for $j = 2, \ldots, n$. Since B is reduced, from (9.4) and the above equation we have that

$$\|g^j\|^2 \geq (\frac{3}{4} - |\mu_{j,j-1}|^2) \|g^{j-1}\|^2 \geq \frac{1}{2} \|g^{j-1}\|^2$$

for $j = 2, \ldots, n$. By induction, it follows that, for $1 \leq k < j \leq n$,

$$\|g^j\|^2 \geq 2^{k-j} \|g^k\|^2. \tag{9.5}$$

Furthermore, since $b^j = g^j + \sum_{k=1}^{j-1} \mu_{jk} g^k$, $j = 1, \ldots, n$,

$$
\begin{aligned}
\|b^j\|^2 &= \|g^j + \sum_{k=1}^{j-1} \mu_{jk} g^k\|^2 \\
&= \|g^j\|^2 + \sum_{k=1}^{j-1} |\mu_{jk}|^2 \|g^k\|^2 \quad \text{(because } g^1, \ldots, g^n \text{ are orthogonal)} \\
&\leq \|g^j\|^2 + \frac{1}{4} \sum_{k=1}^{j-1} \|g^k\|^2 \quad \text{(because } |\mu_{jk}| \leq \frac{1}{2}) \\
&\leq \|g^j\|^2 (1 + \frac{1}{4} \sum_{k=1}^{j-1} 2^{j-k}) \quad \text{(by (9.5))} \\
&\leq 2^{j-1} \|g^j\|^2
\end{aligned}
$$

This implies $\prod_{j=1}^{n} \|b^j\| \leq \prod_{j=1}^{n} 2^{(j-1)/2} \|g^j\| = 2^{n(n-1)/4} \prod_{j=1}^{n} \|g^j\| = 2^{n(n-1)/4} \det(B)$. □

For $a \in \mathbb{R}$, let $\lceil a \rfloor$ denote the closest integer to a. Each iteration of the basis reduction algorithm consists of two steps, a "normalization step" and a "swapping step."

Lovász' Basis Reduction Algorithm

Input: A rational basis B of a lattice Λ.
Output: A reduced basis B of Λ.

Step 1. (Normalization)
For $j := 2$ to n and for $k := j - 1$ down to 1, replace b^j by $b^j - \lceil \mu_{jk} \rfloor b^k$.

Step 2. (Swapping)
If Condition (9.4)(ii) holds, output the basis $b^1, \ldots, b^n$ and stop.
Else, find an index j, $2 \leq j \leq n$, that violates Condition (9.4)(ii) and interchange vectors b^{j-1} and b^j in the basis.

Note that the operations involved in the algorithm are unimodular operations (see Sect. 1.5.2), thus they do not change the lattice generated by B. This follows from Lemma 1.16 and Theorem 9.1. Let us now analyze the effect of the normalization step.

Lemma 9.3. *The Gram–Schmidt basis remains unchanged in Step 1 of the basis reduction algorithm.*

Proof. Let $\tilde{B}$ be the basis obtained from B after Step 1. Since $\tilde{b}^j$ equals b^j plus a linear combination of $b^1, \ldots, b^{j-1}$, it follows that $b^1, \ldots, b^j$ and $\tilde{b}^1, \ldots, \tilde{b}^j$ generate the same vector space L_j, $j = 1, \ldots, n$, and that the projections of b^j and $\tilde{b}^j$ into the orthogonal complement of L_{j-1} are identical. This implies that the Gram–Schmidt basis associated with B and $\tilde{B}$ is the same. □

Theorem 9.4. *When the basis reduction algorithm terminates, the basis $b^1, \ldots, b^n$ is reduced.*

Proof. The algorithm stops in Step 2 when Condition (9.4)(ii) is satisfied. To show that the basis is reduced, it suffices to show that, at each iteration, at the end of Step 1 the current basis always satisfies Condition (9.4)(i).

By Lemma 9.3, the Gram–Schmidt basis $g^1, \ldots, g^n$ remains unchanged in Step 1. Thus, if $\tilde{B}$ denotes the basis obtained from B after Step 1, we can express $\tilde{b}^j$ at the end of Step 1 as $\tilde{b}^j = g^j + \sum_{k=1}^{j-1} \tilde{\mu}_{jk} g^k$, $j = 1, \ldots, n$. We claim that $|\tilde{\mu}_{jk}| \leq \frac{1}{2}$, $1 \leq k < j \leq n$. Indeed, in a given iteration j^*, k^* in Step 1, among the several coefficients μ_{jk} that are modified, note that $\mu_{j^*k^*}$ is modified to $\mu_{j^*k^*} - \lfloor \mu_{j^*k^*} \rceil \leq \frac{1}{2}$. Furthermore, this new coefficient $\mu_{j^*k^*}$ remains unchanged in all subsequent iterations, since they involve $k < k^*$ or $j > j^*$. □

It is not obvious that this algorithm terminates, let alone that it terminates in polynomial time. Before giving the proof, we illustrate the algorithm on an example (Fig. 9.3).

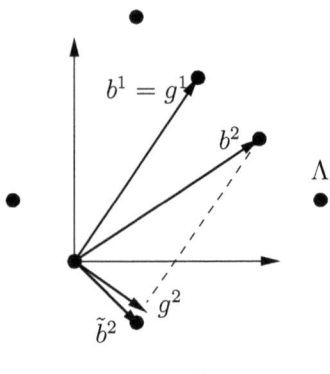

Figure 9.3: A basis (b^1, b^2) in $\mathbb{R}^2$ and a reduced basis $(\tilde{b}^2, b^1)$

9.1. INTEGER PROGRAMMING IN FIXED DIMENSION

Example 9.5. Consider the lattice Λ in $\mathbb{R}^2$ generated by the two vectors $b^1 = \begin{pmatrix} 2 \\ 3 \end{pmatrix}$ and $b^2 = \begin{pmatrix} 3 \\ 2 \end{pmatrix}$. The Gram–Schmidt basis is $g^1 = b^1$ and $g^2 = b^2 - \mu_{21} g^1$. We compute $\mu_{21} = \frac{(b^2)^T g^1}{\|g^1\|^2} = \frac{12}{13}$. Since Condition (9.4)(i) is not satisfied we replace b^2 by $\tilde{b}^2 := b^2 - b^1$ (Step 1 of Lovász' algorithm). $\tilde{b}^2$ is shorter than b^2, as suggested by the name "reduced basis." For the new basis $(b^1, \tilde{b}^2)$ we have $\mu_{21} = \frac{-1}{13}$, therefore Condition (9.4)(i) is satisfied. However Condition (9.4)(ii) is violated: Since $\|\tilde{b}^2\|^2 = 2$ and $\|b^1\|^2 = 13$, we have $\|g^2 + \mu_{21} g^1\|^2 = \|\tilde{b}^2\|^2 < \frac{3}{4}\|b^1\|^2$. Hence we interchange b^1 and $\tilde{b}^2$ putting the shortest vector in front (Step 2 of Lovász's algorithm). The reader can check that now both Conditions (9.4)(i) and (ii) are satisfied. Therefore the basis $\begin{pmatrix} 1 \\ -1 \end{pmatrix}$ and $\begin{pmatrix} 2 \\ 3 \end{pmatrix}$ is a reduced basis for the lattice Λ. ∎

Theorem 9.6. *Lovász' basis reduction algorithm terminates, and it runs in polynomial time in the input size of the original basis.*

Proof. We will prove that the algorithm terminates in a polynomial number of iterations. To prove that it runs in polynomial time, one should also show that the encoding size of the numbers remains polynomially bounded at every iteration; however we will not do it here (see [325] for a proof).

We may assume that B is an integral matrix. Indeed, given an integer number δ such that δB is integral, the basis reduction algorithm applied to δB executes the same iterations as for B, only the matrix produced at each iteration is multiplied by δ. Note that, if we apply the algorithm to an integral matrix, the basis remains integral at every iteration.

To prove finiteness of the algorithm, we use the following potential function associated with a basis $B = (b^1, \ldots, b^n)$, expressed in terms of the Gram–Schmidt basis associated with B:

$$\Phi(B) := \|g^1\|^{2n} \|g^2\|^{2n-2} \cdots \|g^n\|^2.$$

We observed that the Gram–Schmidt basis remains unchanged in Step 1 of the basis reduction algorithm (Lemma 9.3). Therefore the potential function Φ only changes in Step 2 when vectors b^j and b^{j-1} are interchanged.

Let B be the basis at a given iteration, and let us denote by $\tilde{B} = (\tilde{b}^1, \ldots, \tilde{b}^n)$ the basis obtained from B by interchanging b^j and b^{j-1}. We will prove that $\Phi(\tilde{B})/\Phi(B) \le 3/4$ and that the potential is always integer. These two facts imply that the algorithm terminates after $O(\log \Phi(B))$ steps. In particular, the number of iterations is polynomial in the encoding size of the input matrix B. Indeed, if M is the largest absolute value of an entry of B, then $\Phi(B) \leq \det(B)^{2n} \leq (n^n M^n)^{2n} = (nM)^{2n^2}$, hence the number of iterations is $O(n^2 \log(nM))$.

Let $\tilde{g}^1, \ldots, \tilde{g}^n$ be the Gram–Schmidt orthogonalization of $\tilde{B}$. Since $b^h = \tilde{b}^h$ for all $h \neq j-1, j$, $1 \leq h \leq n$, it follows that $g^h = \tilde{g}^h$ for all $h \neq j-1, j$. Furthermore, since $b^j = g^j + \sum_{k=1}^{j-1} \mu_{jk} g^k$, it follows that $g^j + \mu_{j,j-1} g^{j-1}$ is the projection of b^j onto the orthogonal complement of the space generated by $b^1, \ldots, b^{j-2}$, thus $\tilde{g}^{j-1} = g^j + \mu_{j,j-1} g^{j-1}$. We also have that $\|g^1\| \cdots \|g^n\| = |\det(G)| = |\det(\tilde{G})| = \|\tilde{g}^1\| \cdots \|\tilde{g}^n\|$, therefore $\|g^{j-1}\| \|g^j\| = \|\tilde{g}^{j-1}\| \|\tilde{g}^j\|$. It then follows that

$$
\begin{aligned}
\frac{\Phi(\tilde{B})}{\Phi(B)} &= \frac{\|\tilde{g}^{j-1}\|^{2(n-j+2)} \|\tilde{g}^j\|^{2(n-j+1)}}{\|g^{j-1}\|^{2(n-j+2)} \|g^j\|^{2(n-j+1)}} \\
&= \frac{(\|\tilde{g}^{j-1}\| \|\tilde{g}^j\|)^{2(n-j+1)}}{(\|g^{j-1}\| \|g^j\|)^{2(n-j+1)}} \cdot \frac{\|\tilde{g}^{j-1}\|^2}{\|g^{j-1}\|^2} \\
&= \frac{\|g^j + \mu_{j,j-1} g^{j-1}\|^2}{\|g^{j-1}\|^2} < \frac{3}{4},
\end{aligned}
$$

where the last inequality follows from the fact that b^{j-1} and b^j are interchanged at Step 2 if they violate Condition (9.4)(ii).

We finally show that $\Phi(B)$ is an integer for every integral matrix B. Let us denote by B_i the matrix with columns $b^1, \ldots, b^i$, and G_i the matrix with columns $g^1, \ldots, g^i$, $i = 1, \ldots, n$. By (9.3), $B_i = G_i R_i$, where R_i is an $i \times i$ upper-triangular matrix with all diagonal elements equal to 1. It follows that $\det(B_i^T B_i) = \det(G_i^T G_i) = \|g^1\|^2 \cdots \|g^i\|^2$, where the last equality follows from the fact that $g^1, \ldots, g^n$ are pairwise orthogonal. This shows that $\|g^1\|^2 \cdots \|g^i\|^2$ is integer for $i = 1, \ldots, n$. Thus $\Phi(B) = \|g^1\|^{2n} \|g^2\|^{2n-2} \cdots \|g^n\|^2 = \prod_{i=1}^n (\|g^1\|^2 \cdots \|g^i\|^2)$ is integer as well. □

9.1.2 The Flatness Theorem and Rounding Polytopes

Let $K \subseteq \mathbb{R}^n$ be a *convex body*, that is, a closed bounded convex set. Given a vector $d \in \mathbb{R}^n$, we define the *width of K along d* to be

$$w_d(K) = \max_{x \in K} d^T x - \min_{x \in K} d^T x.$$

The *lattice width of K* is defined as the minimum width along any **integral** vector d, that is

$$w(K) = \min_{d \in \mathbb{Z}^n} w_d(K).$$

9.1. INTEGER PROGRAMMING IN FIXED DIMENSION

The fundamental result used in Lenstra's algorithm for integer programming in fixed dimension is Khinchine's *flatness theorem* [237]. The theorem states that any full-dimensional convex body, either contains an integral point, or is "fairly flat," in the sense that its lattice width is bounded by some constant that depends only on the dimension. Figure 9.4 shows that such a constant is greater than 2 in the plane (Hurkens [208] showed that $1 + \frac{2}{\sqrt{3}} \approx 2.155$ is tight).

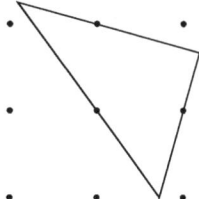

Figure 9.4: A lattice-free triangle with lattice width greater than 2

Theorem 9.7 (Flatness Theorem). *Let $K \subseteq \mathbb{R}^n$ be a full-dimensional convex body. If K does not contain any integral point, then $w(K) \leq k(n)$, where $k(n)$ is a constant depending only on n.*

For rational polytopes, the proof will lead to a polynomial time algorithm that, for any full-dimensional polytope P expressed as a system of rational linear inequalities, outputs either an integral point in P or a *flat direction* for P, that is a vector $d \in \mathbb{Z}^n$ such that $w_d(P) \leq n(n+1)2^{n(n-1)/4}$.

The first step is proving the flatness theorem for ellipsoids. An ellipsoid in $\mathbb{R}^n$ is an affine transformation of the unit ball. That is, an *ellipsoid centered at a* is a set of the form $E(C, a) = \{x \in \mathbb{R}^n : \|C(x-a)\| \leq 1\}$, where $a \in \mathbb{R}^n$ and C is an $n \times n$ nonsingular matrix.

Theorem 9.8 (Flatness Theorem for Ellipsoids). *Let $E \subseteq \mathbb{R}^n$ be an ellipsoid. If E does not contain any integral point, then $w(E) \leq n 2^{n(n-1)/4}$.*

Proof. Let $a \in \mathbb{R}^n$ and $C \in \mathbb{R}^{n \times n}$ be a nonsingular matrix such that $E = E(C, a)$. For any $d \in \mathbb{R}^n$, we first compute $w_d(E)$. Let us view d as a row vector. We have $\max\{dx : \|C(x-a)\| \leq 1\} = da + \max\{dC^{-1}y : \|y\| \leq 1\} = da + \|dC^{-1}\|$, where we have applied the change of variables $y = C(x-a)$, and the maximum is achieved by $y = (dC^{-1})^T/\|dC^{-1}\|$. Therefore, for every $d \in \mathbb{R}^n$,

$$w_d(E) = \max_{x \in E} dx - \min_{x \in E} dx = 2\|dC^{-1}\|. \tag{9.6}$$

Let Λ be the lattice generated by C, and let B be a basis of Λ satisfying $\|b^1\| \cdots \|b^n\| \leq 2^{n(n-1)/4} |\det(B)|$. (While the existence of such a basis B is implied by the basis reduction algorithm when C is rational, the general case follows from a result of Hermite [200] mentioned in Sect. 9.1.1. See inequality (9.2).) Since $|\det(B)|$ is invariant under permuting the columns of B, we may assume that b^n is the element of maximum norm in B, that is $\|b^n\| = \max_{j=1,\ldots,n} \|b^j\|$.

Since C and B are bases of the same lattice, by Theorem 9.1 there exists a unimodular matrix U such that $C = BU$. Let $\lambda = Ua$, and define $\bar{x} = U^{-1}\lfloor \lambda \rfloor$ where $\lfloor \lambda \rfloor = (\lfloor \lambda_1 \rfloor, \ldots, \lfloor \lambda_n \rfloor)$. Note that $\bar{x}$ is integral. Define the vector $d \in \mathbb{Z}^n$ to be the last row of U. We will show that, if $\bar{x} \notin E$, then $w_d(E) \leq n 2^{n(n-1)/4}$. This will conclude the proof of the theorem.

Assume that $\bar{x} \notin E$. Then $\|C(a - \bar{x})\| > 1$, that is, $\|B(\lambda - \lfloor \lambda \rfloor)\| > 1$. Hence

$$1 < \|\sum_{j=1}^{n} (\lambda_j - \lfloor \lambda_j \rfloor) b^j\| \leq \sum_{j=1}^{n} |\lambda_j - \lfloor \lambda_j \rfloor| \|b^j\| \leq \frac{n}{2} \|b^n\|. \qquad (9.7)$$

Consider the Gram–Schmidt orthogonal basis $g^1, \ldots, g^n$ obtained from $b^1, \ldots, b^n$. We have $|\det(B)| = \|g^1\| \cdots \|g^n\|$. Since $\|b^1\| \cdots \|b^n\| \leq 2^{n(n-1)/4} \|g^1\| \cdots \|g^n\|$ and $\|b^j\| \geq \|g^j\|$ for $j = 1, \ldots, n$, it follows that $\|b^n\| \leq 2^{n(n-1)/4} \|g^n\|$. The latter and (9.7) imply

$$\|g^n\| > 2/(n 2^{n(n-1)/4}). \qquad (9.8)$$

We now evaluate $w_d(E)$. Let us denote by v the last row of B^{-1}. Then $w_d(E) = 2\|dC^{-1}\| = 2\|v\|$, since $C^{-1} = U^{-1} B^{-1}$ and d is the last row of U. Since g^n is orthogonal to $g^1, \ldots, g^{n-1}$, it follows from (9.3) that $(g^n)^T b^j = 0$ for $j = 1, \ldots, n-1$ and $(g^n)^T b^n = \|g^n\|^2$. In particular, $v = (g^n)^T / \|g^n\|^2$. Thus, by (9.8), $w_d(E) = 2\|g^n\|/\|g^n\|^2 \leq n 2^{n(n-1)/4}$. $\square$

Note that, when C and a are rational, the elements $\bar{x}$ and d defined in the proof of Theorem 9.8 can be computed in polynomial time, since this amounts to computing a reduced basis of Λ and solving systems of linear equations. The proof shows that $\bar{x} \in E(C, a)$ or $w_d(E(C, a)) \leq n 2^{n(n-1)/4}$. This proves the following.

Remark 9.9. *There is a polynomial-time algorithm that, given $a \in \mathbb{Q}^n$ and a nonsingular matrix $C \in \mathbb{Q}^{n \times n}$, either finds an integral point in the ellipsoid $E(C, a)$, or finds a vector $d \in \mathbb{Z}^n$ such that $w_d(E(C, a)) \leq n 2^{n(n-1)/4}$.*

9.1. INTEGER PROGRAMMING IN FIXED DIMENSION

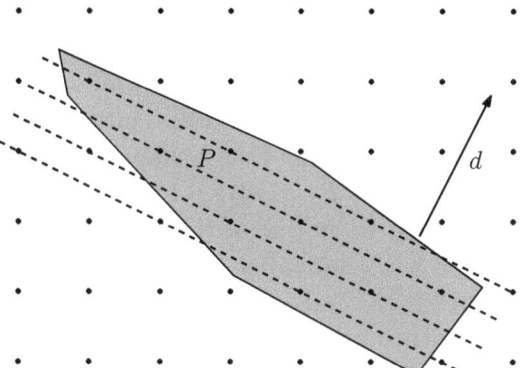

Figure 9.5: The *dashed lines* represent the hyperplanes $d^T x = k$ that intersect the polytope P (where $d = (1, 2)$ and k is an integer). Every integral point in P must lie in one of the four *dashed lines*

Proof of the Flatness Theorem (Theorem 9.7). The proof relies on the flatness theorem for ellipsoids and the following theorem of Löner (reported by Danzer, Grünbaum, and Klee [105]) and John [214].

For every full-dimensional convex body K, there exists an ellipsoid $E(C, a)$ such that $E(nC, a) \subseteq K \subseteq E(C, a)$.

Note that $E(nC, a)$ is obtained by scaling $E(C, a)$ by $1/n$ around its center. Given such an ellipsoid $E(C, a)$, we know from Theorem 9.8 that either $E(nC, a)$ contains an integral point, and so does K because $E(nC, a) \subseteq K$, or there exists $d \in \mathbb{Z}^n$ such that $w_d(E(nC, a)) \leq n2^{n(n-1)/4}$. It then follows that $w_d(K) \leq w_d(E(C, a)) = nw_d(E(nC, a)) \leq n^2 2^{n(n-1)/4}$. The statement of Theorem 9.7 then follows if we choose $k(n) = n^2 2^{n(n-1)/4}$. □

The scaling factor of $1/n$ in Löner and John's theorem is the best possible. A result of Goffin [174] implies that there exists a polynomial time algorithm which, given a full-dimensional polytope $P = \{x \in \mathbb{R}^n : Ax \leq b\}$ expressed by a rational system of linear inequalities, computes an ellipsoid $E(C, a)$ such that $E((n+1)C, a) \subseteq P \subseteq E(C, a)$. This improves an earlier algorithm of Lenstra [256] guaranteeing a scaling factor of $\frac{1}{2}n^{-3/2}$. Following the proof of the flatness theorem, the above discussion and Remark 9.9 imply the following.

Theorem 9.10. *There exists a polynomial time algorithm that, given $A \in \mathbb{Q}^{m \times n}$ and $b \in \mathbb{Q}^m$ such that $P = \{x \in \mathbb{R}^n : Ax \leq b\}$ is a full-dimensional polytope, outputs either an integral point in P, or a vector $d \in \mathbb{Z}^n$ such that $w_d(P) \leq n(n+1)2^{n(n-1)/4}$.*

9.1.3 Lenstra's Algorithm

Theorem 9.10 leads immediately to the basic idea of Lenstra's algorithm: given a rational system $Ax \leq b$ in n variables, either find an integral point in $P := \{x \in \mathbb{R}^n : Ax \leq b\}$, or find a flat direction for P, namely a direction $d \in \mathbb{Z}^n$ such that $w_d(P) \leq n(n+1)2^{n(n-1)/4}$. Since d is integral, every point in $P \cap \mathbb{Z}^n$ must lie in one of the $(n-1)$-dimensional polytopes $P \cap \{x : d^T x = k\}$, for $k = \lceil \min_{x \in P} d^T x \rceil, \ldots, \lfloor \max_{x \in P} d^T x \rfloor$. Since there are at most $n(n+1)2^{n(n-1)/4} + 1$ such polytopes and n is a constant, we need to apply the algorithm recursively to a constant number of polytopes of lower dimension (see Fig. 9.5).

However, Theorem 9.10 applies to *full-dimensional, bounded* polyhedra. We need to address the following two technical issues: what to do if P is not bounded, and what to do if P is not full-dimensional (which will be the case at every iteration, since polytopes of the form $P \cap \{x : d^T x = k\}$ are not full-dimensional).

We first address the issue of non-boundedness of P. By Corollary 4.37, if we denote by L the maximum among the encoding sizes of the coefficients of (A, b), there exists an integer valued function f of n and L such that the encoding size of $f(n, L)$ is polynomially bounded by n and L and such that P has an integral point if and only if $P' := \{x \in P : -f(n, L) \leq x \leq f(n, L)\}$ has an integral point. Thus we only need to check if P' contains an integral point. We have therefore reduced to the case that the input system $Ax \leq b$ defines a polytope.

If P is not full-dimensional, then by Theorem 3.17 the system $Ax \leq b$ must include some implicit equality. An implicit equality can be determined in polynomial time by solving the m linear programs $\beta_i := \min\{a_i x : Ax \leq b\}$, $i = 1, \ldots, m$, where $a_1, \ldots, a_m$ are the rows of A, and checking if $\beta_i = b_i$. Thus, in polynomial time, we can determine a rational equation $\alpha x = \beta$ such that $P \subseteq \{x : \alpha x = \beta\}$. Possibly by multiplying the equation by the greatest denominators of the entries of α, we may assume that α is an integral vector with relatively prime entries. If β is not integer, then P does not contain any integral point and we are done. If β is integer, the next lemma shows how we can reduce to a problem with $n - 1$ variables.

Lemma 9.11. *Let $A \in \mathbb{Q}^{m \times n}$, $b \in \mathbb{Q}^m$, $\alpha \in \mathbb{Z}^n$, $\beta \in \mathbb{Z}$ be such that the entries of α are relatively prime. There exists a matrix $D \in \mathbb{Z}^{n \times (n-1)}$ and a vector $b' \in \mathbb{Q}^m$ such that the system $Ax \leq b$, $\alpha x = \beta$ has an integral solution if and only if the system $ADy \leq b'$ has an integral solution.*

9.1. INTEGER PROGRAMMING IN FIXED DIMENSION

Proof. Since all entries of α are relatively prime, by Corollary 1.9, $\alpha x = \beta$ has an integral solution $\bar{x}$, and there exists a unimodular matrix U such that $\alpha U = e^1$, where e^1 denotes the first unit vector in $\mathbb{R}^n$. If we define D as the $n \times (n-1)$ matrix formed by the last $n-1$ columns of U, we have that $\{x \in \mathbb{Z}^n : \alpha x = \beta\} = \{\bar{x} + Dy : y \in \mathbb{Z}^{n-1}\}$. Therefore, if we let $b' = b - A\bar{x}$, we have

$$\{x \in \mathbb{Z}^n : Ax \le b, \alpha x = \beta\} = \{\bar{x} + Dy : ADy \le b', y \in \mathbb{Z}^{n-1}\}. \quad (9.9)$$

In particular $Ax \le b$, $\alpha x = \beta$ has an integral solution if and only if $ADy \le b'$ has an integral solution. □

Note that the matrix D and vector b' in the statement of Lemma 9.11 can be computed in polynomial time by Remark 1.13. It thus follows from the above discussion that one can reduce to the case where P is full-dimensional. We are now ready to formally present Lenstra's algorithm.

Lenstra's Algorithm

Input: A matrix $A \in \mathbb{Q}^{m \times n}$ and a vector $b \in \mathbb{Q}^m$ such that $P := \{x \in \mathbb{R}^n : Ax \le b\}$ is a full-dimensional polytope.
Output: "Yes" if P contains an integral point, "No" otherwise.

    Apply the algorithm of Theorem 9.10.

    If the outcome is an integral point in P, then output "Yes."

    If the outcome is a vector $d \in \mathbb{Z}^n$ such that $w_d(P) \le n(n+1)2^{(n(n-1)/4)}$, do the following;

        If $\lceil \min_{x \in P} d^T x \rceil > \lfloor \max_{x \in P} d^T x \rfloor$, output "No."
        Else, for $k = \lceil \min_{x \in P} d^T x \rceil, \ldots, \lfloor \max_{x \in P} d^T x \rfloor$,

            Compute a matrix $D \in \mathbb{Z}^{n \times (n-1)}$ and a vector $b' \in \mathbb{Q}^m$ such that the system $Ax \le b$, $d^T x = k$ has an integral solution if and only if the system $ADy \le b'$ has an integral solution.
            Apply Lenstra's algorithm recursively to the system $ADy \le b'$.

Lenstra [256] observed that the above algorithm can be modified to solve mixed integer linear programming problems with a fixed number of integer variables (and arbitrarily many continuous variables) in polynomial time.

9.2 Implementing Branch-and-Cut

Lenstra's algorithm is of great theoretical interest for solving integer programs. However, in practice, the most successful solvers are currently based on the branch-and-cut approach introduced in Chap. 1. This is because many applications have a combinatorial flavor (a large number of 0,1 variables) rather than a number theoretic origin (a small number of interconnected variables that must take integer values, possibly in a wide range). The complexity of Lenstra's algorithm explodes as the number of variables exceeds a few dozens. On the other hand, integer programs with thousands of 0,1 variables are solved routinely in several application areas using software based on the branch-and-cut method.

In this section we return to branch-and-cut. Many implementation issues were left open in Chap. 1: branching strategy, node selection strategy, heuristics for getting feasible solutions, and many others. To better understand the range of questions that arise, consider a mixed integer linear program (MILP) and suppose that we just solved its linear programming relaxation; let $\bar{z}$ be the optimum value and $\bar{x}$ the optimal solution of this linear programming relaxation. What should one do if $\bar{x}_j$ is fractional for at least one of the variables that are required to be integer in the MILP that we are trying to solve? Should one generate cutting planes in the hope of improving the linear programming relaxation, or should one branch? When branching, should one use a strategy in the spirit of Lenstra's algorithm, creating smaller subproblems in parallel hyperplanes along a thin direction, or should one simply branch on one of the integer variables x_j for which $\bar{x}_j$ is fractional, setting $x_j \leq \lfloor \bar{x}_j \rfloor$ on one branch, and $x_j \geq \lceil \bar{x}_j \rceil$ on the other. In this case, which variable should we choose for branching? We should favor choices that help pruning the enumeration tree faster: infeasibility or integrality of the subproblems that we create, and pruning by bounds. The goal of pruning by bounds is best achieved by strategies that generate good upper and lower bounds on the optimal value of the MILP. To obtain these bounds, heuristics and cutting plane generation need to be integrated in the solver. These components are all interconnected. By now, the reader should have realized that building an efficient branch-and-cut solver is a sophisticated affair. Let us discuss some of the key issues.

Consider the MILP

$$\begin{aligned} z_I = \quad & \max cx \\ & Ax \leq b \\ & x \geq 0 \\ & x_j \text{ integer for } j = 1, \ldots, p. \end{aligned} \qquad (9.10)$$

9.2. IMPLEMENTING BRANCH-AND-CUT

The data are a vector $c \in \mathbb{Q}^n$, an $m \times n$ rational matrix A, a vector $b \in \mathbb{Q}^m$, and an integer p such that $1 \leq p \leq n$. The set $I := \{1, \ldots, p\}$ indexes the integer variables whereas the set $C := \{p+1, \ldots, n\}$ indexes the continuous variables.

The branch-and-cut algorithm keeps a list of linear programming problems obtained by relaxing the integrality requirements on the variables and imposing linear constraints such as bounds on the variables: $x_j \leq u_j$ or $x_j \geq l_j$, and/or cutting planes. Each such linear program corresponds to a *node* of the branch-and-cut tree. For a node N_i, let z_i denote the value of the corresponding linear program LP_i. Let $\mathcal{L}$ denote the list of nodes that must still be solved (i.e., that have not been pruned nor branched on). Initially, $\mathcal{L}$ just contains node N_0 corresponding to (MILP). Let $\underline{z}$ denote a lower bound on the optimum value z_I (initially, the bound $\underline{z}$ can be derived from a heuristic solution of (MILP), or it can be set to $-\infty$).

Branch-and-Cut Algorithm

0. **Initialize**

    $\mathcal{L} = \{N_0\}$, $\underline{z} = -\infty$, $x^* = \emptyset$. One may also apply preprocessing (to improve the formulation LP_0) and heuristics (to improve $\underline{z}$ and to obtain a feasible solution x^*).

1. **Terminate?**

    If $\mathcal{L} = \emptyset$, the solution x^* is optimal (If $x^* = \emptyset$, (MILP) is infeasible or unbounded).

2. **Select node**

    Choose a node N_i in $\mathcal{L}$ and delete it from $\mathcal{L}$.

3. **Bound**

    Solve LP_i. If it is unbounded, (MILP) is unbounded, stop. If it is infeasible, go to Step 1. Else, let x^i be an optimal solution of LP_i and z_i its objective value.

4. **Prune**

    If $z_i \leq \underline{z}$, go to Step 1.

    If x^i is feasible to (MILP), let $\underline{z} = z_i$, $x^* = x^i$. Delete from $\mathcal{L}$ any node N_j for which a bound z_j is known and satisfies $z_j \leq \underline{z}$. Go to Step 1.

    If x^i is not feasible to (MILP), go to Step 5.

5. **Add Cuts?**

   Decide whether to strengthen the formulation LP_i or to branch.

   In the first case, strengthen LP_i by adding cutting planes and go back to Step 3.

   In the second case, go to Step 6.

6. **Branch**

   From LP_i, construct $k \geq 2$ linear programs $LP_{i_1}, \ldots, LP_{i_k}$ with smaller feasible regions whose union does not contain (x^i, y^i), but contains all the solutions of LP_i with $x \in \mathbb{Z}^n$. Add the corresponding new nodes $N_{i_1}, \ldots, N_{i_k}$ to $\mathcal{L}$ and go to Step 1.

Various choices have been left open in this algorithm. In particular, five issues need special attention when solving integer programs by branch-and-cut.

- Preprocessing (to decrease the size of the instance whenever possible),
- Heuristics (to find a good lower bound $\underline{z}$ on z_I),
- Cutting plane generation (to reduce the upper bound $\bar{z}$ obtained when solving the linear programming relaxation),
- Branching,
- Node selection.

We discuss branching strategies first, followed by node selection strategies, heuristics, preprocessing, and cutting plane generation.

Branching

Although Step 6 of the branch-and-cut algorithm provides the flexibility of branching into any number $k \geq 2$ of subproblems, the most popular choice is to generate $k = 2$ disjoint subproblems (Exercise 9.14 helps to understand why this is a good strategy). A natural way of generating two disjoint subproblems is by using a split disjunction $\sum_{j=1}^{p} \pi_j x_j \leq \pi_0$ or $\sum_{j=1}^{p} \pi_j x_j \geq \pi_0 + 1$ where $(\pi, \pi_0) \in \mathbb{Z}^{p+1}$ is chosen such that x^i satisfies $\pi_0 < \sum_{j=1}^{p} \pi_j x_j^i < \pi_0 + 1$. The simplest such disjunction is obtained by choosing π to be a unit vector. This branching strategy is called *variable branching* and it is by far the most widely used in integer programming solvers.

9.2. IMPLEMENTING BRANCH-AND-CUT

Specifically, let x_j^i be one of the fractional values for $j = 1, \ldots, p$, in the optimal solution x^i of LP_i (we know that there is such a j, since otherwise N_i would have been pruned in Step 4 on account of x^i being feasible to (MILP)). From problem LP_i, we can construct two linear programs LP_{ij}^- and LP_{ij}^+ that satisfy the requirements of Step 6 by adding the constraints $x_j \leq \lfloor x_j^i \rfloor$ and $x_j \geq \lceil x_j^i \rceil$ respectively to LP_i. An advantage of variable branching is that the number of constraints in the linear programs does not increase, since linear programming solvers treat bounds on variables implicitly.

An important question is: On which variable x_j should the algorithm branch, among the $j = 1, \ldots, p$ such that x_j^i is fractional? To answer this question, it would be very helpful to know the decrease D_{ij}^- in objective value between LP_i and LP_{ij}^-, and D_{ij}^+ between LP_i and LP_{ij}^+. A good branching variable x_j at node N_i is one for which both D_{ij}^- and D_{ij}^+ are relatively large (thus tightening the upper bound z_i, which is useful for pruning). For example, a reasonable strategy is to choose j such that the product $D_{ij}^- \times D_{ij}^+$ is the largest.

The strategy which consists of computing D_{ij}^- and D_{ij}^+ explicitly for each j is called *strong branching*. It involves solving linear programs that are small variations of LP_i by performing dual simplex pivots, for each $j = 1, \ldots, p$ such that x_j^i is fractional. Experiments indicate that strong branching reduces the size of the enumeration tree by one or more orders of magnitude in most cases, relative to a simple branching rule such as branching on the most fractional variable. Thus there is a clear benefit to spending time on strong branching. But the computing time of doing it at each node N_i, for every fractional variable x_j^i, may be too high. This suggests the following idea, based on the notion of *pseudocosts* that are initialized at the root node and then updated throughout the branch-and-bound tree.

Let $f_j^i = x_j^i - \lfloor x_j^i \rfloor$ be the fractional part of x_j^i, for $j = 1, \ldots p$. For an index j such that $f_j^i > 0$, define the *down pseudocost* and *up pseudocost* as

$$P_j^- = \frac{D_{ij}^-}{f_j^i} \quad \text{and} \quad P_j^+ = \frac{D_{ij}^+}{1 - f_j^i}$$

respectively. Benichou et al. [48] observed that the pseudocosts tend to remain fairly constant throughout the branch-and-bound tree. Therefore the pseudocosts need not be computed at each node of the tree. They can be estimated instead. How are they initialized and how are they updated in the tree? A good way of initializing the pseudocosts is through strong branching at the root node or other nodes of the tree when new variables become fractional for the first time (or the first r times, where $r \geq 2$, to get more reliable initial estimates). To update the estimate $\hat{P}_j^-$ of the pseudocost P_j^-, the algorithm averages the observations $\frac{D_{ij}^-}{f_j^i}$ over all the nodes N_i of the tree in which x_j was branched on or in which D_{ij}^- was computed through strong branching. Similarly for the estimate $\hat{P}_j^+$ of the up pseudocost. The decision of which variable to branch on at the current node N_i of the tree is then made based on these estimated pseudocosts $\hat{P}_j^-$ and $\hat{P}_j^+$ as follows. For each $j = 1, \ldots, p$ such that $f_j^i > 0$, compute estimates of D_{ij}^- and D_{ij}^+ by the formula $\hat{D}_{ij}^- = \hat{P}_j^- f_j^i$ and $\hat{D}_{ij}^+ = \hat{P}_j^+ (1 - f_j^i)$. Branch on the variable x_j with the largest value of the product $\hat{D}_{ij}^- \times \hat{D}_{ij}^+$.

Variable branching is not the only branching strategy that is implemented in state-of-the-art solvers. Many integer programs contain constraints of the form

$$\sum_{t=1}^{k} x_{j_t} = 1$$

with $x_{j_t} = 0$ or 1, for $t = 1, \ldots k$. If one or more of these variables is fractional in the current solution x^i, variable branching would pick one such variable, say x_{j^*}, and set $x_{j^*} = 0$ on one branch and $x_{j^*} = 1$ on the other. This leads to unbalanced trees since only one variable is fixed on the first branch but all variables x_{j_t}, $t = 1, \ldots k$, are fixed on the other. A strategy that better balances the tree is to partition the set $\{j_1, \ldots, j_k\}$ into $J' \cup J''$ so that both $\sum_{j_t \in J'} x_{j_t}^i > 0$ and $\sum_{j_t \in J''} x_{j_t}^i > 0$ and the cardinalities of J' and J'' are roughly the same. One then fixes $x_{j_i} = 0$ for all $j_i \in J'$ on one branch and $x_{j_i} = 0$ for all $j_i \in J''$ on the other. This branching scheme is known to reduce the size of the enumeration tree significantly in practice. It is often called GUB branching (GUB stands for Generalized Upper Bound). SOS branching is a variation on this idea (SOS stands for Special Ordered Sets).

Node Selection

How does one choose among the different problems N_i available in Step 2 of the algorithm? Two goals need to be considered: finding a better feasible solution (thus increasing the lower bound z) and proving optimality of the current best feasible solution (by decreasing the upper bound as quickly as possible).

For the first goal, we estimate the value of the best feasible solution in each node N_i. For example, we could use the following estimate:

$$E_i = z_i - \sum_{j=1}^{p} \min(\hat{P}_j^- f_j^i, \hat{P}_j^+ (1 - f_j^i))$$

based on the pseudocost estimates defined earlier. This corresponds to rounding the noninteger solution x^i to a nearby integer solution and using the pseudocosts to estimate the degradation in objective value. We then select a node N_i with the largest E_i. This is the so-called best estimate criterion node selection strategy. A good feasible solution may be found by "diving" from node N_i, computing estimates E_j for its possible sons N_j, selecting a son with the largest estimate, and then repeating from the selected node N_j. Diving heuristics will be revisited in the next section.

For the second goal, the best strategy depends on whether the first goal has been achieved already. If we currently have a very good lower bound $\underline{z}$, it is reasonable to adopt a depth-first search strategy. This is because the linear programs encountered in a depth-first search are small variations of one another. As a result they can be solved faster in sequence, using the dual simplex method initialized with the optimal solution of the father node (they are solved about ten times faster, based on empirical evidence). On the other hand, if no good lower bound is available, depth-first search tends to be wasteful: it might explore many nodes N_i with a value z_i smaller than the optimum z_I. Assuming that we have a good lower bound $\underline{z}$, and that we adopt a depth-first search strategy whenever possible, what should we do when we reach a node of the tree that can be pruned? An alternate to the "best estimate criterion" is the "best bound" node selection strategy, which consists in picking a node N_i with the largest bound z_i. No matter how good a solution of (MILP) is found in other nodes of the branch-and-bound tree, the node with the largest bound z_i cannot be pruned by bounds (assuming no ties) and therefore it will have to be explored eventually. So we might as well explore it first.

The most successful node selection strategy may differ depending on the application. For this reason, most integer programming solvers have several node selection strategies available as options to the user. The default strategy is usually a combination of the "best estimate criterion" (or a variation) and depth-first search. Specifically, the algorithm may dive using depth-first search until it reaches an infeasible node N_i or it finds a feasible solution of (MILP). At this point, the next node might be chosen using the "best estimate criterion" strategy, and so on, alternating between dives in a depth-first search fashion to get feasible solutions at the bottom of the tree and the "best estimate criterion" to select the next most promising node.

Heuristics

Several types of heuristics are routinely applied in integer programming solvers. Heuristics help to improve the bound $\underline{z}$, which is used in Step 4 for pruning the enumeration tree. Heuristic solutions are even more important when the branch-and-cut algorithm has to be terminated before completion, returning a solution of value $\underline{z}$ without a proof of its optimality. We present two successful ideas, *diving* and *local branching*. Each can be applied at any node N_i of the branch-and-cut tree and can be implemented in many different variations. In some applications, even finding a feasible solution might be an issue, in which case heuristics such as the *feasibility pump* are essential.

Diving heuristics can be viewed as depth-first searches in the context of the node selection strategy presented above. One chooses an integer variable x_j that is fractional in the current linear programming solution $\bar{x}$ and adds the constraint $x_j \leq \lfloor \bar{x}_j \rfloor$ or $x_j \geq \lceil \bar{x}_j \rceil$; one then solves the new linear program; the process is repeated until a solution of (MILP) is found or infeasibility is reached. Solvers usually contain several different heuristic rules for choosing the variable x_j and the constraint to add in this procedure. One option is to choose a variable with smallest fractionality $\min(\bar{f}_j, 1-\bar{f}_j)$ among the integer variables x_j that have nonzero fractionality at $\bar{x}$, where $\bar{f}_j = \bar{x}_j - \lfloor \bar{x}_j \rfloor$, and to add the constraint $x_j \leq \lfloor \bar{x}_j \rfloor$ if $\bar{f}_j < \frac{1}{2}$, and $x_j \geq \lceil \bar{x}_j \rceil$ otherwise. Other more sophisticated rules use strong branching or pseudo-cost information to choose the variable x_j to branch on, and move down the branch that corresponds to the largest estimate of the objective value of the integer program at the children nodes, using the function E_i introduced earlier. Diving heuristics can be repeated from a variety of starting points N_i to improve the chance of getting good solutions.

Once a feasible solution x^* is available, *local branching* [140] can be applied to try to improve upon it. For simplicity of exposition, assume that all the integer variables are 0,1 valued. The idea is to define a neighborhood of x^* as follows:

$$\sum_{j=1}^{p} |x_j - x_j^*| \leq k$$

where k is an integer chosen by the user (for example $k = 20$ seems to work well), to then add this constraint to (MILP) and to apply your favorite integer programming solver. Instead of getting lost in a huge enumeration tree, the search is restricted to the neighborhood of x^* by this constraint. Note that the constraint should be linearized before adding it to the formulation, which is easy to do:

$$\sum_{j \in I : x_j^* = 0} x_j + \sum_{j \in I : x_j^* = 1} (1 - x_j) \leq k.$$

If a better solution than x^* is found, the neighborhood is redefined relatively to this new solution, and the procedure is repeated until no better solution is found. This heuristic can be modified into appealing variations, by changing the definition of the neighborhood. One such example is called RINS (which stands for relaxation induced neighborhood search) [101]. RINS needs a feasible solution x^* and a solution $\bar{x}$ of some linear program in the branch-and-cut tree. It fixes the variables that have the same value in x^* and $\bar{x}$. The resulting smaller integer program is then processed by the integer programming solver (for a limited time or a limited number of nodes) in the hope of finding a better feasible solution than x^*. This recursive use of the integer programming solver is a clever feature of local branching heuristics.

For some instances of (MILP), just finding a feasible solution might be an issue. Heuristics such as the *feasibility pump* are specifically designed for this purpose [139]. The main idea is to construct two sequences of points that hopefully converge to a feasible solution of (MILP). Let P denote the polyhedron defined by the linear constraints of (MILP). One sequence consists of points in P, possibly integer infeasible, the other one of integer feasible points, but possibly not in P. These two sequences are produced by alternately rounding a point $\bar{x}^i \in P$ to an integer feasible point x^i, and finding a point $\bar{x}^{i+1}$ in the polyhedron P that is closest to x^i (using the ℓ_1-norm) by solving a linear program. All current integer programming solvers incorporate some variant of this basic idea (see [6] for an improved feasibility pump).

Preprocessing

Integer programming solvers try to tighten and simplify the formulation before launching into a full-fledged branch-and-cut. These preprocessing steps can also be performed when subproblems are created in the course of the enumeration process. They involve simple cleaning operations: identifying infeasibilities or redundancies in the constraints, improving bounds on the variables, improving constraint coefficients, fixing variables, and identifying logical implications. We give a sample of such steps. Let

$$\sum_{j \in B} a_j x_j + \sum_{j \in C} g_j y_j \leq b \qquad (9.11)$$

be a constraint of the integer programming formulation where B denotes the set of 0,1 variables and C denotes the set of remaining variables (both the continuous variables and the general integer variables). Let $B_+ := \{j \in B : a_j > 0\}$, $B_- := \{j \in B : a_j < 0\}$, $C_+ := \{j \in C : g_j > 0\}$ and $C_- := \{j \in C : g_j < 0\}$. Assume that the variables in C are bounded:

$$\ell_j \leq y_j \leq u_j \quad \text{for } j \in C.$$

The smallest and largest values that the LHS of (9.11) can take are, respectively,

$$L_{\min} := \sum_{j \in B_-} a_j + \sum_{j \in C_-} g_j u_j + \sum_{j \in C_+} g_j \ell_j$$

$$L_{\max} := \sum_{j \in B_+} a_j + \sum_{j \in C_-} g_j \ell_j + \sum_{j \in C_+} g_j u_j.$$

We can now perform the following checks:

- If $L_{\min} > b$, infeasibility has been identified,

- If $L_{\max} \leq b$, redundancy has been identified,

- If $u_k > \frac{b - L_{\min} + g_k \ell_k}{g_k}$ for some $k \in C_+$, then the bound u_k can be improved to the RHS value,

- If $\ell_k < \frac{b - L_{\min} + g_k u_k}{g_k}$ for some $k \in C_-$, then the bound ℓ_k can be improved to the RHS value,

- If $L_{\min} + a_k > b$ for some $k \in B_+$, then $x_k = 0$ (variable fixing),

- If $L_{\min} - a_k > b$ for some $k \in B_-$, then $x_k = 1$ (variable fixing),

- If $a_k > L_{\max} - b$, for some $k \in B_+$, then the constraint coefficient a_k and the RHS b can both be reduced by $a_k - (L_{\max} - b)$,

- If $a_k < b - L_{\max}$, for some $k \in B_-$, then the constraint coefficient a_k can be increased to the RHS value $b - L_{\max}$.

Performing these checks on all the constraints in the formulation only takes linear time. Whenever some improvement is discovered, an additional pass through the constraints is performed. Other standard preprocessing steps are typically applied, such as reduced cost fixing (Exercise 9.18), implication based fixing, and identification of clique inequalities (Exercise 9.19). We refer the reader to the presentation of Savelsbergh [321] for additional details. Preprocessing is surprisingly effective in practice as can be seen from the table given in Sect. 5.3.

Cut Pool

When a node N_i is explored, cuts may be generated to strengthen the formulation, thus improving the bound z_i. Some cuts may be local (i.e., valid only at node N_i and its descendants) or global (valid at all the nodes of the branch-and-bound tree). A good strategy is to store cuts in a cut pool instead of adding them permanently to the formulation. The reason is that solving the linear programs tends to be slowed significantly when a large number of cuts are present. Thus, if a cut has become inactive in the linear program at a node N_i, it is moved to the cut pool. At later stages, the algorithm searches the cut pool for those inequalities that are violated by the current fractional point $\bar{x}$, and moves these cuts back into the linear program.

Software

Currently, Gurobi, Cplex, and Xpress are excellent commercial branch-and-cut codes. `Scip` and `cbc` are open source.

9.3 Dealing with Symmetries

Some problems in integer programming have a highly symmetric structure. This means that there are ways of permuting the variables and constraints that leave the formulation invariant. This implies that the feasible region of the linear programming relaxation is also symmetric. This can be a major issue in branch-and-cut algorithms, since they may end up solving many subproblems that are essentially identical. For example, in the operating room scheduling problem (2.17) in Sect. 2.8 with n identical operating rooms, permuting the indices in $\{1, \ldots, n\}$ does not change the structure of the constraints. So, for each solution y of the linear programming relaxation, there are $n!$ equivalent solutions, one for each permutation π, obtained by setting $y'_{ij} = y_{i\pi(j)}$.

Various approaches have been tried to break the symmetry in branch-and-cut, such as perturbing the coefficients in the formulation or adding symmetry breaking constraints. The most successful approach to date is to ensure that only one isomorphic copy of each node is kept in the enumeration tree. A way of efficiently constructing such a tree was presented by Margot [267] in the context of integer programming. We will present some of these ideas in this section. To simplify the treatment, we will only present it in the context of pure 0, 1 programs.

Before doing this we need to introduce some basic terminology about permutations. A *permutation* on n-elements is a bijection $\pi : \{1, \ldots, n\} \to \{1, \ldots, n\}$. We denote by Σ_n the set of all permutations on n elements.

We will represent π by the n-vector $(\pi(1), \ldots, \pi(n))$. Given $S \subseteq \{1, \ldots, n\}$, we denote by $\pi(S) = \{\pi(i) : i \in S\}$ the image of S under π. Given $v \in \mathbb{R}^n$, we denote by $\pi(v)$ the vector $(v_{\pi(1)}, \ldots, v_{\pi(n)})$. Given a matrix $A \in \mathbb{R}^{m \times n}$ and permutations $\pi \in \Sigma_n$, $\sigma \in \Sigma_m$, we denote by $A(\pi, \sigma)$ the matrix obtained by permuting the columns of A according to π and the rows according to σ.

Consider a pure $0, 1$ linear program (BIP)

$$\begin{aligned} \max \quad & cx \\ & Ax \leq b \\ & x \in \{0,1\}^n. \end{aligned} \qquad (9.12)$$

The set of permutations

$$\Gamma = \{\pi \in \Sigma_n : \exists\, \sigma \in \Sigma_m \text{ such that } \pi(c) = c,\ \sigma(b) = b,\ A(\pi, \sigma) = A\} \qquad (9.13)$$

is called the *symmetry group of (BIP)*. It is not difficult to show that Γ is indeed a group, that is, (i) the identity is in Γ, (ii) if $\pi \in \Gamma$, then $\pi^{-1} \in \Gamma$, (iii) if $\pi^1, \pi^2 \in \Gamma$, then $\pi^1 \circ \pi^2 \in \Gamma$. Note that, for every feasible solution $\bar{x}$ of the linear programming relaxation of (9.12), and for every $\pi \in \Gamma$, $\pi(\bar{x})$ is also feasible and it has the same objective value. The vectors $\bar{x}$ and $\pi(\bar{x})$ are said to be *isomorphic solutions*. Observe that the definition of Γ depends on the choice of formulation, and not only on the of geometry of the linear programming relaxation. For example, multiplying a constraint by a positive number, adding or removing redundant constraints may change the symmetry group.

Isomorphism Pruning

Consider a branch-and-bound algorithm for solving (BIP), in which we perform branching on the variables. At any node N_a of the enumeration tree, let F_a^0 and F_a^1 be the set of indices of the variables that have been fixed to 0 and to 1 respectively. Two nodes N_a and N_b of the enumeration tree are *isomorphic* if there exists some permutation $\pi \in \Gamma$ in the symmetry group of (BIP) such that $\pi(F_b^1) = F_a^1$ and $\pi(F_b^0) = F_a^0$. We are interested in strategies that avoid enumerating isomorphic subproblems. One obvious possibility is to check, every time a new node is generated, if there is an isomorphic node already in the tree, but this is computationally impractical because checking isomorphism is too expensive to be performed multiple times at every node.

9.3. DEALING WITH SYMMETRIES

We now present *isomorphism pruning*, which can be performed locally at each node without explicitly checking isomorphism to other nodes of the tree. Despite its simplicity, isomorphism pruning guarantees that we never solve two isomorphic subproblems.

Let N_a be a node of the enumeration tree, and let $p_1, \ldots, p_\ell$ be the sequence of indices of the variables that have been branched on, on the path from the root to node N_a.

Isomorphism Pruning Rule. *If there exists $\pi \in \Gamma$ and $t \in \{1, \ldots, \ell\}$ such that*

$$\begin{aligned} p_i \in F_a^1 &\Rightarrow \pi(p_i) \in F_a^1, \quad i = 1, \ldots, t-1; \\ p_t \in F_a^0, &\quad \pi(p_t) \in F_a^1; \end{aligned} \qquad (9.14)$$

then prune node N_a.

Example 9.12. Consider a 0,1 linear program (BIP) with three variables and assume that $\pi = (3, 1, 2)$ belongs to the symmetry group of (BIP). This implies that, whenever $(\bar{x}_1, \bar{x}_2, \bar{x}_3)$ is feasible to (BIP), $(\bar{x}_3, \bar{x}_1, \bar{x}_2)$ is also feasible to (BIP) and has the same objective value. Consider the enumeration tree of Fig. 9.6. We can apply the isomorphic pruning rule with the above permutation π and $t = 2$: indeed $1 \in F_a^1$ and $\pi(1) = 3 \in F_a^1$; and $2 \in F_a^0$ and $\pi(2) = 1 \in F_a^1$. The isomorphic pruning rule tells us to prune node N_a. Note that it makes sense in this case because node N_b must still be solved and it contains the solution $(1, 1, 0)$ which is isomorphic to the solution $(1, 0, 1)$ that we are pruning in node N_a. ∎

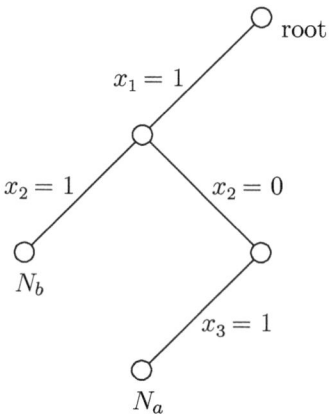

Figure 9.6: Example of isomorphic pruning

Next we explain why the isomorphism pruning rule ensures that we never consider isomorphic subproblems, and that we always keep at least one optimal solution of (BIP). We assume that, when branching on a variable x_k at a given node in the enumeration tree, we create two children, the *left child* obtained by fixing $x_k = 1$, and the *right child* obtained by fixing $x_k = 0$. Given two nodes N_a and N_b in the enumeration tree, we say that N_b is *to the left* of N_a if N_b is a descendant of the left child of the common ancestor N_d of N_a and N_b, and N_a is a descendant of the right child of N_d. We say that a vector $\bar{x}$ is a *solution in node* N_a if it is feasible for (BIP) and $\bar{x}_i = 1$ for all $i \in F_a^1$, $\bar{x}_i = 0$ for all $i \in F_a^0$ (in particular $\bar{x}$ must be a $0,1$ vector).

Proposition 9.13. *If node N_a has an isomorphic node to its left in the enumeration tree, then node N_a is pruned by isomorphism.*

Conversely, if node N_a is pruned by isomorphism, then for every solution $\bar{x}$ in node N_a there is a node N_b to the left of node N_a containing a solution isomorphic to $\bar{x}$.

Proof. Suppose N_a has an isomorphic node N_b to its left. Then there exists a permutation $\pi \in \Gamma$ such that $\pi(F_b^1) = F_a^1$ and $\pi(F_b^0) = F_a^0$. Let N_d be the common ancestor of N_a and N_b and p_t be the index of the branching variable at node N_d. Thus, for $i = 1, \ldots, t-1$, if $p_i \in F_a^1$, then $p_i \in F_b^1$, because the paths from the root to N_a and N_b coincide up to node N_d. By definition of π, this implies that $\pi(p_i) \in F_a^1$ for $i = 1, \ldots, t-1$. Since N_b is a descendant of the left child of N_d, and N_a is a descendant of the right child of N_d, it follows that $p_t \in F_a^0$ and $p_t \in F_b^1$. Again by definition of π, this implies that $\pi(p_t) \in F_a^1$. This shows that π and t satisfy the conditions (9.14). Therefore N_a is pruned by isomorphism.

For the second part of the statement, assume that N_a is pruned by isomorphism. Let π and t satisfy conditions (9.14). Given a solution $\bar{x}$ in N_a, let $\tilde{x} = \pi(\bar{x})$. Consider the minimum index k such that $\bar{x}_{p_k} \neq \tilde{x}_{p_k}$. It follows from the conditions (9.14) that $k \leq t$ and $\bar{x}_{p_k} = 0$, $\tilde{x}_{p_k} = 1$. Thus, if N_d is the kth node on the path from the root to N_a (that is, N_d is the node where we branched on x_{p_k}), it follows that $\tilde{x}$ is feasible for the left child N_b of N_d. Hence $\bar{x}$ is isomorphic to a feasible solution of N_b, which is a node to the left of N_a. □

The above proposition shows that a branch-and-bound algorithm that implements isomorphism pruning will produce an optimal solution of (BIP).

Various procedures are often incorporated in branch-and-bound to fix variables at the nodes, such as reduced cost fixing or implication based

9.3. DEALING WITH SYMMETRIES

fixing. Let x^* be the current best feasible solution found. A variable x_j can be fixed to 0 (resp. to 1) at a node N_a of the enumeration tree when it is determined that no better solution $\bar{x}$ than x^* can exist in N_a with $\bar{x}_j = 1$ (resp. $\bar{x}_j = 0$). We call any such procedure *variable fixing by bounds*. We can describe such a fixing in the enumeration tree by branching at node N_a on variable x_j, and pruning one of the two children by bound. Namely, we prune the left child if no better solution than x^* can exist in N_a with $x_j = 1$, and similarly for the right child. With this convention, all variables fixed in the enumeration tree are branching variables. Therefore these fixing procedures are still valid even when isomorphism pruning is performed.

We also remark that the proof of Proposition 9.13 depends on the set of solutions of (BIP) in a give node N_a, and not on the specific linear programming relaxation at that node. In particular, one can add cutting planes to the formulation at node N_a as long as they are valid for the set of 0,1 solutions in N_a.

The above discussion implies that one can incorporate isomorphism pruning in a general branch-and-cut framework for (BIP).

In order to implement the isomorphism pruning rule, one needs to compute the symmetry group Γ defined in (9.13). Furthermore, at each node of the enumeration tree one needs to compute a permutation π satisfying (9.14), if any exists. The computation of Γ needs to be performed only once. Note that applying isomorphism pruning using a subgroup of Γ instead of Γ itself will still produce a correct algorithm, although in this case there may be isomorphic nodes in the enumeration tree. It is often the case that the user has knowledge of the group Γ, or at least a large subgroup of Γ, as for example in the operating room scheduling problem. The group Γ can be also generated automatically using tools from computational group theory. While the status of computing a set of generators for Γ in polynomial time is an open problem, there is software that runs efficiently in practice, such as nauty by McKay [275]. Furthermore, permutation groups can be represented in a compact form, called the Schreier–Sims representation (see for example [328]). Algorithms that are based on this representation, to detect a permutation π satisfying (9.14) if any exists, are typically efficient in practice [267].

In principle, one would like to work with a larger group of permutations than Γ, namely the group Γ' of all permutations π such that, for any $\bar{x} \in \mathbb{R}^n$, $\bar{x}$ is an optimal solution for (BIP) if and only if $\pi(\bar{x})$ is optimal for (BIP). Using Γ' for isomorphism pruning would still produce a correct algorithm (as this would guarantee that at least one optimal solution is kept in the enumeration tree), while resulting in smaller trees. However, computing Γ' is not possible without "a priori" knowledge of the symmetries of the optimal solutions, and in practice one is able to detect symmetries only if they are "displayed" by the formulation.

Orbital Fixing

We consider a branch-and-cut algorithm that performs the following operations on the enumeration tree: addition of valid cutting planes at nodes of the tree, branching on variables, pruning by bound, by infeasibility, by integrality or by isomorphism, and variable fixing by bound as introduced earlier. Recall that the latter operation can be represented in the enumeration tree by branching on variables and pruning by bound.

In addition to these operations, we describe how symmetry in the formulation can be exploited to fix additional variables.

Given a group G of permutations in Σ_n and an element $i \in \{1, \ldots, n\}$, the *orbit of i under G* is the set $\{\pi(i) : \pi \in G\}$. Given $i, j \in \{1, \ldots, n\}$, i is in the orbit of j if and only if j is in the orbit of i (indeed, $j = \pi(i)$ for $\pi \in G$ if and only if $i = \pi^{-1}(j)$ for $\pi^{-1} \in G$). Therefore the orbits of the elements $1, \ldots, n$ under G form a partition. Given a set $S \subset \{1, \ldots, n\}$, the *stabilizer of S in G* is the set $\mathrm{stabil}(S, G) := \{\pi \in G : \pi(S) = S\}$. One can show that $\mathrm{stabil}(S, G)$ is also a group.

Let N_a be a node of the enumeration tree and let $\mathcal{O}_a$ be the family of orbits of the elements $1, \ldots, n$ under $\mathrm{stabil}(F_a^1, \Gamma)$.

Orbital Fixing Rule. Let Z be the union of all orbits $O \in \mathcal{O}_a$ such that $O \cap F_a^0 \neq \emptyset$ and $O \setminus F_a^0 \neq \emptyset$. If $Z \neq \emptyset$, create a child $N_{a'}$ of N_a and set $F_{a'}^0 := F_a^0 \cup Z$.

We say that the variables in $Z \setminus F_a^0$ have been *fixed to zero by orbital fixing* at node N_a.

Example 9.14. We illustrate orbital fixing on the following 0,1 linear program, where $n \geq 3$.

$$\begin{array}{ll} \max & \sum_{j=1}^n x_j \\ & x_i + x_j \leq 1 \quad \text{for all } i, j \\ & x \in \{0, 1\}^n. \end{array} \quad (9.15)$$

A branch-and-bound algorithm first solves the linear programming relaxation. The optimum of this linear program puts all variables at $\frac{1}{2}$ for an

9.3. DEALING WITH SYMMETRIES

objective value of $\frac{n}{2}$. Branching is done on variable x_1, say. The branch where x_1 is fixed to 1 can be pruned by integrality (in this subproblem, the linear program sets $x_j = 0$ for all $j \neq 1$ and it has objective value 1). Let us now concentrate on the branch where x_1 is fixed to 0. See Fig. 9.7.

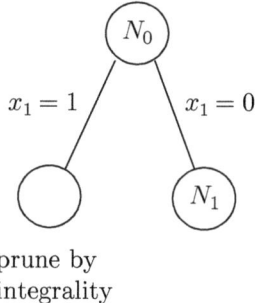

Figure 9.7: Branch-and-bound tree when orbital fixing is applied

Let N_1 denote the corresponding node in the branch-and-bound tree. No variable has been fixed to 1 on the path from N_0 to N_1, therefore $F_1^1 = \emptyset$. In our example, the symmetry group of the formulation (9.15) is Σ_n, the set of all permutations on $1, \ldots, n$. The stabilizer $\mathrm{stabil}(F_1^1, \Sigma_n)$ is Σ_n itself. The family $\mathcal{O}_1$ of orbits of the elements $1, \ldots, n$ under $\mathrm{stabil}(F_1^1, \Sigma_n)$ consists of the single orbit $O = \{1, \ldots, n\}$. Since $F_1^0 = \{1\}$, it follows that $O \cap F_1^0 \neq \emptyset$ and $O \setminus F_1^0 \neq \emptyset$. Therefore the set Z defined in the orbital fixing rule is $Z = O = \{1, \ldots, n\}$. As a consequence we can fix $x_2 = x_3 = \ldots = x_n = 0$ by orbital fixing in node N_1. Since all variables are now fixed in N_1, this node can be pruned and the branch-and-bound algorithm terminates. ∎

Proposition 9.15. *Consider a node N_a of the enumeration tree. If $\bar{x}$ is a solution in node N_a such that there exists an orbit $O \in \mathcal{O}_a$ satisfying $O \cap F_a^0 \neq \emptyset$ and an index $i \in O$ such that $\bar{x}_i = 1$, then there exists a node N_b to the left of node N_a containing a solution isomorphic to $\bar{x}$.*

Proof. Let P be the path of the enumeration tree from the root N_0 to node N_a. We will prove the statement by induction on the length of P. The statement is true when $N_a = N_0$. Now consider a general node N_a. Let $\bar{x}$ satisfy the hypothesis. Let S be the union of all orbits $O \in \mathcal{O}_a$ such that $O \cap F_a^0 \neq \emptyset$ and $\bar{x}_i = 1$ for some $i \in O$. Among all indices in $S \cap F_a^0$, choose h to be the index of a variable x_h that was first fixed to zero going from N_0 to N_a in P. Let O be the orbit in $\mathcal{O}_a$ that contains h. By definition of S, there exists $i \in O$ such that $\bar{x}_i = 1$. Since i, h belong to O, it follows that there exists $\pi \in \text{stabil}(F_a^1, \Gamma)$ such that $\pi(h) = i$. Let $\tilde{x} = \pi(\bar{x})$. Note that $\tilde{x}_h = 1$.

Let N_d be the last node on P where variable x_h was not yet fixed to zero. We next show that $\tilde{x}$ is a solution in N_d. Since $\pi \in \text{stabil}(F_a^1, \Gamma)$, it follows that $\tilde{x}_j = 1$ for every $j \in F_a^1$, and thus $\tilde{x}_j = 1$ for every $j \in F_d^1$ because $F_d^1 \subseteq F_a^1$. Thus, if $\tilde{x}$ is not a solution in node N_d, there exists $j \in F_d^0$ such that $\tilde{x}_j = 1$. This implies that $\bar{x}_{\pi(j)} = 1$. But then j is an index in $S \cap F_a^0$ such that the variable x_j was fixed before x_h, contradicting the choice of h.

Now, variable x_h has been fixed along the path P either by branching or by orbital fixing. Suppose x_h was fixed by branching. By the choice of node N_d, x_h must be the branching variable at node N_d, and N_a is a descendant of the right child of N_d. It follows that $\tilde{x}$ is a solution in the left child N_b of N_d, because $\tilde{x}_h = 1$ and $\tilde{x}$ is a solution in N_d. Thus $\tilde{x}$ is a solution isomorphic to $\bar{x}$ in a node to the left of N_a.

Henceforth, we may assume that x_h was fixed to 0 by orbital fixing at node N_d. Let O' be the orbit of h under $\text{stabil}(F_d^1, \Gamma)$. Since h is fixed by orbital fixing at N_d, it follows that $O' \cap F_d^0 \neq \emptyset$. Since $\tilde{x}_h = 1$ and the length of the path from the root N_0 to N_d is less than the length of P, it follows by induction that there exists a node N_b to the left of N_d containing a solution isomorphic to $\tilde{x}$. Since $\tilde{x}$ is isomorphic to $\bar{x}$, N_b contains a solution isomorphic to $\bar{x}$ and the statement follows. $\square$

It is important to note that Proposition 9.13 remains true even when isomorphism pruning is used in conjunction with orbital fixing. Therefore the two rules can both be added to a general branch-and-cut algorithm for (BIP).

9.4 Further Readings

Integer Programming in Fixed Dimension

Surveys on lattice basis reduction and integer programming in fixed dimension are given by Aardal, Weismantel and Wolsey [4] and by Eisenbrand (in [217, pp. 505–560]).

The first papers to apply basis reduction for solving integer programs were those of Hirschberg and Wong [202], who gave a polynomial algorithm for the knapsack problem in two variables, and Kannan [225], who gave a polynomial algorithm for the two-variable integer programming problem.

9.4. FURTHER READINGS

A question related to basis reduction is to find the shortest nonzero vector in a lattice. Ajtai [9] shows that the shortest vector problem in L2 is NP-hard for randomized reductions, and Micciancio [277] shows that the shortest vector in a lattice is hard to approximate to within some constant.

Kannan [226, 227] provides a variant of Lenstra's algorithm, based on a basis reduction algorithm different from the Lovász' algorithm, which improves the number of hyperplanes to be considered in the recursion to $O(n^{5/2})$. Lovász and Scarf [263] describe an algorithm for integer programming in fixed dimension that avoids computing a rounding ellipsoid in order to find a flat direction. Cook, Rutherford, Scarf and Shallcross [91] report a successful application of Lovász and Scarf's algorithm to a class of small but difficult mixed integer linear programming problems with up to 100 integer variables, arising from a network design application, that could not otherwise be solved by regular branch-and-bound techniques.

Banaszczyk, Litvak, Pajor, and Szarek [35] show that the flatness theorem holds for $k(n) = Cn^{3/2}$, where C is a universal constant. The best choice for $k(n)$ is not known. Note that $k(n) \geq n$ since the simplex $\text{conv}\{0, ne^1, \ldots, ne^n\}$ has width n and its interior does not contain any integer point (e^i denotes the ith unit vector).

John [214] shows that every full-dimensional convex body $K \subseteq \mathbb{R}^n$ contains a unique ellipsoid of maximum volume E^*, and that $K \subseteq nE^*$. If $P = \{x : Ax \leq b\}$ is a full-dimensional polytope and E^* is the maximum-volume inscribed ellipsoid in P, one can compute $E \subseteq P$ such that $(1 + \varepsilon)\text{vol}(E) \geq \text{vol}(E^*)$ in time polynomial in the encoding size of (A, b) and in $\log(\varepsilon^{-1})$ using the shallow cut ellipsoid method. Nesterov and Nemirovsky [291] show how to solve such a problem in polynomial time using interior point methods.

It is NP-hard to compute the volume of a polytope, and of a convex body more generally. However, Dyer, Frieze, and Kannan [122] give a random polynomial-time algorithm for approximating these volumes.

A more general problem than the integer feasibility problem is the question of counting integral points in a polyhedron. Barvinok [38] gave a polynomial-time algorithm for counting integral points in polyhedra when the dimension is fixed.

A topic that we do not cover in this book is that of strongly polynomial algorithms. We just mention two papers: Frank and Tardos [149], and Tardos [334].

Computational Aspects of Branch-and-Bound

Much work has been devoted to branching. Related to the topic of basis reduction, we cite Aardal, Bixby, Hurkens, Lenstra, and Smeltink [1], Aardal and Lenstra [2], Aardal, Hurkens, and Lenstra [3].

Strong branching was proposed by Applegate, Bixby, Chvátal, and Cook [13] in the context of the traveling salesman problem.

Computational studies on branching rules for mixed integer linear programming were performed by Linderoth and Savelsbergh [257], Achterberg, Koch, and Martin [7, 8], Achterberg [5], Patel and Chinneck [306], and Kilinc–Karzan, Nemhauser, and Savelsbergh [238].

One aspect that deserves more attention is the design of experiments to evaluate empirical aspects of algorithms. Hooker [205] makes this case in his paper "Needed: An empirical science of algorithms." We also mention "Experimental analysis of algorithms" by McGeogh [274].

Dealing with Symmetry

While the concept of isomorphism pruning in the context of integer programming was first introduced by Margot [267], similar search algorithms have been independently developed in the computer science and constraint programming communities by, among others, Brown, Finkelstein and Purdom [65] and Fahle, Shamberger, and Sellmann [134].

In the context of packing and covering problems, Ostrowski, Linderoth, Rossi, and Smriglio [293] introduced the idea of *orbital branching*. Consider a node a of the enumeration tree and let Γ^a be the symmetry group of the formulation at node a. Suppose we want to branch on the variable x_k. Orbital branching creates two subproblems, the left problem where x_k is set equal to 1, and the right problem where all variables with index in the orbit of k under the group Γ^a are set to 0. This branching strategy excludes many isomorphic solutions, while guaranteeing that an optimal solution is always present in the enumeration tree. However, there is no guarantee that the enumeration tree will not contain isomorphic subproblems.

A common approach to deal with symmetries in the problem is adding symmetry-breaking inequalities. An interesting example is given by Kaibel and Pfetsch in [222], where they characterize the inequalities defining the *partitioning orbitope*, that is the convex hull of all $0, 1$ $m \times n$ matrices with at most one nonzero entry in each row and whose columns are sorted lexicographically. These inequalities can be used to break symmetries in certain integer programming formulations, such as the graph coloring formulation in which a variable x_{ij} indicates if node i is given color j.

9.5 Exercises

Exercise 9.1. Let $k_1, \ldots, k_n$ be relatively prime integers, and let p be a positive integer. Let $\Lambda \subseteq \mathbb{Z}^n$ be the set of points $x = (x_1, \ldots, x_n)$ satisfying $\sum_{i=1}^{n} k_i x_i \equiv 0 \pmod{p}$. Prove that Λ is a lattice and that $\det(\Lambda) = p$.

Exercise 9.2. Let $b^1, \ldots, b^n$ be an orthogonal basis of $\mathbb{R}^n$. Prove that the shortest vector in the lattice generated by $b^1, \ldots, b^n$ is one of the vectors b^i, for $i = 1, \ldots, n$.

Exercise 9.3. Let A, B be two $n \times n$ nonsingular integral matrices. Assume that the lattices Λ_A and Λ_B generated by these matrices satisfy $\Lambda_A \subseteq \Lambda_B$.

(i) Prove that $|\det B|$ divides $|\det A|$.

(ii) Prove that there exist bases $u^1, \ldots, u^n$ of Λ_A and $v^1, \ldots, v^n$ of Λ_B such that $u^i = k_i v^i$ for some positive integers k_i, $i = 1, \ldots, n$.

Exercise 9.4. Prove that any lattice $\Lambda \subset \mathbb{R}^2$ has a basis b^1, b^2 such that the angle between b^1 and b^2 is between $60°$ and $90°$.

Exercise 9.5. Let $\Lambda \subset \mathbb{R}^n$ be a lattice and $b^1, \ldots, b^n$ a reduced basis of Λ. Prove that

$$\|b^1\| \leq 2^{\frac{n-1}{2}} \min_{u \in \Lambda \setminus \{0\}} \|u\|.$$

Exercise 9.6. Let $\Lambda \subset \mathbb{R}^n$ be a lattice and $b^1, \ldots, b^n$ a reduced basis of Λ. For $u \in \Lambda$, consider $\lambda \in \mathbb{Z}^n$ such that $u = \sum_{i=1}^{n} \lambda_i b^i$. Prove that, if u is a shortest nonzero vector in Λ, then $|\lambda_i| \leq 3^n$ for $i = 1, \ldots, n$.

Exercise 9.7. Let $K \subseteq \mathbb{R}^n$ be a convex body and let $w_d(K)$ denote its width along direction $d \in \mathbb{R}^n$. Prove that $\inf_{d \in \mathbb{Z}^n} w_d(K) = \min_{d \in \mathbb{Z}^n} w_d(K)$, i.e., there exists $d^* \in \mathbb{Z}^n$ such that $w_{d^*}(K) = \inf_{d \in \mathbb{Z}^n} w_d(K)$.

Exercise 9.8. For a matrix $B \in \mathbb{R}^{n \times n}$, let M be the absolute value of the largest entry in B. Prove that $|\det(B)| \leq n^{n/2} M^n$.

Exercise 9.9. Let $K \in \mathbb{R}^n$ be a full-dimensional convex body that does not contain any integral point. Let us call an *open split set* any set of the form $\{x \in \mathbb{R}^n : \pi_0 < \pi x < \pi_0 + 1\}$ where $\pi \in \mathbb{Z}^n$, $\pi_0 \in \mathbb{Z}$. Prove that K is contained in the union of $C(n)$ open split sets, where $C(n)$ is a constant depending only on n.

Exercise 9.10. Modify Lenstra's algorithm so that it computes an integral solution to $Ax \leq b$ if one exists (hint: use (9.9)).

Exercise 9.11. Let $\Lambda \subset \mathbb{Q}^n$ be the lattice generated by a basis $b^1, \ldots, b^n \in \mathbb{Q}^n$. Prove that, for fixed n, the shortest vector in Λ can be found in polynomial time.

Exercise 9.12. Apply Lenstra's algorithm to find a solution to
$$243243x_1 + 244223x_2 + 243334x_3 = 8539262753$$
$$x_1, x_2, x_3 \geq 0 \text{ integer.}$$

Exercise 9.13. Consider a mixed integer linear program with one integer variable. Assume that its linear programming relaxation has a unique optimal solution. Prove that any branch-and-bound algorithm using variable branching and the LP relaxation to compute lower bounds has an enumeration tree of size at most three.

Exercise 9.14. Consider the mixed integer linear program (9.10) where all integer variables are bounded, $0 \leq x_j \leq u_j$ with u_j integer for $j = 1, \ldots, p$. Specialize the branching rule in the branch-and-cut algorithm as follows. Choose a variable x_j, $j = 1, \ldots, p$ such that x_j^i is fractional. From the linear program LP_i, construct $u_j + 1$ subproblems LP_i^t obtained by adding the constraint $x_j = t$ for $t = 0, \ldots, u_j$. Let z_i^t be the objective value of LP_i^t. Let $t_l = \lfloor x_j^i \rfloor$ and $t_u = \lceil x_j^i \rceil$.

(i) Prove that $z_i^{t_l} = \max_{t=0,\ldots,t_l} z_i^t$ and $z_i^{t_u} = \max_{t=t_u,\ldots,u_j} z_i^t$.

(ii) Suppose that the above branching rule is used in conjunction with a node selection rule that chooses subproblem $\text{LP}_i^{t_l}$ or $\text{LP}_i^{t_u}$ before the others. Is there an advantage in using this branching rule rather than the more common $x_j \leq t_l$ or $x_j \geq t_u$?

Exercise 9.15. Consider the symmetric traveling salesman problem on the undirected graph $G = (V, E)$ with costs c_e, $e \in E$. Devise a branch-and-bound algorithm for solving this problem where bounds are obtained by computing Lagrangian bounds based on the 1-tree relaxation (8.7).

Explain each step of your branch-and-bound algorithm, such as branching, and how subproblems are created and solved at the nodes of the enumeration tree.

Exercise 9.16. Let $[l, u] \subset \mathbb{R}$ be an interval. An approach for maximizing a nonlinear function $f : [l, u] \to \mathbb{R}$ is to approximate it by a piecewise linear function g with breakpoints $l = s_1 < s_2 < \ldots < s_n = u$. The formulation is
$$\begin{array}{rl} x = & \sum_{k=1}^n \lambda_k s_k \\ g(x) = & \sum_{k=1}^n \lambda_k f(s_k) \\ & \sum_{k=1}^n \lambda_k = 1 \\ & \lambda_k \geq 0, \text{ for } k = 1, \ldots, n \end{array}$$

together with the condition that no more than two of the λ_k are positive and these are of the form λ_k, λ_{k+1}.

(i) Prove that a valid separation is given by the constraints

$$\sum_{k=1}^{p-1} \lambda_k = 0 \quad \text{or} \quad \sum_{k=p+1}^{n} \lambda_k = 0$$

for some $p = 2, \ldots, n-1$.

(ii) Based on this separation, develop a branch-and-bound algorithm to find an approximate solution to a separable nonlinear program of the form

$$\begin{array}{l} \max \ \sum_{j=1}^{n} f_j(x_j) \\ Ax \leq b \\ l \leq x \leq u \end{array}$$

where $f_j : [l_j, u_j] \to \mathbb{R}$ are nonlinear functions for $j = 1, \ldots, n$.

Exercise 9.17. Denote by (LP) the linear programming relaxation of (9.10). Let $\bar{x}$ denote an optimal solution of (LP). For a split disjunction $\sum_{j=1}^{p} \pi_j x_j \leq \pi_0$ or $\sum_{j=1}^{p} \pi_j x_j \geq \pi_0 + 1$ where $(\pi, \pi_0) \in \mathbb{Z}^{p+1}$ is chosen such that $\bar{x}$ satisfies $\pi_0 < \sum_{j=1}^{p} \pi_j \bar{x}_j < \pi_0 + 1$, let (LP$^-$) and (LP$^+$) be the linear programs obtained from (LP) by adding the constraints $\sum_{j=1}^{p} \pi_j x_j \leq \pi_0$ and $\sum_{j=1}^{p} \pi_j x_j \geq \pi_0 + 1$ respectively. Let z^- and z^+ be the optimum objective values of (LP$^-$) and (LP$^+$) respectively. Compare z^-, z^+ with the value $\hat{z}$ obtained by adding to (LP) a split cut from the above disjunction: Construct an example showing that the differences $\hat{z} - z^-$ and $\hat{z} - z^+$ can be arbitrarily large. Compare z^-, z^+ with the value z^* obtained by adding to (LP) all the split inequalities from the above disjunction.

Exercise 9.18. Consider a pure integer linear program, with bounded variables $0 \leq x_j \leq u_j$ for $j = 1, \ldots, n$, where the objective function z is to be maximized. Assume that we know a lower bound $\underline{z}$ and that the linear programming relaxation has been solved to optimality with the objective function represented as $z = \bar{z} + \sum_{j \in N_0} \bar{c}_j x_j + \sum_{j \in N_u} \bar{c}_j (x_j - u_j)$ where N_0 indexes the nonbasic variables at 0, N_u indexes the nonbasic variables at their upper bound, $\bar{c}_j \leq 0$ for $j \in N_0$, and $\bar{c}_j \geq 0$ for $j \in N_u$. Prove that in any optimal solution of the integer program

$$x_j \leq \lfloor \frac{\bar{z} - \underline{z}}{-\bar{c}_j} \rfloor \quad \text{for } j \in N_0 \text{ such that } \bar{c}_j \neq 0, \quad \text{and}$$

$$x_j \geq u_j - \lceil \frac{\bar{z} - z}{\bar{c}_j} \rceil \quad \text{for } j \in N_u \text{ such that } \bar{c}_j \neq 0.$$

Exercise 9.19. Let B index the set of binary variables in a mixed 0,1 linear program. For $j \in B$, let $\bar{x}_j := 1 - x_j$ and let $\bar{B}$ index the set of binary variables $\bar{x}_j$, $j \in B$. Construct the graph $G = (B \cup \bar{B}, E)$ where two nodes are joined by an edge if and only if the two corresponding variables cannot be 1 at the same time in a feasible solution.

1. Show that, for any clique C of G, the inequality

$$\sum_{C \cap B} x_j - \sum_{C \cap \bar{B}} x_j \leq 1 - |C \cap \bar{B}|$$

   is a valid inequality for the mixed 0,1 linear program.

2. What can you deduce when G has a clique C of cardinality greater than two such that $C \cap B$ and $C \cap \bar{B}$ contain nodes corresponding to variables x_k and $\bar{x}_k$ respectively, for some index k.

3. Construct the graph $(B \cup \bar{B}, E)$ for the 0,1 program

$$\begin{aligned} 4x_1 + x_2 - 3x_4 &\leq 2 \\ 3x_1 + 2x_2 + 5x_3 + 3x_4 &\leq 7 \\ x_2 + x_3 - x_4 &\leq 0 \\ x_1, x_2, x_3, x_4 &\in \{0, 1\}. \end{aligned}$$

   Deduce all the possible maximal clique inequalities and variable fixings.

Exercise 9.20. Consider the following integer linear program

$$\begin{aligned} \min \quad & z \\ 2x_1 + \ldots + 2x_n + z &= n \\ x_1, \ldots, x_n &= 0 \text{ or } 1 \\ z &\geq 0 \end{aligned}$$

where n is an odd integer.

(i) Prove that any branch-and-bound algorithm using variable branching and the linear programming relaxation to compute lower bounds, but no isomorphism pruning, has an enumeration tree of size at least $2^{n/2}$.

(ii) What is the size of the enumeration tree when isomorphism pruning is used?

Exercise 9.21. Given a group G of permutations in Σ_n and a set $S \subset \{1, \ldots, n\}$, show that the stabilizer $\text{stabil}(S, G)$ is a group.

9.5. EXERCISES

Exercise 9.22. Consider the following 0,1 linear program, where $n \geq 3$.

$$\begin{aligned}
\max \quad & \sum_{j=1}^{n} x_j \\
& x_i + x_j + x_k \leq 2 \quad \text{for all } i, j, k \\
& x \in \{0, 1\}^n.
\end{aligned} \qquad (9.16)$$

Provide the complete enumeration tree for a branch-and-bound algorithm that uses isomorphism pruning and orbital fixing.

Exercise 9.23. Formulate as a pure 0,1 linear program the question of whether there exists a $v \times b$ binary matrix with exactly r ones per row, k ones per column, and with a scalar product of λ between any pair of distinct rows. Solve the problem for $(v, b, r, k, \lambda) = (7, 7, 3, 3, 1)$.

The authors working on Chap. 9

One of the authors enjoying a drink under a fig tree

Chapter 10

Semidefinite Bounds

Semidefinite programs are a generalization of linear programs. Under mild technical assumptions, they can also be solved in polynomial time. In certain cases, they can provide tighter bounds on integer programming problems than linear programming relaxations. The first use of semidefinite programming in combinatorial problems dates back to Lovász [260], who introduced a semidefinite relaxation of the stable set polytope, the so-called theta body. A similar idea is also the basis of an elegant, and very tight, approximation algorithm for the max-cut problem, due to Goemans and Williamson [173]. The approach can be generalized to mixed $0,1$ linear programming problems within a framework introduced by Lovász and Schrijver [264]. The idea is to obtain a relaxation of the feasible region of the mixed $0,1$ program as the projection of some higher-dimensional convex set, which is defined by linear and semidefinite constraints. This approach is closely related to the lift-and-project inequalities discussed in Chap. 5, and also to relaxations introduced by Sherali–Adams [330] and by Lasserre [247, 248].

10.1 Semidefinite Relaxations

A matrix $A \in \mathbb{R}^{n \times n}$ is *positive semidefinite* if, for every $x \in \mathbb{R}^n$, $x^T A x \geq 0$. A semidefinite program consists of maximizing a linear function subject to linear constraints, where the variables are the entries of a square symmetric matrix X, with the further (nonlinear) constraint that X is positive semidefinite. The condition that X is symmetric and positive semidefinite will be denoted by $X \succeq 0$.

Formally, a *semidefinite program* is an optimization problem, defined by symmetric $n \times n$ matrices $C, A_1, \ldots, A_m$ and a vector $b \in \mathbb{R}^m$, of the following form

$$\begin{aligned} \max \quad & \langle C, X \rangle \\ & \langle A_i, X \rangle = b_i \quad i = 1, \ldots, m \\ & X \succeq 0. \end{aligned} \qquad (10.1)$$

Recall that $\langle \cdot, \cdot \rangle$ applied to two $n \times n$ matrices A, B is defined as $\langle A, B \rangle := \sum_k \sum_j a_{kj} b_{kj}$. We observe that, in the special case where the matrices $C, A_1, \ldots, A_m$ are diagonal matrices, (10.1) reduces to a linear program, because only the diagonal entries of X appear in the optimization and the condition $X \succeq 0$ implies that these variables are nonnegative. Note also that the feasible region of (10.1) is convex, since any convex combination of symmetric positive semidefinite matrices is also symmetric positive semidefinite.

Similar to linear programming, a natural notion of dual exists for semidefinite programming problems. The *dual* problem of (10.1) is defined by

$$\begin{aligned} \min \quad & ub \\ & \sum_{i=1}^{m} u_i A_i - C \succeq 0. \end{aligned} \qquad (10.2)$$

It is not difficult to show that the value of a primal feasible solution X in (10.1) never exceeds the value of a dual solution u in (10.2). Despite their similarity with linear programs, semidefinite programs do not always admit an optimal solution, even when the objective function in (10.1) is bounded. Furthermore strong duality might not hold, namely the optimal values of the primal and dual problems might not coincide. The following theorem gives sufficient conditions for strong duality to hold [290]. A matrix $A \in \mathbb{R}^{n \times n}$ is *positive definite* if $x^T A x > 0$ for all $x \in \mathbb{R}^n \setminus \{0\}$. The notation $A \succ 0$ expresses the property that A is a symmetric positive definite matrix. The primal problem (10.1) is *strictly feasible* if it has a feasible solution X such that $X \succ 0$, while the dual problem (10.2) is strictly feasible if there exists $u \in \mathbb{R}^m$ such that $\sum_{i=1}^{m} u_i A_i - C \succ 0$.

Theorem 10.1. *If the primal problem (10.1) (resp. the dual problem (10.2)) is strictly feasible and bounded, then the dual (resp. the primal) problem admits an optimal solution, and the optimal values of (10.1) and (10.2) coincide.*

See Theorem 1.4.2 (3a) in [49] for a proof. When optimal solutions exist, they might have irrational components, even if the data in (10.1) are all rational. Thus one cannot hope to compute an optimal solution exactly, in general. However, semidefinite programs can be solved in polynomial time to any desired precision. This can be done either with the ellipsoid method [186], or more efficiently using interior point methods [291]. Interior point methods typically require that both primal and dual problems are strictly feasible, while ellipsoid methods require the problem to satisfy other technical conditions that we do not discuss here.

Symmetric positive semidefinite matrices have several equivalent characterizations, which will be useful in the remainder. Given a symmetric matrix $A \in \mathbb{R}^{n \times n}$, a *principal submatrix* of A is a matrix B whose rows and columns are indexed by a set $I \subseteq \{1, \ldots, n\}$, so that $B = (a_{ij})_{i,j \in I}$. Clearly, if $A \succeq 0$, then $B \succeq 0$ for every principal submatrix B.

Proposition 10.2. *Let $A \in \mathbb{R}^{n \times n}$ be a symmetric matrix. The following are equivalent.*

(i) *A is positive semidefinite.*

(ii) *There exists $U \in \mathbb{R}^{d \times n}$, for some $d \leq n$, such that $A = U^T U$.*

(iii) *All principal submatrices of A have nonnegative determinant.*

The proof can be found in [207] for example.

10.2 Two Applications in Combinatorial Optimization

10.2.1 The Max-Cut Problem

Given a graph $G = (V, E)$ with edge weights $w \in \mathbb{Q}_+^E$, the *max-cut problem* consists of finding a cut $C \subseteq E$ of maximum weight $\sum_{e \in C} w_e$. This is an NP-hard problem [232]. By contrast, we have seen in Sect. 4.3.3 that the problem of finding a cut of minimum weight is polynomially solvable.

Goemans and Williamson [173] give a semidefinite relaxation of the max-cut problem and they show that the upper bound defined by the optimum value is always less than 14 % away from the maximum value of a cut. We present this result. Let $n := |V|$. The max-cut problem can be formulated as the following quadratic integer program.

$$z_I := \max \quad \frac{1}{2} \sum_{ij \in E} w_{ij}(1 - x_i x_j)$$
$$x \in \{+1, -1\}^n.$$

For any column vector $x \in \{+1,-1\}^n$, the $n \times n$ matrix $Y := xx^T$ is symmetric, positive semidefinite, has rank 1, and $y_{jj} = 1$ for all $j = 1, \ldots, n$. Relaxing the rank 1 condition, we get the following semidefinite relaxation of the max-cut problem.

$$z_{\text{sdp}} := \max \; \frac{1}{2} \sum_{ij \in E} w_{ij}(1 - y_{ij})$$

$$y_{jj} = 1 \text{ for } j = 1, \ldots, n,$$
$$Y \succeq 0.$$

Clearly $z_{\text{sdp}} \geq z_I$. One can show that both the above semidefinite program and its dual admit a strictly feasible solution, therefore they both admit an optimal solution. Furthermore, the value z_{sdp} can be computed in polynomial time to any desired precision by interior point methods. The next theorem demonstrates that the quality of the upper bound z_{sdp} is very good.

Theorem 10.3 (Goemans and Williamson [173]). $\dfrac{z_I}{z_{\text{sdp}}} > 0.87856.$

Proof. Consider an optimal solution Y to the semidefinite relaxation. Since $Y \succeq 0$, by Proposition 10.2 we can write $Y = U^T U$ where U is a $d \times n$ matrix for some $d \leq n$. Let $u_j \in \mathbb{R}^d$ denote the jth column of U. Note that $\|u_i\|^2 = u_i^T u_i = y_{ii} = 1$ for $i = 1, \ldots, n$. Thus, $u_1, \ldots, u_n$ are points on the surface of the unit sphere in $\mathbb{R}^d$.

Generate a vector $r \in \mathbb{R}^d$ with $\|r\|_2 = 1$ uniformly at random, and let $S := \{i \in V : u_i^T r > 0\}$. Consider the cut $C := \delta(S)$ in the graph G.

If H_r denotes the hyperplane normal to r and going through the origin, an edge ij is in the cut C exactly when u_i is on one side of H_r and u_j on the other. Therefore the probability that an edge ij is in the cut C is proportional to the angle between the vectors u_i and u_j, hence it is $\dfrac{\arccos(u_i^T u_j)}{\pi}$. It follows that the expected total weight of the cut C is

$$\mathbb{E}(\sum_{ij \in C} w_{ij}) = \sum_{ij \in E} w_{ij} \frac{\arccos(u_i^T u_j)}{\pi} = \frac{1}{2} \sum_{ij \in E} w_{ij}(1 - u_i^T u_j) \frac{2 \arccos(u_i^T u_j)}{\pi(1 - u_i^T u_j)}.$$

It can be computed that $\min_{0 < \theta < \pi} \frac{2\theta}{\pi(1-\cos\theta)} > 0.87856$. Since $w \geq 0$ and $1 - u_i^T u_j \geq 0$ (because $|u_i^T u_j| \leq \|u_i\|_2 \|u_j\|_2 = 1$), we get

$$\mathbb{E}(\sum_{ij \in C} w_{ij}) > 0.87856 \, (\frac{1}{2} \sum_{ij \in E} w_{ij}(1 - u_i^T u_j)) = 0.87856 \, z_{\text{sdp}}.$$

Now the theorem follows from the fact that $\mathbb{E}(\sum_{ij \in C} w_{ij}) \leq z_I$. □

10.2. TWO APPLICATIONS IN COMBINATORIAL...

The proof of the theorem provides a randomized 0.87856-approximation algorithm for max-cut, that is, a polynomial-time algorithm which returns a solution to the max-cut problem whose expected value is at least 0.87856 times the optimum.

10.2.2 The Stable Set Problem

Consider a graph $G = (V, E)$ and let $n := |V|$. Recall that a stable set in G is a set of nodes no two of which are adjacent. Thus a vector $x \in \{0,1\}^n$ is the characteristic vector of a stable set if and only if it satisfies $x_i + x_j \leq 1$ for all $ij \in E$. Let STAB(G) denote the stable set polytope of G, namely the convex hull of the characteristic vectors of the stable sets of G. A common linear relaxation for STAB(G) is the so-called *clique relaxation*, defined by the clique inequalities relative to all cliques of G, that is

$$\text{QSTAB}(G) := \{x \geq 0 : \sum_{j \in C} x_j \leq 1 \text{ for all cliques } C \text{ of } G\}. \quad (10.3)$$

Clearly STAB(G) $\subseteq$ QSTAB(G). However, optimizing a linear function over QSTAB(G) is NP-hard in general (this follows by Theorem 7.26 because the separation problem for QSTAB(G) is a maximum weight clique problem, which is NP-hard [158]). Lovász [260] introduced the following semidefinite relaxation of STAB(G). Define the *theta body* of G, denoted by TH(G), as the set of all $x \in \mathbb{R}^n$ for which there exists a matrix $Y \in \mathbb{R}^{(n+1)\times(n+1)}$ satisfying the following constraints (see Fig. 10.1)

$$\begin{aligned} y_{00} &= 1 \\ y_{0j} = y_{j0} = y_{jj} &= x_j \quad j \in V \\ y_{ij} &= 0 \quad ij \in E \\ Y &\succeq 0. \end{aligned} \quad (10.4)$$

The theta body has the advantage that one can optimize a rational linear function cx over TH(G) in polynomial time with an arbitrary precision using semidefinite programming. Furthermore, as shown next, the set TH(G) is sandwiched between STAB(G) and QSTAB(G).

Theorem 10.4. *For any graph G, STAB(G) $\subseteq$ TH(G) $\subseteq$ QSTAB(G).*

Proof. Let x be the characteristic (column) vector of a stable set. Define $Y := \begin{pmatrix} 1 \\ x \end{pmatrix} (1 \; x^T)$. Then Y satisfies all the properties needed in (10.4). In particular $y_{jj} = x_j^2 = x_j$ since $x_j = 0$ or 1, and $y_{ij} = x_i x_j = 0$ when $ij \in E$. This shows that STAB(G) $\subseteq$ TH(G).

$$Y := \begin{pmatrix} 1 & x_1 & \cdots & x_i & \cdots & x_j & \cdots & x_n \\ x_1 & x_1 & & & & & & \\ \vdots & & \ddots & & & & & \\ x_i & & & x_i & & y_{ij} & & \\ \vdots & & & & \ddots & & & \\ x_j & & & y_{ij} & & x_j & & \\ \vdots & & & & & & \ddots & \\ x_n & & & & & & & x_n \end{pmatrix}$$

Figure 10.1: The matrix Y in (10.4)

Consider $x \in \mathrm{TH}(G)$ and let Y be a matrix satisfying (10.4). Let C be any clique of G. Consider the principal submatrix Y_C of Y whose rows and columns are indexed by $C \cup \{0\}$. Note that all entries of Y_C are zero except $y_{00} = 1$ and, possibly, $y_{0j} = y_{j0} = y_{jj} = x_j$ for $j \in C$. Since $Y \succeq 0$, also $Y_C \succeq 0$ by Proposition 10.2(iii), therefore $v^T Y_C v \geq 0$ for every $v \in \mathbb{R}^{|C|+1}$. In particular, choosing v such that $v_0 = 1$ and $v_j = -1$ for $j \in C$, we get $1 - \sum_{j \in C} x_j \geq 0$. This proves $x \in \mathrm{QSTAB}(G)$. □

The above semidefinite relaxation has an interesting connection to perfect graphs. A graph G is *perfect* if, in G and any of its node-induced subgraphs, the chromatic number is equal to the size of a largest clique. It is known, as a consequence of results of Lovász [259], Chvátal [74] and Fulkerson [155], that $\mathrm{STAB}(G) = \mathrm{QSTAB}(G)$ if and only if G is a perfect graph. In this case $\mathrm{STAB}(G) = \mathrm{TH}(G)$, therefore one can find the stability number of a perfect graph in polynomial time and, more generally, solve the maximum weight stable set problem in a perfect graph in polynomial time. Interestingly, no other direct polynomial-time algorithm to compute the stability number of a perfect graph is currently known.

10.3 The Lovász–Schrijver Relaxation

The approach described in the two previous examples can be viewed in a more general framework, that extends to general mixed 0,1 programming problems. Consider a polyhedron $P := \{x \in \mathbb{R}_+^{n+p} : Ax \geq b\}$ and the mixed 0,1 linear set $S := \{x \in \{0,1\}^n \times \mathbb{R}_+^p : Ax \geq b\}$. Without loss of generality, we assume that the constraints $Ax \geq b$ include $x_j \geq 0$ for $j = 1, \ldots, n+p$, and $x_j \leq 1$ for $j = 1, \ldots, n$. Lovász and Schrijver [264] study the following "lift-and-project" procedure.

10.3. THE LOVÁSZ–SCHRIJVER RELAXATION

Lovász–Schrijver Procedure

Step 1. Generate the nonlinear system

$$x_j(Ax - b) \geq 0$$
$$(1 - x_j)(Ax - b) \geq 0 \qquad j = 1, \ldots, n. \qquad (10.5)$$

Step 2. Linearize the system (10.5) as follows: Substitute y_{ij} for $x_i x_j$, for all $i = 1, \ldots, n + p$ and $j = 1, \ldots, n$ such that $j < i$, and substitute x_j for x_j^2 for all $j = 1, \ldots, n$.

Denote by $Y = (y_{ij})$ the symmetric $(n+1) \times (n+1)$ matrix such that $y_{00} = 1$, $y_{0j} = y_{j0} = y_{jj} = x_j$ for $j = 1, \ldots, n$, and $y_{ij} = y_{ji}$ for $1 \leq i < j \leq n$.

Denote by $M_+(P)$ the convex set in $\mathbb{R}_+^{\frac{n(n+1)}{2}+np}$ defined by all (x, y) that satisfy the linearized inequalities, and such that $Y \succeq 0$.

Step 3. Project $M_+(P)$ onto the x-space. Call the resulting convex set $N_+(P)$.

Whenever there is no ambiguity, we will refer to $M_+(P)$ and $N_+(P)$ simply as M_+ and N_+.

Observe that $S \subseteq N_+$ because, for any $x \in S$, choosing $y_{ij} = x_i x_j$ for $1 \leq j \leq n$, $j < i \leq n + p$, produces a feasible solution for M_+. Indeed the diagonal elements of this matrix Y are equal to x_i since $x \in S$ implies

$$x_i^2 = x_i = 0 \text{ or } 1 \text{ for } i = 1, \ldots, n, \text{ and } Y = \begin{pmatrix} 1 \\ x_1 \\ \vdots \\ x_n \end{pmatrix} (1, x_1, \ldots, x_n), \text{ therefore}$$

Y is symmetric, positive semidefinite, and it has rank 1.

We remark that S is the projection onto the x-space of the set of elements $(x, y) \in M_+$ for which the matrix Y in Fig. 10.1 has rank 1. In particular, N_+ is a relaxation of S since the condition that Y has rank 1 is waived.

Since the problem of optimizing a linear function over M_+ is a semidefinite program, one can optimize over the Lovász–Schrijver relaxation with arbitrary precision in polynomial time.

Example 10.5. Let $P := \{x \in [0,1]^4 : x_1 - 2x_2 + 4x_3 + 5x_4 \geq 3\}$, and $S := P \cap \mathbb{Z}^4$. Constructing a nonlinear system as in Step 1 of the Lovász–Schrijver procedure and then linearizing, we obtain

$$x_1(x_1 - 2x_2 + 4x_3 + 5x_4 - 3) \geq 0$$
$$(1 - x_1)(x_1 - 2x_2 + 4x_3 + 5x_4 - 3) \geq 0$$
$$x_2(x_1 - 2x_2 + 4x_3 + 5x_4 - 3) \geq 0$$
$$(1 - x_2)(x_1 - 2x_2 + 4x_3 + 5x_4 - 3) \geq 0$$
$$x_3(x_1 - 2x_2 + 4x_3 + 5x_4 - 3) \geq 0$$
$$(1 - x_3)(x_1 - 2x_2 + 4x_3 + 5x_4 - 3) \geq 0$$
$$x_4(x_1 - 2x_2 + 4x_3 + 5x_4 - 3) \geq 0$$
$$(1 - x_4)(x_1 - 2x_2 + 4x_3 + 5x_4 - 3) \geq 0$$
$$(1 - x_i)(1 - x_j) \geq 0$$
$$x_i(1 - x_j) \geq 0$$
$$(1 - x_i)x_j \geq 0$$
$$(1 - x_i)(1 - x_i) \geq 0$$
$$x_i x_j \geq 0$$

$$-2x_1 - 2y_{12} + 4y_{13} + 5y_{14} \geq 0$$
$$3x_1 - 2x_2 + 4x_3 + 5x_4 + 2y_{12} - 4y_{13} - 5y_{14} \geq 3$$
$$-5x_2 + y_{12} + 4y_{23} + 5y_{24} \geq 0$$
$$x_1 + 3x_2 + 4x_3 + 5x_4 - y_{12} - 4y_{23} - 5y_{24} \geq 3$$
$$x_3 + y_{13} - 2y_{23} + 5y_{34} \geq 0$$
$$x_1 - 2x_2 + 3x_3 + 5x_4 - y_{13} + 2y_{23} - 5y_{34} \geq 3$$
$$2x_4 + y_{14} - 2y_{24} + 4y_{34} \geq 0$$
$$x_1 - 2x_2 + 4x_3 + 3x_4 - y_{14} + 2y_{24} - 4y_{34} \geq 3$$
$$x_i + x_j - y_{ij} \leq 1$$
$$y_{ij} \leq x_i$$
$$y_{ij} \leq x_j$$
$$x_i \leq 1$$
$$y_{ij} \geq 0$$

where $1 \leq i < j \leq 4$. The convex set M_+ is the set of all $(x_1, x_2, x_3, y_{12}, y_{13}, y_{23}) \in \mathbb{R}^6$ satisfying the above linear inequalities and the semidefinite constraint

$$Y := \begin{pmatrix} 1 & x_1 & x_2 & x_3 & x_4 \\ x_1 & x_1 & y_{12} & y_{13} & y_{14} \\ x_2 & y_{12} & x_2 & y_{23} & y_{24} \\ x_3 & y_{13} & y_{23} & x_3 & y_{34} \\ x_4 & y_{14} & y_{24} & y_{34} & x_4 \end{pmatrix} \succeq 0.$$

∎

Example 10.6. The relaxation TH(G) of the stable set polytope defined in Sect. 10.2.2 is related to the Lovász–Schrijver relaxation $N_+(\text{FRAC}(G))$: Let us apply the Lovász–Schrijver procedure to $P := \text{FRAC}(G) = \{x \in [0,1]^n : x_i + x_j \leq 1 \text{ for all } ij \in E\}$ and $S := P \cap \{0,1\}^n$.

For every $i,j \in V$, linearizing $x_i x_j \geq 0$ we obtain $y_{ij} \geq 0$. For every $ij \in E$, linearizing $x_i(1 - x_i - x_j) \geq 0$ we obtain $y_{ij} \leq 0$, thus implying $y_{ij} = 0$ for all $ij \in E$.

Therefore $N_+(\text{FRAC}(G)) \subseteq \text{TH}(G)$. The inclusion is strict in general (Exercise 10.8).

Similarly, the semidefinite bound used by Goemans and Williamson in Sect. 10.2.1 can also be viewed in the framework of the Lovász–Schrijver relaxation, by transforming the ± 1 variables into $\chi_i := \frac{1}{2}(1 + x_i)$ to get $\chi \in \{0,1\}^n$, and set $P := [0,1]^n$ (see Exercise 10.4). ∎

10.3.1 Semidefinite Versus Linear Relaxations

Consider the variation of the Lovász–Schrijver procedure where the constraint "$Y \succeq 0$" is removed in Step 2. Thus the linearization in Step 2 simply gives a polyhedron M in $\mathbb{R}_+^{\frac{n(n+1)}{2}+np}$. Its projection onto the x-space is a polyhedron $N \subseteq \mathbb{R}^{n+p}$. Clearly, $N_+ \subseteq N$. This relaxation was first considered by Sherali and Adams [330]; it is also studied in Lovász and Schrijver [264]. We will write $N(P)$ instead of just N when it is important to specify the set P to which the procedure is applied.

We remark that $N \subseteq P$, since each inequality $a^i x \leq b_i$ of $Ax - b \geq 0$ can be obtained by summing the linearizations of $x_j(a^i x - b_i) \geq 0$ and $(1 - x_j)(a^i x - b_i) \geq 0$. Thus we have the following inclusions.

Lemma 10.7. $\mathrm{conv}(S) \subseteq N_+ \subseteq N \subseteq P$.

How tight is the Lovász–Schrijver relaxation N_+ compared to N? From a theoretical point of view, the Goemans and Williamson [173] result given in Sect. 10.2.1 shows that, at least for the max-cut problem, the semidefinite relaxation is strikingly strong. From a practical perspective the size of the semidefinite program creates a tremendous challenge: the number of variables has been multiplied by n and the number of constraints as well! Burer and Vandenbussche [66] solve it using an augmented Lagrangian method and they report computational results on three classes of combinatorial problems, namely the maximum stable set problem, the quadratic assignment problem and the following problem of Erdös and Turan: Calculate the maximum size of a subset of numbers in $\{1,\ldots,n\}$ such that no three numbers are in arithmetic progression. In all three cases, the Lovász–Schrijver bound given by N_+ is substantially tighter than the bound given by N. To illustrate this, we give the results obtained in [66] for the size of a maximum stable set (graphs with more than 100 nodes):

Name	Nodes	Edges	Optimum	N_+	N
Brock200-1	200	5,066	21	27.9	66.6
c-fat200-1	200	18,366	12	14.9	66.6
Johnson16-2-4	120	1,680	8	10.2	23.3
Keller4	171	5,100	11	15.4	57.0
Rand-200-05	200	982	64	72.7	75.1
Rand-200-50	200	10,071	11	17.1	66.6

10.3.2 Connection with Lift-and-Project

Next, we relate the relaxation N defined in Sect. 10.3.1 to the lift-and-project relaxations P_j introduced in Sect. 5.4. Consider the following "lift-and-project" procedure.

Lift-and-Project Procedure

Step 0. Select $j \in \{1,\ldots,n\}$.
Step 1. Generate the nonlinear system $x_j(Ax - b) \geq 0$, $(1 - x_j)(Ax - b) \geq 0$.
Step 2. Linearize the system by substituting y_i for $x_i x_j$, $i \neq j$, and x_j for x_j^2. Call this polyhedron Q_j.
Step 3. Project Q_j onto the x-space. Let P_j be the resulting polyhedron.

Theorem 10.9 below will show that P_j obtained in this way coincides with the definition given in Sect. 5.4.

Lemma 10.8. $N \subseteq \bigcap_{j=1}^{n} P_j \subseteq P$

Proof. The linear inequalities defining Q_j are a subset of those defining M. Therefore $N \subseteq P_j$ for $j = 1,\ldots,n$.

The inclusion $P_j \subseteq P$ follows by observing that the inequalities $Ax \geq b$ can be obtained by summing up the constraints defining Q_j. □

The inclusion $N \subseteq \bigcap_{j=1}^{n} P_j$ can be strict (see Exercise 10.9). This is because Step 2 of the Lovász–Schrijver procedure takes advantage of the fact that $y_{ij} = y_{ji}$ whereas this is not the case for the different Q_js used in generating $\bigcap_{j=1}^{n} P_j$.

Theorem 10.9. $P_j = \operatorname{conv}\{(P \cap \{x : x_j = 0\}) \cup (P \cap \{x : x_j = 1\})\}$

Proof. The linear system produced at Step 2 of the lift-and-project procedure is

$$\begin{aligned} Ax^1 &\geq \lambda b \\ x_j^1 &= \lambda \\ Ax^2 &\geq (1-\lambda)b \\ x_j^2 &= 1-\lambda \\ x^1 + x^2 &= x \\ 0 \leq \lambda &\leq 1 \end{aligned}$$

where we introduced the additional variable $y_j = x_j$ so that the vector y is now in $\mathbb{R}^{n+p}$, and we defined $x^1 := y$, $x^2 := x - y$, $\lambda := x_j$.

Since P_j is the projection onto the x variables of the polyhedron defined by the above inequalities, the statement now follows from Lemma 4.45. □

10.3. THE LOVÁSZ–SCHRIJVER RELAXATION

10.3.3 Iterating the Lovász–Schrijver Procedure

Let $N^1 := N(P)$ be the polyhedron obtained by the Sherali–Adams procedure of Sect. 10.3.1 applied to $P := \{x \in \mathbb{R}_+^{n+p} : Ax \geq b\}$ and the mixed 0,1 set $S := \{x \in \{0,1\}^n \times \mathbb{R}_+^p : Ax \geq b\}$. For any integer $t \geq 2$, define $N^t := N(N^{t-1})$ as the polyhedron obtained by the Sherali–Adams procedure of Sect. 10.3.1 applied to N^{t-1}.

Theorem 10.10. $P \supseteq N^1 \supseteq N^2 \supseteq \ldots \supseteq N^n = \mathrm{conv}(S)$

Proof. The inclusions follow from Lemma 10.7.

As a consequence of Lemma 10.8, we have $N^1 \subseteq P_1$, $N^2 \subseteq P_2(P_1)$, ..., $N^n \subseteq P_n(\ldots P_2(P_1))$. By Theorem 10.9, $P_j = \mathrm{conv}\{(Ax \geq b, x_j = 0) \cup (Ax \geq b, x_j = 1)\}$. By Balas' sequential convexification theorem, $P_n(\ldots P_2(P_1)) = \mathrm{conv}(S)$. It follows that $N^n \subseteq \mathrm{conv}(S)$. By Lemma 10.7, $\mathrm{conv}(S) \subseteq N^n$. Therefore $N^n = \mathrm{conv}(S)$. □

Because N_+ is not a polyhedron, one cannot iterate the Lovász–Schrijver procedure presented in Sect. 10.3. However the procedure can be extended to general convex bodies. We give the extension in the pure binary case for simplicity of notation. Let $K \subseteq [0,1]^n$ be a convex body and let $S := K \cap \{0,1\}^n$.

We denote by

$$\tilde{K} := \{\lambda \begin{pmatrix} 1 \\ x \end{pmatrix} : x \in K, \lambda \geq 0\}$$

the *homogenization cone* of K, where the additional coordinate is indexed by 0. Note that, given $x \in \{0,1\}^n$, we have that $x \in K$ if and only if, for $j = 1, \ldots, n$, $x_j \begin{pmatrix} 1 \\ x \end{pmatrix} \in \tilde{K}$ and $(1 - x_j) \begin{pmatrix} 1 \\ x \end{pmatrix} \in \tilde{K}$. In particular, for any $x \in \{0,1\}^n$, the matrix $Y := \begin{pmatrix} 1 \\ x \end{pmatrix} (1, x^T)$ is a symmetric positive semidefinite matrix such that

$$y_{00} = 1, \quad y_{0j} = y_{j0} = y_{jj} \quad \text{for } j = 1, \ldots, n, \tag{10.6}$$

and $x \in K$ if and only if Y satisfies

$$Ye^j, \quad Ye^0 - Ye^j \in \tilde{K} \quad \text{for } j = 1, \ldots, n \tag{10.7}$$

where $e^0, e^1, \ldots, e^n$ denote the unit vectors in $\mathbb{R}^{n+1}$.

Let

$$M_+ := \{Y \in \mathbb{R}^{(n+1) \times (n+1)} : Y \text{ satisfies } (10.6), (10.7), Y \succeq 0\}.$$

We define

$$N_+ := \{x \in \mathbb{R}^n : \begin{pmatrix} 1 \\ x \end{pmatrix} = Ye^0 \text{ for some } Y \in M_+\}. \qquad (10.8)$$

It follows from the above discussion that $K \cap \{0,1\}^n \subseteq N_+$.

When K is a polytope, the set N_+ defined in (10.8) is identical to the set N_+ obtained by the Lovász–Schrijver procedure presented earlier (see Exercise 10.19). We can now iterate the Lovász–Schrijver relaxation, by defining $N_+^1 := N_+(K)$ and $N_+^t := N_+(N_+^{t-1})$ for any integer $t \geq 2$.

In the next section, we present the Sherali–Adams and Lasserre [247] hierarchies, which are stronger than N^t and N_+^t respectively, yet one can still optimize a linear function over them in polynomial time for fixed t. However these relaxations are substantially more computationally demanding than N and N_+.

10.4 The Sherali–Adams and Lasserre Hierarchies

In this section, we will restrict our attention to pure 0,1 programs. Let $P := \{x \in \mathbb{R}_+^n : Ax \geq b\}$ and $S := P \cap \{0,1\}^n$. As in the previous section we assume that the constraints $Ax \geq b$ include $x_j \geq 0$ and $x_j \leq 1$ for $j = 1, \ldots, n$.

10.4.1 The Sherali–Adams Hierarchy

Instead of multiplying the constraints $Ax \geq b$ by x_j and $1 - x_j$ only, as done in the previous section, Sherali–Adams [330] propose to multiply them by all the possible products $\prod_{i \in I} x_i \prod_{j \in J}(1 - x_j)$ where I, J are disjoint sets of indices such that $1 \leq |I| + |J| \leq t$. Here t is a fixed integer, $1 \leq t \leq n$, which defines the tth level of the hierarchy. Let S_t be the relaxation of S provided by the tth level of the Sherali–Adams hierarchy. Note that $S_1 = N$. Formally, S_t is obtained by the following lift-and-project procedure. Let us denote by $\mathcal{P}_n$ the family of all 2^n subsets of $\{1, \ldots, n\}$.

10.4. THE SHERALI–ADAMS AND LASSERRE HIERARCHIES

Sherali–Adams Procedure

Step 1. Generate the nonlinear system

$$(Ax-b)\prod_{i \in I} x_i \prod_{j \in J}(1-x_j) \geq 0 \quad \text{for all } I, J \subseteq \{1,\ldots,n\} \text{ s.t. } 1 \leq |I|+|J| \leq t,\ I \cap J = \emptyset.$$
(10.9)

Step 2. Linearize system (10.9) by first substituting x_j for x_j^2 for all $j = 1, \ldots, n$ and then substituting y_I for $\prod_{i \in I} x_i$ for all $I \in \mathcal{P}_n$, $I \neq \emptyset$. Set $y_\emptyset = 1$ and call R_t the polyhedron in $\mathbb{R}^{\mathcal{P}_n}$ defined by these linear inequalities.

Step 3. Project R_t onto the x-space. Call the resulting polytope S_t.

Note that the system produced at Step 2 only involves variables y_I for all $I \in \mathcal{P}_n$ such that $|I| \leq t+1$. However, it will be more convenient to consider R_t as a polyhedron in the whole space $\mathbb{R}^{\mathcal{P}_n}$, where the variables relative to y_I, $|I| > t+1$, are unconstrained. In particular, $R_1 \supseteq R_2 \supseteq \ldots \supseteq R_n$.

Theorem 10.11. $S_t \subseteq N^t$.

Proof. We already observed that $S_1 = N^1$. We prove the theorem by induction. Assume $S_{t-1} \subseteq N^{t-1}$ for some $t \geq 2$. We have $N(S_{t-1}) \subseteq N^t$. Therefore, to prove the theorem, it suffices to show that $S_t \subseteq N(S_{t-1})$.

Let $\tilde{C}y \geq \tilde{d}$ denote the linear system of inequalities defining R_{t-1} and $\tilde{A}x \geq \tilde{b}$ the system defining its projection S_{t-1}.

Consider a valid inequality $\alpha x \geq \beta$ for $N(S_{t-1})$. It is implied by a nonnegative combination of the inequalities defining $M(S_{t-1})$, each of which is a linearization of inequalities from $x_j(\tilde{A}x - \tilde{b}) \geq 0$, $(1 - x_j)(\tilde{A}x - \tilde{b}) \geq 0$, for $j = 1, \ldots, n$. The inequalities in $\tilde{A}x - \tilde{b} \geq 0$ are themselves nonnegative combinations of inequalities in $\tilde{C}y - \tilde{d} \geq 0$.

It follows that $\alpha x \geq \beta$ is implied by a nonnegative combination of the linearization of inequalities of the form $x_j(\tilde{C}y - \tilde{d}) \geq 0$ and $(1-x_j)(\tilde{C}y - \tilde{d}) \geq 0$. But such inequalities are valid for R_t. This implies $S_t \subseteq N(S_{t-1})$. □

We leave the proof of the following proposition as an exercise.

Proposition 10.12. *Let $y \in R_t$. Then*

(i) $0 \leq y_I \leq y_J \leq 1$ for all $I, J \subseteq \{1, \ldots, n\}$ such that $J \subseteq I$ and $|I| \leq t+1$.

(ii) Given $I \subseteq \{1, \ldots, n\}$ such that $|I| \leq t+1$, if $\{x \in P : x_i = 1, i \in I\} = \emptyset$, then $y_I = 0$.

10.4.2 The Lasserre Hierarchy

The Lasserre hierarchy strengthens the Sherali–Adams hierarchy by adding semidefinite constraints. The variables y_I for $I \subseteq \{1, \ldots, n\}$ can be organized in a so-called moment matrix $M(y)$. This is a square matrix whose rows (and columns) are indexed by all 2^n subsets of $\{1, \ldots, n\}$. The entry in row $I \subseteq \{1, \ldots, n\}$ and column $J \subseteq \{1, \ldots, n\}$ is $y_{I \cup J}$. Figure 10.2 shows an example for $n = 3$. For simplicity, we write $y_{i \ldots j}$ instead of $y_{\{i, \ldots, j\}}$.

$$M(y) := \begin{pmatrix} y_\emptyset & y_1 & y_2 & y_3 & y_{12} & y_{13} & y_{23} & y_{123} \\ y_1 & y_1 & y_{12} & y_{13} & y_{12} & y_{13} & y_{123} & y_{123} \\ y_2 & y_{12} & y_2 & y_{23} & y_{12} & y_{123} & y_{23} & y_{123} \\ y_3 & y_{13} & y_{23} & y_3 & y_{123} & y_{13} & y_{23} & y_{123} \\ y_{12} & y_{12} & y_{12} & y_{123} & y_{12} & y_{123} & y_{123} & y_{123} \\ y_{13} & y_{13} & y_{123} & y_{13} & y_{123} & y_{13} & y_{123} & y_{123} \\ y_{23} & y_{123} & y_{23} & y_{23} & y_{123} & y_{123} & y_{23} & y_{123} \\ y_{123} & y_{123} & y_{123} & y_{123} & y_{123} & y_{123} & y_{123} & y_{123} \end{pmatrix}$$

Figure 10.2: Moment matrix of the vector y, and submatrix $M_2(y)$

For a fixed integer t such that $1 \leq t \leq n$, we define the matrix $M_t(y)$ to be the square submatrix of $M(y)$ that contains the rows and columns indexed by sets of cardinality at most t. The submatrix of $M(y)$ highlighted in Fig. 10.2 represents $M_2(y)$ for $n = 3$.

The *Lasserre relaxation* $L_t \subseteq \mathbb{R}^n$ of P is defined as the projection of a higher dimensional convex set $K_t \subseteq \mathbb{R}^{\mathcal{P}_n}$, which is obtained from the Sherali–Adams relaxation R_t by further imposing that $M_{t+1}(y) \succeq 0$, that is,

$$K_t := \{y \in \mathbb{R}^{\mathcal{P}_n} : y \in R_t, M_{t+1}(y) \succeq 0\},$$
$$L_t := \{x \in \mathbb{R}^n : \text{there exists } y \in K_t \text{ s.t. } x_j = y_j \; j = 1, \ldots, n\}.$$

For example, when $n = 3$, $x \in L_1$ and $y \in K_1$ must satisfy $M_2(y) \succeq 0$, where $M_2(y)$ has the following form

$$M_2(y) := \begin{pmatrix} 1 & x_1 & x_2 & x_3 & y_{12} & y_{13} & y_{23} \\ x_1 & x_1 & y_{12} & y_{13} & y_{12} & y_{13} & y_{123} \\ x_2 & y_{12} & x_2 & y_{23} & y_{12} & y_{123} & y_{23} \\ x_3 & y_{13} & y_{23} & x_3 & y_{123} & y_{13} & y_{23} \\ y_{12} & y_{12} & y_{12} & y_{123} & y_{12} & y_{123} & y_{123} \\ y_{13} & y_{13} & y_{123} & y_{13} & y_{123} & y_{13} & y_{123} \\ y_{23} & y_{123} & y_{23} & y_{23} & y_{123} & y_{123} & y_{23} \end{pmatrix}.$$

10.4. THE SHERALI–ADAMS AND LASSERRE HIERARCHIES

When the column vector y is defined by $y_I = \prod_{i \in I} x_i$, $I \in \mathcal{P}_n$, for some $0,1$ vector x, then we have that $M(y) = yy^T$, and $M(y)$ is symmetric and positive semidefinite. It follows that, in this case, all principal submatrices $M_t(y)$, $t = 1, \ldots, n$, are positive semidefinite. In particular, this shows that $S \subseteq L_t$ and therefore $\operatorname{conv}(S) \subseteq L_t$ for $t = 1, \ldots, n-1$.

Note that $M_1(y)$ is the matrix Y in the Lovász–Schrijver procedure. Furthermore, if $M_2(y) \succeq 0$, then also $M_1(y) \succeq 0$ since $M_1(y)$ is a principal submatrix of $M_2(y)$. Therefore, $L_1 \subseteq N_+$.

Our definition of the Lasserre hierarchy L_t is nonstandard. The usual definition imposes positive semidefiniteness conditions on m additional moment matrices obtained from each row $a^k x - b_k$ of the system $Ax - b \geq 0$. Thus the relaxation L_t that we defined above is weaker than the usual Lasserre relaxation. Nevertheless our presentation preserves key properties of Lasserre's hierarchy while being a little simpler. In his papers [247, 248], Lasserre also gives an algebraic perspective to his hierarchy of relaxations, based on representing nonnegative polynomials as sums of squares. We do not elaborate on this point of view here.

In the remainder of this section, we will present two results from Laurent [249] (Theorems 10.14 and 10.15 below). The first relates the Lasserre hierarchy L_t to the Lovász–Schrijver hierarchy N_+^t, while the second is an application to the stable set problem. We first state some basic properties of the Lasserre relaxation.

Proposition 10.13. *Let $y \in K_t$. Then*

(i) *Given $I, J \subseteq \{1, \ldots, n\}$ such that $|I|, |J| \leq t + 1$, if $y_I = 0$, then $y_{I \cup J} = 0$.*

(ii) *Given $I, J \subseteq \{1, \ldots, n\}$ such that $|I|, |J| \leq t + 1$, if $y_I = 1$, then $y_{I \cup J} = y_J$.*

(iii) *Given $I \subseteq \{1, \ldots, n\}$ such that $|I| \leq 2t + 2$, if $x_i = 1$ for all $i \in I$, then $y_I = 1$.*

Proof. (i) If $y_I = 0$, then the 2×2 principal submatrix of $M_{t+1}(y)$ indexed by I, J has determinant $-y_{I \cup J}^2$, which must be nonnegative since $M_{t+1}(y) \succeq 0$. Thus $y_{I \cup J} = 0$.

(ii) Let $I, J \subseteq \{1, \ldots, n\}$ such that $|I|, |J| \leq t+1$ and $y_I = 1$. The principal 3×3 submatrix of $M_{t+1}(y)$ indexed by $\{\emptyset, I, J\}$ has determinant $-(y_J - y_{I \cup J})^2$, which must be nonnegative since $M_{t+1}(y)$ is positive semidefinite. It follows that $y_J = y_{I \cup J}$.

We prove (iii) by induction on $|I|$, the statement being trivial if $|I| \leq 1$. If $|I| \geq 2$, then I can be partitioned into nonempty sets I_1, I_2 such that $|I_1|, |I_2| \leq t+1$. By induction, $y_{I_1} = y_{I_2} = 1$. Now the result follows from (ii). □

Theorem 10.14. $L_t \subseteq N_+^t$.

Proof. We already observed that $L_1 \subseteq N_+$. To prove $L_t \subseteq N_+^t$ for $t \geq 2$, we will show by induction the stronger claim that $L_t \subseteq N_+(L_{t-1})$. Let $\hat{x} \in L_t$. Then there exists $\hat{y} \in K_t$ such that $\hat{x}_j = \hat{y}_j$, $j = 1 \ldots, n$. In order to show that $\hat{x} \in N_+(L_{t-1})$, we need to provide a matrix $Y \in M_+(L_{t-1})$ such that $Ye^0 = \binom{1}{\hat{x}}$ (recall the definition of N_+ in (10.8)). We will show that the matrix $Y := M_1(\hat{y})$ satisfies such properties. Clearly $Ye^0 = \binom{1}{\hat{x}}$ and $Y \succeq 0$ since it is a principal submatrix of $M_{t+1}(\hat{y}) \succeq 0$. We need to show that Y satisfies (10.7). That is, we need to show that, given $h \in \{1, \ldots, n\}$, both Ye^h and $Y(e^0 - e^h)$ belong to $\tilde{L}_{t-1}$, the homogenization cone of L_{t-1}.

If $\hat{y}_h = 0$, then, by Proposition 10.12(i), $Ye^h = 0 \in \tilde{L}_{t-1}$ and $Y(e^0 - e^h) = \binom{1}{\hat{x}} \in \tilde{L}_{t-1}$. If $\hat{y}_h = 1$, then, by Proposition 10.13(ii), $Ye^h = Ye^0$, and thus $Ye^h = \binom{1}{\hat{x}} \in \tilde{L}_{t-1}$ and $Y(e^0 - e^h) = 0 \in \tilde{L}_{t-1}$. Thus we assume that $0 < \hat{y}_h < 1$.

Define the points $\bar{x}, \tilde{x} \in \mathbb{R}^n$ by $\bar{x}_j = \hat{y}_h^{-1} \hat{y}_{jh}$ and $\tilde{x}_j = (1-\hat{y}_h)^{-1}(\hat{y}_j - \hat{y}_{jh})$, $j = 1, \ldots, n$. From the definition, $Ye^h, Y(e^0 - e^h) \in \tilde{L}_{t-1}$ if and only if $\bar{x}, \tilde{x} \in L_{t-1}$.

To show the latter, we need to give vectors $\bar{y}, \tilde{y} \in K_{t-1}$ such that $\bar{y}_j = \bar{x}_j$ and $\tilde{y}_j = \tilde{x}_j$ for all $j = 1 \ldots, n$. Let us define $\bar{y} := \hat{y}_h^{-1} M(\hat{y}) e^h$ and $\tilde{y} := (1-\hat{y}_h)^{-1} M(\hat{y})(e^0 - e^h)$. Clearly $\bar{y}_j = \bar{x}_j$ and $\tilde{y}_j = \tilde{x}_j$ for $j = 1, \ldots, n$. Note that, for every $S \subseteq \{1, \ldots, n\}$, $\bar{y}_S = \hat{y}_h^{-1} \hat{y}_{S \cup \{h\}}$ and $\tilde{y}_S = (1-\hat{y}_h)^{-1}(\hat{y}_S - \hat{y}_{S \cup \{h\}})$. We need to show that $\bar{y}, \tilde{y} \in K_{t-1}$. In other words, we need to show $\bar{y}, \tilde{y} \in R_{t-1}$ and $M_t(\bar{y}) \succeq 0$, $M_t(\tilde{y}) \succeq 0$. We will show these results in Claims 1 and 2 below.

Claim 1: $\bar{y}, \tilde{y} \in R_{t-1}$.

Consider any valid inequality for R_{t-1} generated in Step 2 of the Sherali–Adams procedure

$$\sum_{S \in \mathcal{P}_n} c_S y_S \geq 0. \tag{10.10}$$

It is obtained by linearizing an inequality $(a^k x - b_k) \prod_{i \in I} x_i \prod_{j \in J} (1 - x_j) \geq 0$ defined by some row k of the system $Ax - b \geq 0$ and disjoint sets $I, J \subseteq$

10.4. THE SHERALI–ADAMS AND LASSERRE HIERARCHIES

$\{1, \ldots, n\}$ such that $1 \leq |I| + |J| \leq t - 1$. To prove $\bar{y}, \tilde{y} \in R_{t-1}$ we will show that $\bar{y}$ and $\tilde{y}$ satisfy (10.10). Let

$$\sum_{S \in \mathcal{P}_n} \bar{c}_S y_S \geq 0, \quad \sum_{S \in \mathcal{P}_n} \tilde{c}_S y_S \geq 0 \tag{10.11}$$

denote the linearized inequalities obtained from $(a^k x - b_k) \prod_{i \in I \cup \{h\}} x_i \prod_{j \in J} (1 - x_j) \geq 0$ and $(Ax - b) \prod_{i \in I} x_i \prod_{j \in J \cup \{h\}} (1 - x_j) \geq 0$, respectively.

By definition, both inequalities in (10.11) are valid for K_t, and thus they are satisfied by $\hat{y}$. Note that, for all $S \subseteq \{1, \ldots, n\}$, $\bar{c}_S = 0$ if $h \notin S$, while $\bar{c}_S = c_S + c_{S \setminus \{h\}}$ if $h \in S$. Since the first inequality in (10.11) is satisfied by $\hat{y}$, and since $\bar{y}_S = \hat{y}_h^{-1} \hat{y}_{S \cup \{h\}}$, it is routine to verify that $\bar{y}$ satisfies (10.10). Analogously, one can show that, since $\hat{y}$ satisfies the second inequality in (10.11), $\tilde{y}$ satisfies (10.10).

Claim 2: $M_t(\bar{y})$ and $M_t(\tilde{y})$ are positive semidefinite.

Define the sets $\mathcal{P}_1 := \{I \subseteq \{1, \ldots, n\} : h \notin I, |I| \leq t - 1\}$, $\mathcal{P}_2 := \{I \subseteq \{1, \ldots, n\} : h \notin I, |I| = t\}$, $\mathcal{P}'_1 := \{I \subseteq \{1, \ldots, n\} : h \in I, |I| \leq t\}$, $\mathcal{P}'_2 := \{I \subseteq \{1, \ldots, n\} : h \in I, |I| = t + 1\}$. The principal submatrix W of $M_{t+1}(\hat{y})$ indexed by $\mathcal{P}_1 \cup \mathcal{P}_2 \cup \mathcal{P}'_1 \cup \mathcal{P}'_2$ is of the form

$$W = \begin{array}{c} \\ \mathcal{P}_1 \\ \mathcal{P}_2 \\ \mathcal{P}'_1 \\ \mathcal{P}'_2 \end{array} \begin{pmatrix} \mathcal{P}_1 & \mathcal{P}_2 & \mathcal{P}'_1 & \mathcal{P}'_2 \\ A & E & C & F \\ E^T & B & F^T & D \\ C & F & C & F \\ F^T & D & F^T & D \end{pmatrix}$$

Observe that, for every $S, T \subset \{1, \ldots, n\}$, $|S|, |T| \leq t$, the S, T entry of $M_t(\bar{y})$ is $\bar{y}_{S \cup T} = \hat{y}_h^{-1} \hat{y}_{S \cup (T \cup \{h\})}$. Analogously, the S, T entry of $M_t(\tilde{y})$ is $\tilde{y}_{S \cup T} = (1 - \hat{y})_h^{-1} (\hat{y}_{S \cup T} - \hat{y}_{S \cup (T \cup \{h\})})$. We conclude that $M_t(\bar{y}) = \hat{y}_h^{-1} \bar{W}$ and $M_t(\tilde{y}) = (1 - \hat{y}_h)^{-1} \tilde{W}$, where

$$\bar{W} = \begin{array}{c} \\ \mathcal{P}_1 \\ \mathcal{P}_2 \\ \mathcal{P}'_1 \end{array} \begin{pmatrix} \mathcal{P}_1 & \mathcal{P}_2 & \mathcal{P}'_1 \\ C & F & C \\ F^T & D & F^T \\ C & F & C \end{pmatrix}, \quad \tilde{W} = \begin{array}{c} \\ \mathcal{P}_1 \\ \mathcal{P}_2 \\ \mathcal{P}'_1 \end{array} \begin{pmatrix} \mathcal{P}_1 & \mathcal{P}_2 & \mathcal{P}'_1 \\ A - C & E - F & 0 \\ E^T - F^T & B - D & 0 \\ 0 & 0 & 0 \end{pmatrix}.$$

Note that, for all column vectors u, v, w of appropriate dimension,

$$(u^T, v^T, w^T) \bar{W} \begin{pmatrix} u \\ v \\ w \end{pmatrix} = (u^T + w^T, v^T) \begin{pmatrix} C & F \\ F^T & D \end{pmatrix} \begin{pmatrix} u + w \\ v \end{pmatrix} \geq 0 \text{ where the}$$

latter inequality follows from the fact that $\begin{pmatrix} C & F \\ F^T & D \end{pmatrix} \succeq 0$, since it is a principal submatrix of $M_{t+1}(\hat{y})$. It follows that $M_t(\bar{y}) \succeq 0$.

Finally, $(u^T, v^T) \begin{pmatrix} A - C & E - F \\ E^T - F^T & B - D \end{pmatrix} \begin{pmatrix} u \\ v \end{pmatrix} = (u^T, v^T, -u^T, -v^T)$

$W \begin{pmatrix} u \\ v \\ -u \\ -v \end{pmatrix} \geq 0$, where the last inequality follows from the fact that $W \succeq 0$.

It follows that $M_t(\tilde{y}) \succeq 0$. $\square$

An Application to the Stable Set Problem

Let $G = (V, E)$ be a graph. In this section we describe the Lasserre relaxation of the edge formulation of the stable set polytope, $\text{FRAC}(G) := \{x \in [0,1]^n : x_i + x_j \leq 1 \text{ for all } ij \in E\}$.

Theorem 10.15. $K_t(\text{FRAC}(G))$ *is the set of vectors* $y \in \mathbb{R}^{\mathcal{P}_n}$ *satisfying*

$$\begin{aligned} M_{t+1}(y) &\succeq 0 \\ y_I &= 0 \quad I \subseteq V, \ |I| \leq 2t+2, \ I \text{ not a stable set.} \end{aligned} \quad (10.12)$$

We will need the following lemma.

Lemma 10.16. *Let S be a finite set. Then* $\displaystyle\sum_{\substack{H, K \subseteq S \\ H \cup K = S}} (-1)^{|H|+|K|} = (-1)^{|S|}$.

Proof. By induction on $|S|$, the statement being trivial for $|S| = 0$. Let $S \neq \emptyset$, and let $a \in S$. Let $\mathcal{F} := \{(H, K) : H \cup K = S \setminus \{a\}\}$. Then

$$\sum_{\substack{H, K \subseteq S \\ H \cup K = S}} (-1)^{|H|+|K|} = \sum_{(H,K) \in \mathcal{F}} (-1)^{|H \cup \{a\}|+|K|} + (-1)^{|H|+|K \cup \{a\}|} + (-1)^{|H \cup \{a\}|+|K \cup \{a\}|}$$

$$= -\sum_{(H,K) \in \mathcal{F}} (-1)^{|H|+|K|} = (-1)^{|S|},$$

where the last equation follows from induction. $\square$

Proof of Theorem 10.15. For the "if" direction, let $\bar{y} \in K_t$, and let $I \subset \{1, \ldots, n\}$ such that I is not stable and $|I| \leq 2t+2$. Then there exist I_1, I_2 such that $I_1 \cup I_2 = I$, $|I_1|, |I_2| \leq t+1$, and such that I_1 is not a stable set. It follows that $\{x \in FR(G) : x_i = 1, i \in I_1\} = \emptyset$, thus $y_{I_1} = 0$ by Proposition 10.12(ii). By Proposition 10.13(i), $y_I = 0$.

10.4. THE SHERALI–ADAMS AND LASSERRE HIERARCHIES

For the "only if" direction, let $\bar{y}$ satisfying (10.12). To show that $\bar{y} \in K_t$, we only need to prove that $\bar{y}$ satisfies all the linearized constraints defining K_t in the Lasserre procedure. Consider the inequality

$$(1 - x_a - x_b) \prod_{i \in I} x_i \prod_{j \in J}(1 - x_j) \geq 0,$$

where $ab \in E$ and I, J are disjoint and $|I| + |J| \leq t$. The resulting linearized inequality is

$$\sum_{I \subseteq S \subseteq I \cup J} (-1)^{|S \setminus I|}(y_S - y_{S \cup \{a\}} - y_{S \cup \{b\}}) \geq 0. \tag{10.13}$$

Note that, if $I \subseteq S \subseteq I \cup J$ and S contains a or b, then, by (10.12), $\bar{y}_S - \bar{y}_{S \cup \{a\}} - \bar{y}_{S \cup \{b\}} = 0$.

Define $\mathcal{P}_0 := \{H : I \subseteq H \subseteq I \cup J \setminus \{a,b\}\}$ and, for $c = a, b$, $\mathcal{P}_c := \{H \cup \{c\} : H \in \mathcal{P}_0\}$. The principal submatrix of $M_{t+1}(\bar{y})$ indexed by $\mathcal{P}_0 \cup \mathcal{P}_a \cup \mathcal{P}_b$ is of the form

$$Z = \begin{array}{c} \\ \mathcal{P}_0 \\ \mathcal{P}_a \\ \mathcal{P}_b \end{array} \begin{pmatrix} \mathcal{P}_0 & \mathcal{P}_a & \mathcal{P}_b \\ A & B & C \\ B & B & 0 \\ C & 0 & C \end{pmatrix}.$$

Define the vector $u \in \mathbb{R}^{\mathcal{P}_0}$ by $u_H = (-1)^{|H \setminus I|}$, $H \in \mathcal{P}_0$. Since Z is positive semidefinite,

$$0 \leq (-u^T, u^T, u^T) Z \begin{pmatrix} -u \\ u \\ u \end{pmatrix} = u^T(A - B - C)u$$

$$= \sum_{H, K \in \mathcal{P}_0} (-1)^{|H \setminus I| + |K \setminus I|} (\bar{y}_{H \cup K} - \bar{y}_{H \cup K \cup \{a\}} - \bar{y}_{H \cup K \cup \{b\}})$$

$$= \sum_{S \in \mathcal{P}_0} (-1)^{|S \setminus I|} (\bar{y}_S - \bar{y}_{S \cup \{a\}} - \bar{y}_{S \cup \{b\}}),$$

where the last equation follows from Lemma 10.16. Thus $\bar{y}$ satisfies (10.13).

Finally, we need to consider the linearized constraints corresponding to $x_a \geq 0$, $a \in V$. Any such constraint is of the form

$$\sum_{I \subseteq S \subseteq I \cup J} (-1)^{|S \setminus I|} y_{S \cup \{a\}} \geq 0.$$

where I, J are disjoint and $|I|+|J| \le t$. Let W be the submatrix of $M_{t+1}(\bar{y})$ indexed by the family $\mathcal{P} := \{H \cup \{a\} : I \subseteq H \subseteq I \cup J\}$. Let $u \in \mathbb{R}^{\mathcal{P}}$ be defined by $u_{H \cup \{a\}} = (-1)^{|H \setminus I|}$, for $I \subseteq H \subseteq I \cup J$. Since W is positive semidefinite, it follows that

$$0 \le u^T W u = \sum_{I \subseteq H, K \subseteq I \cup J} (-1)^{|H \setminus I|+|K \setminus I|} \bar{y}_{H \cup K \cup \{a\}} = \sum_{I \subseteq S \subseteq I \cup J} (-1)^{|S \setminus I|} \bar{y}_{S \cup \{a\}},$$

where the last equation follows from Lemma 10.16. $\square$

10.5 Further Readings

Interior point methods, developed for linear programming by Karmarkar [229], were extended to semidefinite programming problems by Nesterov and Nemirovski [289] (see also [291]), who developed a general approach to adapt interior point methods for solving convex optimization problems, and independently by Alizadeh [10], who also reports several applications of semidefinite programming to combinatorial problems.

Lovász's seminal work on the Shannon capacity [260] initiated the use of semidefinite relaxations to obtain stronger bounds for combinatorial problems, and it motivated subsequent works in the field. Goemans and Williamson [173] were the first to adopt these ideas to design polynomial-time approximation algorithms for combinatorial optimization problems. Their technique was later applied to other problems by several authors, such as Frieze and Jerrum's work on bisection and max k-cut [151], and Karger, Motwani, Sudan's work on graph coloring [231] and Nesterov's $\frac{\pi}{2}$ theorem [288]. We refer the reader to Laurent and Rendl [250] for a survey on the applications of semidefinite programming to integer programming. Chlamtac and Tulsiani [70] survey the use of positive semidefinite hierarchies for the design of approximation algorithms in theoretical computer science.

Goemans and Tunçel [172] study the relative strength of the N and N_+ operators, and show several cases for which the two operators provide the same relaxation. They also show that, in the worst case, n iterations of the N_+ operator are indeed needed to obtain the convex hull (see Theorem 10.10).

Bienstock and Zuckerberg [52] defined a lift operator that uses subset algebra. It is one of the strongest known, neither dominated by nor dominating the Lasserre operator.

Gouveia, Parrilo, and Thomas [181] study positive semidefinite hierarchies for combinatorial problems based on a sum of squares" perspective.

10.5. FURTHER READINGS

Burer and Vandenbussche [66] apply an augmented Lagrangian approach to optimize over the Lovász–Schrijver relaxation, and report computational experiments on the strength of the bounds.

In this chapter, we focused on semidefinite relaxations, but there are other promising nonlinear approaches to integer programming. Ideas coming from algebra and geometry are presented in the new book of De Loera, Hemmecke, and Koppe [110]. The Hilbert's Nullstellensatz is an interesting example.

Infeasibility Certificates Using Hilbert's Nullstellensatz Feasibility of certain combinatorial problems can be expressed through systems of polynomial equations. For example, De Loera, Lee, Malkin, and Margulies [109] consider the graph k-coloring problem. Given an undirected graph $G = (V, E)$ on n vertices, a k-coloring of G is a function from V to the set of colors $\{1, \ldots, k\}$ such that every pair of adjacent vertices are assigned different colors. Instead of assigning to each vertex a number from 1 to k, one can assign a kth root of the unity, so that adjacent vertices are assigned different roots. We observe that G has a k-coloring if and only if the following system of polynomials has a solution in $\mathbb{C}^n$.

$$x_i^k - 1 = 0 \quad i \in V \quad (10.14)$$
$$x_i^{k-1} + x_i^{k-2}x_j^1 + \cdots + x_i x_j^{k-2} + x_j^{k-1} = 0 \quad ij \in E. \quad (10.15)$$

Indeed, given a k-coloring c of G and a kth root of the unity $\beta \neq 1$, then setting $\bar{x}_i = \beta^{c(i)}$ for all $i \in V$ gives a solution to the above system. Conversely, given a solution $\bar{x}_i$ to the above system, then such solution defines a k-coloring. If not, then there exist two adjacent nodes $i, j \in V$ that are assigned the same kth complex root of the unity, say $\bar{x}_i = \bar{x}_j = \beta$, but then (10.15) is violated as $\beta^{k-1} + \beta^{k-2}\beta^1 + \cdots + \beta\beta^{k-2} + \beta^{k-1} = k\beta^{k-1} \neq 0$.

Note that the above proof shows a useful fact. Let q be a positive integer, relatively prime with k, and denote by $\mathbb{F}_q$ the finite field on q elements, and by $\bar{\mathbb{F}}_q$ its algebraic closure. The graph G has a k-coloring if and only if (10.14), (10.15) has a solution in $\bar{\mathbb{F}}_q^n$. Indeed, $k\beta^{k-1} \neq 0$ in $\bar{\mathbb{F}}_q$, because k and q are relatively prime.

Hilbert's Nullstellensatz gives a certificate of infeasibility for systems of polynomial equations, much in the spirit of the Fredholm alternative for systems of linear equations or Farkas's lemma for systems of linear inequalities.

Theorem 10.17 (Hilbert's Nullstellensatz). *Let $\mathbb{K}$ be a finite field and $\bar{\mathbb{K}}$ its algebraic closure. Given $f_1, \ldots, f_m \in \mathbb{K}[x_1, \ldots, x_n]$, the system of* polynomial equations $f_1(x) = 0, \ldots, f_m(x) = 0$ has no solution in $\bar{\mathbb{K}}^n$ if and only if there exist $g_1, \ldots, g_m \in \mathbb{K}[x_1, \ldots, x_n]$ such that $g_1 f_1 + \cdots + g_m f_m = 1$.

Note that the "if" direction of the statement is trivial, since given $g_1, \ldots, g_m \in \mathbb{K}[x_1, \ldots, x_n]$ such that $g_1 f_1 + \cdots + g_m f_m = 1$, then $f_1, \ldots, f_m$ cannot have a common root.

Hilbert's Nullstellensatz can therefore be used to provide certificates of non colorability, based on the following observation. Suppose we are given a positive integer d and we want to check if there exists a Nullstellensatz certificate of infeasibility $g_1, \ldots, g_m$ for $f_1, \ldots, f_m$ such that the degree of the polynomials $g_i f_i$ is bounded by d for $i = 1, \ldots, m$. The condition $g_1 f_1 + \cdots + g_m f_m = 1$ can be expressed as a linear system of equations, where the variables are the coefficients of the monomials defining $g_1, \ldots, g_m$.

For example, consider the system of polynomial equations $x_1^2 + x_2^2 - 1 = 0$, $x_1 + x_2 - 1 = 0$, $x_1 - x_2 = 0$. We want to find if there exists a Nullstellensatz certificate of degree 2. Thus, we are looking for coefficients $\alpha_0, \ldots, \alpha_6$ such that

$$\alpha_0(x_1^2 + x_2^2 - 1) + (\alpha_1 x_1 + \alpha_2 x_2 + \alpha_3)(x_1 + x_2 - 1) + (\alpha_4 x_1 + \alpha_5 x_2 + \alpha_6)(x_1 - x_2) = 1.$$

Grouping the monomials in the above expression, we obtain

$$\begin{aligned} 1 = {} & -(\alpha_0 + \alpha_3) + (-\alpha_1 + \alpha_3 + \alpha_6)x_1 + (-\alpha_2 + \alpha_3 - \alpha_6)x_2 + \\ & + (\alpha_0 + \alpha_1 + \alpha_4)x_1^2 + (\alpha_0 + \alpha_2 - \alpha_5)x_2^2 + (\alpha_1 + \alpha_2 - \alpha_4 + \alpha_5)x_1 x_2. \end{aligned}$$

Thus we need to solve the system of inequalities

$$\begin{aligned} -\alpha_0 - \alpha_3 &= 1 & \alpha_0 + \alpha_1 + \alpha_4 &= 0 \\ -\alpha_1 + \alpha_3 + \alpha_6 &= 0 & \alpha_0 + \alpha_2 - \alpha_5 &= 0 \\ -\alpha_2 + \alpha_3 - \alpha_6 &= 0 & \alpha_1 + \alpha_2 - \alpha_4 + \alpha_5 &= 0. \end{aligned}$$

A solution is given by $\alpha_0 = -2$, $\alpha_1 = 1$, $\alpha_2 = 1$, $\alpha_3 = 1$, $\alpha_4 = 1$, $\alpha_5 = -1$, $\alpha_6 = 0$. Thus the polynomials $g_1 = -2$, $g_2 = x_1 + x_2 + 1$, $g_3 = x_1 - x_2$, are a Nullstellensatz certificate of infeasibility for the initial system of polynomials.

This gives rise to the following algorithm to test for feasibility of a system of polynomial equations $f_i(x) = 0$, $i = 1, \ldots, n$. Start with $d = \max_{i=1, \ldots, m} \deg(f_i)$, and decide if a Nullstellensatz certificate of degree d exists; if such a certificate exists, then the system is infeasible, else increase d by one and repeat. There are upper-bounds on the maximum degree that a Nullstellensatz certificate can have, thus in principle one can carry on the above procedure until d exceeds such upper-bound, in which case one concludes that the system is feasible. Unfortunately such upper-bounds are doubly exponential in the number of polynomial equations in the system and in their degree. As the "target degree" d increases, the size of the linear systems to solve increases exponentially, thus in general the method described can be effective only for proving infeasibility, and only for problems for which there exists a Nullstellensatz certificate of low degree.

10.6 Exercises

Exercise 10.1. Show that the value of any feasible solution for the primal problem (10.1) is less than or equal to the value of any feasible solution to the dual problem (10.2).

Write the dual of the semidefinite relaxation of the max cut problem, and show that both the primal and the dual are strictly feasible.

Exercise 10.2. Prove Proposition 10.2.

Exercise 10.3. Prove that if A and B are two symmetric positive semidefinite matrices, then $\langle A, B \rangle \geq 0$.

Exercise 10.4. The max cut problem can be formulated using $0, 1$ node variables as follows

$$z_I := \max \ \sum_{ij \in E} w_{ij}(\chi_i + \chi_j - 2\chi_i\chi_j)$$
$$\chi \in \{0,1\}^n.$$

1. Show that the following is a relaxation of the max cut problem

$$z'_{\text{sdp}} := \max \ \sum_{ij \in E} w_{ij}(\chi_i + \chi_j - 2z_{ij})$$
$$z_{jj} = z_{0j} = z_{j0} = \chi_j \quad j = 1, \ldots, n$$
$$z_{00} = 1$$
$$Z \succeq 0.$$

2. Show that z'_{sdp} is equal to the value z_{sdp} of the Goemans–Williamson relaxation discussed in Sect. 10.2.1.

Exercise 10.5. Let $G = (V, E)$ be a graph with nonnegative edge weights w_e, $e \in E$. Show that, if G is bipartite, then $z_I = z_{\text{sdp}}$.

Exercise 10.6. Let $G = (V, E)$ be a complete graph on n nodes, with edge-weights $w_e = 1$, $e \in E$. Show that $z_{\text{sdp}} = \left(\frac{n}{2}\right)^2$.

Exercise 10.7. Compute the best lower bound you can on the value of z_{sdp} for the 5-cycle $G = (V, E)$ with $V := \{v_1, \ldots, v_5\}$ and $E := \{v_1v_2, \ldots v_4v_5, v_5v_1\}$ and $w_e = 1$ for all $e \in E$.

Exercise 10.8. Let $G = (V, E)$ be the 5-cycle with $V := \{v_1, \ldots, v_5\}$ and $E := \{v_1v_2, \ldots v_4v_5, v_5v_1\}$.

1. Prove that both inclusions $\text{STAB}(G) \subseteq \text{TH}(G) \subseteq \text{QSTAB}(G)$ are strict.

2. Compute the lift-and project set P_j obtained from $P := \text{FRAC}(G)$ for some $j = 1, \ldots, 5$.

3. Show that the inclusion $N_+(\text{FRAC}(G)) \subseteq \text{TH}(G)$ is strict.

Exercise 10.9. Let $P := \{x \in \mathbb{R}^n_+ : Ax \geq b\}$ and $S := P \cap \{0,1\}^n$. Assume that the constraints $Ax \geq b$ include $x_j \geq 0$ and $x_j \leq 1$ for $j = 1, \ldots, n$. Give an example showing that the inclusion $N \subseteq \cap_{j=1}^n \text{conv}\{(Ax \geq b, x_j = 0) \cup (Ax \geq b, x_j = 1)\}$ can be strict.

Exercise 10.10. Consider a polyhedron $P := \{x \in \mathbb{R}^{2+p}_+ : Ax \geq b\}$ and the mixed 0,1 linear set with two 0,1 variables $S := \{x \in \{0,1\}^2 \times \mathbb{R}^p_+ : Ax \geq b\}$. Without loss of generality, we assume that the constraints $Ax \geq b$ include $x_j \geq 0$ for $j = 1, \ldots, 2 + p$, and $x_j \leq 1$ for $j = 1, 2$. Prove that $N_+ = N$.

Exercise 10.11. Let $P := \{x \in [0,1]^5 : x_i + x_j \leq 1 \text{ for all } i \neq j\}$ and $S := P \cap \{0,1\}^5$. Compute the sets $\text{conv}(S)$, N_+, N and $\cap_{j=1}^n P_j$.

Exercise 10.12. Show that all vertices of $N([0,1]^n)$ are half-integral.

Exercise 10.13. Let F be a face of a polytope $P \subseteq [0,1]^n$. Show that $N(F) = N(P) \cap F$.

Exercise 10.14. Let $G = (V, E)$ be a graph and let C be an odd cycle in G. Show that the inequality $\sum_{j \in V(C)} x_j \leq \frac{|C|-1}{2}$ is valid for $N(\text{FRAC}(G))$.

Exercise 10.15. Let $G = (V, E)$ be a graph. Show that, for each antihole H (see Exercise 7.19), the odd antihole inequality $\sum_{j \in V(H)} x_j \leq 2$ is valid for $N_+(\text{FRAC}(G))$.

Exercise 10.16. Let $P := \{x \in [0,1]^n : \sum_{j=1}^n x_j \geq \frac{1}{2}\}$ and $S := P \cap \mathbb{Z}^n$.

(i) Show that the point $(\frac{1}{2n-k}, \ldots, \frac{1}{2n-k})$ belongs to N_+^k for $k \leq n$.

(ii) Show that $N_+^k \neq \text{conv}(S)$ for $k < n$.

Exercise 10.17. Let $n \geq 2$, $P := \{x \in \mathbb{R}^n : \sum_{j \in J} x_j + \sum_{j \notin J}(1-x_j) \geq 1$ for all $J \subseteq \{1,\ldots,n\}\}$ and $S := P \cap \{0,1\}^n$.

(i) Show that $(\frac{1}{2},\ldots,\frac{1}{2}) \in N_+^{n-1}$.

(ii) Show that $N_+^k \neq \operatorname{conv}(S)$ for $k < n$.

Exercise 10.18. Given a polytope $P \subseteq [0,1]^n \times \mathbb{R}^n$, show that, for every integer $t \geq 0$ and every $I \subseteq \{1,\ldots,n\}$ such that $|I| \leq t$, the tth Lovász–Schrijver closure N^t of P is contained in $\operatorname{conv}(\{x \in P : x_j \in \{0,1\}, j \in I\})$.

Exercise 10.19. Show that, when K is a polytope, the definition of N_+ given in (10.8) is equivalent to the one produced by the Lovász–Schrijver procedure presented earlier in Sect. 10.3.

Exercise 10.20. Prove Proposition 10.12.

This is the end of the book. We hope you enjoyed the journey!

Bibliography

[1] K. Aardal, R.E. Bixby, C.A.J. Hurkens, A.K. Lenstra, J.W. Smeltink, Market split and basis reduction: towards a solution of the Cornuéjols–Dawande instances. INFORMS J. Comput. **12**, 192–202 (2000) (Cited on page 382.)

[2] K. Aardal, A.K. Lenstra, Hard equality constrained integer knapsacks. Math. Oper. Res. **29**, 724–738 (2004); Erratum: Math. Oper. Res. **31**, 846 (2006) (Cited on page 382.)

[3] K. Aardal, C. Hurkens, A.K. Lenstra, Solving a system of diophantine equations with lower and upper bounds on the variables. Math. Oper. Res. **25**, 427–442 (2000) (Cited on page 382.)

[4] K. Aardal, R. Weismantel, L.A. Wolsey, Non-standard approaches to integer programming. Discrete Appl. Math. **123**, 5–74 (2002) (Cited on page 380.)

[5] T. Achterberg, Constraint Integer Programming. Ph.D. thesis, ZIB, Berlin, 2007 (Cited on pages 74 and 382.)

[6] T. Achterberg, T. Berthold, Improving the feasibility pump. Discrete Optim. **4**, 77–86 (2007) (Cited on page 371.)

[7] T. Achterberg, T. Koch, A. Martin, Branching rules revisited. Oper. Res. Lett. **33**, 42–54 (2005) (Cited on page 382.)

[8] T. Achterberg, T. Koch, A. Martin, MIPLIB 2003. Oper. Res. Lett. **34**, 361–372 (2006) (Cited on page 382.)

[9] M. Ajtai, The shortest vector problem in L2 is NP-hard for randomized reductions, in *Proceedings of the 30th Annual ACM Symposium on Theory of Computing (STOC-98)*, (1998), pp. 10–19 (Cited on page 381.)

[10] F. Alizadeh, Interior point methods in semidefinite programming with applications to combinatorial optimization. SIAM J. Optim. **5**, 13–51 (1995) (Cited on page 408.)

[11] K. Andersen, G. Cornuéjols, Y. Li, Split closure and intersection cuts. Math. Program. A **102**, 457–493 (2005) (Cited on pages 201 and 228.)

[12] K. Andersen, Q. Louveaux, R. Weismantel, L.A. Wolsey, Inequalities from two rows of a simplex tableau, in *Proceedings of IPCO XII*, Ithaca, NY. Lecture Notes in Computer Science, vol. 4513 (2007), pp. 1–15 (Cited on page 274.)

[13] D. Applegate, R.E. Bixby, V. Chvátal, W.J. Cook, *The Traveling Salesman Problem. A Computational Study* (Princeton University Press, Princeton, 2006) (Cited on pages 74, 302, 305, 307, 312, and 382.)

[14] S. Arora, B. Barak, *Complexity Theory: A Modern Approach* (Cambridge University Press, Cambridge, 2009) (Cited on page 37.)

[15] A. Atamtürk, Strong formulations of robust mixed 0–1 programming. Math. Program. **108**, 235–250 (2006) (Cited on page 186.)

[16] A. Atamtürk, G.L. Nemhauser, M.W.P. Savelsbergh, Conflict graphs in solving integer programming problems. Eur. J. Oper. Res. **121**, 40–55 (2000) (Cited on page 52.)

[17] R.K. Ahuja, T.L. Magnanti, J.B. Orlin, *Network Flows* (Prentice Hall, New Jersey, 1993) (Cited on pages 140 and 184.)

[18] G. Averkov, On maximal S-free sets and the Helly number for the family of S-convex sets. SIAM J. Discrete Math. **27**(3), 1610–1624 (2013) (Cited on page 275.)

[19] G. Averkov, A. Basu, On the unique lifting property, in *IPCO 2014*, Bonn, Germany, Lecture Notes in Computer Science, **8494**, pp. 76–87 (2014) (Cited on page 275.)

BIBLIOGRAPHY

[20] D. Avis, K. Fukuda, A pivoting algorithm for convex hulls and vertex enumeration of arrangements and polyhedra. Discrete Comput. Geom. **8**, 295–313 (1992) (Cited on pages 121 and 123.)

[21] A. Bachem, R. von Randow, Integer theorems of Farkas lemma type, in *Operations Research Verfahren/ Methods of Operations Research 32, III Symposium on Operations Research*, Mannheim 1978, ed. by W. Oettli, F. Steffens (Athenäum, Königstein, 1979), pp. 19–28 (Cited on page 37.)

[22] E. Balas, Intersection cuts—a new type of cutting planes for integer programming. Oper. Res. **19**, 19–39 (1971) (Cited on pages 240 and 274.)

[23] E. Balas, Integer programming and convex analysis: intersection cuts from outer polars. Math. Program. **2** 330–382 (1972) (Cited on page 274.)

[24] E. Balas, Disjunctive programming: properties of the convex hull of feasible points, GSIA Management Science Research Report MSRR 348, Carnegie Mellon University (1974); Published as invited paper in Discrete Appl. Math. **89**, 1–44 (1998) (Cited on pages 166, 166, 168, 223, and 227.)

[25] E. Balas, Facets of the knapsack polytope. Math. Program. **8**, 146–164 (1975) (Cited on page 285.)

[26] E. Balas, Disjunctive programming and a hierarchy of relaxations for discrete optimization problems. SIAM J. Algebr. Discrete Methods **6**, 466–486 (1985) (Cited on pages 166, 166, and 168.)

[27] E. Balas, A modified lift-and-project procedure. Math. Program. **79**, 19–31 (1997) (Cited on page 229.)

[28] E. Balas, P. Bonami, Generating lift-and-project cuts from the LP simplex tableau: open source implementation and testing of new variants. Math. Program. Comput. **1**, 165–199 (2009) (Cited on page 230.)

[29] E. Balas, S. Ceria, G. Cornuéjols, A lift-and-project cutting plane algorithm for mixed 0–1 programs. Math. Program. **58**, 295–324 (1993) (Cited on pages 225 and 229.)

[30] E. Balas, S. Ceria, G. Cornuéjols, R.N. Natraj, Gomory cuts revisited. Oper. Res. Lett. **19**, 1–9 (1996) (Cited on page 220.)

[31] E. Balas, R. Jeroslow, Strengthening cuts for mixed integer programs. Eur. J. Oper. Res. **4**, 224–234 (1980) (Cited on pages 229 and 271.)

[32] E. Balas, M. Perregaard, A precise correspondence between lift-and-project cuts, simple disjunctive cuts and mixed integer Gomory cuts for 0–1 programming. Math. Program. B **94**, 221–245 (2003) (Cited on page 230.)

[33] E. Balas, W.R. Pulleyblank, The perfectly matchable subgraph polytope of an arbitrary graph. Combinatorica **9**, 321–337 (1989) (Cited on page 186.)

[34] E. Balas, A. Saxena, Optimizing over the split closure. Math. Program. **113**, 219–240 (2008) (Cited on pages 222 and 229.)

[35] W. Banaszczyk, A.E. Litvak, A. Pajor, S.J. Szarek, The flatness theorem for nonsymmetric convex bodies via the local theory of Banach spaces. Math. Oper. Res. **24** 728–750 (1999) (Cited on page 381.)

[36] F. Barahona, R. Anbil, The volume algorithm: producing primal solutions with a subgradient method. Math. Program. **87**, 385–399 (2000) (Cited on page 342.)

[37] I. Barany, T.J. Van Roy, L.A. Wolsey, Uncapacitated lot-sizing: the convex hull of solutions. Math. Program. **22**, 32–43 (1984) (Cited on page 186.)

[38] A. Barvinok, A polynomial time algorithm for counting integral points in polyhedra when the dimension is fixed. Math. Oper. Res. **19**, 769–779 (1994) (Cited on page 381.)

[39] A. Barvinok, *A Course in Convexity*. Graduate Studies in Mathematics, vol. 54 (American Mathematical Society, Providence, 2002) (Cited on pages 249, 250, and 353.)

[40] A. Basu, M. Campelo, M. Conforti, G. Cornuéjols, G. Zambelli, On lifting integer variables in minimal inequalities. Math. Program. A **141**, 561–576 (2013) (Cited on pages 272 and 275.)

[41] A. Basu, M. Conforti, G. Cornuéjols, G. Zambelli, Maximal lattice-free convex sets in linear subspaces. Math. Oper. Res. **35**, 704–720 (2010) (Cited on pages 242 and 250.)

BIBLIOGRAPHY

[42] A. Basu, M. Conforti, G. Cornuéjols, G. Zambelli, Minimal inequalities for an infinite relaxation of integer programs. SIAM J. Discrete Math. **24**, 158–168 (2010) (Cited on page 274.)

[43] A. Basu, R. Hildebrand, M. Köppe, M. Molinaro, A (k+1)-Slope Theorem for the k-Dimensional Infinite Group Relaxation. SIAM J. Optim. **23**(2), 1021–1040 (2013) (Cited on page 274.)

[44] A. Basu, R. Hildebrand, M. Köppe, Equivariant perturbation in Gomory and Johnson infinite group problem III. Foundations for the k-dimensional case with applications to the case $k = 2$. www.optimization-online.org (2014) (Cited on page 274.)

[45] D.E. Bell, A theorem concerning the integer lattice. Stud. Appl. Math. **56**, 187–188 (1977) (Cited on page 250.)

[46] R. Bellman, *Dynamic Programming* (Princeton University Press, Princeton, 1957) (Cited on page 37.)

[47] J.F. Benders, Partitioning procedures for solving mixed variables programming problems. Numerische Mathematik **4**, 238–252 (1962) (Cited on page 343.)

[48] M. Bénichou, J.M. Gauthier, P. Girodet, G. Hentges, G. Ribière, O. Vincent, Experiments in mixed-integer linear programming. Math. Program. **1**, 76–94 (1971) (Cited on page 367.)

[49] A. Ben-Tal, A.S. Nemirovski, *Lectures on Modern Convex Optimization: Analysis, Algorithms, and Engineering Applications*. MPS/SIAM Series in Optimization (SIAM, Philadelphia, 2001) (Cited on page 390.)

[50] C. Berge, Two theorems in graph theory. Proc. Natl. Acad. Sci. USA **43**, 842–844 (1957) (Cited on page 147.)

[51] D. Bertsimas, R. Weismantel, *Optimization over Integers* (Dynamic Ideas, Belmont, 2005) (Cited on pages 36 and 38.)

[52] D. Bienstock, M. Zuckerberg, Subset algebra lift operators for 0–1 integer programming. SIAM J. Optim. **15**, 63–95 (2004) (Cited on page 408.)

[53] L.J. Billera, A. Sarangarajan, All 0,1 polytopes are traveling salesman polytopes. Combinatorica **16**, 175–188 (1996) (Cited on page 305.)

[54] S. Binato, M.V.F. Pereira, S. Granville, A new Benders decomposition approach to solve power transmission network design problems. IEEE Trans. Power Syst. **16**, 235–240 (2001) (Cited on page 341.)

[55] J. R. Birge, F. Louveaux, *Introduction to Stochastic Programming* (Springer, New York, 2011) (Cited on page 341.)

[56] R.E. Bixby, S. Ceria, C.M. McZeal, M.W.P. Savelsbergh, An updated mixed integer programming library: MIPLIB 3.0. Optima **58**, 12–15 (1998) (Cited on page 229.)

[57] R.E. Bixby, M. Fenelon, Z. Gu, E. Rothberg, R. Wunderling, Mixed integer programming: a progress report, in *The Sharpest Cut: The Impact of Manfred Padberg and His Work*, ed. by M. Grötschel. MPS/SIAM Series in Optimization (SIAM, 2004), pp. 309–326 (Cited on pages 37, 220, 220, and 229.)

[58] P. Bonami, On optimizing over lift-and-project closures. Math. Program. Comput. **4**, 151–179 (2012) (Cited on pages 198 and 227.)

[59] P. Bonami, M. Conforti, G. Cornuéjols, M. Molinaro, G. Zambelli, Cutting planes from two-term disjunctions. Oper. Res. Lett. **41**, 442–444 (2013) (Cited on page 229.)

[60] P. Bonami, G. Cornuéjols, S. Dash, M. Fischetti, A. Lodi, Projected Chvátal-Gomory cuts for mixed integer linear programs. Math. Program. **113**, 241–257 (2008) (Cited on pages 229 and 229.)

[61] P. Bonami, F. Margot, Cut generation through binarization, IPCO 2014, eds. by J. Lee, J. Vygen. LNCS, vol 8494 (2014) pp. 174–185 (Cited on page 225.)

[62] J.A. Bondy, U.S.R. Murty, *Graph Theory* (Springer, New York, 2008) (Cited on page 73.)

[63] V. Borozan, G. Cornuéjols, Minimal valid inequalities for integer constraints. Math. Oper. Res. **34**, 538–546 (2009) (Cited on page 274.)

[64] O. Briant, C. Lemaréchal, Ph. Meurdesoif, S. Michel, N. Perrot, F. Vanderbeck, Comparison of bundle and classical column generation. Math. Program. **113**, 299–344 (2008) (Cited on page 342.)

[65] C.A. Brown, L. Finkelstein, P.W. Purdom, Backtrack Searching in the Presence of Symmetry, Nordic J. Comput. **3**, 203–219 (1996) (Cited on page 382.)

BIBLIOGRAPHY

[66] S. Burer, D. Vandenbussche, Solving lift-and-project relaxations of binary integer programs. SIAM J. Optim. **16**, 726–750 (2006) (Cited on pages 397, 397, and 409.)

[67] A. Caprara, M. Fischetti, $\{0, \frac{1}{2}\}$ Chvátal–Gomory cuts. Math. Program. **74**, 221–235 (1996) (Cited on page 228.)

[68] A. Caprara, A.N. Letchford, On the separation of split cuts and related inequalities. Math. Program. B **94**, 279–294 (2003) (Cited on page 203.)

[69] R.D. Carr, G. Konjevod, G. Little, V. Natarajan, O. Parekh, Compacting cuts: new linear formulation for minimum cut. ACM Trans. Algorithms **5**, 27:1–27:6 (2009) (Cited on page 184.)

[70] E. Chlamtac, M. Tulsiani, Convex relaxations and integrality gaps, in *Handbook on Semidefinite, Conic and Polynomial Optimization, International Series in Operations Research and Management Science*, Springer, vol. 166 (Springer, 2012), pp. 139–169 (Cited on page 408.)

[71] M. Chudnovsky, G. Cornuéjols, X. Liu, P. Seymour, K. Vušković, Recognizing Berge graphs. Combinatorica **25**, 143–186 (2005) (Cited on page 184.)

[72] M. Chudnovsky, N. Robertson, P. Seymour, R. Thomas, The strong perfect graph theorem. Ann. Math. **164**, 51–229 (2006) (Cited on page 184.)

[73] V. Chvátal, Edmonds polytopes and a hierarchy of combinatorial optimization. Discrete Math. **4**, 305–337 (1973) (Cited on pages 208, 208, 210, and 228.)

[74] V. Chvátal, On certain polytopes associated with graphs. J. Combin. Theory B **18**, 138–154 (1975) (Cited on page 394.)

[75] V. Chvátal, W. Cook, M. Hartmann, On cutting-plane proofs in combinatorial optimization. Linear Algebra Appl. **114/115**, 455–499 (1989) (Cited on pages 224 and 228.)

[76] M. Conforti, G. Cornuéjols, A. Daniilidis, C. Lemaréchal, J. Malick, Cut-generating functions and S-free sets, Math. Oper. Res. http://dx.doi.org/10.1287/moor.2014.0670 (Cited on page 274.)

[77] M. Conforti, G. Cornuéjols, G. Zambelli, A geometric perspective on lifting. Oper. Res. **59**, 569–577 (2011) (Cited on page 275.)

[78] M. Conforti, G. Cornuéjols, G. Zambelli, Equivalence between intersection cuts and the corner polyhedron. Oper. Res. Lett. **38**, 153–155 (2010) (Cited on page 247.)

[79] M. Conforti, G. Cornuéjols, G. Zambelli, Extended formulations in combinatorial optimization. 4OR **8**, 1–48 (2010) (Cited on pages 184 and 186.)

[80] M. Conforti, G. Cornuéjols, G. Zambelli, Corner polyhedron and intersection cuts. Surv. Oper. Res. Manag. Sci. **16**, 105–120 (2011) (Cited on page 273.)

[81] M. Conforti, M. Di Summa, F. Eisenbrand, L.A. Wolsey, Network formulations of mixed-integer programs. Math. Oper. Res. **34**, 194–209 (2009) (Cited on page 184.)

[82] M. Conforti, L.A. Wolsey, Compact formulations as unions of polyhedra. Math. Program. **114**, 277–289 (2008) (Cited on page 186.)

[83] M. Conforti, L.A. Wolsey, G. Zambelli, Split, MIR and Gomory inequalities (2012 submitted) (Cited on page 227.)

[84] S.A. Cook, The complexity of theorem-proving procedures, in *Proceedings 3rd STOC* (Association for Computing Machinery, New York, 1971), pp. 151–158 (Cited on pages 20 and 37.)

[85] W.J. Cook, Fifty-plus years of combinatorial integer programming, in *50 Years of Integer Programming 1958–2008*, ed. by M. Jünger et al. (Springer, Berlin, 2010), pp. 387–430 (Cited on page 312.)

[86] W.J. Cook, *In Pursuit of the Traveling Salesman: Mathematics at the Limits of Computation* (Princeton University Press, Princeton, 2012) (Cited on pages 74 and 312.)

[87] W.J. Cook, W.H. Cunningham, W.R. Pulleyblank, A. Schrijver, *Combinatorial Optimization* (Wiley, New York, 1998) (Cited on page 183.)

[88] W.J. Cook, S. Dash, R. Fukasawa, M. Goycoolea, Numerically accurate Gomory mixed-integer cuts. INFORMS J. Comput. **21**, 641–649 (2009) (Cited on page 221.)

[89] W.J. Cook, J. Fonlupt, A. Schrijver, An integer analogue of Carathéodory's theorem. J. Combin. Theory B **40**, 63–70 (1986) (Cited on page 185.)

[90] W.J. Cook, R. Kannan, A. Schrijver, Chvátal closures for mixed integer programming problems. Math. Program. **47**, 155–174 (1990) (Cited on pages 196, 202, 203, 203, and 228.)

[91] W.J. Cook, T. Rutherford, H.E. Scarf, D. Shallcross, An implementation of the generalized basis reduction algorithm for integer programming. ORSA J. Comput. **5**, 206–212 (1993) (Cited on page 381.)

[92] G. Cornuéjols, *Combinatorial Optimization: Packing and Covering.* SIAM Monograph, CBMS-NSF Regional Conference Series in Applied Mathematics, vol. 74 (SIAM, 2001) (Cited on page 184.)

[93] G. Cornuéjols, M.L. Fisher, G.L. Nemhauser, Location of bank accounts to optimize float: an analytic study of exact and approximate algorithms. Manag. Sci. **23**, 789–810 (1977) (Cited on page 324.)

[94] G. Cornuéjols, Y. Li, On the rank of mixed 0,1 polyhedra. Math. Program. A **91**, 391–397 (2002) (Cited on page 224.)

[95] G. Cornuéjols, Y. Li, A connection between cutting plane theory and the geometry of numbers. Math. Program. A **93**, 123–127 (2002) (Cited on page 203.)

[96] G. Cornuéjols, R. Tütüncü, *Optimization Methods in Finance* (Cambridge University Press, Cambridge, 2007) (Cited on page 73.)

[97] A.M. Costa, A survey on Benders decomposition applied to fixed-charge network design problems. Comput. Oper. Res. **32**, 1429–1450 (2005) (Cited on page 344.)

[98] H. Crowder, M.W. Padberg, Solving large-scale symmetric travelling salesman problems to optimality. Manag. Sci. **26**, 495–509 (1980) (Cited on page 312.)

[99] H. Crowder, E. Johnson, M.W. Padberg, Solving large scale zero-one linear programming problems. Oper. Res. **31**, 803–834 (1983) (Cited on page 312.)

[100] R.J. Dakin, A tree-search algorithm for mixed integer programming problems. Comput. J. **8**, 250–255 (1965) (Cited on page 37.)

[101] E. Danna, E. Rothberg, C. Le Pape, Exploring relaxation induced neighborhoods to improve MIP solutions. Math. Program. A **102**, 71–90 (2005) (Cited on page 371.)

[102] G.B. Dantzig, Maximization of a linear function of variables subject to linear inequalities, in *Activity Analysis of Production and Allocation*, ed. by T.C. Koopmans (Wiley, New York, 1951), pp. 339–347 (Cited on pages v and 18.)

[103] G. Dantzig. R. Fulkerson, S. Johnson, Solution of a large-scale traveling-salesman problem. Oper. Res. **2**, 393–410 (1954) (Cited on pages vi, 37, 62, 136, 302, and 311.)

[104] G.B. Dantzig, P. Wolfe, Decomposition principle for linear programs. Oper. Res. **8**, 101–111 (1960) (Cited on page 343.)

[105] L. Danzer, B. Grünbaum, V. Klee, Helly's theorem and its relatives, in *Convexity*, ed. by V. Klee (American Mathematical Society, Providence, 1963), pp. 101–180 (Cited on page 361.)

[106] S. Dash, S.S. Dey, O. Günlük, Two dimensional lattice-free cuts and asymmetric disjunctions for mixed-integer polyhedra. Math. Program. **135**, 221–254 (2012) (Cited on page 275.)

[107] S. Dash, O. Günlük, A. Lodi, in *On the MIR Closure of Polyhedra, IPCO 2007*, ed. by M. Fischetti, D.P. Williamson. LNCS, vol. 4513 (Springer, 2007), pp. 337–351 (Cited on pages 202, 222, and 228.)

[108] R. Dechter, *Constraint Processing* (Morgan Kaufmann, San Francisco, 2003) (Cited on page 61.)

[109] J.A. De Loera, J. Lee, P.N. Malkin, S. Margulies, Computing infeasibility certificates for combinatorial problems through Hilbert's Nullstellensatz. J. Symb. Comput. **46**, 1260–1283 (2011) (Cited on page 409.)

[110] J.A. De Loera, R. Hemmecke, M. Köppe, *Algebraic and Geometric Ideas in the Theory of Discrete Optimization*. MOS-SIAM Series on Optimization, vol. 14 (SIAM, 2012) (Cited on pages 36, 38, and 409.)

[111] R. de Wolf, Nondeterministic quantum query and communication complexities. SIAM J. Comput. **32**, 681–699 (2003) (Cited on page 187.)

[112] A. Del Pia, R. Weismantel, Relaxations of mixed integer sets from lattice-free polyhedra. 4OR **10**, 221–244 (2012) (Cited on page 273.)

[113] A. Del Pia, R. Weismantel, On convergence in mixed integer programming. Math. Program. **135**, 397–412 (2012) (Cited on page 228.)

[114] J. Desrosiers, F. Soumis, M. Desrochers, Routing with time windows by column generation. Networks **14**, 545–565 (1984) (Cited on page 343.)

[115] S.S. Dey, Q. Louveaux, Split rank of triangle and quadrilateral inequalities. Math. Oper. Res. **36**, 432–461 (2011) (Cited on page 247.)

[116] S. S. Dey, D.A. Morán, On maximal S-free convex sets. SIAM J. Discrete Math. **25**(1), 379–393 (2011) (Cited on page 275.)

[117] S.S. Dey, J.-P.P. Richard, Y. Li, L.A. Miller, On the extreme inequalities of infinite group problems. Math. Program. A **121**, 145–170 (2010) (Cited on page 260.)

[118] S.S. Dey, L.A. Wolsey, *Lifting Integer Variables in Minimal Inequalities Corresponding to Lattice-Free Triangles, IPCO 2008*, Bertinoro, Italy. Lecture Notes in Computer Science, Springer, vol. 5035 (Springer, 2008), pp. 463–475 (Cited on pages 272 and 275.)

[119] S.S. Dey, L.A. Wolsey, Constrained infinite group relaxations of MIPs. SIAM J. Optim. **20**, 2890–2912 (2010) (Cited on page 274.)

[120] E.A. Dinic, Algorithm for solution of a problem of maximum flow in networks with power estimation. Soviet Math. Dokl. **11**, 1277–1280 (1970) (Cited on page 144.)

[121] J.-P. Doignon, Convexity in cristallographical lattices. J. Geom. **3**, 71–85 (1973) (Cited on page 250.)

[122] M. Dyer, A. Frieze, R. Kannan, A random polynomial-time algorithm for approximating the volume of convex bodies. J. ACM **38**, 1–17 (1991) (Cited on page 381.)

[123] J. Edmonds, Paths, trees, and flowers. Can. J. Math. **17**, 449–467 (1965) (Cited on pages vi, 37, 37, and 147.)

[124] J. Edmonds, Maximum matching and a polyhedron with 0,1-vertices. J. Res. Natl. Bur. Stand. B **69**, 125–130 (1965) (Cited on pages 151 and 184.)

[125] J. Edmonds, Systems of distinct representatives and linear algebra. J. Res. Natl. Bur. Stand. B **71**, 241–245 (1967) (Cited on pages vi and 37.)

[126] J. Edmonds, Submodular functions, matroids, and certain polyhedra, in *Combinatorial Structures and Their Applications*, ed. by R. Guy, H. Hanani, N. Sauer, J. Schönheim. (Gordon and Breach, New York, 1970), pp. 69–87 (Cited on page 185.)

[127] J. Edmonds, D.R. Fulkerson, Bottleneck extrema. J. Combin. Theory **8**, 299–306 (1970) (Cited on page 54.)

[128] J. Edmonds, R. Giles, A min-max relation for submodular functions on graphs. Ann. Discrete Math. **1**, 185–204 (1977) (Cited on page 184.)

[129] J. Edmonds, R.M. Karp, Theoretical improvements in algorithmic efficiency for network flow problems. J. ACM **19**, 248–264 (1972) (Cited on page 144.)

[130] F. Eisenbrand, On the membership problem for the elementary closure of a polyhedron. Combinatorica **19**, 297–300 (1999) (Cited on page 208.)

[131] F. Eisenbrand, G. Shmonin, Carathéodory bounds on integer cones. Oper. Res. Lett. **34**, 564–568 (2006) (Cited on pages 342 and 343.)

[132] F. Eisenbrand, A.S. Schulz, Bounds on the Chvátal rank of polytopes in the 0/1 cube. Combinatorica **23**, 245–261 (2003) (Cited on pages 211 and 228.)

[133] D. Erlenkotter, A dual-based procedure for uncapacitated facility location. Oper. Res. **26**, 992–1009 (1978) (Cited on page 342.)

[134] T. Fahle, S. Shamberger, M. Sellmann, Symmetry Breaking, CP 2001. LNCS, vol. 2239 (Springer, 2001), pp. 93–107 (Cited on page 382.)

[135] Gy. Farkas, On the applications of the mechanical principle of Fourier, Mathematikai és Természettudományi Értesötö **12**, 457–472 (1894) (Cited on page v.)

[136] S. Fiorini, S. Massar, S. Pokutta, H.R. Tiwary, R. de Wolf, Linear vs. semidefinite extended formulations: exponential separation and strong lower bounds, in *STOC 2012* (2012) (Cited on pages 177, 179, 181, 182, and 187.)

[137] S. Fiorini, V. Kaibel, K. Pashkovich, D.O. Theis Combinatorial bounds on the nonnegative rank and extended formulations. Discrete Math. **313**, 67–83 (2013) (Cited on page 187.)

[138] M.L. Fischer, The Lagrangian relaxation method for solving integer programming problems. Manag. Sci. **27**, 1–18 (1981) (Cited on page 341.)

[139] M. Fischetti, F. Glover, A. Lodi, The feasibility pump. Math. Program. **104**, 91–104 (2005) (Cited on page 371.)

[140] M. Fischetti, A. Lodi, Local branching. Math. Program. B **98**, 23–47 (2003) (Cited on page 370.)

[141] M. Fischetti, A. Lodi, Optimizing over the first Chvátal closure. Math. Program. **110**, 3–20 (2007) (Cited on page 229.)

[142] M. Fischetti, A. Lodi, A. Tramontani, On the separation of disjunctive cuts. Math. Program. A **128**, 205–230 (2011) (Cited on page 229.)

[143] M. Fischetti, D. Salvagnin, C. Zanette, A note on the selection of Benders' cuts. Math. Program. B **124**, 175–182 (2010) (Cited on pages 343 and 344.)

[144] R. Fortet, Applications de l'algèbre de Boole en recherche opérationnelle. Revue Française de Recherche Opérationnelle **4**, 17–26 (1960) (Cited on page 72.)

[145] J.B.J. Fourier, Solution d'une question particulière du calcul des inégalités. Nouveau Bulletin des Sciences par la Société Philomatique de Paris (1826), pp. 317–319 (Cited on pages v and 85.)

[146] L.R. Ford Jr., D.R. Fulkerson, Maximal flow through a network. Can. J. Math. **8**, 399–404 (1956) (Cited on page 55.)

[147] L.R. Ford Jr., D.R. Fulkerson, *Flows in Networks* (Princeton University Press, Princeton, 1962) (Cited on page 184.)

[148] A. Frank, Connections in combinatorial optimization, in *Oxford Lecture Series in Mathematics and Its Applications*, vol. 38 (Oxford University Press, Oxford, 2011) (Cited on pages 185 and 185.)

[149] A. Frank, E. Tardos, An application of simultaneous Diophantine approximation in combinatorial optimization. Combinatorica **7**, 49–65 (1987) (Cited on page 381.)

[150] R. M. Freund, J.B. Orlin, On the complexity of four polyhedral set containment problems. Math. Program. **33**, 139–145 (1985) (Cited on page 123.)

[151] A.M. Frieze, M. Jerrum, Improved approximation algorithms for MAX k-CUT and MAX BISECTION. Algorithmica **18**, 67–81 (1997) (Cited on page 408.)

[152] K. Fukuda, Frequently Asked Questions in Polyhedral Computation. Research Report, Department of Mathematics, and Institute of Theoretical Computer Science ETH Zurich, available online (2013) (Cited on page 123.)

[153] K. Fukuda, Lecture: Polyhedral Computation. Research Report, Department of Mathematics, and Institute of Theoretical Computer Science ETH Zurich, available online (2004) (Cited on page 120.)

[154] D.R. Fulkerson, Blocking and anti-blocking pairs of polyhedra. Math. Program. **1**, 168–194 (1971) (Cited on page 184.)

[155] D.R. Fulkerson, Anti-blocking polyhedra. J. Combin. Theory B **12**, 50–71 (1972) (Cited on page 394.)

[156] D.R Fulkerson, Blocking polyhedra, in *Graph Theory and Its Applications*, edited by B. Harris (Academic, New York, 1970), pp. 93–112 (Cited on page 55.)

[157] D.R. Fulkerson, G.L. Nemhauser, L.E. Trotter, Two computationally difficult set covering problems that arise in computing the 1-width of incidence matrices of Steiner triples. Math. Program. Study **2**, 72–81 (1974) (Cited on page 58.)

[158] M.R. Garey, D.S. Johnson, *Computers and Intractability: A Guide to the Theory of NP-Completeness* (W.H. Freeman and Co., New York, 1979) (Cited on pages 138 and 393.)

[159] R.S. Garfinkel, G. Nemhauser, *Integer Programming* (Wiley, New York, 1972) (Cited on pages 36 and 213.)

[160] C.F. Gauss, *Theoria Motus Corporum Coelestium in Sectionibus Conicis Solem Ambientium* (F. Perthes & J.H. Besser, Hamburg, 1809) (Cited on page v.)

[161] A.M. Geoffrion, Generalized Benders decomposition. J. Optim. Theory Appl. **10**, 237–260 (1972) (Cited on page 343.)

[162] A.M. Geoffrion, Lagrangean relaxation for integer programming. Math. Program. Study **2**, 82–114 (1974) (Cited on page 341.)

[163] A.M. Geoffrion, G.W. Graves, Multicommodity distribution design by Benders' decomposition. Manag. Sci. **20**, 822–844 (1974) (Cited on page 343.)

[164] A.M.H. Gerards, A short proof of Tutte's characterization of totally unimodular matrices. Linear Algebra Appl. **114/115**, 207–212 (1989) (Cited on page 183.)

[165] A. Ghouila-Houri, Caractérisation des matrices totalement unimodulaires. Comptes Rendus Hebdomadaires des Scéances de l'Académie des Sciences (Paris) **254**, 1192–1194 (1962) (Cited on page 133.)

[166] F.R. Giles, W.R. Pulleyblank, Total dual integrality and integer polyhedra. Linear Algebra Appl. **25**, 191–196 (1979) (Cited on pages 184 and 185.)

[167] P.C. Gilmore, Families of sets with faithful graph representation. IBM Research Note N.C., vol. 184 (Thomas J. Watson Research Center, Yorktown Heights, 1962) (Cited on page 53.)

[168] P.C. Gilmore, R.E. Gomory, A linear programming approach to the cutting-stock problem. Oper. Res. **9**, 849–859 (1961) (Cited on pages 50 and 343.)

[169] M.X. Goemans, Worst-case comparison of valid inequalities for the TSP. Math. Program. **69**, 335–349 (1995) (Cited on page 312.)

[170] M.X. Goemans, Smallest compact formulation for the permutahedron. Math. Program. Ser. A (2014) doi 10.1007/s101007-014-0757-1 (Cited on page 186.)

[171] M.X. Goemans, T. Rothvoß, Polynomiality for bin packing with a constant number of item types. arXiv:1307.5108 [cs.DS] (2013) (Cited on page 343.)

[172] M.X. Goemans, L. Tunçel, When does the positive semidefiniteness constraint help in lifting procedures. Math. Oper. Res. **26**, 796–815 (2001) (Cited on page 408.)

[173] M.X. Goemans, D.P. Williamson, Improved approximation algorithms for maximum cut and satisfiability problems using semidefinite programming. J. ACM **42**, 1115–1145 (1995) (Cited on pages vi, 57, 389, 391, 392, 397, and 408.)

[174] J.L. Goffin, Variable metric relaxation methods, part II: the ellipsoid method. Math. Program. **30**, 147–162 (1984) (Cited on page 361.)

[175] R.E. Gomory, Outline of an algorithm for integer solutions to linear programs. Bull. Am. Math. Soc. **64**, 275–278 (1958) (Cited on pages v, 14, 37, 212, and 213.)

[176] R.E. Gomory, An algorithm for the mixed integer problem. Tech. Report RM-2597 (The Rand Corporation, 1960) (Cited on pages 205, 205, 216, 219, and 227.)

[177] R.E. Gomory, An algorithm for integer solutions to linear programs, in *Recent Advances in Mathematical Programming*, ed. by R.L. Graves, P. Wolfe (McGraw-Hill, New York, 1963), pp. 269–302 (Cited on pages 212, 213, 214, 219, and 229.)

[178] R.E. Gomory, Some polyhedra related to combinatorial problems. Linear Algebra Appl. **2**, 451–558 (1969) (Cited on pages 236 and 274.)

[179] R.E. Gomory, E.L. Johnson, Some continuous functions related to corner polyhedra I. Math. Program. **3**, 23–85 (1972) (Cited on pages 254, 254, 256, 271, and 274.)

[180] R.E. Gomory, E.L. Johnson, T-space and cutting planes. Math. Program. **96**, 341–375 (2003) (Cited on pages 260, 261, 274, and 274.)

[181] J. Gouveia, P. Parrilo, R. Thomas, Theta bodies for polynomial ideals. SIAM J. Optim. **20**, 2097–2118 (2010) (Cited on page 408.)

[182] J. Gouveia, P. Parrilo, R. Thomas, Lifts of convex sets and cone factorizations. Math. Oper. Res. **38**, 248–264 (2013) (Cited on page 187.)

[183] M. Grötschel, *Polyedrische Charackterisierungen kombinatorischer Optimierungsprobleme* (Anton Hain, Meisenheim/Glan, 1977) (Cited on page 312.)

[184] M. Grötschel, On the symmetric travelling salesman problem: solution of a 120-city problem. Math. Program. Study **12**, 61–77 (1980) (Cited on pages 302 and 312.)

[185] M. Grötschel, M. Jünger, G. Reinelt, A cutting plane algorithm for the linear ordering problem. Oper. Res. **32**, 1195–1220 (1984) (Cited on page 312.)

[186] M. Grötschel, L. Lovász, A. Schrijver, The ellipsoid method and its consequences in combinatorial optimization. Combinatorica **1**, 169–197 (1981) (Cited on pages 307, 308, 312, 313, 314, and 391.)

[187] M. Grötschel, L. Lovász, A. Schrijver, Geometric methods in combinatorial optimization, in *Progress in Combinatorial Optimization*, ed. by W.R. Pulleyblank (Academic, Toronto, 1984), pp. 167–183 (Cited on page 309.)

[188] M. Grötschel, L. Lovász, A. Schrijver, *Geometric Algorithms and Combinatorial Optimization* (Springer, New York, 1988) (Cited on pages 36, 121, 308, 309, 309, 311, and 313.)

[189] M. Grötschel, M.W. Padberg, On the symmetric travelling salesman problem I: inequalities. Math. Program. **16**, (1979) 265–280 (Cited on page 304.)

[190] B. Grünbaum, *Convex Polytopes* (Wiley-Interscience, London, 1967) (Cited on page 120.)

[191] Z. Gu, G.L. Nemhauser, M.W.P. Savelsbergh, Lifted flow covers for mixed 0–1 integer programs. Math. Program. **85**, 439–467 (1999) (Cited on pages 298 and 312.)

[192] Z. Gu, G.L. Nemhauser, M.W.P. Savelsbergh, Sequence independent lifting in mixed integer programming. J. Combin. Optim. **1**, 109–129 (2000) (Cited on pages 290, 297, 312, and 312.)

[193] C. Guéret, C. Prins, M. Servaux, *Applications of Optimization with Xpress* (Dash Optimization Ltd., London, 2002) (Cited on page 73.)

[194] M. Guignard, S. Kim, Lagrangean decomposition for integer programming: theory and applications. RAIRO **21**, 307–323 (1987) (Cited on page 341.)

[195] O. Günlük, Y. Pochet, Mixing mixed-integer inequalities. Math. Program. **90**, 429–458 (2001) (Cited on page 163.)

[196] W. Harvey, Computing two-dimensional integer hulls. SIAM J. Comput. **28**, 2285–2299 (1999) (Cited on page 24.)

[197] M. Held, R.M. Karp, The traveling-salesman problem and minimum spanning trees. Oper. Res. **18**, 1138–1162 (1970) (Cited on pages 325 and 331.)

[198] M. Held, R.M. Karp, The traveling-salesman problem and minimum spanning trees: part II. Math. Program. **1**, 6–25 (1971) (Cited on page 325.)

[199] I. Heller, C.B. Tompkins, An extension of a theorem of Dantzig's, in *Linear Inequalities and Related Systems*, ed. by H.W. Kuhn, A.W. Tucker (Princeton University Press, Princeton, 1956), pp. 247–254 (Cited on page 134.)

[200] Ch. Hermite, Extraits de lettres de M. Ch. Hermite à M. Jacobi sur différents objets de la théorie des nombres. Journal für dei reine und angewandte Mathematik **40**, pp. 261–277 (1850) (Cited on pages 353 and 360.)

[201] J.-B. Hiriart-Urruty, C. Lemaréchal. *Fundamentals of Convex Analysis* (Springer, New York, 2001) (Cited on page 248.)

[202] D.S. Hirschberg, C.K. Wong, A polynomial algorithm for the knapsack problem in two variables. J. ACM **23**, 147–154 (1976) (Cited on pages 24 and 380.)

[203] A.J. Hoffman, A generalization of max-flow min-cut. Math. Program. **6**, 352–259 (1974) (Cited on page 184.)

[204] A.J. Hoffman, J.B. Kruskal, Integral boundary points of polyhedra, in *Linear Inequalities and Related Systems*, ed. by H.W. Kuhn, A.W. Tucker (Princeton University Press, Princeton, 1956), pp. 223–246 (Cited on page 108.)

[205] J.N. Hooker, Needed: an empirical science of algorithms. Oper. Res. **42**, 201–212 (1994) (Cited on page 382.)

[206] J. Hooker, *Integrated Methods for Optimization*. International Series in Operations Research and Management Science (Springer, New York, 2010) (Cited on page 74.)

[207] R.A. Horn, C.R. Johnson, *Matrix Analysis* (Cambridge University Press, Cambridge, 2013) (Cited on page 391.)

[208] C.A.J. Hurkens, Blowing up convex sets in the plane. Linear Algebra Appl. **134**, 121–128 (1990) (Cited on page 359.)

[209] S. Iwata, L. Fleischer, S. Fujishige, A combinatorial, strongly polynomial-time algorithm for minimizing submodular functions. J. ACM **48**, 761–777 (2001) (Cited on page 185.)

[210] R.G. Jeroslow, There cannot be any algorithm for integer programming with quadratic constraints. Oper. Res. **21**, 221–224 (1973) (Cited on page 20.)

[211] R.G. Jeroslow, Representability in mixed integer programming, I: characterization results. Discrete Appl. Math. **17**, 223–243 (1987) (Cited on page 185.)

[212] R.G Jeroslow, On defining sets of vertices of the hypercube by linear inequalities. Discrete Math. **11**, 119–124 (1975) (Cited on page 170.)

[213] R.G Jeroslow, J.K. Lowe, Modelling with integer variables. Math. Program. Stud. **22**, 167–184 (1984) (Cited on page 171.)

[214] F. John, Extremum problems with inequalities as subsidiary conditions, in *Studies and Essays Presented to R. Courant on his 60th Birthday, January 8, 1948* (Interscience Publishers, New York, 1948), pp. 187–204 (Cited on pages 361 and 381.)

[215] E.L. Johnson, On the group problem for mixed integer programming. Math. Program. Study **2**, 137–179 (1974) (Cited on page 268.)

[216] E.L. Johnson, Characterization of facets for multiple right-hand choice linear programs. Math. Program. Study **14**, 112–142 (1981) (Cited on page 274.)

[217] M. Jünger, T. Liebling, D. Naddef, G. Nemhauser, W. Pulleyblank, G. Reinelt, G. Rinaldi, L. Wolsey (eds.), *50 Years of Integer Programming 1958-2008* (Springer, Berlin, 2010) (Cited on pages 31, 36, and 380.)

[218] M. Jünger, D. Naddef (eds.), *Computational Combinatorial Optimization. Optimal or provably near-optimal solutions*. Lecture Notes in Computer Science, vol. 2241 (Springer, Berlin, 2001) (Cited on page 36.)

[219] V. Kaibel, Extended formulations in combinatorial optimization. Optima **85**, 2–7 (2011) (Cited on page 186.)

[220] V. Kaibel, K. Pashkovich, Constructing extended formulations from reflection relations, in *Proceedings of IPCO XV O. Günlük*, ed. by G. Woeginger. Lecture Notes in Computer Science, vol. 6655 (Springer, Berlin, 2011), pp. 287–300 (Cited on page 186.)

[221] V. Kaibel, K. Pashkovich, D.O. Theis, Symmetry matters for sizes of extended formulations. SIAM J. Discrete Math. **26**(3), 1361–1382 (2012) (Cited on page 186.)

[222] V. Kaibel, M.E. Pfetsch, Packing and partitioning orbitopes. Math. Program. **114**, 1–36 (2008) (Cited on page 382.)

[223] V. Kaibel, S. Weltge, A short proof that the extension complexity of the correlation polytope grows exponentially. arXiv:1307.3543 (2013) (Cited on pages 178 and 187.)

[224] V. Kaibel, S. Weltge, Lower bounds on the sizes of integer programs without additional variables. arXiv:1311.3255 (2013) (Cited on page 64.)

[225] R. Kannan, A polynomial algorithm for the two-variable integer programming problem. J. ACM **27**, 118–122 (1980) (Cited on page 380.)

[226] R. Kannan, Improved algorithms for integer programming and related problems, in *Proceedings of the 15th Annual ACM Symposium on Theory of Computing (STOC-83)* (1983), pp. 193–206 (Cited on page 381.)

[227] R. Kannan, Minkowski's convex body theorem and integer programming. Math. Oper. Res. **12**, 415–440 (1987) (Cited on page 381.)

[228] R. Kannan, A. Bachem, Polynomial algorithms for computing the Smith and Hermite normal forms of an integer matrix. SIAM J. Comput. **8**, 499–507 (1979) (Cited on pages 31, 31, and 37.)

[229] N. Karmarkar, A new polynomial-time algorithm for linear programming. Combinatorica **4**, 373–395 (1984) (Cited on pages vi, 121, and 408.)

[230] D.R. Karger, Global min-cuts in RNC, and other ramifications of a simple min-cut algorithm, in *Proceedings of SODA* (1993), pp. 21–30 (Cited on page 184.)

[231] D.R. Karger, R. Motwani, M. Sudan, Approximate graph coloring by semidefinite programming. J. ACM **45**, 246–265 (1998) (Cited on page 408.)

[232] R.M. Karp, Reducubility among combinatorial problems, in *Complexity of Computer Computations* (Plenum Press, New York, 1972), pp. 85–103 (Cited on page 391.)

[233] R.M. Karp, C.H. Papadimitriou, On linear characterizations of combinatorial optimization problems. SIAM J. Comput. **11**, 620–632 (1982) (Cited on pages 308 and 313.)

[234] H. Kellerer, U. Pferschy, D. Pisinger, *Knapsack Problems* (Springer, Berlin, 2004) (Cited on page 73.)

[235] L.G. Khachiyan, A polynomial algorithm in linear programming. Soviet Math. Dokl. **20**, 191–194 (1979) (Cited on pages vi, 120, 308, and 313.)

[236] L. Khachiyan, E. Boros, K. Borys, K. Elbassioni, V. Gurvich, Generating all vertices of a polyhedron is hard. Discrete Comput. Geom. **39**, 174–190 (2008) (Cited on page 123.)

[237] A. Khinchine, A quantitative formulation of Kronecker's theory of approximation (in russian). Izvestiya Akademii Nauk SSR Seriya Matematika **12**, 113–122 (1948) (Cited on pages 24 and 359.)

[238] F. Kilinc-Karzan, G.L. Nemhauser, M.W.P. Savelsbergh, Information-based branching schemes for binary linear mixed integer problems. Math. Program. Comput. **1**, 249–293 (2009) (Cited on page 382.)

[239] D. Klabjan, G.L. Nemhauser, C. Tovey, The complexity of cover inequality separation. Oper. Res. Lett. **23**, 35–40 (1998) (Cited on page 283.)

[240] V. Klee, G.J. Minty, How good is the simplex algorithm? in *Inequalities, III*, ed. by O. Shisha (Academic, New York, 1972), pp. 159–175 (Cited on pages 18 and 120.)

[241] M. Köppe, Q. Louveaux, R. Weismantel, Intermediate integer programming representations using value disjunctions. Discrete Optim. **5**, 293–313 (2008) (Cited on page 225.)

[242] M. Köppe, R. Weismantel, A mixed-integer Farkas lemma and some consequences. Oper. Res. Lett. **32**, 207–211 (2004) (Cited on page 37.)

[243] B. Korte, J. Vygen, *Combinatorial Optimization: Theory and Algorithms* (Springer, Berlin/Hidelberg, 2000) (Cited on pages 36 and 183.)

[244] J.B. Kruskal Jr., On the shortest spanning subtree of a graph and the traveling salesman problem. Proc. Am. Math. Soc. **7**, 48–50 (1956) (Cited on page 155.)

[245] H.W. Kuhn, The Hungarian method for the assignment problem. Naval Res. Logistics Q. **2**, 83–97 (1955) (Cited on pages 5 and 149.)

[246] A.H. Land, A.G. Doig, An automatic method of solving discrete programming problems. Econometrica **28**, 497–520 (1960) (Cited on page 37.)

[247] J.B. Lasserre, An Explicit Exact SDP Relaxation for Nonlinear 0–1 Programs. Lecture Notes in Computer Science, vol. 2081 (Springer, 2001), pp. 293–303 (Cited on pages 389, 400, and 403.)

[248] J.B. Lasserre, Global optimization with polynomials and the problem of moments. SIAM J. Optim. **11**, 796–817 (2001) (Cited on pages 389 and 403.)

[249] M. Laurent, A comparison of the Sherali-Adams, Lovász-Schrijver and Lasserre relaxations for 0–1 programming. SIAM J. Optim. **28**, 345–375 (2003) (Cited on page 403.)

[250] M. Laurent, F. Rendl, Semidefinite programming and integer programming, in *Handbook on Discrete Optimization*, ed. by K. Aardal, G.L. Nemhauser, R. Weimantel (Elsevier, Amsterdam, 2005), pp. 393–514 (Cited on page 408.)

[251] E. L. Lawler, Covering problems: duality relations and a method of solution. SIAM J. Appl. Math. **14**, 1115–1132 (1966) (Cited on page 54.)

[252] E. L. Lawler, *Combinatorial Optimization: Networks and Matroids* (Holt, Rinehart and Winston, New York, 1976) (Cited on page 183.)

[253] E.L. Lawler, J.K. Lenstra, A.H.G. Rinnooy Kan, D.B. Shmoys (eds.), *The Traveling Salesman Problem: A Guided Tour of Combinatorial Optimization* (Wiley, New York, 1985) (Cited on page 74.)

[254] A. Lehman, On the width-length inequality. Math. Program. **17**, 403–417 (1979) (Cited on page 55.)

[255] A.K. Lenstra, H.W. Lenstra, L. Lovász, Factoring polynomials with rational coefficients. Math. Ann. **261**, 515–534 (1982) (Cited on page 353.)

[256] H.W. Lenstra, Integer programming with a fixed number of variables. Math. Oper. Res. **8**, 538–548 (1983) (Cited on pages vi, 37, 351, 353, 361, and 363.)

[257] J.T. Linderoth, M.W.P. Savelsbergh, A computational study of search strategies for mixed integer programming. INFORMS J. Comput. **11**, 173–187 (1999) (Cited on page 382.)

[258] Q. Louveaux, L.A. Wolsey, Lifting, superadditivity, mixed integer rounding and single node flow sets revisited. 4OR **1**, 173–207 (2003) (Cited on page 312.)

[259] L. Lovász, Normal hypergraphs and the perfect graph conjecture. Discrete Math. **2**, 253–267 (1972) (Cited on pages 184 and 394.)

[260] L. Lovász, On the Shannon capacity of a graph. IEEE Trans. Inf. Theory **25**, 1–7 (1979) (Cited on pages 389, 393, and 408.)

[261] L. Lovász, Geometry of numbers and integer programming, in *Mathematical Programming: Recent Developments and Applications*, ed. by M. Iri, K. Tanabe (Kluwer, Dordrecht, 1989), pp. 177–201 (Cited on pages 249 and 250.)

[262] L. Lovász, M.D. Plummer, *Matching Theory* (Akadémiai Kiadó, Budapest, 1986) [Also: North Holland Mathematics Studies, vol. 121 (North Holland, Amsterdam)] (Cited on page 184.)

[263] L. Lovász, H.E. Scarf, The generalized basis reduction algorithm. Math. Oper. Res. **17**, 751–764 (1992) (Cited on page 381.)

[264] L. Lovász, A. Schrijver, Cones of matrices and set-functions and 0–1 optimization. SIAM J. Optim. **1**, 166–190 (1991) (Cited on pages vi, 389, 394, and 397.)

[265] T.L. Magnanti, R.T. Wong, Accelerated Benders decomposition: algorithmic enhancement and model selection criteria. Oper. Res. **29**, 464–484 (1981) (Cited on page 343.)

[266] H. Marchand, L.A. Wolsey, Aggregation and mmixed integer rounding to solve MIPs. Oper. Res. **49**, 363–371 (2001) (Cited on page 229.)

[267] F. Margot, Pruning by isomorphism in branch-and-cut. Math. Program. **94**, 71–90 (2002) (Cited on pages 373, 377, and 382.)

[268] S. Martello, P. Toth, *Knapsack Problems: Algorithms and Computer Implementations* (Wiley, Chichester, 1990) (Cited on page 73.)

[269] R.K. Martin, Generating alternative mixed integer programming models using variable definition. Oper. Res. **35**, 820–831 (1987) (Cited on page 186.)

[270] R.K. Martin, Using separation algorithms to generate mixed integer model reformulations. Oper. Res. Lett. **10**(3), 119–128 (1991) (Cited on pages 174 and 186.)

[271] R.K. Martin, R.L. Rardin, B.A. Campbell, Polyhedral characterization of discrete dynamic programming. Oper. Res. **38**, 127–138 (1990) (Cited on page 186.)

[272] J.F. Maurras, Bon algorithmes, vieilles idées, Note E.d.F. HR 32.0320 (1978) (Cited on page 120.)

[273] J.F. Maurras, K. Truemper, M. Agkül, Polynomial algorithms for a class of linear programs. Math. Program. **21**, 121–136 (1981) (Cited on page 120.)

[274] C.C. McGeogh, Experimental analysis of algorithms. Notices Am. Math. Assoc. **48**, 204–311 (2001) (Cited on page 382.)

[275] B.D. McKay, Practical graph isomorphism. Congressus Numerantium **30**, 45–87 (1981) (Cited on page 377.)

[276] R.R. Meyer, On the existence of optimal solutions to integer and mixed integer programming problems. Math. Program. **7**, 223–235 (1974) (Cited on page 159.)

[277] D. Micciancio, The shortest vector in a lattice is hard to approximate to within some constant, in *Proceedings of the 39th Annual Symposium on Foundations of Computer Science (FOCS-98)* (1998), pp. 92–98 (Cited on page 381.)

BIBLIOGRAPHY

[278] C.E. Miller, A.W. Tucker, R.A. Zemlin, Integer programming formulation of traveling salesman problems. J. ACM **7**, 326–329 (1960) (Cited on pages 63 and 136.)

[279] H. Minkowski, Geometrie der Zahlen (Erste Lieferung) (Teubner, Leipzig, 1896) (Cited on pages v and 96.)

[280] T.S. Motzkin, H. Raiffa, G.L. Thompson, R.M. Thrall, The double description method, in *Contributions to Theory of Games*, vol. 2, ed. by H.W. Kuhn, A.W. Tucker (Princeton University Press, Princeton, 1953) (Cited on page 121.)

[281] J. Munkres, Algorithms for the assignment and transportation problems. J. SIAM **5**, 32–38 (1957) (Cited on page 5.)

[282] H. Nagamochi, T. Ibaraki, Computing edge-connectivity in multiple and capacitated graphs. SIAM J. Discrete Math. **5**, 54–66 (1992) (Cited on page 184.)

[283] G.L. Nemhauser, L.E. Trotter Jr., Properties of vertex packing and independence system polyhedra. Math. Program. **6**, 48–61 (1974) (Cited on page 312.)

[284] G.L. Nemhauser, L.E. Trotter Jr., Vertex packings: structural properties and algorithms. Math. Program. **8**, 232–248 (1975) (Cited on page 312.)

[285] G.L. Nemhauser, L.A. Wolsey, *Integer and Combinatorial Optimization* (Wiley, New York, 1988) (Cited on page 36.)

[286] G.L. Nemhauser, L.A. Wolsey, A recursive procedure to generate all cuts for 0–1 mixed integer programs. Math. Program. **46**, 379–390 (1990) (Cited on pages 206, 206, and 228.)

[287] Y.E. Nesterov, Smooth minimization of non-smooth functions. Math. Program. A **103**, 127–152 (2005) (Cited on page 329.)

[288] Y.E. Nesterov, Semidefinite relaxation and nonconvex quadratic optimization. Optim. Methods Softw. **12**, 1–20 (1997) (Cited on page 408.)

[289] Y.E. Nesterov, A.S. Nemirovski, Self-concordant functions and polynomial time methods in convex programming. Technical report, Central Economical and Mathematical Institute, U.S.S.R (Academy of Science, Moscow, 1990) (Cited on page 408.)

[290] Y.E. Nesterov, A.S. Nemirovski, Conic formulation of a convex programming problem and duality. Optim. Methods Softw. **1**, 95–115 (1992) (Cited on page 390.)

[291] Y.E. Nesterov, A.S. Nemirovski, *Interior Point Polynomial Algorithms in Convex Programming* (SIAM, Philadelphia, 1994) (Cited on pages 381, 391, and 408.)

[292] J. Ostrowski, J.T. Linderoth, F. Rossi, S. Smriglio, Solving large Steiner triple covering problems. Oper. Res. Lett. **39**, 127–131 (2011) (Cited on page 58.)

[293] J. Ostrowski, J. Linderoth, F. Rossi, S. Smriglio, Orbital branching. Math. Program. **126**, 147–178 (2011) (Cited on page 382.)

[294] J.H. Owen, S. Mehrotra, A disjunctive cutting plane procedure for general mixed-integer linear programs. Math. Program. A **89**, 437–448 (2001) (Cited on page 228.)

[295] J.H. Owen, S. Mehrotra, On the value of binary expansions for general mixed-integer linear programs. Oper. Res. **50**, 810–819 (2002) (Cited on page 225.)

[296] J. Oxley, *Matroid Theory* (Oxford University Press, New York, 2011) (Cited on page 183.)

[297] M.W. Padberg, On the facial structure of set packing polyhedra. Math. Program. **5**, 199–215 (1973) (Cited on pages 312 and 312.)

[298] M.W. Padberg, A note on zero-one programming. Oper. Res. **23**, 833–837 (1975) (Cited on page 312.)

[299] M.W. Padberg, M.R. Rao, The Russian method for linear programming III: bounded integer programming. Research Report 81-39, Graduate School of Business Administration, New York University (1981) (Cited on pages 308 and 313.)

[300] M.W. Padberg, M.R. Rao, Odd minimum cut-sets and b-matchings. Math. Oper. Res. **7**, 67–80 (1982) (Cited on page 304.)

[301] M.W. Padberg, G. Rinaldi, Optimization of a 532-city symmetric traveling salesman problem by branch and cut. Oper. Res. Lett. **6**, 1–7 (1987) (Cited on pages 37 and 302.)

BIBLIOGRAPHY

[302] M.W. Padberg, G. Rinaldi, A branch-and-cut algorithm for the resolution of large-scale symmetric traveling salesman problems. SIAM Rev. **33**, 60–100 (1991) (Cited on page 37.)

[303] M.W. Padberg, T.J. Van Roy, L.A. Wolsey, Valid linear inequalities for fixed charge problems. Oper. Res. **33**, 842–861 (1985) (Cited on page 291.)

[304] J. Pap, Recognizing conic TDI systems is hard. Math. Program. **128**, 43–48 (2011) (Cited on page 185.)

[305] C.H. Papadimitriou, On the complexity of integer programming. J. ACM **28**, 765–768 (1981) (Cited on page 37.)

[306] J. Patel, J.W. Chinneck, Active-constraint variable ordering for faster feasibility of mixed integer linear programs. Math. Program. **110**, 445–474 (2007) (Cited on page 382.)

[307] J. Petersen, Die Theorie der regulären graphs. Acta Matematica **15**, 193–220 (1891) (Cited on page 147.)

[308] Y. Pochet, L.A. Wolsey, Polyhedra for lot-sizing with Wagner–Whitin costs. Math. Program. **67**, 297–324 (1994) (Cited on page 186.)

[309] Y. Pochet, L.A. Wolsey, *Production Planning by Mixed-Integer Programming*. Springer Series in Operations Research and Financial Engineering (Springer, New York, 2006) (Cited on page 73.)

[310] B.T. Poljak, A general method for solving extremum problems. Soviet Math. Dokl. **8**, 593–597 (1967) (Cited on pages 328, 328, 329, and 329.)

[311] C.H. Papadimitriou, M. Yannakakis, On recognizing integer polyhedra. Combinatorica **10**, 107–109 (1990) (Cited on page 183.)

[312] M. Queyranne, A.S. Schulz, Polyhedral approaches to machine scheduling. Preprint (1994) (Cited on page 74.)

[313] A. Razborov, On the distributional complexity of disjointness. Theor. Comput. Sci. **106**(2), 385–390 (1992) (Cited on pages 178 and 187.)

[314] J. Renegar, A polynomial-time algorithm based on Newton's method for linear programming. Math. Program. **40**, 59–93 (1988) (Cited on page 18.)

[315] J.-P.P. Richard, S.S. Dey (2010). The group-theoretic approach in mixed integer programming, in *50 Years of Integer Programming 1958–2008*, ed. by M. Jünger, T. Liebling, D. Naddef, G. Nemhauser, W. Pulleyblank, G. Reinelt, G. Rinaldi, L. Wolsey (Springer, New York, 2010), pp. 727–801 (Cited on page 273.)

[316] R.T. Rockafellar, *Convex Analysis* (Princeton University Press, Princeton, 1969) (Cited on pages 248, 248, and 249.)

[317] T. Rothvoß, Some 0/1 polytopes need exponential size extended formulations. Math. Program. A **142**, 255–268 (2012) (Cited on page 186.)

[318] T. Rothvoß, The matching polytope has exponential extension complexity, in *Proceedings of the 46th Annual ACM Symposium on Theory of Computing (STOC 2014)* (2014), pp. 263–272 (Cited on pages 177, 181, and 187.)

[319] T. Rothvoß, L. Sanitá, $0-1$ polytopes with quadratic Chvátal rank, in *Proceedings of the 16th IPCO Conference*. Lecture Notes in Computer Science, vol. 7801 (Springer, New York, 2013) (Cited on page 211.)

[320] J.-S. Roy, Reformulation of bounded integer variables into binary variables to generate cuts. Algorithmic Oper. Res. **2**, 810–819 (2007) (Cited on page 225.)

[321] M.P.W. Savelsbergh, Preprocessing and probing techniques for mixed integer programming problems. ORSA J. Comput. **6**, 445–454 (1994) (Cited on page 372.)

[322] H.E. Scarf, An observation on the structure of production sets with indivisibilities. Proc. Natl. Acad. Sci. USA **74**, 3637–3641 (1977) (Cited on page 250.)

[323] A. Schrijver, On cutting planes. Ann. Discrete Math. **9**, 291–296 (1980) (Cited on page 210.)

[324] A. Schrijver, On total dual integrality. Linear Algebra Appl. **38**, 27–32 (1981) (Cited on page 184.)

[325] A. Schrijver, *Theory of Linear and Integer Programming* (Wiley, New York, 1986) (Cited on pages 31, 36, 120, 165, 165, 183, 184, 213, and 357.)

[326] A. Schrijver, A combinatorial algorithm minimizing submodular functions in strongly polynomial time. J. Combin. Theory Ser. B **80**, 346–355 (2000) (Cited on page 185.)

[327] A. Schrijver, *Combinatorial Optimization: Polyhedra and Efficiency* (Springer, Berlin, 2003) (Cited on pages 36, 156, 183, 184, and 184.)

[328] Á. Seress, *Permutation Group Algorithms, Cambridge Tracts in Mathematics*, vol. 152 (Cambridge University Press, Cambridge, 2003) (Cited on page 377.)

[329] P.D. Seymour, Decomposition of regular matroids. J. Combin. Theory B **28**, 305–359 (1980) (Cited on page 183.)

[330] H. Sherali, W. Adams, A hierarchy of relaxations between the continuous and convex hull representations for zero-one programming problems. SIAM J. Discrete Math. **3**, 311–430 (1990) (Cited on pages 389, 397, and 400.)

[331] H. Sherali, W. Adams, *A Reformulation-Linearization Technique for Solving Discrete and Continuous Nonconvex Problems*, Chap. 4 (Kluwer Academic Publishers, Norwell, 1999) (Cited on page 225.)

[332] N. Z. Shor, Cut-off method with space extension in convex programming problems. Cybernetics **13**, 94–96 (1977) (Cited on pages 120 and 313.)

[333] M. Stoer, F. Wagner, A simple min-cut algorithm. J. ACM **44**, 585–591 (1997) (Cited on page 184.)

[334] E. Tardos, A strongly polynomial algorithm to solve combinatorial linear programs. Oper. Res. **34**, 250–256 (1986) (Cited on page 381.)

[335] R.E. Tarjan, Depth-first search and linear graph algorithms. SIAM J. Comput. **1**, 146–160 (1972) (Cited on page 303.)

[336] S. Tayur, R.R. Thomas, N.R. Natraj, An algebraic geometry algorithm for scheduling in presence of setups and correlated demands. Math. Program. **69**, 369–401 (1995) (Cited on page 38.)

[337] P. Toth, D. Vigo, *The Vehicle Routing Problem*. Monographs on Discrete Mathematics and Applications (SIAM, Philadelphia, 2001) (Cited on page 74.)

[338] K. Truemper, *Matroid Decomposition* (Academic, Boston, 1992) (Cited on page 183.)

[339] W.T. Tutte, A homotopy theorem for matroids I, II. Trans. Am. Math. Soc. **88**, 905–917 (1958) (Cited on page 183.)

[340] T.J. Van Roy, L.A. Wolsey, Solving mixed integer programming problems using automatic reformulation. Oper. Res. **35**, 45–57 (1987) (Cited on page 312.)

[341] M. Van Vyve, The continuous mixing polyhedron. Math. Oper. Res. **30**, 441–452 (2005) (Cited on page 186.)

[342] F. Vanderbeck, L.A. Wolsey, Reformulation and decomposition of integer programs, in *50 Years of Integer Programming 1958–2008*, ed. by M. Jünger, T. Liebling, D. Naddef, G. Nemhauser, W. Pulleyblank, G. Reinelt, G. Rinaldi, L. Wolsey (Springer, New York, 2010), pp. 431–502 (Cited on page 186.)

[343] R.J. Vanderbei, *Linear Programming: Foundations and Extentions*, 3rd edn. (Springer, New York, 2008) (Cited on page 121.)

[344] S. Vavasis, On the complexity of nonnegative matrix factorization. SIAM J. Optim. **20**, 1364–1377 (2009) (Cited on page 177.)

[345] V.V. Vazirani, *Approximation Algorithms* (Springer, Berlin, 2003) (Cited on page 37.)

[346] J.P. Vielma, A constructive characterization of the split closure of a mixed integer linear program. Oper. Res. Lett. **35**, 29–35 (2007) (Cited on page 228.)

[347] J.P. Vielma, Mixed integer linear programming formulation techniques. SIAM Rev. (2014, to appear) (Cited on page 185.)

[348] H. Weyl, The elementary theory of convex polyhedra, in *Contributions to the Theory of Games I*, ed. by H.W. Kuhn, A.W. Tucker (Princeton University Press, Princeton, 1950), pp. 3–18 (Cited on page 96.)

[349] D.P. Williamson, D.B. Shmoys, *The Design of Approxiamtion Algorithms* (Cambridge University Press, Cambridge, 2011) (Cited on page 37.)

[350] L.A. Wolsey, Further facet generating procedures for vertex packing polytopes. Math. Program. **11**, 158–163 (1976) (Cited on page 312.)

[351] L.A. Wolsey, Valid inequalities and superadditivity for 0–1 integer programs. Math. Oper. Res. **2**, 66–77 (1977) (Cited on page 312.)

[352] L.A. Wolsey, Heuristic analysis, linear programming, and branch and bound. Math. Program. Stud. **13**, 121–134 (1980) (Cited on page 312.)

[353] L.A. Wolsey, *Integer Programming* (Wiley, New York, 1999) (Cited on pages 36 and 74.)

[354] R.T. Wong, Dual ascent approach for Steiner tree problems on directed graphs. Math. Program. **28**, 271–287 (1984) (Cited on page 342.)

[355] M. Yannakakis, Expressing combinatorial optimization problems by linear programs. J. Comput. Syst. Sci. **43**, 441–466 (1991) (Cited on pages 174, 177, 182, and 186.)

[356] D. B. Yudin, A. S. Nemirovski, Evaluation of the information complexity of mathematical programming problems. Ekonomika i Matematicheskie Metody **12**, 128–142 (1976) (in Russian). English Translation: Matekon **13**, 3–45 (1976) (Cited on pages 120 and 313.)

[357] G.M. Ziegler, *Lectures on Polytopes* (Springer, New York, 1995) (Cited on page 120.)

Index

$(\cdot)^+$: $a^+ = \max\{0, a\}$ $(a \in \mathbb{R})$, 199
$(\cdot)^-$: $a^- = \max\{0, -a\}$ $(a \in \mathbb{R})$, 199
$A^<$, 98
$A^=$, 98
$E[\cdot]$, 151
M-alternating path, 146
N_+, 393, 398
O notation, 18
$P(B)$, 236
P^{Ch}, 208
P^{MIR}, 206
P_n^{corr}, 180
P^{cut}, 179
P^{mix}, 161
P^{split}, 197
P^{lift}, 222
P_{tsp}, 300
S_n^{even}, 170
$V(\cdot)$, 56
Σ_n, 372
$\mathbb{Z}^p$-free convex set, 242
$\mathbb{Z}^p \times \mathbb{R}^{n-p}$-free convex set, 249
aff$(\cdot)$, 92
α-critical edge, 318
cone$(\cdot)$, 93
conv$(\cdot)$, 93
corner(B), 236
$\delta(\cdot)$, 55
$\delta^+(\cdot), \delta^-(\cdot)$, 62
dim$(\cdot)$, 92
intcone, 160
$\langle \cdot \rangle$, 91
$\langle \cdot, \cdot \rangle$, 180
$\lceil \cdot \rceil$, 6
$\lfloor \cdot \rceil$, 353
$\lfloor \cdot \rfloor$, 6
lin$(\cdot)$, 97
FRAC$(\cdot)$, 394
MIX, 66
QSTAB$(\cdot)$, 391
STAB$(\cdot)$, 106, 181, 391
TH$(\cdot)$, 391
stab$(\cdot)$, 51
det(Λ), 350
proj$_x(\cdot)$, 116
rec$(\cdot)$, 97
stabil$(\cdot, \cdot)$, 376
$\subset$, 3
$\subseteq$, 3
$\succ$, 388
$\succeq$, 387
$n(\cdot)$, 59
s, t-cut, 55, 140
s, t-flow, 139
s, t-path, 55, 137
$x(\cdot)$, 158
z_{LD}, 320
$z_{LR}(\cdot)$, 320

\alldifferent{} , 61
1-tree, 323

affine
    combination, 92
    hull, 92
    independence, 92
    space, 92
aggregated formulation, 69
algorithm, 18
antihole, 318
approximation algorithm, 37
assignment
    polytope, 99
    problem, 4, 149
asymmetric traveling salesman problem, 61
atomic proposition, 58
augmenting path
    relative to a flow, 143
    relative to a matching, 146
auxiliary digraph, 148

Balas' sequential convexification theorem, 223
basic
    feasible solution, 115
    solution, 115
basis, 115
    dual feasible basis, 115
    Gram-Schmidt, 351
    of a lattice, 350
    of a linear space, 91
    of an affine space, 92
    optimal basis, 115
    primal feasible basis, 115
    reduction algorithm, 351
Bellman-Ford algorithm, 138

Benders
    decomposition, 338
    reformulation, 338
    subproblem, 338
best
    bound criterion, 367
    estimate criterion, 367
better
    formulation, 46
    representation, 47
big M formulation, 67
big O notation, 18
binary encoding, 17
block diagonal structure, 330, 339
blocker, 54
blocking pair, 54
blossom inequality
    for the matching polytope, 151
    for the TSP, 304
Bonami's lemma, 198
branch-and-bound algorithm, 10
branch-and-cut algorithm, 15, 363
branch-and-price, 335
branching, 9, 364
breakpoint, 259

Camion's theorem, 183
capacity of a cut, 140
Carathéodory's theorem, 113
cdd, 121
certificate, 19
characteristic vector, 51
chordless cycle, 317
chromatic number, 77
Chvátal
    closure, 208
    inequality, 207
    rank of a polyhedron, 209
    rank of an inequality, 209

INDEX 449

circuit, 135
    simple, 135
circulation, 135
    cone, 135
circumscribed convex set, 313
clause, 58
clique, 52
    inequality, 52
    matrix, 53
    maximal, 52
    relaxation, 391
closure
    Chvátal, 208
    split, 197
clutter, 54
coloring, 407
column generation, 332
comb inequality, 303
combinatorial auction, 78
combinatorial optimization, 36
compact extended formulation, 117
complementary slackness, 90
complexity class
    Co-NP, 19
    NP, 19
    P, 18
cone, 93
    circulation, 135
    finitely generated, 94
    of a nonempty set, 93
    ray of a cone, 93
conflict graph, 52
conic combination, 93
conjunction of literals, 58
conjunctive normal form, 59
constraint
    irredundant, 104
    redundant, 104
constraint programming, 61
continued fraction, 24

continuous infinite relaxation, 264
convex
    analysis, 248
    body, 356
    combination, 21, 92
    cone, 93
    hull, 20, 93
    set, 20, 93
Cook-Kannan-Schrijver
    example, 203
    theorem, 203
corner polyhedron, 236
correlation polytope, 180
counting integral points in a
    polyhedron, 379
cover, 46, 282
    inequality, 282
    minimal, 46, 282
covering, 51
crew scheduling, 57
criterion
    best bound, 367
    best estimate, 367
cut
    in a graph, 55
    polytope, 110, 179
    pool, 371
cut-generating linear program, 228,
    342
cutting plane, 12
    algorithm, 12
cutting stock problem, 49, 331,
    340
cycle, 56
    odd, 56

Dantzig-Fulkerson-Johnson
    formulation, 62, 64, 136,
    299

Dantzig-Wolfe
    reformulation, 328
    relaxation, 328
decision problem, 19
decomposition theorem for
        polyhedra, 111
defining a face, 102
degree
    constraint, 62, 64, 145
    of a node, 55
delayed column generation, 306
depth-first search, 367
determinant of a lattice, 350
digraph, 61
dimension
    of a linear space, 91
    of a set, 92
Diophantine equations, 25
directed s,t-path, 137
disaggregated formulation, 69
disjunction of literals, 58
disjunctive constraint, 70, 169
diving heuristic, 368
dominated inequality, 202, 241
double description method, 121
down pseudocost, 365
dual
    feasible basis, 115
    of a linear program, 89
    semidefinite program, 388
duality
    strong, 89
    theorem, 89
dynamism of a cut, 221

edge of a polyhedron, 109
ellipsoid, 357
    algorithm, 120, 308
encoding size, 17, 112

enumeration tree, 8
equitable bicoloring, 133
Euclidean algorithm, 26
extended cover, 316
extended formulation, 117
extreme ray, 110
extreme valid function, 260

face, 102
    proper face, 102
facet, 102
    enumeration for a polytope, 123
facet-defining inequality, 102
facility location problem, 67
Farkas' lemma, 88
feasibility cut, 338
feasibility pump, 369
feasible
    flow, 140
    solution, 2
finite support, 254
finitely generated cone, 94
fixed
    charge, 66
    cost, 67
flat direction, 349
flatness theorem, 356
flow
    feasible, 140
    integral, 140
flow cover, 291
    inequality, 291
Fortet's linearization, 72
Fourier elimination, 85
fractional cut, 14, 212
Fredholm alternative, 35
Frobenius product, 180
full-dimensional polyhedron, 99

function
    valid for G_f, 256
    valid for M_f, 254
    valid for R_f, 264

gauge, 248
gcd, 26
generalized assignment problem, 65
generalized set covering, 58
generators
    of a cone, 94
    of a linear space, 91
Ghouila-Houri's theorem, 133
Goemans-Williamson theorem, 390
Gomory
    fractional cut, 14, 212
    function, 245
    lexicographic method, 213
    mixed integer cut, 216
    mixed integer inequality, 205, 216
Gomory-Johnson theorem, 256
Gram-Schmidt
    basis, 351
    procedure, 351
graph k-coloring problem, 407
graphical traveling salesman polyhedron, 306
Graver test set, 38
greatest common divisor, 26
greedy algorithm, 158
GUB branching, 366

Hadamard's inequality, 351
Hamiltonian tour
    directed, 61
    undirected, 64

Hamiltonian-path polytope, 101
head of an arc, 134
Heller-Tompskins characterization, 134
Helly's theorem, 114
Hermite normal form, 30
heuristic for integer programs, 368
Hilbert basis, 185
Hilbert's Nullstellensatz, 407
Hoffman-Kruskal theorem, 132
Hungarian method, 5, 149

ideal matrix, 184
implicit equality, 98
incidence matrix, 51
    of a digraph, 134
    of a graph, 145
incomparable representations, 47
inequality
    Chvátal, 207
    defining a face, 102
infeasibility certificate, 407
infinite relaxation, 254
instance, 17
integer
    decomposition property, 189
    Farkas lemma, 36
    feasibility problem, 349
    program, 2
integral
    convex set, 130
    flow, 140
    polyhedron, 22
    vector, 2
integrality gap, 229
intersection cut, 240
intersection graph, 51
interval lemma, 261
irredundant constraint, 104

isomorphic
    nodes, 372
    solutions, 372
isomorphism pruning, 373

j-cut, 226

Karmarkar's algorithm, 18
Khinchine's flatness theorem, 357
knapsack
    polytope, 100
    problem, 45
    set, 45
Kronecker approximation theorem, 36
Kruskal's algorithm, 155
kth split closure, 203

Löner-John theorem, 359
Lagrangian
    dual, 320
    relaxation, 320
laminar, 188
Lasserre
    hierarchy, 400
    relaxation, 400
lattice, 350
    basis, 350
    determinant, 350
    generated by a matrix, 350
    reduced basis, 352
    width, 356
lattice-free convex set, 242
left child, 374
length of a path, 137
Lenstra's algorithm, 361
lexicographically
    larger vector, 214
    optimal basis, 214
lift-and-project
    closure, 222
    inequality, 222
lifted flow cover inequality, 295
lifting, 283
    a valid function for R_f, 271
    function, 288
    function for flow cover, 295
    sequence independent, 287
    sequential, 284
lineality space, 97
linear
    combination, 91
    independence, 91
    relaxation, 3
    space, 91
linear programming, 3, 17, 89
    bounding, 9
    duality, 90
    relaxation, 3
linearly isomorphic polytopes, 180
literal, 58
LLL algorithm, 351
local branching, 368
local cut, 307
logical inference, 59
lot sizing, 78
Lovász' basis reduction algorithm, 351
Lovász-Scarf algorithm, 379
Lovász-Schrijver
    procedure, 393
    relaxation, 392
lrs, 121

machine scheduling, 74
master problem, 332, 338
matching, 145
    polytope, 146, 150, 181
max-cut problem, 56, 389
max-flow min-cut theorem, 141
maximal clique, 52

INDEX

maximal lattice-free convex set, 249
maximum cardinality matching, 145
maximum flow problem, 140
membership problem, 208
metaheuristic, 37
Meyer's theorem, 159
Miller-Tucker-Zemlin formulation, 63, 136
MILP feasibility problem, 164
min-cut problem, 56, 140
minimal
    cover, 46, 282
    cover inequality, 69, 282
    face, 108
    lifting of a function, 271
    representation of a polyhedron, 105
    transversal, 54
    valid function for G_f, 256
    valid function for M_f, 255
    valid function for R_f, 264
minimum weight
    s,t-cut problem, 55
    perfect matching problem, 149
Minkowski function, 248
Minkowski sum, 95
Minkowski-Weyl theorem, 96
    for cones, 94
    for polyhedra, 95
    for polytopes, 97
MIX, 66
mixed 0, 1 linear
    program, 3
    set, 2
mixed integer
    infinite relaxation, 268
    linear program, 2
    linear representable set, 171
    linear set, 2
    rounding closure, 206

rounding inequality, 206
mixing
    inequalities, 163
    set, 66, 161
moment matrix, 400

natural linear relaxation, 3
natural linear programming relaxation, 3
negative-length circuit, 138
neighborly polytope, 110
network design, 69
node
    covered by a matching, 145
    of an enumeration tree, 10, 363
    selection, 366
    to the left of another node, 374
nonnegative rank, 174
nontrivial inequality, 239
normalization constraint, 228
notation
    big O, 18
NP, 19
NP-complete, 20
NP-hard, 20

octahedron, 117
odd cut inequality, 151
odd cycle, 56
    inequality, 317
one-tree, 323
operating rooms scheduling problem, 65, 371
optimal basis, 115
optimality cut, 338
optimization problem, 307
oracle-polynomial time, 314
orbit, 376
orbital fixing, 376

orthogonality defect, 351
orthogonal complement, 119

packing, 51
partitioning, 51
perfect
    formulation, 22, 129
    graph, 53, 392
    matching, 145
    matching polytope, 146, 151
    matrix, 184
periodic function, 256
permutahedron, 100, 106
permutation, 371
piecewise-linear function, 266
    in $[0, 1]$, 259
pointed polyhedron, 97
polar, 119
polyhedral
    approach, 37
polyhedral cone, 94
polyhedron, 94
    integral, 130
polynomial algorithm, 18
polynomial-time algorithm, 18
polynomially bounded function, 17
polytope, 95
    verification, 123
Porta, 121
positive
    definite, 388
    semidefinite, 387
positively homogeneous, 248
potential function, 355
premise, 59
preprocessing, 369
pricing problem, 333
primal feasible basis, 115
primal method, 38
principal submatrix, 389

problem, 17
program
    mixed integer linear, 2
    pure integer linear, 1
projection, 116
    cone, 118
proper
    cut, 55
propositional logic, 58
pruning, 363
    by bound, 7
    by infeasibility, 7
    by integrality, 7
    by isomorphism, 373
pseudocost, 365
pure integer infinite relaxation, 256
pure integer linear
    program, 1
    set, 2

QSTAB, 391
quadratic assignment problem, 72

Radon's theorem, 114
rank
    Chvátal, 209
    split, 204
rational polyhedron, 94
ray
    of a cone, 93
    of a polyhedron, 97
recession cone, 97
rectangle
    cover, 178
    covering number, 178
reduced basis, 352
reduced cost, 115, 333
redundant constraint, 104
relative interior, 249
relatively prime, 28

INDEX

relaxation, 3
representation of a polyhedron, 104
residual digraph, 142
right child, 374
RINS, 369
round of cuts, 220
rounding polytopes, 356

s,t-cut, 55
s,t-flow, 139
s,t-path, 55
SAT, 59
satisfiability problem, 59
satisfied clause, 58
scheduling, 70
Schreier-Sims representation, 375
semidefinite program, 388
separation problem, 12, 307
sequence independent lifting, 287
sequential convexification theorem, 223
sequential lifting, 284
set
    covering, 51, 53
    packing, 51
    packing family, 51
    partitioning, 51
Sherali-Adams
    hierarchy, 398
    procedure, 399
shortest augmenting path
    algorithm, 144
shortest path problem, 137
simple
    circuit, 135
simple rounding inequality, 206
simplex method, 18
single-node flow set, 291
size of an instance, 17
skeleton of a polytope, 110

slack matrix, 175
sliding objective, 309
slope, 259
Smith normal form, 37
software, 371
solution
    basic, 115
    basic feasible, 115
SOS branching, 366
spanning tree, 153
    polytope, 153
split, 196
    closure, 197
    disjunction, 196
    inequality, 196
    rank, 204
STAB, 106, 181, 391
stab, 51
stability number, 318
stabilizer, 376
stable set, 51
    polytope, 106, 181
    problem, 391, 404
standard equality form, 115
Steiner triple system, 58
stochastic integer program, 38
stochastic programming, 339
strictly feasible, 388
strong
    branching, 365
    duality, 89
subadditive, 248
subgradient, 325
    algorithm, 326
sublinear, 248
submodular
    function, 157
    polyhedron, 157

subtour elimination
    constraints, 62, 64, 300
    polytope, 302
Sudoku game, 60
superadditive function, 288
support of a vector, 54
supporting hyperplane, 102
symmetric traveling salesman
        problem, 61, 299, 323
symmetry
    breaking constraint, 371
    condition, 256
    group, 372
    in integer programming, 371

tail of an arc, 134
TDI, 155
TH, 391
theorem of the alternative, 89
theta body, 391
tightness of formulation, 47
total dual integrality, 155
totally unimodular matrix, 131
tour, 61
trace of a square matrix, 180
transversal, 54
traveling salesman
    polytope, 182, 300
    problem, 61
trivial inequality, 239
trivial lifting, 271
truth assignment, 58
two-slope theorem, 263

uncapacitated facility location, 68, 322
undecidable problem, 20

undominated inequality, 202
unimodular
    matrix, 31, 350
    operations, 30, 353
union
    of polyhedra, 166
    of polytopes, 70
unique disjointness matrix, 178
up pseudocost, 365

valid
    function, 254, 256, 264
    inequality, 12, 101
variable
    branching, 9, 364
    cost, 67
variable fixing, 370
    by bounds, 375
vehicle routing, 74
vertex
    cover, 190
    enumeration for a polytope, 123
    of a polyhedron, 109

weak
    optimization oracle, 313
    optimization problem, 313
    separation oracle, 313
    separation problem, 313
web, 318
well-described family of polyhedra, 310
wheel, 315
witdh of a convex body, 356

Yannakakis' theorem, 176

MIX
Papier aus verantwortungsvollen Quellen
Paper from responsible sources
FSC® C105338

If you have any concerns about our products,
you can contact us on
ProductSafety@springernature.com

In case Publisher is established outside the EU,
the EU authorized representative is:
**Springer Nature Customer Service Center GmbH
Europaplatz 3, 69115 Heidelberg, Germany**

Printed by Libri Plureos GmbH
in Hamburg, Germany